PARTICLE PHYSICS

PERSPECTIVES AND OPPORTUNITIES

Report of the DPF Committee on Long-Term Planning

PARTICLE PHYSICS

PERSPECTIVES AND OPPORTUNITIES

Report of the DPF Committee on Long-Term Planning

Editors

Roberto Peccei
University of California, Los Angeles

Michael E. Zeller
Yale University, New Haven

David G. Cassel
Cornell University, Ithaca

Jonathan A. Bagger
Johns Hopkins University, Baltimore

Robert N. Cahn
Lawrence Berkeley Laboratory, Berkeley

Paul D. Grannis
State University of New York, Stony Brook

Frank J. Sciulli
Columbia University, Irvington-on Hudson

World Scientific
Singapore • New Jersey • London • Hong Kong

Published by

World Scientific Publishing Co. Pte. Ltd.
P O Box 128, Farrer Road, Singapore 9128
USA office: Suite 1B, 1060 Main Street, River Edge, NJ 07661
UK office: 57 Shelton Street, Covent Garden, London WC2H 9HE

British Library Cataloguing-in-Publication Data
A catalogue record of this book is available from the British Library.

PARTICLE PHYSICS — PERSPECTIVES AND OPPORTUNITIES
Report of the DPF Committee on Long-Term Planning

ISBN 981-02-2271-8

Preface

The SSC was the scientific instrument which would allow us to study one of our fundamental open questions, the origin of mass, and to search for new and unexpected phenomena. When the survival of the SSC became uncertain in the summer of 1993, the Executive Committee of the Division of Particles and Fields (DPF) of the American Physical Society began serious discussions of steps that would best help our community face the challenges of the future. These discussions were strongly influenced by the loss of the SSC and the subsequent appointment of the Drell Subpanel of the Department of Energy's High Energy Physics Advisory Panel (HEPAP) to help provide a vision for the future of the field. Since the Drell Subpanel had a specific charge and a relatively short time for deliberation, the DPF Executive Committee felt that the high-energy physics community and the public on whose support it relies would profit by having an additional broad-based and in-depth study of the field. This study, called the Committee on Long-Term Planning, was initiated in early 1994 and its fruits are the reports contained in this volume.

This study of high-energy physics was organized on the basis of working groups. Each of the working groups was led by conveners chosen among established leaders in the area under consideration. The conveners then assembled their groups by inviting other active workers in the field to participate. Most of the working groups also gathered further input from the community by holding various specialized meetings and workshops. The working groups presented their preliminary findings at a Workshop at Johns Hopkins University in May 1994, and their final reports at the DPF general meeting in Albuquerque the following August. Draft written versions of these reports were critically discussed at a retreat of all conveners with the members of the DPF Executive Committee at the Lawrence Berkeley Laboratory in October 1994. The corrected draft versions of the reports were examined one more time by an editorial board appointed by the DPF Executive Committee.

We would like to take this opportunity to thank our hosts at Johns Hopkins University, the University of New Mexico, and Lawrence Berkeley Laboratory for hospitality during the events used to prepare this report. We also wish to express our gratitude to the Department of Energy, National Science Foundation, Brookhaven National Laboratory, Fermi National Accelerator Laboratory, Cornell Laboratory of Nuclear Studies, and Stanford Linear Accelerator Center for support of this effort.

Roberto Peccei
Michael E. Zeller
David G. Cassel
Jonathan Bagger
Robert N. Cahn
Paul D. Grannis
Frank J. Sciulli

Table of Contents

Preface v

The Character of Particle Physics **1**

R. Peccei, M. E. Zeller, D. G. Cassel, J. A. Bagger, R. N. Cahn, P. D. Grannis and F. J. Sciulli

Executive Summary **5**

R. Peccei, M. E. Zeller, D. G. Cassel, J. A. Bagger, R. N. Cahn, P. D. Grannis and F. J. Sciulli

1 Introduction 5
2 Precision Tests of Electroweak Theory 6
3 Quantum Chromodynamics 7
4 Heavy Flavor Physics and CP Violation 8
5 Neutrino Mass and Mixing 10
6 Electroweak Symmetry Breaking and New Physics at the TeV Scale 11
7 Astrophysics, Cosmology, and Unification of Forces 12
8 Exploratory Theory 14
9 Accelerator Physics, Technology, and Facilities 14
10 Detectors 15
11 Computing 16
12 Structural Issues in High-Energy Physics 16
13 Conclusions 17

Precision Tests of Electroweak Physics: Current Status and Prospects for the Next Two Decades **19**

F. S. Merritt, H. Montgomery, A. Sirlin and M. Swartz

1 Introduction 19
2 Standard Electroweak Model 19
3 Expected Schedule of Future Measurements 24
4 Measurements of Electroweak Couplings Near the Z Pole 25
5 Measurements of Electroweak Couplings at low Q^2 27
6 Measurements of the W-boson Mass 29
7 The Top Quark 31
8 Global Analysis of Electroweak Data 36
9 Direct Searches for the Higgs Boson 45
10 Couplings of the Gauge Bosons 48
11 Conclusions 48
Acknowledgements 49
References 49

Quantum Chromodynamics 55

A. H. Mueller, B. Müller, C. Rebbi and W. H. Smith

1 Preface ... 56
2 Introduction and Executive Summary ... 56
3 Experimental Studies of QCD ... 58
4 Theoretical Studies of QCD ... 66
5 QCD and Dense Matter ... 70
6 Lattice QCD ... 75
References ... 84

Heavy Flavor Physics and *CP* Violation 91

J. P. Cumalat, E. J. Eichten, H. R. Quinn, M. P. Schmidt, E. H. Thorndike and Y. W. Wah

1 Overview ... 91
2 Standard Model Parameters and Tests ... 92
3 *CP* Violation and *T* Violation ... 97
4 Rare and Forbidden Decays ... 103
5 Strong Interaction Effects in Heavy Flavor Physics ... 106
6 Facilities ... 111
7 Conclusions ... 115
Acknowledgements ... 115
References ... 115

Neutrino Mass and Mixing 119

P. Langacker, R. Rameika and H. Robertson

1 Introduction ... 119
2 Basic Properties of Neutrinos ... 120
3 Neutrino Counting ... 123
4 Cosmology and Astrophysics ... 124
5 Supernova Constraints on Neutrino Masses ... 124
6 Direct Mass Measurements ... 126
7 Double Beta Decay ... 129
8 Neutrino Oscillations ... 130
9 Masses Below 10^{-2} eV (Solar Neutrinos) ... 131
10 Atmospheric Neutrinos ... 136
11 Reactor Neutrinos ... 140
12 Accelerator Neutrinos ... 142
13 Summary ... 147
References ... 148

Electroweak Symmetry Breaking and New Physics at the TeV Scale 153

T. Barklow, S. Dawson, H. E. Haber and J. Siegrist

1 Introduction ... 153
2 Weakly-Coupled Higgs Bosons ... 156
3 Low Energy Supersymmetry: Implications of Models ... 163

4 Low Energy Supersymmetry: Phenomenology ... 166
5 Strongly Coupled ESB Sector: Model-Independent Approaches ... 170
6 Strongly Coupled ESB Sector: Implications of Models ... 172
7 New Gauge Bosons ... 173
8 New Particles and Interactions ... 176
9 Anomalous Gauge Boson Couplings ... 179
10 Top Quarks as a Window to Electroweak Symmetry Breaking ... 182
11 Virtual Effects of New Physics ... 184
12 Experimental Issues at Hadron Colliders ... 186
13 Experimental Issues at e^+e^- Linear Colliders ... 187
14 Conclusions ... 189
References ... 190

Astrophysics, Cosmology, and Unification of Forces **195**
B. C. Barish, M. S. Turner and F. Wilczek
1 Executive Summary ... 195
2 Cosmology ... 195
3 Unification ... 200
4 Gravity ... 203
5 Interface with Cosmic Rays, Nuclear Physics, and Astrophysics ... 204
6 Experiment ... 205
7 Structural Issues ... 208
Selected Bibliography ... 209

The State of Exploratory Theory Beyond the Standard Model **211**
D. J. Gross, G. Kane and E. Witten
1 Introduction ... 211
2 Experimental Opportunities to Detect Supersymmetric Physics ... 213
3 Recent Theoretical Progress ... 214
4 Theoretical Challenges ... 215

Accelerator Physics, Technologies, and Facilities **217**
D. L. Hartill, S. D. Holmes, R. B. Palmer and R. D. Ruth
1 Introduction ... 217
2 Hadron Facilities ... 219
3 Electron Linear Collider Facilities ... 228
4 Electron–Positron Circular Collider Facilities ... 251
5 Advanced Accelerators ... 253
6 Conclusions ... 258
Acknowledgments ... 258
References ... 258

Detectors **261**

J. R. Bensinger, A. J. Lankford and D. R. Marlow

1 Introduction 261
2 Tracking 262
3 Calorimetry 266
4 Muon Detectors 269
5 Particle Identification 271
6 Large-Area Detectors for Non-Accelerator Physics 275
7 Electronics 278
8 Conclusions and Recommendations 280
References 282

Computing **287**

K. McFarlane

1 Introduction 287
2 Past Progress and Current Status 287
3 Mode of Operation 289
4 Technology Needs 289
5 Recommendations 292

Structural Issues in High-Energy Physics **293**

R. L. Brock, R. N. Cahn, A. J. Smith and M. Barnett

1 Introduction 293
2 Governance and Advising 293
3 Career Issues 295
4 Education and Outreach 296
Appendix A 300
Appendix B 300
References 301

PARTICLE PHYSICS

PERSPECTIVES AND OPPORTUNITIES

Report of the DPF Committee
on Long-Term Planning

The Character of Particle Physics

Roberto Peccei

University of California, Los Angeles, CA 90024

Michael E. Zeller

Yale University, New Haven, CT 06511

David G. Cassel

Cornell University, Ithaca, NY 14853

Jonathan A. Bagger

Johns Hopkins University, Baltimore, MD 21218

Robert N. Cahn

Lawrence Berkeley Laboratory, Berkeley, CA 94720

Paul D. Grannis

State University of New York, Stony Brook, NY 11794

Frank J. Sciulli

Columbia University, Irvington-on Hudson, NY 10533

From the time humans began to ask questions about themselves and their world, people have wondered what the world is made of and how it behaves. Over and over, in different ages and with different methods, people have tried to answer the questions, "What is the smallest possible piece of matter? What are the fundamental forces of nature?" From the time of the Greeks, with their designation of the elements Earth, Air, Fire, and Water, this ongoing quest has yielded extraordinary advances in knowledge that make our universe more understandable. Today, the fundamental science of high-energy, or particle, physics continues to pursue answers to these most ancient, and most modern, of questions.

Over the centuries of thought and experiment, there has gradually arisen a powerful belief in a simple framework underlying the seeming complexity of the universe. For example, in the 19th century, the periodic table of elements born of the Mendeleevian revolution gave simple family relationships to the welter of new elements that chemists had discovered, and led to the recognition that matter is made of atoms. In the early 20th century, further simplification came from the identification of the electron, proton and neutron as the constituent parts of atoms.

In the 1950's and 1960's, the discovery of hundreds of additional "fundamental" brothers and sisters of these three atomic constituents appeared to complicate these simple patterns. The query "Who needs so many building blocks?" led to a new level of understanding of the fundamental structure of matter. To explain the myriad particles, physicists conjectured even more elementary particles called quarks – the quarks that have now, in turn, been observed in experiments. Many studies world-wide have affirmed our current picture of the fundamental world of matter: all the particles of matter are composed of varying arrangements of six quarks (and their antimatter partners) and a corresponding set of six leptons, grouped into three pairings, called generations, with similar properties.

At the time the quarks and leptons were proposed, it was by no means understood that they came in three generations. In the original conjecture, three quark types – named 'up', 'down', and 'strange' – sufficed to explain the observations. Four leptons were also known. In

1974, experiments at Stanford Linear Accelerator Center (SLAC) and at Brookhaven National Laboratory identified a fourth quark. The mass of the new 'charm' quark was noteworthy – five times that of its strange quark partner in the second generation. For the moment, the symmetry of two generations of quarks and leptons seemed to signal the end of the road of particle discovery. Then, in 1976, an additional pair of leptons was found at SLAC. The following year, experiments at Fermi National Accelerator Laboratory found one more quark, the 'bottom', with a mass about three times that of the charm quark.

From theory we expect that the number of quarks and leptons should be equal, and several experiments implied that there are but six leptons. This symmetry thus strongly suggested that a final quark should exist – the 'top' quark, the partner to the bottom. The hunt for the top quark began in a series of ever more sensitive experiments at laboratories around the world. Near the end of the DPF study process, in early 1995, a dramatic confirmation of this picture was made with the discovery at Fermilab that the top quark does indeed exist. Indeed, the working group reports of this study contain many plaintive references to the need for the knowledge of the top quark's existence, its mass and its properties.

But the completion of the "periodic table" of matter comes with a twist: the top quark finally weighs in with a mass of about 40 times that of the bottom quark! And so, along with the satisfaction of completing the search for the elementary bits that everything is made of, come intriguing new questions. Is there a deep reason for the extraordinary massiveness of the top? Does it hold some clue to the underlying reason for nature's provision of mass to matter's building blocks? Why are there three full generations of quarks and leptons, and not one, two or more? The world we actually see and touch is in fact made of quarks and leptons of only the first generation. Why are there more?

Our understanding of the fundamental forces has evolved along with our growing knowledge of the particles of matter. Early observations identified the gravitational, electric, and magnetic forces as distinct, but the multitude of everyday phenomena seemed to be governed by a long list of idiosyncratic forces, each operating in special circumstances. A large step toward simplification came in the mid-19th century with Maxwell's unification of the electric and magnetic forces into a single electromagnetic entity. Fifty years later came the recognition that the newly discovered atoms are also governed by the same force. By the late 1800's, all commonly observed phenomena could be understood with only the electromagnetic and gravitational forces.

At the turn of the century, new experiments began to explore the subatomic world. They revealed the necessity for a very short-range 'strong' force to bind together the elements of the nucleus, and a similarly short-range 'weak' force to account for radioactive decays. All four forces – electromagnetic, gravitational, strong, and weak – became more understandable when the development of quantum mechanics and the quantum theory of fields allowed physicists to postulate the existence of the force-carrying particles called bosons. In our current understanding, the electromagnetic and weak forces are further unified as the electroweak force, carried by combinations of the ordinary photon of light and three very heavy bosons called W^+, W^-, and Z^0. The W and Z bosons were discovered in 1983 at European Center for Nuclear Research (CERN), and demonstrated that at very large energies, the weak and electromagnetic forces merge into a single entity.

The understanding of the strong force that binds the nucleus has also undergone stages of simplification over the past 30 years. Early descriptions focused on dozens of force-carrying particles and a tangle of difficult-to-determine parameters. Again a single force-carrier operating at the level of quarks was postulated. This theory, called quantum chromodynamics or QCD was based on the properties of electromagnetic theory (QED). QCD postulates the existence of massless gluons, bosons which mediate the strong force; these were discovered in the 1970's experiments at the Deutsches Elektronen-Synchroton (DESY) in Germany.

Of all the forces, gravity remains the odd man out. We expect there to be a force-carrying graviton, but gravity is such a weak force that observation of the graviton has thus far eluded experiment.

However the satisfying simplicity and harmony of the picture of the forces again evokes questions. Having gone from many forces to just three, we ask whether further unification cannot be made. It seems plausible to join the electroweak and QCD forces. This would satisfy our intuitive guess that nature is at root simple and elegant. In our dreams, we envision still further unification incorporating quantum gravity as well. We even know roughly at which energy the various forces should reach equal strength and thus exhibit their underlying unity. But present experiments have also shown us that with only the three generations of quarks and leptons now known, it would seem unlikely that the forces we observe in our labs today could evolve to the unification scale. The price to be paid for tidiness and the dream of further unification of forces may thus be a further revolution in our picture of the fundamental constituents!

Particle physics has brought answers to many questions that once seemed all but unanswerable. Through the study of particles and forces, we have found crucial keys to the way the universe itself developed in the first moments following the burst of energy called the Big Bang. We see that the pattern of particles uncovered in our present-day laboratories influenced the initial conditions of the universe so as to produce the particular world that we experience today. Among the most mysterious links between cosmology and particle physics are two seemingly disparate observations. First, by observing the heavens, we see that the universe contains a preponderance of matter over antimatter. We thus conclude that the laws of nature must somehow discriminate between these opposite forms. Second, our seemingly unrelated study of particles tells us that the simplest pattern of quarks that allows for such matter-antimatter distinction requires at least three generations of quarks. In the past two years, striking new progress in understanding the distinction between matter and antimatter has come from experiments at the Cornell Electron Storage Ring.

Thus, in a broad, and perhaps unsatisfying anthropomorphic manner, we can suggest that in order for matter – that which now forms stars, people and proteins – to exist without having been annihilated in the very early universe, we need the three pairs of quarks and leptons that had on first sight seemed redundant. In this surprising turn, our own existence seems to depend upon the existence of the newly discovered top quark, whose life lasts but a trillionth of a trillionth of a second and whose earthly utility is hard to imagine!

We recognize that our present understanding of the constituents of matter and fundamental forces, though conceptually pleasing and consistent with all our observations so far, is likely to be only a stage on the way to deeper appreciation of the structure of our world. We know the taxonomy of particle and forces, but we know little as yet of what underlies our classification system. Nevertheless, the questions we can now think to ask were unthinkable a few years ago. "What is the origin of the mysterious attribute of matter called mass?" "What causes the peculiar pattern of quarks and leptons with their strong family relationships but differing individual characteristics?" "How did the particles we see in the laboratory influence the character of the universe at large?" "Can we find connections between such seemingly disparate phenomena as black holes and fundamental particles, as twists in space-time?"

The reports of this DPF study describe many opportunities for progress in the exploration of these fundamental questions, both with the facilities now in operation and through the building of new accelerators and experimental detectors based on proven technologies. The study is the product of many of the active particle physicists in the U.S, a group consisting of some 2000 individuals ranging from university faculty and laboratory scientists to postdoctoral and predoctoral students. It is an assessment of the promise and possibilities for scientific progress from the point of view of the participants – not necessarily those of the the existing laboratories or governmental funding agencies – and addresses the field from the perspective of global physics opportunities.

The global character of particle physics should be underlined. The U.S. community of about 2000 individuals is strongly interconnected and is embedded within the world community of about 6000 particle physicists, both in coordinated attack on inter-related questions and through a web of mutual collaborations in experimental and theoretical studies.

Much of the experimental research is conducted at particle accelerators at Fermilab, SLAC, Brookhaven and Cornell in the U.S. and at facilities at CERN and DESY as well as those in Japan, Russia, China, and Italy. These accelerators bring a variety of particles into collision – electrons and positrons, protons and antiprotons, and nuclei – with differing combinations, energies and intensities. Each has particular advantages for studies of elementary structure and forces. Together they give complementary views of phenomena.

There are some questions for which the accelerators do not provide the relevant capability. Fundamental questions such as the possible instability of the proton; observation of neutrinos produced by our sun, in the earth's atmosphere or by supernova collapses; or studies of rare nuclear decays require specialized facilities, often in deep underground or underwater experiments shielded from the bombardment of cosmic rays. Particle physics also has very strong interconnections with astrophysics and cosmology through experiments directed at the search for cosmological dark matter candidates, for gravitational waves, and studies of such astronomical phenomena as supernovae, black holes and distributions of galaxies in the universe.

The experimental particle physics activity is intertwined with the theoretical community, interacting daily at the universities and Laboratories, with each sub-community stimulating advances within the other. The theorists play dual roles in interpreting and correlating the experimental findings and in developing new frameworks and conjectures which guide further experimental study. The health of the field depends crucially on the close and supportive roles of theory and experiment.

Finally, there is a crucial group of physicists focussing upon the development of more incisive instruments. The research into accelerators, with both experimental and theoretical components, is a crucial fulcrum upon which progress depends. Similarly, the development of experimental detectors, electronics, computing and data-handling techniques is a prerequisite to the field's ability to meet the challenges ahead. Again the interfaces between these endeavors and the other sub-communities are critical. In many cases, individual physicists contribute to more than one of the three main sub-communities.

Our report first discusses the current status and future prospects of particle physics and then issues related to infrastructure and the tools required for further progress with our research. The physics discussions are separated into those topics which build upon our present paradigm called the Standard Model, and those which look beyond it. Each is the product of the appropriate mix of experimentalists, theorists, and instrumentalists and accelerator physicists.

The topics relating to the physics of the Standard Model are divided into four chapters: the electroweak force properties of the W and Z bosons together with the interplay of these with the top quark and the Higgs boson; the study of the constituents of the proton and the elucidation of the strong QCD force; the properties of the heavy quarks (strange, charm and bottom) and the violation of matter-antimatter symmetry; and the exploration of the properties and interactions of neutrinos both in accelerator experiments and underground detectors.

The topics beyond the Standard Model include three chapters: new physics which could serve to unify all forces and explain some of the mysterious patterns observed in current data; the connections between particle physics and astrophysics; and the outlook for the theoretical investigations which extend and simplify presently disparate observations.

The discussion of the structural aspects of our discipline comprises four chapters: the prospects for the development of new accelerator techniques; the possibilities and means for advancing new detector technologies; the evolution of computing for acquiring and analyzing the data from particle experiments; and the structures for governing the progress of the field, nurturing the young physicists and educating the broader public.

This overview has focused on the evolution of particle physics to the present level of understanding, and on the nature of the questions that we are poised to explore. It is indeed true that the basic urge to pursue such studies originates in our innate desire to understand the character of our world. Analogies are often made with our urge to create and enjoy music, art or literature, as a fundamental aspect of our humanity. However, this is not to gainsay the pragmatic benefits of scientific research. Every basic scientific field of study has enriched society by the new understanding gained. The ability to produce the materials to clothe, feed and serve us results from the understanding of atoms, molecules and the chemical bond. The understanding of the enormously complex couplings of organic macromolecules has led to tremendous improvements in our ability to heal and prolong life. Our electric power and transportation industries are dependent on esoteric studies of electric, magnetic, and thermal phenomena of a century ago. Our computer-based information age rests upon the intellectual foundation of the quantum revolution early in this century. All scientists are proud that our studies can simultaneously serve to illuminate the fundamental character of the world that we inhabit and contain a spark that, somewhere down the road, enriches our individual and collective lives.

From the present vantage point, we can appreciate both the panorama of past achievements and the next range of peaks to climb. It is with pride that we review in this study the recent path that we have scaled in understanding the constituents of matter. We find it an achievement of impressive proportions that we have glimpsed the regularities and order inherent in the physical world.

In addition to describing our field we have attempted to present and explain, as well as we can foresee, what opportunities exist for the next ten years, and where the best chances for real progress lie. In the year that we have made these studies, we have already seen discoveries that open new vistas. We have begun to plan for the next stage at new accelerators in this country and abroad. We find reason to hope for increasing internationalization of an already global community, and look for the experience in particle physics to encourage other world-wide collaborative efforts. Most of all, we are delighted that the range of absorbing and fundamental questions, far from ending with the discovery of the "last" quark, can be expected to lead to new understanding that enriches our culture and benefits our society.

Executive Summary

Roberto Peccei

University of California, Los Angeles, CA 90024

Michael E. Zeller

Yale University, New Haven, CT 06511

David G. Cassel

Cornell University, Ithaca, NY 14853

Jonathan A. Bagger

Johns Hopkins University, Baltimore, MD 21218

Robert N. Cahn

Lawrence Berkeley Laboratory, Berkeley, CA 94720

Paul D. Grannis

State University of New York, Stony Brook, NY 11794

Frank J. Sciulli

Columbia University, Irvington-on Hudson, NY 10533

1 Introduction

Particle physics comprises the study of the most basic components of matter and the forces that govern their interactions. These subjects are the bedrock upon which the physical sciences rest, and ultimately govern such diverse phenomena as the evolution of the universe from the moment of the Big Bang, the generation of energy in the stars, the properties of atoms and nuclei, and the understanding of the forces that bind protons and neutrons and stimulate radioactive decays.

Over the past two decades, substantial progress has been made in identifying the basic constituents of matter, their patterns and relationships, and in understanding and partially unifying the fundamental forces between them. These advances are embedded in the paradigm called the Standard Model of particles and forces. This model makes predictions that agree with virtually all existing experimental data. However, the Standard Model has many components and parameters that are arbitrarily inserted to fit the observations, and it offers little insight into the deepest questions of why the observed arrangements of particles exist and how the forces governing them may be more completely unified.

A rich program of investigations can shed light on these fundamental issues and the instruments available to study them are quite varied. The Tevatron at Fermilab, currently the highest energy accelerator, can study the properties of the newly discovered top quark and can search for new particles and phenomena in a heretofore unexplored energy range. At the other extreme, experiments attempt to determine if neutrinos have mass, an issue that relates intimately to our understanding of the future of the universe. For the next decade there are facilities under construction and in operation in the United States that will continue these studies and that offer great opportunity for expanding our knowledge.

While the program that has been set in place is diverse and has great potential, there are major advances that can only be made through accelerator-based studies at energies about ten to twenty times those available at present facilities. Compelling scientific arguments persuaded particle physicists to try to reach these energies by constructing the Superconducting Super Collider

(SSC). The SSC will not be built, however, and the aspirations of the particle-physics community for understanding these fundamental issues now rest upon the recently approved Large Hadron Collider (LHC) at CERN in Switzerland and upon accelerators, yet to be proposed, that might be built in the future.

When the survival of the SSC became uncertain in the summer of 1993, the Executive Committee of the Division of Particles and Fields (DPF) of the American Physical Society began serious discussions of steps that would best help our community face the challenges of the future. These discussions were strongly influenced by the loss of the SSC and the subsequent appointment of the Drell Subpanel of the Department of Energy's High Energy Physics Advisory Panel (HEPAP) to help provide a vision for the future of the field. Since the Drell Subpanel had a specific charge and a relatively short time for deliberation, the DPF Executive Committee felt that the high-energy-physics community and the public on whose support it relies would profit by having an additional broad-based and in-depth study of the field. This study, called the Committee on Long-Term Planning, was initiated in early 1994 and its fruits are the reports contained in this volume.

This study of high-energy physics was organized on the basis of working groups. Of the eleven working groups in the study, seven dealt with specific physics topics and three with more technical areas (accelerators, particle detectors, and computers). One concentrated on structural and educational issues of the field. Broadly speaking, four of the physics working groups dealt with physics issues connected with the Standard Model, our present paradigm for particle physics, while the other three working groups focused on questions beyond our present knowledge, but whose answers are crucial for a deeper understanding of nature. The charge to each of the physics working groups was to articulate the broad range of physics questions in the group's particular area and to discuss the means by which these questions might best be addressed, in the context of existing and future facilities in the United States and in the rest of the world. Each of these working groups was asked to identify important areas of emphasis for future study.

The accelerator physics working group was charged to undertake a study of the current state of the art and of foreseeable advances in accelerator facilities for high-energy physics (HEP) research. The particle detectors working group and the computing working group were asked, respectively, to assess the likely detector technology and computational needs of ongoing and future high-energy physics experiments, considering the impact of existing and emerging technologies. Finally, the structural and educational issues working group was charged with examining various issues in our community that are of broad concern. These include the adequacy of the present advisory structure in high-energy physics; suggestions for possible additional mechanisms to more adequately plan for the future; and whether our community should seek a structured public education role and, if so, how broad should such an activity be.

Each of the working groups was led by conveners chosen among established leaders in the area under consideration. The conveners then assembled their groups by inviting other active workers in the field to participate. Most of the working groups also gathered further input from the community by holding various specialized meetings and workshops. The working groups presented their preliminary findings at a Workshop at Johns Hopkins University in May 1994, and their final reports at the DPF general meeting in Albuquerque the following August. Draft written versions of these reports were critically discussed at a retreat of all conveners with the members of the DPF Executive Committee at the Lawrence Berkeley Laboratory in October 1994. The corrected draft versions of the reports were examined one more time by an editorial board appointed by the DPF Executive Committee.

The reports of the working groups, along with their conclusions and recommendations are summarized in the following sections.

2 Precision Tests of Electroweak Theory

The theory of electroweak interactions unites the electromagnetic and weak interactions and is the cornerstone of the Standard Model. In this theory, the electroweak forces arise from interactions of four bosons, the photon, Z, and $W^{\pm}$, with the quarks and leptons of matter. Gauge symmetry is a fundamental ingredient of this theory. The gauge symmetry of the electroweak interactions implies that the masses of the photon, Z, $W^{\pm}$, and the quarks and leptons all vanish. Some mechanism is required to break this symmetry and provide the observed masses of particles. In the standard electroweak theory, the simplest way to produce this symmetry breaking is via interaction with an additional particle, called the Higgs boson.

The resulting electroweak theory depends on three primary parameters, which can be taken to be the electromagnetic fine structure constant, α, the Fermi constant of weak interactions G_F, and the mass M_Z of the Z boson. It is then possible to predict many other important quantities, such as the mass M_W of the $W^{\pm}$ bosons, in terms of these parameters. Actually, such predictions also depend in detail on quantities like the masses of the quarks and the Higgs boson. Very extensive tests of the electroweak theory have been carried out in a variety of circumstances, most notably in the production and decay of the Z boson. The agreement of these tests with the theory is remarkably good, often at the level of 1%

or better. In addition, from the slight dependence of predictions on the mass of the top quark, it is possible to estimate its mass. This prediction of approximately 175 GeV is in excellent agreement with recent direct measurements of the top mass by the CDF and DØ collaborations at Fermilab.

Despite these notable successes, more stringent tests of the electroweak interactions are important. There may well be new interactions outside the Standard Model, or new particles that are too massive to be produced and detected at current accelerators. These phenomena can cause small deviations from expectations for measurable quantities and precision measurements can provide circumstantial evidence for their existence. There is an active experimental program planned worldwide that will provide significant improvement in precision, with a substantial reduction in the errors for key electroweak parameters.

Forthcoming data from LEP at CERN and from polarized Z's at the SLC at SLAC will sharpen our knowledge of electroweak interactions at the Z. Further improvements in the accuracy with which we know the W boson's mass will provide powerful complementary information. This will come from data now being gathered at the Fermilab Tevatron Collider and from future running with the Main Injector, as well as from data from LEP 2 at CERN, a machine that is scheduled to begin operation in 1996. The ability to pin down the value of the top quark mass with high precision will also be essential for more stringent tests of the electroweak theory. Only Fermilab will have sufficient energy to explore directly the properties of the top quark until the turn-on of the LHC.

Current investigations will surely deepen our understanding of electroweak interactions and may presage new physical phenomena, whose full exploration will need higher energies. A case in point is the Higgs boson, which can be directly detected at LEP 2 if its mass is less than about 90 GeV. This search could perhaps be extended if the Fermilab Collider luminosity is upgraded significantly, but above this mass range the detection of the Higgs boson will require the LHC or a high-energy electron-positron (e^+e^-) collider. Nevertheless, precision electroweak measurements may help to further narrow the expected mass range for this possible particle.

The theory of the electroweak interactions is one of the two pillars which underlies the Standard Model, the paradigm for how the fundamental constituents of matter interact. Continued probing of this theory by experiments of increasing precision remains central to particle physics and has two principal goals:

- to refine our knowledge of the parameters that characterize the Standard Model and the breaking of electroweak symmetry, (requiring improvements in the accuracy with which we know the masses of the W bosons and of the top quark) and
- to look for small discrepancies between experiment and theory that, if found, will point to new physics beyond our present understanding.

3 Quantum Chromodynamics

Quantum chromodynamics (QCD), the theory of the strong forces between partons (quarks and gluons), is the second central ingredient of the Standard Model. In QCD, the force between quarks is carried by bosons called gluons. Indeed, an important feature of QCD is the fact that there would be a strong force even without the quarks because gluons interact with each other. The strength of the interaction, called α_S, is not a constant but depends on the energy scale at which it is measured. At the mass of the Z boson its value is about 0.12 – about a factor of 16 larger than α, the corresponding strength of the electromagnetic interaction.

Since its formulation in the early 1970's, QCD has developed into a mature theory that is well verified in a variety of experiments. While the evidence is compelling that QCD is the correct theory of the strong interactions, we cannot say that we understand all phenomena completely. This is not an entirely new situation: although we believe that quantum mechanics and electromagnetism describe completely our everyday world, we cannot yet explain quantitatively many complex phenomena. Some processes in QCD can be understood with perturbation theory (expansions of the interactions between the partons in low orders of α_S); others, such as the formation of bound states of quarks, called hadrons, cannot. In particular, at present in QCD there is still a very incomplete understanding of how perturbative regimes are connected with non-perturbative regimes; of quark confinement (the experimental absence of free quarks); of the high temperature and high density phases of quarks and gluons; and of the reason for the absence of strong CP violation.

There is a large and active experimental program worldwide pursuing different aspects of QCD. When explored at high energies, protons are seen to contain additional quark-antiquark pairs and gluons, besides the u and d quarks of the simple quark model of hadron structure. The measurements of the distributions of these parton densities has been an important thrust of high-energy physics for nearly thirty years and continues at both hadron and electron machines. Data obtained at the new electron-proton (ep) collider HERA in Hamburg show a large increase in the quark and gluon density at small values of the Bjorken scaling variable x, and may be the first indication of an eventual saturation of parton densities (a new non-perturbative domain of QCD at very high parton densities). Experiments at CERN and SLAC are giving new information on how the proton spin is shared among the constituents. Studies of jet decays of the Z's produced in e^+e^- collisions at LEP and SLAC have given very accurate understanding of the coupling and underlying structure of QCD processes, and provide insights on the differences between quark and gluon jets.

Experiments at hadron machines are a rich source of complementary information about QCD. Whereas in e^+e^- and ep collisions, one proceeds from a clean and well understood initial state to a final state involving parton jets, in hadron collisions one may study simple final states (γ, W, and Z) free from fragmentation effects but with partonic initial states that are less well understood. This complementarity is important because it allows cross checks of the effects of proton structure and fragmentation into hadrons. The very high energy of hadron colliders enables sensitive studies of QCD and the measurement of α_S at short distances from high transverse momentum production of jets and vector bosons, and their angular correlations. The increased energy, masses, and phase space available also permits new tests of gluon radiation calculations "resummed" to all orders. Further insights into the transition from a perturbative to a non-perturbative regime of QCD can come from studies of diffractive scattering and events with large kinematic gaps between clusters of particles in ep and hadron-hadron collisions.

Comparison of observations depending on the same underlying distributions provided by different measurements give important cross-checks and enhance our understanding of QCD. The high precision jet data now being gathered warrant refined theoretical calculations beyond those that currently exist, and an ongoing interaction between theorists and experimenters is both helpful and necessary for continuing progress.

Among the phenomena that may reveal surprises are systems with a high density of quarks and gluons. Such systems can be created by colliding nuclei at very high energy, which will become possible soon at the Relativistic Heavy Ion Collider (RHIC) at Brookhaven. Heavy ion collisions offer the possibility of studying a new phase of QCD where the separate nucleons coalesce into a plasma of quarks and gluons. A major challenge here is to understand how, at early times in the collision, the transition from hadronic degrees of freedom to an equilibrated quark-gluon phase takes place. This is a field where nuclear physics and particle physics merge, and where perturbative physics, non-perturbative physics and phenomenology are essential in order to extract the desired information that these collisions can provide on high density and high temperature QCD.

Because of confinement, quarks and antiquarks are bound together by a force which increases with the distance of their separation. As a result, the properties of hadronic bound states cannot be determined by straightforward theoretical techniques. Lattice gauge calculations, treating space-time as discrete, are an important tool for probing the implications of QCD in the strong coupling regime. These methods are just now beginning to produce reliable results for low energy phenomena. With more computing power and further progress on the development of algorithms, the dream of a reliable calculation of the hadron mass spectrum and of low energy hadronic matrix elements may be close. Lattice calculations for α_S are now already competitive with perturbative derivations and should eventually provide the most precise determination of this fundamental quantity.

Even though the present experimental evidence for QCD is compelling, QCD deserves further intense experimental scrutiny in diverse circumstances to probe areas of the theory where our understanding is far from complete. These include:

- continued studies at Fermilab of the production and correlations of jets, W's, Z's, γ's, and b quarks at large transverse momentum and all angles;
- studies of hadronic final states in e^+e^- collisions with well-defined initial conditions;
- measurements aiming at elucidating the proton's structure at high energy and low x in ep collisions at HERA;
- studies of how the spin of the hadrons is carried by the underlying partons;
- exploration of diffractive phenomena in both ep and hadron-hadron colliders;
- searches for states (*e.g.*, glueballs, hybrids, and strangelets) allowed by QCD but not yet observed; and
- searches for indications of a quark-gluon plasma in high-energy collisions between nuclei at RHIC.

4 Heavy Flavor Physics and CP Violation

Our tangible world is made of electrons and up (u) and down (d) quarks, which are constituents of the proton and neutron. Neutrinos (ν's) are intangible, but also present. These four basic particles appear to be sufficient to explain ordinary matter. Nonetheless, there are two additional sets of four particles that follow the same pattern: the "second generation" of charm (c) and strange (s) quarks, with the muon (μ) and its neutrino, and the "third generation" of top (t) and bottom (b) quarks, with the tau (τ) and its neutrino. These three generations of quarks and leptons and the associated mechanism by which these particles decay into each other are an essential ingredient of the Standard Model. The reason for the triplication is unknown, but it allows the Standard Model to accommodate matter-antimatter asymmetry due in part to the phenomenon known as CP violation.

The weak interaction can turn one kind ("flavor") of quark into another. In nuclear β decay a d quark becomes a u quark while an electron and an antineutrino are emitted. There is a general propensity for quarks to remain within one family. Thus a c quark prefers to change to an s quark rather than a d quark. These propensities can be described quantitatively in terms of a unitary matrix, V, called the Cabibbo-Kobayashi-Maskawa (CKM) matrix. In this description, V_{cs} is a factor in the amplitude for a transition from a c quark to an s quark, while V_{cd} is the smaller factor associated with a transition from a c quark to a d quark.

In the Standard Model the CKM matrix arises from the mismatch between the quarks as seen by the strong and the weak interactions. Unitarity and the fact that only relative phases are observable imply that four independent real parameters determine the nine complex elements of the matrix. It is possible to test this expectation by measuring more than four of the matrix elements and seeing if they are related in the anticipated way. To do this, many different weak decay processes involving many different kinds of particles must be measured carefully.

Understanding the CKM matrix is important for two reasons. First, it seems likely that the CKM matrix is intimately connected to the process through which the quarks get their masses, that is, to the electroweak symmetry breaking mechanism. Second, the CKM matrix may also play an important role in CP violation. CP is short-hand for the joint application of the operations of replacing particles by their antiparticles (C), and replacing right handed particles and processes by left handed ones (P). The violation of CP symmetry has been seen in the decays of K^0 mesons, one type of particle containing strange quarks.

CP violation is an essential element in understanding the observed dominance of matter over antimatter in the universe. In the Big Bang, matter and antimatter should have been produced in equal amounts. If CP were an exact symmetry, this primordial matter and antimatter would have annihilated each other shortly after they were produced.

Although the CP violation that can be described by the CKM matrix may not be fully responsible for the dominance of matter in the universe, it is still important to learn all we can about it. For thirty years, the decays of K^0's has been the only source of our knowledge of CP violation. Crucial experiments at Fermilab, CERN, and Frascati in Italy, are now being readied to explore further CP violation in this system, looking for tiny effects that should help elucidate whether its origin can be traced to the CKM matrix.

The decays of B mesons, particles containing b quarks, can provide new and more complete understanding of the CKM matrix. The decays of B mesons are already being studied very effectively at CESR, the e^+e^- storage ring at Cornell, as well as at LEP and at the Tevatron collider. These studies provide direct measurement of two CKM matrix elements, V_{cb} and V_{ub}, and indirect measurements of some others. However, a full understanding of the role of the CKM matrix in CP violation requires intricate and difficult measurements on many more B decays than have been observed so far. B Factories, especially designed e^+e^- colliders with high intensities and unequal beam energies, are being built at SLAC and at KEK in Japan to facilitate these measurements.

High-energy hadron machines, including the Tevatron Collider, which collide strongly interacting particles, are more prolific sources of B mesons. However, B's are produced in only a tiny fraction of the vast number of events that reach the detectors and the B's are produced in a much more complex environment, making the desired particles and correlations more difficult to observe. Full exploitation of these high event rates will thus require advances in detector technology, particularly in the ability to "trigger" on B events (selecting B events quickly and efficiently at some stage well before events are fully processed in a computer). The high production rate of b quarks in high-energy hadron-hadron collisions has led many to suggest there should be a dedicated b physics facility based on a hadron collider. The desirability of such a facility deserves careful consideration, with attention to the additional reach it would provide beyond that given by the e^+e^- B Factories. A vigorous and dedicated b physics facility at a hadron collider may also be able to fill the gap which will result when the LEP b program ceases with the turn-on of LEP 2.

It is an old adage of particle physics that everything that is not forbidden is compulsory. But what is absolutely forbidden? A decay like $\pi^+ \to e^+\gamma$, which would violate conservation of angular momentum, certainly seems to be forbidden. Some decays which have no such fundamental problem, like $K^0 \to e^+\mu^-$, have never been observed. From the absence of such processes we infer that electron number and muon number are separately conserved. These and similar conservation "laws" are included in the Standard Model, even though no compelling theoretical reason for their existence has been found. Whether these conservation laws are absolute, however, is an experimental question which can be investigated especially well with K beams, as in the current program at Brookhaven National Laboratory. The observation of violation of any of these lepton-number rules would have profound consequences. Because we are unable to predict in which sector (s, c, b, t, or τ) evidence for such new physics might appear, our studies need to remain as broad as possible. In addition, rare decays of heavy quarks provide alternative possibilities for studying CP violation and may reveal signs of CP violation beyond that predicted by the CKM matrix.

The present U.S. experimental program in heavy flavor physics is healthy and extensive and includes:

- CLEO at Cornell (b, c, and τ),
- the Fermilab collider program (b and t),
- the Fermilab fixed target program (c and s), and
- rare K decay studies at Brookhaven (s).

In addition, there is substantial U.S. involvement in the LEP experiments at CERN (b, τ), and in BES in Beijing (c, τ).

The approved upgrades to existing U.S. facilities (Phase III of CESR plus CLEO III and the Fermilab Main Injector plus CDF and DØ upgrades) and approved new facilities (SLAC B Factory) assure a continued healthy program. However, the long-range future of K physics is less certain as Brookhaven's role shifts with the advent of RHIC, although new opportunities may arise with the turn-on of the Main Injector at Fermilab.

The mysteries of mass, flavor, and CP violation, which are likely to have deep connections to each other, compel us to continue a vigorous experimental and theoretical program of studying heavy flavor decay. The two most important goals of this research are:

- to provide enough measurements to unambiguously overconstrain the CKM matrix and clarify its role in CP violation, and
- to search for rare and forbidden decays in order to exploit their sensitivity to physics beyond the Standard Model, at mass scales exceeding those that can be directly probed at accelerators.

Continued progress towards these goals requires more joint experimental and theoretical effort in understanding the interplay of the strong, weak, and electromagnetic interactions in decay processes.

5 Neutrino Mass and Mixing

The field of neutrino physics is diverse and is carried out in experiments at university laboratories, deep underground sites, and accelerators. The scientists come from particle and nuclear physics, astrophysics, and cosmology and are comprised of both experimentalists and theorists. Experiments range from mass measurements of neutrinos in decays of particles and nuclei, to observations of neutrinos produced in our sun and in supernovae, to studies of neutrinos produced by accelerators and in reactors.

Forty years after the first direct detection of neutrino interactions with matter, our knowledge of the three families of neutrinos and their interactions has improved dramatically. However, crucial aspects of our understanding are still lacking. Many fundamental properties, particularly neutrino masses and mixing of the neutrino states, are known only as experimental upper limits. Indeed, the issue of neutrino mass is central to the question of whether the neutrino is a Dirac particle (with a distinct antiparticle) or a Majorana particle (being its own antiparticle). In the search for physics beyond the Standard Model, new information in the neutrino sector may well play an essential role.

The question of neutrino mass and mixing not only is important for particle physics, but has also enormous relevance to astrophysics and cosmology. For example, a nonzero neutrino mass would contribute to the nonbaryonic dark matter, and perhaps significantly to the total mass of the universe. If neutrinos are massive, it is likely that they mix in a manner similar to that seen in neutral K mesons. A neutrino produced as a partner of a muon will initially possess muonic nature, but as time passes, would evolve a component of electron-type or tau-type neutrino. Such changes can be observed in a variety of ways by observing neutrino interactions at a distance from the source. Several substantial indications are emerging that these neutrino oscillations might exist.

One hint is provided by the solar neutrino problem. The intensity of neutrinos produced in the interior of the sun may be calculated from known nuclear processes, which ultimately produce the radiant energy visible at the solar surface. All measurements of this electron-type neutrino intensity at the earth are considerably lower than the rates expected from the standard model of energy production from the sun. This is hard to accommodate even by highly nonstandard astrophysical or nuclear physics mechanisms.

Another hint comes from cosmic rays; measurements of neutrinos produced by the interaction of cosmic rays in our atmosphere indicate that the relative number of electron-type and muon-type neutrinos differs from expectations by about a factor of two. Neutrino oscillations of muon-type neutrinos are a serious possibility to explain this discrepancy.

Finally, very recently, another indication comes from an accelerator-based experiment at Los Alamos, where an anomalous signal might be interpreted in terms of neutrino oscillations.

Although oscillations between different neutrino species may provide an attractive explanation for some or all of these data, a full resolution of these issues with further experiments is essential. For the solar neutrino problem the next generation of solar-neutrino detectors, the Sudbury Neutrino Observatory (SNO) in Canada, SuperKamiokande in Japan and Borexino in Italy, should be able to look for signatures of solar neutrino interactions and isolate effects of electron-type neutrino oscillations independently of solar models.

Accelerator experiments may be the best way to resolve the hints of muon-type neutrino oscillation provided by the atmospheric neutrino results. Long baseline experiments sensitive to oscillations by muon-type neutrinos are being proposed for the KEK, Fermilab, and

Brookhaven accelerators and could begin taking data in the not too distant future. Oscillations of electron-type neutrinos in the same parameter range should be visible in experiments beginning shortly at the San Onofre and Chooz reactors.

New information about the nature of neutrinos is likely to make important, perhaps vital, contributions to our understanding of the universe and of what lies beyond the Standard Model. For this reason, it is important that innovative experiments exploring neutrino mass, couplings, and other properties be encouraged and that indications of nonzero mass and/or flavor mixing be vigorously pursued. Fortunately, an active program of neutrino experiments is underway or planned:

- Direct mass measurements of electron-type neutrinos in β decay are probing masses of cosmological interest. Other experiments seeking evidence for double β decay with no neutrinos may provide the first demonstration of the Majorana identity of the neutrino.
- Indications of neutrino oscillations from solar neutrinos will be probed by at least three new experiments with significant U.S. participation (SNO, SuperKamiokande, Borexino). These experiments should be able to resolve the issue of whether the observed deficit of solar neutrinos has an astrophysical explanation or instead requires neutrino mixing.
- Short baseline experiments presently underway at CERN and planned for Fermilab have the potential to explore cosmologically relevant mass ranges for ν_μ oscillations to ν_τ (or ν_e), with excellent mixing sensitivity.
- Full exploration of the hints of oscillations seen in atmospheric neutrino interactions awaits direct measurements by long baseline experiments using accelerator-generated neutrino beams.

6 Electroweak Symmetry Breaking and New Physics at the TeV Scale

The electroweak theory is based on the realization that the quantum of light, the photon, and the quanta of β decay, the $W^\pm$ bosons, are intimately related. Just as isospin, a symmetry of strong interactions, identifies the neutron and proton as partners, a new symmetry, weak isospin, identifies the electron (or rather, its left-handed piece) and its neutrino as partners.

The electroweak symmetry is far from exact. The W and Z bosons are among the heaviest known elementary particles, while the photon is the lightest, though they are related by this symmetry. Similarly, the neutrino and the electron can hardly be confused, even though they are partners.

How is the electroweak symmetry broken? The rotational symmetry of an atom can be broken by applying a magnetic field, which explicitly distinguishes the direction of the field. Theory requires that electroweak symmetry not be broken in this way, but more subtly. Without the symmetry breaking, the $W^\pm$ and Z, and all the quarks and leptons would be massless. If any progress is to be made in understanding these masses, the source of electroweak symmetry breaking must be discovered. This is one of the core questions in high-energy physics today.

In one approach to electroweak symmetry breaking, a number of scalar (spinless) particles are introduced. Some of these particles are absorbed by the W and Z bosons when they acquire masses. The remaining scalars appear as new particles. In the simplest version, there is just one such particle, the Higgs boson.

A second possibility, called supersymmetry, predicts the existence of many new particles, among them a number of scalars like the Higgs boson. While there is no direct evidence for supersymmetry, there is strong theoretical motivation for it. There is also some supporting circumstantial evidence from extrapolating the electroweak and strong couplings to high energy, where the three couplings coalesce – if supersymmetric effects are included – as they should if there is a unification of these forces at high energy (grand unification).

A third possibility, referred to as strongly coupled electroweak symmetry breaking, introduces no new scalar particles but requires that their roles be played by bound pairs of new, heavy analogs of the usual quarks, dubbed techniquarks.

Which of these theories, if any, is the correct description of nature can be determined only by experiment. Finding the Higgs boson or any of the other new particles that appear in models of electroweak symmetry breaking would be a major step forward. The negative results of current experiments bound the Higgs boson mass to be above about 60 GeV. This range will be extended to about 90 GeV at LEP 2. Supersymmetric analogs of the Higgs boson could also be discovered at this accelerator if they were light enough.

While we cannot predict the mass of the Higgs boson we do have confidence that whatever the source of electroweak symmetry breaking, there must be indications for it below about 1000 GeV (1 TeV). Therefore, it is crucial to explore the TeV energy scale to elucidate the dynamics of electroweak symmetry breaking and the associated fundamental particle spectrum. This program necessarily will be a primary focus of particle physics research in the future and, at present, can only be contemplated at the LHC.

An extensive study of the electroweak symmetry breaking sector will require a robust experimental program at the LHC. However, finding the source of electroweak symmetry breaking is not the only goal of the LHC. Our current picture of particle interactions appears incomplete so that there may well be new phenomena within the LHC's range. One possibility is that there may be more particles like the W and Z bosons, which transmit new forces. Moreover, due to its mass reach, the LHC is in an excellent position to test for low-energy supersymmetry by looking for the signatures characteristic of the decays of strongly-interacting supersymmetric particles, missing transverse momentum and multi-lepton final states. Full exploitation of the potential for these searches requires the energy and luminosity of the LHC, but the substantial luminosity improvements now being proposed for the Fermilab collider should also permit substantial progress in the next decade. For example, the upgraded Fermilab collider with $q\bar{q}$ initial states (as opposed to the gluon-gluon initial states predominating at the LHC) could allow the observation of a Higgs boson with mass up to 130 GeV in association with a W or Z.

An electron positron linear collider with total energy between 0.5 to 1.5 TeV and sufficient luminosity (beam intensity) would provide lower backgrounds and a better controlled initial state than the LHC for similar explorations. Such a machine would have greater sensitivity to weaker signals, such as Higgs bosons in supersymmetric models and low-mass weakly interacting supersymmetric particles. An electron-positron collider of appropriate energy would be an ideal laboratory for the detailed study of the properties of Higgs bosons, and a 1.5 TeV linear collider could probe a strongly coupled electroweak symmetry breaking sector. Thus, such a linear collider would be in many ways complementary to the LHC.

Physics beyond the Standard Model may appear in other ways. Unequivocal evidence for neutrino oscillations, or CP violation outside the CKM structure, or decay modes that are forbidden by the Standard Model, would have dramatic consequences. At very high energies we may find new fundamental fermions, part of a fourth generation, or extensions of the three known generations. However, this is only a speculation; such particles may or may not exist in nature. In contrast, although we do not know precisely the origin of electroweak symmetry breaking, at least we know where and at which energy clues await us.

Exploration of the TeV scale is essential for the long-term vitality of particle physics. This requires a new generation of colliders and detectors beyond those that are currently available or those that can be attained with modest upgrades. This leads us to two conclusions:

- Participation of the United States in the LHC program is important and should be vigorously supported.
- An e^+e^- linear collider with sufficient energy and luminosity could provide important opportunities for discovery at this scale, while enhancing our ability to interpret evidence for new physics from the LHC.

7 Astrophysics, Cosmology, and Unification of Forces

The fields of astroparticle physics, cosmology, and unification of forces are growing areas of research that are at the intellectual frontier of particle physics. Because the energy reach of theory has exceeded that of terrestrial experiment, the pursuit of fundamental physics has more and more come to involve astrophysics and cosmology. Today many particle physicists, both theorists and experimentalists, are working at the interface of particle physics with astrophysics and cosmology, and this trend is likely to increase.

The boundaries of particle physics now touch on cosmology, astrophysics and gravity. Our best attempts to understand some of the most basic features of the universe – the cosmic asymmetry between matter and antimatter, the tiny primordial inhomogeneities that seed the growth of all galactic structure in the universe, the mysterious dark matter, and the large-scale smoothness and flatness of the universe – involve fundamental physics. Conversely, the universe has become an important laboratory for testing our boldest ideas about the laws of Nature – supersymmetry, grand unification, and superstrings.

Most of the basic features of the universe can be explained by postulating a brief inflationary instant during which an enormous burst of expansion allows a tiny, smooth piece of the universe to grow and encompass all that we can see today. This tremendous expansion also allows quantum-mechanical fluctuations on microscopic scales to evolve into variations in the energy density on cosmological scales, explaining the origin of the inhomogeneities needed to seed structure. Inflation requires that the total mass density of the universe be equal to the critical density, balanced between a universe that expands forever and one that ultimately ceases to expand and then collapses. However, the mass density in stars is far less than this, and the study of primordial nucleosynthesis shows that the mass density in ordinary matter, luminous or not, is far less than the critical density, indicating the existence of a new form of matter, called dark matter.

Even setting aside the inflationary prediction, there is strong evidence that much, if not most, of the mass density of the universe is something other than the familiar baryons comprising atomic nuclei. No problem more dramatically demonstrates the confluence of particle physics and cosmology than that of dark matter. It appears likely that the bulk of the dark matter exists in the form of elementary particles remaining from the earliest moments. These particles might be neutrinos of mass between 5 and 25 eV, or something more exotic, like particles called axions of mass ranging from 10^{-6} to 10^{-4} eV or supersymmetric particles of mass between 10

GeV and 1 TeV.

The dark matter problem provides further impetus for the search for evidence of neutrino mass and for new, long-lived particles. Conversely, direct searches for particle dark matter in our own neighborhood, using highly innovative detectors, have the potential to discover new elementary particles.

Inflation and particle dark matter have led to the most promising explanation for how the universe evolved from a very smooth beginning to the abundance of structure that exists today – galaxies, clusters of galaxies, superclusters, voids, and great walls. This theory, involving cold (or mixed) dark matter, holds that most of the mass in the universe exists in the form of slowly moving elementary particles and that the density perturbations that seeded all the structure arose during inflation. Cold dark matter is being put to the test in many different ways: measurements of the temperature variations in the cosmic background radiation, which probe the primeval density perturbations; redshift surveys, which map the inhomogeneity that exists today; deep images of the universe taken by the Hubble Space Telescope and Keck Telescope, which reveal the recent evolution of galaxies and clusters; x-ray studies of clusters of galaxies, which allow a determination of the ratio of baryons to exotic matter; and the search for the cold dark matter particles themselves. In addition to its obvious cosmological significance, the cold dark matter theory provides a window to physics at energies up to the unification scale.

A fundamental feature of the universe is the absence of antimatter and the tiny ratio of matter to radiation (about one baryon per three billion photons). One of the great triumphs of particle cosmology is baryogenesis, which holds that the small net baryon number of the universe evolved through interactions that do not respect baryon-number conservation or CP symmetry and occurred out of thermal equilibrium. The details of baryogenesis are not yet understood, and it is possible that it involved electroweak physics. Through baryogenesis the cosmic asymmetry between matter and antimatter and CP violation are intimately linked: advances made in understanding CP violation in the laboratory will further our understanding of the baryon asymmetry of the universe and vice versa.

Neutrino experiments with astrophysical and cosmological significance range from table-top searches for nuclear decays with the emission of two electrons and no neutrinos to arrays of photomultipliers monitoring vast domains of water or ice in order to detect very-high-energy cosmic neutrino sources. Neutrinos from the supernovae of 1987 (SN 1987a) in the Large Magellanic Cloud were detected in devices intended to observe proton decay, illustrating how intertwined astrophysics and particle physics have become.

Measurements of the flux of solar neutrinos are of special astrophysical interest. The solar neutrino deficit has been a puzzle for many years and new powerful detectors are beginning to address the problem. It may be that a resolution of the solar-neutrino problem is near and will advance our understanding of both neutrino physics and the details of the solar interior.

Particle physics began with the observation of cosmic rays. Remarkably, we still understand little about the origin and propagation of very-high-energy cosmic rays. Investigations high in the atmosphere, at the surface of the earth, and in underground detectors are addressing these important questions. Larger detectors, borrowing significantly from technology developed in high-energy physics and involving high-energy physicists, are being designed.

Because astrophysical and cosmological phenomena occur over such a wide range of circumstances – from the cosmic rays to the early universe – they can be exploited to study fundamental physics in regimes beyond the reach of terrestrial laboratories. The knowledge derived has been very useful: hadronic cross sections at very high energies; nucleosynthesis limits to the number of light neutrino species (and other light particle species); cosmological limits to neutrino masses and other properties; constraints on the axion based upon neutron stars, white dwarfs, and red giant stars; and a host of limits to neutrino properties based upon SN 1987A.

The unification of the strong, weak and electromagnetic forces seems to require the inclusion of gravity and to demand its quantization. There is much important theoretical activity, ranging from the study of quantum aspects of tiny black holes and to understanding the implications of superstring theory for particle physics, gravitation, and cosmology. On the experimental side, a new era will dawn soon when the laser interferometric gravity-wave detectors being built in the U.S. (LIGO) and in Europe (VIRGO) are commissioned. These devices have the potential to detect astrophysical sources as well as sources connected with the early universe.

The closely related fields of astrophysics, cosmology, and unification of forces are flourishing. Thanks to the wealth and diversity of experimental and theoretical effort, the field is poised for major advances in our understanding of:

- solar neutrinos and the sun,
- how structure formed in the universe,
- the earliest moments of the universe, and
- the nature of the mysterious dark matter.

Despite the rosy scientific outlook, a difficult structural problem needs to be addressed:

- Because this work is intrinsically interdisciplinary, it is hard to weigh properly quite different research proposals against each other or find appropriate funding, since no agency has well-defined responsibility for the field.

8 Exploratory Theory

The remarkable success of the Standard Model is important not just for the questions it answers, but for the questions it lets us ask. Are the electroweak and strong forces, themselves, expressions of a single, unified force? Is there a yet more complete unification, one that includes gravity? What determines the number of quarks and leptons, their masses and the CKM quark mixing matrix? Speculating about these issues and exploring subjects seemingly remote from experiment are important intellectual frontiers for particle physics. Furthermore, if the past is any guide, it is likely that these deep questions will be eventually answered by the discovery of important new organizing principles.

Theoretical speculations, no matter how attractive, must ultimately be validated by experiment. Thus, although the idea of unifying the electroweak and strong interactions is compelling, the absence of proton decay at the predicted levels has served to rule out the simplest grand unified theories. Moreover, the predicted coalescence of the couplings of the electroweak and strong interactions at a single high energy also did not seem to be fulfilled in these theories. Nevertheless the idea of unification of all forces continues to be a theoretical beacon. There are attractive alternatives to the simplest grand unified models involving the possibility that nature has a symmetry, supersymmetry, that connects bosons with fermions.

If the speculation that supersymmetry exists in nature is correct, then every known particle has a yet-to-be-discovered partner with a different spin. While no supersymmetric partners have yet been found, there is some indirect circumstantial evidence that these particles exist, but have TeV-scale masses. When their effects are considered, the couplings of the electroweak and strong interactions do coalesce at high energy, as required by unification. Furthermore, supersymmetric grand unified theories also lead to longer proton lifetimes. Finally, supersymmetry also helps us understand why some particle masses are much smaller than the scale set by gravity. If these speculations are correct, supersymmetry should reveal itself directly at the energy scales which will be probed by the LHC, if not at lower energy accelerators.

Supersymmetry has additional motivation. It arises from the ambitious efforts to encompass all interactions, including gravity, in a single theory. String theory provides the hope of answering many of the hard, deep questions that face particle physics. Over the past decade, string theory has developed into a thriving discipline, with applications both inside and outside particle physics. In particular, string theory has suggested plausible extensions of the Standard Model and has motivated phenomenologically interesting interrelations among low energy parameters.

Nevertheless, string theory remains speculative, separated from the rest of particle theory and experiment. It is important to bridge this gap and find evidence for or against these speculations. Indeed, perhaps the most important attribute of exploratory theory is that it demands that even the wildest speculations should ultimately be testable experimentally.

9 Accelerator Physics, Technology, and Facilities

In the past, progress in particle physics has depended strongly on developments in accelerator physics and technology. As we reach for higher energies and the increased particle fluxes necessary to extend the frontier, continued advances in accelerator physics will become ever more important. This study has concentrated particularly on three aspects of accelerator physics: hadron colliders, electron linear colliders, and novel technologies. It has led to several conclusions concerning the need for accelerator R&D in the U.S.

Hadron Colliders

Design considerations for luminosity and energy upgrades of the $\bar{p}p$ Collider at Fermilab are important to the near-term prospects for hadron colliders. Antiproton production is the key to high luminosity in the Fermilab Tevatron. Reviews of various options revealed no insurmountable technical barriers to achieve higher luminosity, but the necessary trade-offs, costs, and the potential impact on the Main Injector program need to be evaluated carefully. Thus we recommend that:

- Continued studies and design work on schemes to increase antiproton production at Fermilab should be vigorously pursued.

Superconducting magnets are the enabling technology for the highest energy machines. If the U.S. is to remain in the forefront of accelerator physics, reassembly of a U.S. superconducting magnet R&D program is critical. Thus we also recommend that:

- Support for R&D in the area of superconducting magnet technology should be strengthened.

In this respect, significant participation in the LHC project could be a valuable and important component of a U.S. program for hadron collider R&D. Such participation would be most valuable if it were targeted toward those areas which represent the greatest challenges in moving toward the future, such as superconducting magnet and beam tube vacuum technology.

New ideas for overall systems design of a collider facility in the 30 TeV per beam energy range, which could be constructed sometime following completion of the LHC, were also explored. Although the challenges are enormous, some of the ideas seem feasible and should be explored further.

Electron Linear Colliders

The SLC has demonstrated that linear colliders can be made to work and can provide a way to reach the TeV scale in electron-positron accelerators. As a result, interest in this technique has grown considerably over the last few years and there is considerable R&D underway. Snapshots of the worldwide design and prototype research for future e^+e^- linear colliders (CLIC, JLC, NLC, SBLC, TESLA, and VLEPP) are described in the study and their critical issues evaluated. Although the basic features of these linear colliders are fairly generic, the work has led to several possible approaches.

For electron linear colliders the enabling technologies are the high gradient acceleration system, radio frequency power sources and accelerator structures. The technology required to support a next generation linear collider has made significant advances in the recent past and is expected to reach maturity during the next few years. Therefore, we recommend that:

- Facilities in various parts of the globe that test different technical issues and different technological options for linear colliders should be completed as expeditiously as possible, so that a conceptual design for a next generation linear collider based on proven technologies can be developed.

Novel Technologies

Cost played a pivotal role in the cancellation of the SSC, and is a major issue in the success of the LHC. The importance of cost will increase further due to the technical difficulties that will be encountered in producing higher energies or luminosities. Thus we are challenged both economically and technically to find new acceleration techniques in order to reach new frontiers. This implies that:

- Continued investment in generic accelerator research is required to continue to open new experimental opportunities at expanding energy and luminosity frontiers.

Wakefield, laser near-field, inverse-Cerenkov, plasma, and free-electron laser accelerators are among the methods considered in the report. Alternate particle colliders include $\gamma\gamma$ and $\mu^+\mu^-$ colliders.

10 Detectors

Progress in particle physics requires state-of-the-art detectors to exploit the increasing energy and beam intensities at accelerators. Collider detectors at hadron and electron machines have met the demands of increased data rates and higher precision through innovations in tracking, calorimetry, electronics, and computing.

Innovations in detector technology have had a major impact outside particle physics. A particularly important example is medical imaging, which benefits from the high-resolution, high-speed, high-efficiency x-ray detection pursued at high-energy physics laboratories around the world.

Accelerators now being designed will demand improvements in many aspects of particle detectors. Increased beam intensities will require the use of radiation-hard materials and electronics. The increased data rates will demand higher bandwidth data acquisition systems. Increased particle densities from very high-energy collisions will necessitate higher density, lower cost electronics. The tagging of particles with picosecond lifetimes will require improved vertex detectors and fast vertex triggers. And the use of high precision electromagnetic calorimetry will call for fast, bright, radiation-hard crystals.

There is no question that detector research and development is a vital part of the U.S. HEP program. Until recently, U.S. particle physicists were able to carry out detector R&D using funding provided by the SSC Laboratory. University physicists also benefited from funding provided by the Texas National Research Laboratory Commission. This R&D program stimulated rapid progress and brought together experimentalists from a variety of subdisciplines.

Unfortunately, the cancellation of the SSC brought these programs to an end. Much promising work was stopped or significantly slowed. At the same time, further shortfalls have been created by cutbacks in traditional sources of detector R&D funding, such as laboratory discretionary funds and university seed money. Additional pressures result from the scale of most modern detector projects, which is so large that managers are reluctant to risk adopting new and unproven technologies. Although new technologies often enjoy some support in the early phases of these projects, there is inevitably pressure to focus limited resources on a single technology.

These considerations lead us to recommend that independent support be provided for detector development work. Such an R&D program must be modest in light of tremendous pressures on the HEP budget. The administrative aspects of this program are best determined in consultation with the funding agencies. However, a few general features are desirable:

- The primary aim of the program should be to provide an incubation period for new ideas.
- Funding decisions should be based on review by a peer-review committee.
- Most projects should be limited to three years of funding, after which time their support should come from the detector projects or other sources.

- The general level of funding should total about \$3M per year. This would allow about six new starts of projects with annual budgets in the \$50,000 to \$250,000 range.

Detector development is a vital component of high-energy physics. It is essential that the field invest in new technologies to prepare for the challenging environments of future experimental facilities.

11 Computing

Historically, high-energy physicists have made extensive use of state-of-the-art techniques in computer and communication technology. For example, particle physics experiments pioneered the use of real-time computer techniques. As collaborations grew in size and scope, particle physicists were among the first to make effective use of computer networks. Indeed, the popular World-Wide Web application was developed by high-energy physicists at CERN.

Particle physicists have long employed advanced computer analysis and simulation techniques and have made productive use of the most powerful computers available. However, as the use of computers becomes increasingly part of the mainstream, particle physicists will need to change the way in which they use and develop software. They must make the transition from being major developers of computer technology to being just one set of users among many areas of application.

Although this change may introduce some short-term problems, on balance the effect should be positive because much of the software development work that formerly fell to particle physicists will now be done by specialists. Ultimately, this will allow more time to be devoted to science. This has already long been the case for computer hardware, where the HEP community has consistently been able to exploit commercial products.

The main challenge for the future is to learn how to make good use of commercial software products in a cost-effective way. The following set of recommendations could help the community meet this challenge:

- The U.S. HEP community should formulate a coherent approach to the use of computing technology.
- Computing at U.S. HEP institutions (universities and laboratories) should be re-capitalized in a way that is compatible with current and anticipated industrial developments toward open systems.
- The U.S. should provide centers of CPU power and storage for simulation, reconstruction, and analysis, that are accessible to any HEP institution.
- Modest funding should be made available to carry out R&D in aid of this new direction.

12 Structural Issues in High-Energy Physics

This working group addressed three areas of concern: Governance and Advising, Career Issues, and Education and Outreach.

Governance and Advising

One of the important problems for the field to address is that of governance, advising, and planning. At present HEPAP plays the role of advising the government, specifically the Department of Energy, about issues concerning high-energy physics and, to some extent, helps in the planning for the field. Historically this is the only body that substantively considers the field as a whole; so far the DPF has played only a minimal role. Concerns have been raised throughout the community recently about the effectiveness, limitations, and flaws of this approach in an era of shrinking budgets and increased demand on available funds. In response to these concerns, it is important to consider additional mechanisms which might improve planning the future of the field.

This question was discussed at two open meetings held during the study as well as at town meetings held in conjunction with the work of the Drell subpanel. From those gatherings, and through deliberations among the working group and with the DPF Executive Committee, the following recommendation was formulated:

- The DPF Executive Committee should commission workshops and studies on issues of great importance to the field.
- These studies, which in general should be of short duration, would be carried out by ad-hoc panels appointed by the Executive Committee in consultation with the funding agencies and the national laboratories.
- The reports resulting from these studies should then be disseminated promptly to the HEP community and the public.

Career Issues

The cancellation of the SSC put more than 100 physicists into the job market. That, coupled with reductions in funding and the concomitant loss of employment opportunities emphasizes the importance of studying career issues.

Reliable demographic data are a prerequisite for a meaningful study. These data are now being gathered by the Particle Data Group under the auspices of both the DOE and the NSF. Obtaining accurate data, however, is only the first step toward understanding career opportunities open to high-energy physicists and devising ways to help improve career paths in the discipline. One early conclusion is that it is important for high-energy physicists to be aware of all the opportunities open to them both within and outside the discipline. Within the profession, young physicists should be educated to the needs of employers in government and industrial laboratories, as well as in universities. Conversely, it is important that potential employers be aware of how high-energy physicists can contribute to their enterprises. To foster these goals, we recommend that the DPF Executive Committee:

- should organize a Subcommittee on Careers to arrange meetings, prepare literature, and devise mechanisms to educate young colleagues about employment opportunities and the means of enhancing their prospects for obtaining jobs; and
- should consider the appropriateness of, and possible improvements in, the education of graduate students, with an eye towards enlarging the employment opportunities for those with training in high-energy physics.

Outreach and Education

The cancellation of the SSC made it particularly clear that we must increase our efforts to educate the public and those in government about the societal benefits of the pursuit of research in high-energy physics. Already substantial efforts are underway, both by individuals and by the laboratories. These form a base upon which to build.

As a result of the study, a sub-group was formed to examine ongoing outreach and education efforts and seek proposals for new directions. Named POET, for Public Outreach and Education Team, this group has begun to sensitize the high-energy-physics community to its existence, and to solicit ideas about possible future activities. To encourage these efforts we recommend that:

- The DPF Executive Committee should support the activities of the POET group as a means of enhancing the outreach and education efforts of particle physicists.

13 Conclusions

This study has highlighted the vitality and intellectual breadth of the U.S. program in the context of the international high-energy physics effort. We hope that the study has also helped to delineate for a wider audience the key elements of worldwide high-energy physics research into the first decade of the next century.

In accord with the Drell Subpanel report, we believe that the LHC program will define the high-energy physics frontier and that participation by U.S. physicists in this effort is essential. In addition, we expect that the Fermilab Main Injector, the SLAC B Factory, and Phase III of CESR, as well as the RHIC accelerator at the Brookhaven National Laboratory, will provide frontier capabilities in other fundamental areas of research.

The picture of the future U.S. accelerator-based program beyond these facilities is much less in focus. Specific groups are considering possibilities as diverse as room temperature or superconducting linear e^+e^- colliders, substantial upgrades in the luminosity or energy of the Fermilab Tevatron, and muon colliders. Although many aspects of these options were studied by the working groups, this effort did not include substantial comparisons nor was there any attempt to set priorities.

We believe that now is the appropriate time to begin to develop a coherent plan for the longer term future, based on comparison of the physics potentials of possible new facilities. It is also time to move toward a consensus for the U.S. program beyond the LHC and current upgrades to U.S. facilities, in the context of the international program.

To accomplish this, members of the U.S. HEP community should meet with individuals from other nations to compare the capabilities, feasibilities, and relative strengths of these initiatives with those of the LHC and other facilities that will likely operate in the LHC era. This type of evaluation has traditionally taken place in DPF-sponsored Snowmass workshops. Thus we propose a similar workshop to be held in the summer of 1996 to begin to address these crucial issues.

Precision Tests of Electroweak Physics: Current Status and Prospects for the Next Two Decades

Frank S. Merritt

University of Chicago, Chicago, IL 60637

Hugh Montgomery

Fermi National Accelerator Laboratory, Batavia, IL 60510

Alberto Sirlin

New York University, New York, NY 10003

Morris Swartz

Stanford Linear Accelerator Laboratory, Stanford, California 94309

1 Introduction

We present a brief review of the present status and prospects for progress in electroweak measurements leading to precision tests of the Standard-Model Electroweak Theory. Our emphasis is on the measurements of electroweak couplings, of the masses of the W and Z gauge bosons, and of the searches for and mass measurements of the top quark and the Higgs boson.

Section 2 contains a brief review of the standard theory, and Section 3 summarizes the schedule of the major experimental programs expected to run during the next 15 years. Sections 4 and 5 cover measurements of electroweak couplings at high and low q^2, respectively. Section 6 is devoted to precision measurements of the W mass at Fermilab and LEP 2. Section 7 describes the status of the top quark searches at Fermilab, and the anticipated future improvements in direct top mass measurement. Section 8 describes the electroweak fits of all of these data to the Standard Model. For the first time, these measurements are beginning to give significant constraints on the Higgs mass. The precision of the Standard-Model predictions of the Higgs mass based on indirect measurements provides a useful benchmark for evaluating the experimental program. We have attempted to project how the resolution of these predictions will evolve over the next 15 years, and to see where significant improvements might be made. The ultimate test of the Standard Electroweak Theory may lie in the comparison of these indirect determinations of the Higgs mass with its direct observation. In Section 9 we review the direct searches for the Higgs anticipated in future years. The measurement of triboson couplings is briefly discussed in Section 10. A summary and conclusions are presented in Section 11.

We have drawn heavily on the work of others, particularly in workshop proceedings, design reports, and summary talks. The field of precision electroweak measurements is evolving rapidly, both in terms of new measurements from Fermilab, CERN, and SLAC, and also in terms of new studies relating to future experiments and facilities. We have attempted to include new information as it has become available, but it is clear that new developments will continue to emerge over the next few months in several important areas. These include:

- final measurements of M_W from Run 1a data from both CDF and DØ,
- improved estimations of LHC detector performance (from ATLAS and CMS TDR's),
- improved projections of LEP 2 energy and schedule, based on continuing RF studies,
- better estimates of ultimate LEP 1 errors and running plans for 1995, and
- better understanding of theoretical and experimental limitations in determining $\alpha(M_Z)$.

2 Standard Electroweak Model.

This section is based on a more extensive overview [1] prepared for this working group.

2.1 Overview

Because of its success in describing low energy phenomenology and its relative economy in the number of fundamental fields, the $SU(3)_c \times SU(2)_L \times U(1)$ theory of strong and electroweak interactions, based on the principle of non-abelian gauge invariance, has become the Standard Model (SM). $SU(3)_c$ embodies the current theory of the strong interactions, Quantum Chromodynamics, and is deemed to be an unbroken symmetry of nature. The $SU(2)_L \times U(1)$ sector is the basis of the Standard Electroweak Model and is spontaneously broken at a mass scale $v = (\sqrt{2}G_\mu)^{-1/2} = 246$ GeV[a] into $U(1)_Q$, the abelian gauge group of electromagnetism. The theory is endowed with the fundamental properties of renormalization and asymptotic freedom of the strong interactions. The minimal version involves three generations of quarks and leptons and one Higgs doublet.

The SM has been very successful phenomenologically. It has provided the theoretical framework for the description of a very rich phenomenology spanning a wide range of energies, from the atomic scale up to M_Z. It is being tested at the level of a few tenths of a percent, both at very low energies (superallowed Fermi transitions, low energy decays) and at high energies (CERN, Fermilab, SLC). The aim of these studies is to test the theory at the level of its quantum corrections and to search for deviations that may signal the presence of "new physics".

There are two basic missing pieces in the SM: the top quark and the Higgs boson. Very suggestive, but not definitive, evidence for the existence of the top quark with a mass $M_t = (174 \pm 10^{+13}_{-12})$ GeV has recently been presented [2] by the CDF experiment. There are compelling theoretical reasons for believing that the top quark exists:

- It is known [3] that the left-handed b has quantum numbers corresponding to the lower component of an isodoublet and, therefore, its partner must exist; and
- The top quark is necessary to complete the cancellation of anomalies, a basic requirement for renormalizability [4].

The analysis of the precision electroweak data leads at present to the prediction that M_t lies in the vicinity of 175±30 GeV, as discussed in Section 8; this constraint arises from the top quark contribution to the electroweak radiative corrections.

The most fundamental discovery to be made in the framework of the SM would be the detection of the Higgs boson or, more generally, the elucidation of the Higgs sector of the theory. In fact, this may lead us to understand the nature of electroweak symmetry breaking and the generation of mass. Perturbative partial wave unitarity [5] in the $J = 0, I = 0$ channel for $W_L^+ W_L^- \to W_L^+ W_L^-$ requires $|5\lambda/16\pi| < 1/2$, where λ is the quartic Higgs coupling. Using $M_H^2 = 2\lambda v^2$, this implies $M_H \leq 780$ GeV. This roughly represents the mass scale at which the Higgs sector becomes strongly interacting. For larger values of M_H, the unitarization of the S-matrix requires some manifestations of "new physics" and one expects the formation of resonances such as ρ-like vector mesons. There is also an interesting lower bound from the requirement that the SM vacuum be stable [6]:

$$M_H(\text{GeV}) > 136.3 + 2.2(M_t - 174) - \tfrac{4.5}{0.007}(\alpha_S(M_Z) - 0.118) \quad (1)$$

This bound assumes that the SM is valid up to mass scales $\simeq 10^{15} - 10^{19}$ GeV.

There are strong arguments to expect that, sooner or later, "new physics" will be revealed. The SM contains a plethora of adjustable parameters, most of them associated with the mass matrices of the matter fields, leptons and quarks. If the neutrinos are massless, we have three gauge couplings, M_Z, M_H, three lepton masses, six quark masses, three mixing angles and one CP-violating phase in the CKM matrix, and one CP and P violating Θ parameter, a total of 19 constants (we assume that the Planck mass $M_P \simeq 1.2 \times 10^{19}$ GeV defines the unit of mass). If the neutrinos have mass there are seven additional parameters. The SM also fails to provide a reason for the observed quantization of electric charge and the number of generations. Particularly suspect is the mechanism of symmetry breaking involving a single Higgs doublet. As the self-energies of the scalar fields are quadratically divergent, one has $M_H^2 = M_{0H}^2 - c_\lambda \Lambda^2$, where M_{0H} is the bare mass, Λ is the ultraviolet cutoff, and c_λ is a constant of $\mathcal{O}(\lambda)$. If one assumes that a natural cutoff will emerge at the Planck mass, we have $\Lambda \approx M_p$, and this means that M_{0H} must be fine-tuned to an unnatural precision to maintain M_H at the v scale. The same hierarchy problem emerges if one embeds the SM in a GUT involving scales $\Lambda \gg M_H$. Two main schools of thought have emerged to circumvent this problem. One approach is dynamical symmetry breaking via fermion-antifermion condensation, $\langle \bar{\psi}_R \psi_L + h.c \rangle \neq 0$. Examples include technicolor [7] and $t\bar{t}$ condensation models [8]. The fact that the top quark is the only fundamental fermion with $M \sim v$ may be a strong indication that it is involved in the mechanism of symmetry breaking. Another scenario is supersymmetry. Here Higgs scalars occur as fundamental fields, but the quadratically divergent contributions to their self-energies are cancelled by the fermionic counterparts. The minimal supersymmetric extension of the SM (MSSM) [9] involves two Higgs doublets. There are five Higgs bosons (two neutral scalars, 1 neutral pseudoscalar, and 2 charged) plus a rich spectrum of supersymmetric partners. The lighter Higgs boson satisfies the important constraint $M_H \lesssim 130$ GeV. Recently, the SUSY approach has received support from the observation, derived from the precision electroweak analysis, that in the framework of the MSSM the three gauge couplings unify [10] at energies $\sim 10^{16}$ GeV.

[a] We choose units such that $c = 1$, and consequently masses are expressed in GeV rather than GeV/c^2 throughout this report.

2.2 *Input Parameters*

There are three accurately determined physical parameters [11] that play an important role as inputs in electroweak physics:

1. The fine-structure constant, measured most precisely from $(g\text{-}2)_e$, is

$$\alpha = 1/137.03599\ldots \quad (\delta\alpha/\alpha = 0.045 \text{ ppm}). \quad (2)$$

2. The muon decay constant G_μ, which may be regarded as the modern definition of the Fermi constant. It is determined, by definition, from the muon lifetime, by applying the radiative corrections of the local V-A theory, which in this case are convergent. The current value is

$$G_\mu = 1.16639(1) \times 10^{-5} \text{ GeV}^{-2} \quad (8.5 \text{ ppm}). \quad (3)$$

3. The Z mass, determined from the 1993 LEP data, is

$$M_Z = 91.1888 \pm 0.0044 \text{ GeV} \quad (48 \text{ ppm}). \quad (4)$$

For a long time it has been known that large radiative corrections, associated with the running of α, play an important role in electroweak physics. In particular, $\alpha(M_Z)$ is of special interest. This concept is scheme-dependent. A frequently employed definition is $\alpha(M_Z) = \alpha/(1-\Delta\alpha)$, where $\Delta\alpha$ is the fermionic contribution to the conventionally renormalized vacuum polarization function of QED, evaluated at $q^2 = {M_Z}^2$. We discuss the determination of $\Delta\alpha(M_Z)$ in a later subsection.

2.3 *Basic Radiative Corrections*

Knowing α, G_μ, M_Z, one can evaluate M_W and $\sin^2\theta_W$. At the tree level, this can be done from the basic natural relations:

$$M_W = M_Z \cos\theta_W, \quad (5)$$

$$e = g \sin\theta_W, \quad (6)$$

$$G_\mu/\sqrt{2} = g^2/8M_W^2, \quad (7)$$

where e is the positron charge and the weak mixing angle θ_W is given by $\tan\theta_W = g'/g$ (g and g' are the gauge couplings of $SU(2)_L$ and U(1), respectively). Equation (4) holds if additional doublets and singlets of Higgs scalars are introduced in the theory, but it is not generally valid for other representations, such as triplets. In order to carry out the analysis in the presence of radiative corrections, it is necessary to specify the precise definition of the parameters involved. It is convenient to identify M_W with the physical or pole mass, in analogy with M_Z.

Regarding the weak mixing angle θ_W, a number of possibilities have been employed in the literature. Defining $\sin^2\theta_W = 1 - M_W^2/M_Z^2$, one finds [12]

$$s^2c^2 = \frac{A^2}{M_Z^2(1-\Delta r)}, \quad (8)$$

where s^2 and c^2 are abbreviations for $\sin^2\theta_W$ and $\cos^2\theta_W$, $A^2 = \pi\alpha/(\sqrt{2}G_\mu) = (37.2802 \text{ GeV})^2$ and Δr is the radiative correction. Another possibility is to identify the weak mixing angle with the $\overline{MS}$ parameter $\sin^2\hat{\theta}_W(M_Z)$, evaluated at the M_Z scale. In this case one obtains [13]

$$\hat{s}^2\hat{c}^2 = \frac{A^2}{M_Z^2(1-\Delta\hat{r})}. \quad (9)$$

The quantities Δr and $\Delta\hat{r}$ are basic corrections of the electroweak theory. In particular, the definition of $\sin^2\theta_W$ and Eq. (6) imply that Δr is a physical observable, like g-2, for example. The parameter $\hat{s}^2$ is very convenient to describe physics at the Z peak and can be obtained almost directly from the on-resonance asymmetries with the application of very small radiative corrections. It is crucial for GUTs studies. The two most frequently employed renormalization methods lead precisely to the above definitions. The on-shell approach [12], which employs physical observables as renormalized parameters, leads in a natural manner to $\sin^2\theta_W = 1 - M_W^2/M_Z^2$. In the $\overline{MS}$ method [13], the counterterms are chosen to cancel the pole terms of dimensional regularization.

The dominant contributions to Δr and $\Delta\hat{r}$ arise from fermion loops and involve large logarithms of the form $\ln(M_Z/M_f)$, that can be associated with the running of α (M_f represents a generic fermion mass), and contributions from the t-b isodoublet which are very sensitive to M_t. There are also conceptually very important bosonic electroweak corrections involving virtual $W^\pm$, Z, γ and H. They affect self-energies, vertex and box diagrams but, in four fermion processes, they are numerically smaller than their fermionic counterparts. The leading asymptotic behaviors for large M_t and M_H are quadratic in M_t and logarithmic in M_H. They have opposite signs in both Δr and $\Delta\hat{r}$, which partially accounts for the fact that, in global analyses of the electroweak data, small (large) M_H values favor relatively small (large) values of M_t. The terms proportional to M_t^2 are analogous to those occurring in the fermionic correction [14]

$$(\Delta\rho)_f = \frac{3G_\mu m_t^2}{\sqrt{2}\,8\pi^2} \quad (10)$$

to the ρ-parameter, defined as the ratio of effective neutral to charged-current couplings at $q^2 = 0$. Because of an additional factor c^2/s^2 multiplying the M_t^2 contributions, Δr is much more sensitive to M_t than $\Delta\hat{r}$ or

$\Delta\rho$. Aside from Δr, $\Delta\hat{r}$, and $\Delta\rho$, other important radiative corrections include the electroweak form factor [15] $k(q^2)((k'(q^2))$ that multiplies $s^2(\hat{s}^2)$ in neutral current amplitudes, and the vertex correction to $Z \to b\bar{b}$ which exhibits an intrinsic M_t^2 dependence.

Complete one-loop calculations for various processes have been carried out by several groups. Although full two-loop calculations do not yet exist, leading contributions of $\mathcal{O}(\alpha^n \ln^n(M_Z/M_f))$, $\mathcal{O}(\alpha^2 \ln(M_Z/M_f))$ [12], $\mathcal{O}(\alpha^2(M_t^2/M_W^2)^2)$ [16], and $\mathcal{O}(\alpha\alpha_s, \alpha\alpha_s^2)$ are incorporated. The latter affect mainly the terms proportional to M_t^2 in $\Delta\rho$ [17] and in the $Z \to b + \bar{b}$ vertex [18], and depend sensitively on how M_t is defined. In current analyses one employs the pole-mass as this seems to be the parameter that can most readily be identified with kinematic measurements of M_t.

There is another definition of the weak mixing angle, $\sin^2\theta_W^{\text{eff}}$ (also called $\sin^2\theta_{\text{eff}}^{\text{lept}}$), which is employed in the LEP analysis. Calling g_{V_ℓ} and g_{A_ℓ} the effective vector and axial vector couplings in $Z \to \ell\bar{\ell}$ at resonance (g_{V_ℓ} and g_{A_ℓ} absorb vertex and self-energy corrections), one defines $1 - 4\sin^2\theta_W^{\text{eff}} = \text{Re}\, g_{V_\ell}/g_{A_\ell}$. It has recently been found [19] that $\hat{s}^2 = \sin^2\theta_W^{\text{eff}} - 0.0003$.

The on-shell and $\overline{MS}$ schemes have been systematically applied to a number of fundamental observables, spanning $0 \leq |q^2| \leq M_Z^2$: atomic parity violation, νl scattering, $\nu_\mu N$ deep inelastic scattering and the various asymmetries, widths, and cross sections measured at LEP and SLC.

2.4 Evidence for Quantum Corrections

It has been known for at least two decades that radiative corrections play an important role in the analysis of the SM. For example, if only the Fermi-Coulomb function were included in the analysis of the superallowed Fermi transitions (as in ^{14}O decay), there would be a large violation of the unitarity of the CKM matrix and the SM would be placed in severe jeopardy. Fortunately, there are large corrections in semileptonic decays which literally rescue the theory from obvious contradiction. Furthermore, these corrections are divergent in the local V-A theory. Thus, the renormalizability of the SM and the fact that the resulting corrections are of the correct sign and order of magnitude play a crucial role in the tenability of the new theory [20].

In high-energy phenomena, the dominant radiative corrections involve virtual fermions. They are responsible for the large logarithms that can be absorbed in the running of α, and the contributions of the tb isodoublet, from which the M_t constraints are derived. One way to quantify this question is to "measure" $(\Delta r)_{\text{res}}$, the residual part of Δr after extracting the corrections contained in $\alpha(M_Z)$. We have

$$\frac{\alpha}{1-\Delta r} = \frac{\alpha(M_Z)}{(1-(\Delta r)_{\text{res}})}. \tag{11}$$

Inserting the direct collider determination of $M_W = 80.23 \pm 0.18$ GeV in Equation 6, one finds $\Delta r = 0.0442 \pm 0.0104$. Using $(\alpha(M_Z))^{-1} \approx 128.84$, we have $(\Delta r)_{\text{res}} = -0.0165 \pm 0.0111$. This differs from 0 by only $\approx 1.4\sigma$. Thus, if one employs only the direct M_W measurement, the evidence for $(\Delta r)_{\text{res}} \neq 0$ is rather weak. The determination of Δr and $(\Delta r)_{\text{res}}$ becomes much sharper if one employs all the direct and indirect information obtained by fitting the data to the full SM, with its plethora of radiative corrections and interlocking relations. In that case one currently obtains $M_W = 80.31 \pm 0.07 \pm 0.01$ GeV. Choosing $M_W = 80.32 \pm 0.07$ GeV (which corresponds to $M_H = 60$ GeV, the most unfavorable case) one finds $\Delta r = 0.0390 \pm 0.0041$ and $(\Delta r)_{\text{res}} = -0.0221 \pm 0.0044$, which now differs from zero by 5σ!

Another interesting question is whether there is at present evidence in the high-energy observables for the conceptually important bosonic electroweak corrections, mediated by virtual $W^\pm$, γ, Z, and H. It has been recently shown [21] that two different determinations of $\sin^2\hat{\theta}_W(M_Z)$, namely from the on-resonance asymmetries and from Eq. (7), differ sharply if these corrections are subtracted and the complete fermionic corrections retained. The difference is 4σ for $M_t = 131$ GeV and reaches 7σ for $M_t = 180$ GeV. Thus, this simple argument gives strong evidence for the contribution of bosonic electroweak corrections in the SM.

2.5 S, T, and U Parameters

"New physics," *i.e.*, physics beyond the SM, may affect the quantum corrections. If the new physics is associated with a high-mass scale and affects mainly the self-energies, the idea has been proposed to parametrize its contributions in terms of three amplitudes [22] S_W, S_Z, and T. If the mass scale is much larger than M_Z, $(\alpha/4\hat{s}^2)S_W$ and $(\alpha/4\hat{s}^2\hat{c}^2)S_Z$ are essentially the contributions of the new physics to the W and Z wave function renormalizations, respectively, while αT represents the corresponding contributions to $\Delta\rho$. Alternatively, one defines $S = S_Z$, $U = S_W - S_Z$. T and U are primarily sensitive to the isodoublet mass splittings (generally, $U \ll T$), while S probes contributions from degenerate $SU(2)_L$ doublets. If a direct measurement of M_t is not available, the usual procedure is to evaluate the radiative corrections of the SM at standard values of M_t and M_H and fit S, T, and U from the data. In that case S, T, and U parametrize both effects of unknown particles, as well as departures of M_t and M_H from the standard values. In more recent discussions, it has become useful to introduce a fourth parameter, δ_b, that describes the departure of the $Z_0 \to b\bar{b}$ width from the SM evaluation [23]. If a directly measured value of M_t is employed, it is sufficient to specify M_H. As the dependence of the radiative corrections on M_H is mild, it can be treated as part of the theoretical error. In this case S, T, U, and δ_b represent just the contributions from new physics. A very recent global analysis [24] that incorporates the CDF M_t value and the 1993 LEP data, gives

$$\begin{aligned} S &= -0.15 \pm 0.25 \ {}^{-0.08}_{+0.17} \\ T &= -0.08 \pm 0.32 \ {}^{+0.18}_{-0.11} \\ U &= -0.56 \pm 0.61 \\ \delta_b &= 0.031 \pm 0.014. \end{aligned} \qquad (12)$$

Except for δ_b (a $\sim$2 σ deviation), there are no indications of departures from the SM. An alternative formulation involves the ϵ_i(i=1,2,3,b) parameters [25], defined in terms of the physical observables M_W, Γ_l, $A^{(l)}_{FB}$ and $\Gamma_{b\bar{b}}$. It is worthwhile to note that S_W, S_Z, and T are actually parts of Δr and $\Delta\hat{r}$, which also include vertex and box diagrams in a gauge invariant manner. Therefore, instead of S_W, S_Z, and T, one can also employ $\Delta r = (\Delta r)_{SM} + (\Delta r)_{new}$, $\Delta\hat{r} = (\Delta\hat{r})_{SM} + (\Delta\hat{r})_{new}$, together with effective parameters $G_\mu(1+\hat{\alpha}T)$ and δ_b linked to the Z widths.

Since the early determinations, the data have favored $S < 0$. On the other hand, a fourth generation of degenerate chiral fermions gives $S \simeq 4/6\pi = 0.21$, while technicolor models roughly contribute $S \simeq (0.05 - 0.10)N_T N_D + 0.12$, where N_T and N_D are the number of technicolors and technidoublets, respectively (the 0.12 arises from a heavy effective Higgs scalar in the TC framework [26]). For one generation with $N_T = N_D = 4$, this leads to $S \simeq 0.9$ to 1.7. Thus, the current precision measurements disfavor the most typical technicolor models. On the other hand, supersymmetric theories don't introduce additional chiral fermions and tend to give very small S values. More generally, in the MSSM, the additional radiative corrections are small provided that the supersymmetric partners are much heavier than M_Z. In that limit, the radiative corrections become those of the SM with a light Higgs, $M_H \leq 130$ GeV [27]. Even if the supersymmetric partners are not regarded as very heavy, the decoupling limit provides a bound for the MSSM predictions.

2.6 Determination of $\alpha(M_Z)$

The renormalization of the electromagnetic coupling constant α to its value, $\alpha(M_Z)$, at the mass of the Z is a necessary step in the evaluation of many electroweak quantities. The theoretical error in the calculation is small [28,29] so that the error arising from the loops containing the leptons is small.

Table 1: Hadronic contributions to the determination of $\Delta\alpha$ and their errors.

Region	Contribution	Uncertainty
Resonances	0.00556	0.00018
Continuum 1.0 - 2.3 GeV	0.00198	0.00039
Continuum 2.3 - 9.0 GeV	0.00721	0.00072
Continuum 9.0 - 12.0 GeV	0.00169	0.00017
Continuum 12.0 - ∞ GeV	0.01237	0.00037
Total	0.0288	0.0009

In principle an analogous calculation of the relevant loops could be performed for the quarks, u, d, s, c, and b; however, the masses of the quarks are ill defined quantities. The usual approach is to express the calculation in terms of a dispersion integral over the electron-positron annihilation cross section into hadrons (scaled to that into muons). This ratio is often called R. The procedures used to make this evaluation have been discussed extensively [30,31]. The integral is evaluated in two parts, the resonances are parametrized and evaluated separately from the continuum regions. The contributions and the errors for the different regions are summarized in Table 1. The change in α, including leptonic contributions, is $\Delta\alpha = 0.0595 \pm 0.0009$ [31], and leads to $(\alpha(M_Z))^{-1} = 128.84 \pm 0.10$. The uncertainty is almost entirely due to the hadronic contributions listed in the table.

There are several experiments which might address individual resonance regions; for example there is a program at Novosibirsk [32] which will address the region from the ρ and ω to above the ϕ. New results are expected soon. The Daphne storage rings at Frascati will also operate at the ϕ resonance. However, the biggest individual contribution to the error comes from the region between 2.3 and 9.0 GeV, between charm and bottom thresholds. The data were taken from a single experiment and a systematic error assigned [31] to take account of differences between the results of that experiment and others for other measurements than R.

A reanalysis of existing data has recently been carried out (motivated by this DPF review) which has produced a result [33] differing from the old one by about two sigma. We have used the old value for all results in this paper in order to be consistent with other studies and with the currently accepted value. But this new result and similar work [34] underscore the importance of further theoretical and experimental studies directed toward improving the determination $\Delta\alpha$. Without this, in a few years the uncertainty in $\Delta\alpha$ will be one of the most limiting factors in improving the precision of the determination of the mass of the Higgs from indirect measurements, as we show in Section 8.6 below.

3 Expected Schedule of Future Measurements

In this section we will briefly review the schedule and status of the experimental programs which are most important for precision electroweak measurements. The experimental measurements will be discussed in the following sections, and the combined electroweak tests are discussed in Section 8.

3.1 LEP and SLC

LEP has been running near the Z since 1989. The run of 1995 is scheduled to be the last year of LEP 1 data taking. Final integrated statistics are expected to be improved by perhaps a factor of 3 over that of most current analyses (based on data taken through the end of 1993).

With the installation of new superconducting RF cavities after the 1995 run, the LEP energy will be raised by about a factor of 2 and LEP 2 will begin data-taking in 1996. It is now expected that the LEP 2 energy will be 180 GeV. This is based on studies with the first superconducting cavities and allowing, as a safety margin, for up to 3 non-operating klystrons. LEP 2 will run for a minimum of three or four years, with the goal of attaining an integrated luminosity of 500 pb^{-1}. The most important physics goals for electroweak studies are the precision measurement of M_W and the extension of the Higgs boson search to higher masses.

It is technically possible to increase the LEP 2 E_{cm} by about 15 GeV by producing more superconducting rf cavities at a cost of about 130 MSF. This decision must be made by summer of 1995.

SLC has been running for two years using polarized beams, which allow the measurement of A_{LR}, and has accumulated about 50K events. Beam polarization has now been improved from 63% to 80%. It is expected that polarized running will continue for at least three more years, reaching a total of at least 500K events.

3.2 Fermilab

The Fermilab collider has completed Run 1a (25 pb^{-1}) and is now taking data in Run 1b. This is expected to last through 1995 and to accumulate at least 100 pb^{-1}. The run is going very well now with record luminosities being achieved; Run 1b may well exceed 150 pb^{-1} if this continues, but in this report we will continue to assume 100 pb^{-1}.

The fixed target run will begin in 1996 and will last for approximately one and a half years, producing an improved measurement of $\sin^2\theta_W$ from the deep-inelastic neutrino experiment E815. During this time, extensive upgrades of both the collider detectors and of the accelerator will be carried out. The Main Injector, which is crucial for high-luminosity collider running, should be completed in 1998.

In 1998 or 1999, Tevatron Run 2 will begin with the Main Injector and with upgraded collider detectors. This run will provide about a factor of 5 improvement in luminosity, and is expected to accumulate at least 1000 pb^{-1} of data within two or three years.

It should be noted that all of the direct information about the top quark obtained over the next decade will come from the Fermilab collider program. This is a crucial part of electroweak physics. Improvements in statistics in both Run 1b and Run 2 could give significant improvements in measurements of M_t and M_W, and consequent improvements in electroweak tests.

3.3 LHC

This accelerator is expected[b] to begin operation in about the year 2004, using the existing LEP tunnel with high-field magnets to collide pp beams at $E_{cm} = 14$ TeV. The CERN Council formally approved the project in December 1994. Two experiments, ATLAS and CMS, have been given first-stage approval and their collaborations are completing design reports which will be submitted later this year. Extensive studies of LHC physics and detector design have been carried out and presented in the "Letters of Intent" of these experiments.

The LHC is essential for improved electroweak tests in several important areas. For precision electroweak tests, it should provide the best measurement of the top mass and extensive information about top production and decays, and will improve the precision of measurements of triboson couplings. Most importantly, it will extend the search for the Higgs boson over the mass region 80-800 GeV, which may provide the ultimate test of the electroweak theory.

3.4 Other Possibilities

There have been extensive studies [35,36] of the potential for an e^+e^- linear collider with energy $E_{cm} \geq 500$ GeV (the "Next Linear Collider", or NLC). In several sections of this report we make allusions to some of this potential. However we have not considered the time frame associated with such a collider to be the primary subject of this report. The treatment of the NLC is therefore not to be

[b]The CERN Council voted unanimously to approve the LHC on December 16, 1994, after this report had been completed. Because of financial consideration, the approval requires staging of the project unless there is significant financial participation by non-member states, and especially by the United States. In this report, we have only considered the machine parameters and schedule that were originally proposed, which we hope in the end will turn out to be the correct ones.

considered complete. For further discussion, see also the reports of the Electroweak Symmetry Breaking and the Accelerator subgroups of this DPF study.

During the past year there have been discussions of possible upgrades to the Tevatron Collider complex at Fermilab. These include a luminosity upgrade to $L = 10^{33}$ cm^{-2}sec^{-1} (TeV*) and a combined luminosity and energy upgrade (to $E_{cm} = 4$ TeV). Such possibilities are discussed by the Accelerator Working Group in a different chapter of this report. An understanding of the extent to which such upgrades can contribute to Electroweak Physics has not been fully documented and while reference is made to some possible opportunities the case is not developed in this chapter. Indeed it is one of the recommendations of our report that such studies be pursued.

It may be that some amount of free time will emerge at CERN between the LEP 2 and LHC schedules. If so, this could be used for

- continued LEP 2 running, perhaps at higher energy,
- continued LEP 1 running to gain higher statistics, or
- measurements of A_{LR} using longitudinal polarization at the Z.

Although we allude to some of these possibilities below, the relative strengths of these three possible programs will be determined by the developing physics measurements and machine studies which are carried out over the next few years.

4 Measurements of Electroweak Couplings Near the Z Pole

In this and the following three sections, we will describe the current status of electroweak measurements, their errors, and the prospects for further improvements. A more complete discussion of the physics implications is given in Section 8.

4.1 LEP 1 Results

Over the last four years, the LEP experiments have produced stringent tests of the electroweak theory. Energy scans over the Z peak were carried out in 1990, 1991, and 1993. The 1994 run, like the 1992 run, was devoted to collecting high-statistics data at the Z peak. Under the present schedule, 1995 will be the last year of LEP 1 running before the energy is doubled for LEP 2 running. The specific running plans for 1995 will be determined in late 1994.

The most important tests of the electroweak theory at LEP come from the lineshape parameters (M_Z, Γ_Z, σ_h^0), leptonic and hadronic branching ratios, leptonic forward-backward asymmetries, τ polarization measurements, and branching ratios and asymmetries for $b\bar{b}$ and $c\bar{c}$ events. Within the Standard Model, all LEP measurements can be predicted at tree-level in terms of a single parameter (*e.g.*, $\sin^2\theta_W$ or M_Z). When loop diagrams are included, there are small correction terms depending on M_t (and *very* small corrections depending on M_H). Additional uncertainties arise from the uncertainties in α_s and α_{em} (discussed in Section 2 above) at the Z-pole.

4.1.1 Lineshape

The most precisely measured quantity is M_Z, now measured to a precision of 5×10^{-5}. M_Z is determined from the lineshape, measured from hadronic events taken during energy scans within $\pm$ 3 GeV of the Z peak. The four LEP experiments typically have efficiencies of $> 99\%$ for hadronic events, with an experimental systematic error of $< 0.20\%$. The measurements of both M_Z and Γ_Z are limited by the uncertainty in the LEP beam energy, common to all four experiments. This has been greatly improved during the 1993 scan by exploiting resonant spin depolarization to calibrate the energy at each running point [37]. A precision of about 1 MeV can be attained; this leads to a measurement of E_{cm} to $\pm$ 3 MeV at each running point, limited by the interpolation between calibrations and the greater uncertainty in e^+ energy. The current LEP averages for these quantities are:

$$\begin{aligned} M_Z &= 91.1888 \pm 0.0018 \pm 0.0040 \text{ GeV} \\ \Gamma_Z &= 2.4974 \pm 0.0027 \pm 0.0029 \text{ GeV} \end{aligned} \tag{13}$$

where the errors are statistical and systematic, respectively. The systematic errors are both dominated by the LEP beam energy, which contributes $\pm$ 4 MeV to M_Z and $\pm$ 2.7 MeV to Γ_Z; these errors are now being reduced through work on understanding machine systematics in the 1993 run, and may be further reduced with a possible new scan in 1995.

Normalization of the LEP data, and consequently the measurements of σ^0, Γ_{ee}, and Γ_{inv}, depend on the precision measurement of LEP luminosity; luminosity depends on interaction region as well as time, and is measured separately by each of the experiments. ALEPH, OPAL, and L3 have installed silicon-based luminosity monitors which have reduced the experimental systematic error in luminosity measurement to the level of 0.1%. Each luminosity measurement is currently dominated by the common theoretical uncertainty of 0.25%. This is expected to be improved to the level of 0.10% in the near future, giving a total LEP luminosity error of about 0.15%. The current LEP average of the peak hadronic cross section is [38]:

$$\sigma_h^0 = 41.49 \pm 0.05(\text{exp}) \pm 0.10(\text{theory}) \text{ nb} \tag{14}$$

4.1.2 Leptonic Partial Widths and A_{FB} Asymmetries

The partial widths of the charged leptons at the Z are related to the weak coupling constants by

$$\Gamma_{ll} = \frac{G_F M_Z^3}{6\pi\sqrt{2}}(g_{A_\ell}^2 + g_{V_\ell}^2) \qquad (15)$$

Since $g_{A_\ell} \approx -0.5$ and $g_{V_\ell} \approx -0.03$, the magnitude of the partial widths is strongly dominated by g_{A_ℓ}. The forward-backward asymmetries $A_{FB}^{0,l}$ at the Z pole for each lepton flavor l, on the other hand, are sensitive to g_{V_ℓ}, with

$$A_{FB}^{0,l} = \frac{\sigma_F - \sigma_B}{\sigma_F + \sigma_B} = \frac{3}{4}\mathcal{A}_e\mathcal{A}_\ell \qquad (16)$$

and

$$\mathcal{A}_\ell = \frac{2g_{V_\ell}g_{A_\ell}}{(g_{V_\ell})^2 + (g_{A_\ell})^2} \qquad (17)$$

The leptonic partial widths are experimentally determined from Γ_Z and the leptonic peak cross section by

$$\sigma = \frac{12\pi}{M_Z^2}(\frac{\Gamma_{ee}}{\Gamma_Z})^2 \qquad (18)$$

and consequently the error on Γ_{ll} is dominated by the error on Γ_Z. A better quantity to use in global fits to the data is $R_l = \Gamma_{had}/\Gamma_{ll}$; the errors on R_l are smaller and are uncorrelated with those of other parameters.

Both the relative leptonic partial ratios R_l and the leptonic forward-backward asymmetries A_{FB}^l are dominated by statistics in each experiment and in the combined LEP averages. The systematics in the A_{FB}^l measurements are dominated by charge misidentification and angular resolutions (not correlated between the experiments), and these will improve with better statistics.

The leptonic partial widths of the Z are equal within errors, each measured to a precision of $< 0.5\%$ by combining the LEP experiments. The average of Γ_e, Γ_μ, and Γ_τ gives a mean leptonic partial width Γ_l measured to a precision of 0.2% with a χ^2/df of 0.2/2.

The measurements of A_{FB}^l are also consistent with lepton universality, but here the agreement is not spectacular, giving a χ^2/df of 5.2/2 . The discrepancy between A_{FB}^μ and A_{FB}^τ is about 2.2 σ; we take this to be consistent with lepton universality.

4.1.3 Tau Polarization

Leptonic couplings can also be measured from τ polarization studies. Because of the V-A coupling in τ decay, the energy and angular distributions of the decay products reflect the polarization of the τ, which is determined by the V and A couplings to the Z of both the initial-state e^+e^- and the final-state $\tau^+\tau^-$. Lepton polarization as a function of production angle is given by

$$\mathcal{P}_\ell = \frac{\sigma_R - \sigma_L}{\sigma_R + \sigma_L} = -\frac{\mathcal{A}_\ell(1+\cos^2\Theta) + 2\mathcal{A}_e\cos\Theta}{(1+\cos^2\Theta) + 2\mathcal{A}_\ell\mathcal{A}_e\cos\Theta} \qquad (19)$$

$\mathcal{P}_\tau$ can be measured in many of the exclusive τ decay channels from the energy distribution of the secondaries (including $e\bar{\nu}_e\nu_\tau$, $\mu\bar{\nu}_\mu\nu_\tau$, $\pi\nu_\tau$, $\rho\nu_\tau$, and $a_1\nu_\tau$). Since the polarization depends on production angle, the data can be used to extract both $\mathcal{A}_e$ and $\mathcal{A}_\tau$. Both measurements are dominated by statistical error, especially $\mathcal{A}_e$; the correlation between the measurements of $\mathcal{A}_e$ and $\mathcal{A}_\tau$ is very small ($\leq 4\%$).

Each of the LEP experiments has combined its own measurements of the various τ decay modes, including correlated systematic errors, to obtain $\mathcal{A}_e$ and $\mathcal{A}_\tau$. In some cases the correlations between modes are small, and in others they are significant. The LEP Electroweak Group has considered systematics common to the experiments (including theoretical uncertainty for the a_1 mode and radiative corrections for hadronic modes), and has found these to be small. V-A coupling for the charged-current decays of the τ has been assumed.

4.1.4 Heavy Flavor Couplings

Since the coupling constants for the heavy quarks c and b are different from those of the leptons (as well as from each other) in the SM, the measurement of the electroweak parameters $A_{FB}^{b\bar{b}}$, $A_{FB}^{c\bar{c}}$, $\Gamma_{b\bar{b}}$, and $\Gamma_{c\bar{c}}$ provides an important test of the Standard Model. The ability of each of the LEP experiments to separately tag $b\bar{b}$ and $c\bar{c}$ events allows these quantities to be directly measured.

Heavy quarks are tagged using leptons, displaced vertices, and event-shape parameters. Because of the high mass of the b and c quarks, their semileptonic decays give leptons which have a hard momentum spectrum and a large p_t with respect to the parent direction; the larger p_t of the b events allows them to be separated from the c events. Since the long lifetime of hadrons containing b and c quarks gives them decay lengths of a few millimeters, events can also be tagged through the detection of a displaced secondary vertex. New silicon vertex detectors have improved vertex tagging to the point where it is competitive with lepton tagging, especially in measuring $R_b = \Gamma_{b\bar{b}}/\Gamma_{had}$ and $R_c = \Gamma_{c\bar{c}}/\Gamma_{had}$. Vertex tagging of b-quarks typically gives an efficiency of 20% with a purity of 90%. Neither lepton nor vertex tagging provides a clean sample of c events. However, a relatively pure sample of c events can be selected by requiring a $D^{*\pm}$ carrying a large fraction of the beam energy.

Double-tagging methods compare the numbers of events with only one hemisphere tagged with the number in which both hemispheres are tagged, and in this way the tagging efficiency is determined from the data. These methods, using tags from leptons, vertices, or both, currently provide the best measurements of R_b.

The LEP Electroweak Working Group has performed a simultaneous fit [38] to R_b, R_c, A_{FB}^b, and A_{FB}^c using data from all experiments. Correlations due to errors in BR(b→ l), BR(c→l), and the mixing parameter $\bar{\chi}$ are fully taken into account. Both R_b and R_c are dominated by systematic errors, while $A_{FB}^{b\bar{b}}$ and $A_{FB}^{c\bar{c}}$ are dominated by statistics. The full heavy flavor electroweak fit gives [39]:

$$\begin{aligned} R_b &= 0.2202 \pm 0.0020 \\ R_c &= 0.1583 \pm 0.0098 \end{aligned} \tag{20}$$

and

$$\begin{aligned} A_{FB}^{0,b} &= 0.0967 \pm 0.0038 \\ A_{FB}^{0,c} &= 0.0760 \pm 0.0091 \end{aligned} \tag{21}$$

The fit gives a correlation coefficient of -0.384 between R_b and R_c; other correlations are negligible. The b parameters have a much higher sensitivity than the c parameters in measuring $\sin^2\theta_W$.

4.1.5 Prospects for Future Improvements

One can anticipate improvements in both statistical and systematic errors in all of the LEP measurements over the next two years. A scan in 1995, with reduced interpolation errors between energy measurements and a better understanding of e^+ energy, should reduce the errors on both M_Z and Γ_Z to less than 2.5 MeV. The new silicon-based luminosity monitors, combined with improved theoretical understanding of the low-angle Bhabha production, should reduce the error in σ_{had}^0 by a factor of 2.

The total LEP lepton sample, including 1993 data, consists of 700,000 events. One can expect to increase this number by a factor of 2-3 with the 1994 and 1995 runs, with a corresponding improvement in A_{FB}^l. The τ polarization results do not yet include most of the 1993 data, so one can expect a larger improvement in statistics (perhaps a factor of 4). The relative statistical improvement in $b\bar{b}$ and $c\bar{c}$ statistics should be similar. Systematic uncertainties already dominate the R_b measurement, but it is reasonable to expect some reduction in these with more data.

4.2 SLC Measurement of A_{LR}

The quantity $\mathcal{A}_e$ which is determined by the LEP experiments from the forward-backward asymmetry of electron final states and from the angular dependence of the τ final state polarization is directly measured at the SLC with polarized initial state electrons. The left-right asymmetry A_{LR} formed from the total Z production cross sections with left-handed and right-handed electrons, σ_L and σ_R, is simply the product of the beam polarization and $\mathcal{A}_e$,

$$A_{LR} \equiv \frac{\sigma_L - \sigma_R}{\sigma_L + \sigma_R} = P_e \cdot \mathcal{A}_e. \tag{22}$$

The measurement makes use of all Z final states except e^+e^- pairs which have a large cross section from t-channel photon exchange. The principal experimental challenge is to accurately measure the polarization of the electron beam. The SLD Collaboration has used a Compton scattering polarimeter to measure the beam polarization to thc 1% level. Since essentially all Z final states constitute the signal, rather simple event selection criteria suppress the backgrounds to a negligible level (0.1%). The measured quantity is independent of acceptance and resolution effects and differs by a small correction (2% for electroweak interference) from the extracted quantity $\mathcal{A}_e$.

The current SLD result [40],

$$\mathcal{A}_e = 0.1637 \pm 0.0071 \text{ (stat)} \pm 0.0028 \text{ (syst)}, \tag{23}$$

is dominated by the 1993 measurement which is based upon a sample of 50,000 events logged with 63% beam polarization. The 1994 run is expected to produce a sample of at least 100,000 events with an average beam polarization of 80%. The program is currently expected to run until 1998 and accumulate a total of 500,000 events. The uncertainty on $\mathcal{A}_e$ should decrease by a factor of two at the end of the 1994 run (in Spring 1995) and by an additional factor of two at the end of the program.

5 Measurements of Electroweak Couplings at low Q^2

5.1 Neutrino Measurements of $\sin^2\theta_W$

The first measurements of $\sin^2\theta_W$ came from fixed-target neutrino experiments, which still provide important tests of electroweak theory. Deep-inelastic ν and $\bar{\nu}$ scattering experiments, which measure the ratio R_ν of neutral current (NC) events to charged current (CC) events, have been carried out using narrow-band neutrino beams as well as wide-band beams, both at CERN and at Fermilab. The Q^2 involved in these processes is much smaller than M_Z^2. In the Standard Model, R_ν gives $\sin^2\theta_W = 1 - M_W^2/M_Z^2$; loop-level corrections depending on M_t are quite small [41], and consequently the νN measurement of $\sin^2\theta_W$ is sometimes presented as a measurement of M_W.

5.1.1 Present Status

The most precise measurements have been carried out by CHARM [42], CDHS [43], and CCFR [44]. The dominant systematic error comes from the uncertainty in the effective mass of the charmed quark, which affects the CC cross section through slow-rescaling effects. The CCFR experiment has experimentally measured [45] this effect using opposite-sign dimuon events, obtaining $m_c = 1.31 \pm 0.24$ GeV. When all three experiments are corrected to use this value, the results are quite consistent and give a world average [44] of

$$\sin^2\theta_W = 0.2256 \pm 0.0047 \quad (\nu N) \tag{24}$$

or expressed in terms of the W mass,

$$M_W = 80.24 \pm 0.25 \text{ GeV} \qquad (\nu\text{N}) \qquad (25)$$

This is in excellent agreement with the direct measurement of M_W from the collider experiments presented below.

The CHARM II experiment has recently reported the measurement [46] of low-q^2 leptonic neutral current couplings from $\nu_\mu e$ scattering. They obtain

$$\sin^2\theta_W = 0.2324 \pm 0.0058 \pm 0.0059 \qquad (\nu_\mu e) \qquad (26)$$

where the first error is statistical and the second systematic. This is a purely leptonic process; the results agree well with high-q^2 leptonic processes ($e^+e^- \to l^+l^-$) at LEP.

5.1.2 Future Prospects

E815 (using a modified CCFR detector with a new collaboration) is scheduled to take high-statistics νN data in the next Fermilab fixed target run. By running alternately with ν and $\bar{\nu}$ beams, the collaboration hopes to improve the measurement of $\sin^2\theta_W$ to a precision of ± 0.0032 (equivalent to a resolution of ± 165 MeV on M_W), and also to measure ρ to a precision of 0.5%.

5.2 Atomic Parity Violation Measurements

Atomic parity violation experiments were among the first searches for the weak neutral current. Since it does not conserve parity, the weak neutral current can mix atomic states of (nominally) opposite parity. This implies that electric dipole (E1) transitions can occur between atomic states which are nominally same-parity eigenstates. The amplitudes for such forbidden transitions $\mathcal{E}_{PNC}$ can be expressed as

$$\mathcal{E}_{PNC} = C_{PNC} \cdot Q_W({}^A_Z N), \qquad (27)$$

where the factor C_{PNC} contains the detailed atomic physics and the factor Q_W is the weak-charge of the atomic nucleus ${}^A_Z N$,

$$Q_W({}^A_Z N) \equiv \rho \left\{ Z - N - 4Z\sin^2\theta_W \right\}, \qquad (28)$$

where Z and N are the proton and neutron numbers of the nucleus, respectively, and $A \equiv N + Z$ is the mass number.

The electroweak radiative corrections to the weak charges of heavy nuclei have the interesting property [47] that the corrections to the ρ parameter nearly cancel. The remaining corrections depend mostly upon the S parameter which makes this class of measurements somewhat unique. Unfortunately, direct determinations of Q_W are limited by the precision of the calculations for the C_{PNC} factors.

A number of researchers have discussed eliminating the sensitivity to these factors by measuring the PNC amplitudes for several isotopes of the same atom [48]. Measurement of the ratio $\mathcal{E}_{PNC}({}^A_Z N)/(\mathcal{E}_{PNC}({}^{A'}_Z N) - \mathcal{E}_{PNC}({}^A_Z N))$ cancels the factor $C_{PNC}\rho$ which eliminates the atomic physics uncertainty but also destroys the approximate T cancellation. The SM information (S-T trajectory) determined by the ratio is identical to that provided by the measurement of $\sin^2\theta_W^{eff}$ at the Z pole.

5.2.1 Present Status

At the current time, the most precise Standard Model information is extracted from the 2% measurement of the 7S-6S transition in ${}^{133}_{55}$Cs [49,50,51],

$$\begin{aligned} Q_W({}^{133}_{55}\text{Cs}) &= -71.04 \pm 1.58 \pm 0.88 \\ S &= -3.0 \pm 2.1 \pm 1.2, \end{aligned} \qquad (29)$$

where the first error is experimental and the second is theoretical; the latter is due to the 1% uncertainty in atomic physics calculations for Cs [50]. The measured value of S is consistent, within the rather large errors, with the Standard Model expectation of $S \approx 0$.

Very recently, parity-non-conserving amplitudes for E1 transitions in Bi, Pb, and Th have been measured [52] with 1% precision. Unfortunately, these atomic systems are more complex than cesium and comparably precise calculations of the C_{PNC} factors do not exist, but are being pursued for the thallium system [53].

5.2.2 Future Prospects

In the near term future, the Cs measurement at Colorado is expected to improve by at least a factor of four [54]. The work on lead and thallium at Washington and Oxford is continuing with measurements on pure isotopes. New measurements on francium [55] and dysprosium [56] are being undertaken at Stony Brook and Berkeley, respectively. An improved calculation of the C_{PNC} factor for the Cs transition is difficult but may yield a result with a 0.2-0.4% precision [53]. The situation for the francium system is similar except that it will be necessary for the experimentalists to measure a number of atomic properties as cross checks. The field is therefore at something of a crossroads. In the next few years, it is expected that S can be measured to the level of $\pm$ 0.5 .

In the longer term, progress is likely to come from measurements of multiple isotopes (although as discussed above these measurements introduce a T dependence). The ultimate precision of such measurements is not clear at the current time.

6 Measurements of the W-boson Mass

A critical test of the Standard Model is the comparison of M_W predicted by the quark and lepton couplings with the direct measurement of M_W. As discussed in Section 8, this provides a direct test of electroweak quantum corrections, potentially more exacting than that provided by the top quark. In order to provide the most stringent test, the resolution in the direct measurement of M_W should match the precision of LEP measurements; this requires a resolution of ~40 MeV,

6.1 Present Status: Tevatron Measurements

Precise measurements of M_W have been made by UA2, CDF, and DØ . Event selection for both CDF and DØ consists of requiring a high-p_t lepton ($p_t > 25$ GeV/c) and large missing transverse energy ($\not{E}_t > 25$ GeV). Event selection is quite clean, with backgrounds at the level of a few percent. The measurements from both CDF and DØ are based on a maximum likelihood fit of the transverse mass distribution of W candidates. The errors in the mass measurements quoted by both experiments consist of three components: statistics, systematics, and scale.

The CDF scale error is based on a measurement of the J/ψ mass with J/$\psi \to \mu^+\mu^-$ decays using the magnetic field. That mass is then scaled to the particle data group value for the mass of the J/ψ. A check on the higher momenta is provided by the Υs and Zs but is not used to scale the value. For muons, this results in a scale error of 60 MeV. For electrons, calorimetric energy is used, with the energy scale determined from the tracking scale and the measured E/p distributions. The scale error for electrons is then 130 MeV. The dominant non-scale systematics include ±120 MeV from momentum resolution, ±90 MeV from the uncertainty in the p_t of the underlying event (based on Z production studies), and ±100 MeV from uncertainties in structure functions. At the present time, these errors are preliminary and are expected to be reduced before final publication [57]. The CDF results are:

$$\begin{aligned} M_W^e &= 80.47 \pm 0.15 \pm 0.21 \pm 0.13 \text{ GeV} \\ M_W^\mu &= 80.29 \pm 0.20 \pm 0.22 \pm 0.06 \text{ GeV} \end{aligned} \qquad (30)$$

where the three errors are due respectively to statistics, non-scale systematics, and scale uncertainties. These measurements are combined, taking into account common errors, to give:

$$M_W^{e+\mu} = 80.38 \pm 0.23 \text{ GeV} \quad \text{(CDF, 1994)} \qquad (31)$$

DØ has presented [58] an M_W measurement based on $W \to e\nu$ events only; with no central magnetic field, their electron data provide the best measurement of M_W. The DØ scale error is based entirely on Z events within the same fiducial region of the calorimeter. A fit allowing for a slope and constant term in the electron energy response gives an error of ±260 MeV. Inclusion of lower mass resonances, the J/ψ and π^0, reduces this to 185 MeV. This reduction and others will be reflected in the final published result for 1994. Other systematics, comparable in size to those of CDF, give a non-scale systematic error of ±160 MeV.

$$\begin{aligned} M_W^e &= 79.86 \pm 0.16 \pm 0.16 \pm 0.26 \text{ GeV} \\ &= 79.86 \pm 0.35 \text{ GeV} \quad \text{(DØ, 1994)} \end{aligned} \qquad (32)$$

The errors in the first line are due respectively to statistics, non-scale systematics, and scale uncertainties.

The world average of these measurements, combined with the earlier UA2 [59] and CDF [60] measurements is [61]

$$M_W = (80.23 \pm 0.18) \text{ GeV} \quad \text{(World Av., 1994)} \qquad (33)$$

where the combination assumes common errors of ±100 MeV.

6.2 Future Prospects

6.2.1 The Tevatron

Each of the three Tevatron measurements of M_W from Run 1a has a statistical error in the range 150-200 MeV. Scaling to 100 pb^{-1} and 1000 pb^{-1} gives statistical errors of ~90 MeV and ~30 MeV for Run 1b and Run 2 respectively for each measurement. In addition, DØ will expand their η coverage to gain a factor of 1.5 in W statistics.

The scale errors in both tracking and calorimetry rely on calibration to a well known resonance, and in all cases the errors should improve approximately as $1/\sqrt{N}$ with higher statistics. For example, higher J/ψ statistics improves the checking of the spatial variations in the field integral for CDF, and higher Z statistics improves the calibration of the DØ calorimetry.

The systematic errors comprise a list of more than ten effects each of which is already smaller than 100 MeV. About three quarters of these effects are limited by the statistics in the data for W's and Z's and are instrumental in origin. The remaining errors are quasi-theoretical in nature and are associated with the details of the production process. In the case of the W p_t spectrum, the Z spectrum is expected to provide the appropriate check. The structure function effects are controlled in part through the measurement of the W asymmetry. The use of this constraint [62] will be implemented in future analyses, and may give a significant reduction in the systematic error from structure functions. It is an important step away from reliance on measurements which need to be carried over from other processes and evolved to the appropriate scale.

The most important point to realize is that no "brick walls" have yet been identified above 30 MeV. We therefore assume that systematic errors will continue to decrease with increasing statistics, but not as fast as do statistical errors. After 100 pb^{-1}, three measurements of M_W with an expected precision of about 150 MeV each and common errors of 70 MeV lead us to expect an overall error of about 110 MeV. For a 1 fb^{-1} exposure with the Main Injector, one would estimate an error of approximately 60 MeV per measurement, of which 35 MeV are common. With four measurements, two from each experiment, a resultant precision of approximately 50 MeV is possible.

Reasonable expectations for the error on M_W, from CDF and DØ combined, are 110 ± 20 Mev after 100 pb^{-1} and 50 ± 20 MeV after 1000 pb^{-1} from the Tevatron measurements.

6.2.2 LEP 2

The center of mass energy of LEP 200 is expected to be at least 10 GeV higher than W pair threshold. The studies referred to in this section assumed 190 GeV and an integrated luminosity of approximately 500 pb^{-1} from about 3 years or so of running.

In principle the mass of the W boson can be measured at LEP 200 by either reconstruction of the final state of hadronic and leptonic decays of the W bosons, or through measurement of the excitation curve for the process. The procedure common to both measurements is that of event selection. W pair production is by no means the dominant process and a careful selection must be made to reduce the backgrounds from conventional annihilation into hadrons and other processes. Care must be taken to handle the radiative effects which rob the initial state of energy but for which the radiated photon appears in the detector. Although the different techniques give comparable precision, it is unlikely that the excitation curve approach will be used since it requires substantial running very low on the threshold curve and hence with diminished event yield.

Reconstruction of events in which one of the W's decays leptonically, as well as those in which both decay hadronically, may be used in the analysis. At some level the leptonic reconstruction and the hadronic reconstruction give independent measurements. According to Monte Carlo studies [63], in either case, reconstruction of the masses results in values lower than that generated. This is attributed to energy lost during reconstruction. The typical shifts are a few GeV and the distributions are of order 10 GeV wide. In order to correct for this, use is made of the beam energy constraint; the resultant masses are rescaled by the ratio of $E_{beam}/E_{W,Recons}$. The resultant distributions are much narrower and are about 200-300 MeV lower than the input mass.

The final correction, of the order of a few hundred MeV, must be made using careful Monte Carlo studies. One component is due to radiative corrections and is well understood. That due to mis-reconstruction will have an error estimated to be about 30 MeV. In addition there is a final error of about 20 MeV, common to all methods and to all experiments, from knowledge of the beam energy. This is the major contributor to an expected overall common error of about 30 MeV. Thus there are potentially 8 measurements each with about 50 MeV overall error. With the common error taken into account, a best estimate for an overall LEP 2 precision based on these Monte Carlo studies would be $\delta M_W = 40$ MeV.

6.2.3 Other Possibilities (TeV*, LHC, NLC)

As we describe in Section 8.6 below, an improvement in δM_W from 40 MeV to 20 MeV would provide a better match to the expected resolutions on M_t and other electroweak measurements expected over the next decade. This would also significantly improve the resolution of the predictions of the Standard Model Higgs mass from global electroweak fits. It is not clear that an M_W resolution as low as 20 MeV can be achieved, but with a dedicated effort one can probably do better than the 40 MeV we have estimated. At LEP 2 this might be possible through improvements in beam energy measurement (using resonant depolarization closer to the beam energy [64]) and increased integrated luminosity in an extended LEP 2 program. Both LHC and the Tevatron may also be able to surpass 40 MeV, but it will be exceedingly difficult to achieve the necessary improvements in systematic errors.

The determination of M_W has been discussed [65] for an e^+e^- linear collider. The beam energy precision is not as good as at LEP 2 so that particular constraint is not available; the statistics are however rather high. At 500 GeV with 100 fb^{-1}, approximately 80K events of the type $e^+e^- \to e\nu W$ would be reconstructed. The statistical error is then less than 20 MeV. Calibration would use about 40k $e^+e^- \to \gamma Z$ events in the same fiducial region in conjunction with a calibration run at the Z pole.

6.3 Other Aspects of W Physics

The couplings of the W to other particles are additional parameters which are predicted by the Standard Model and which can be determined from the measured partial widths of the W to specific final states. Deviations from the SM predictions could signal new decay modes (*e.g.*, $W \to t\bar{b}$ for a light top) or loop corrections involving new particles or new couplings. Both Γ_W and $B(W \to l\nu) =$

$\Gamma(W \to l\nu)/\Gamma_W$ have been measured in hadron collider experiments.

Indirect measurements [66,67,68] of $B(W \to l\nu)$ have been carried out by UA1, UA2, CDF, and DØ using the measured event ratio $R = \sigma B(W \to l\nu)/\sigma B(Z \to l^+l^-)$, multiplied by the theoretical production cross-section ratio $\sigma(p\bar{p} \to Z)/\sigma(p\bar{p} \to W)$ and the LEP measurement of the branching ratio $B(Z \to l^+l^-)$. The published world average [11] for the leptonic branching ratio is $B(W \to l\nu) = 0.107 \pm 0.005$, to be compared with the Standard Model expectation [69] of 0.1084 ± 0.0002. The agreement between the measurement and theory has been used by both CDF and DØ to exclude a light top ($M_t < 62$ GeV) in a way which is independent of any assumptions about top decay modes. In the future, the error in $B(W \to l\nu)$ is likely to be limited at about the 1% level by the theoretical uncertainty in the production cross-section ratios (note that the quantity actually measured here is $B(W \to l\nu) \times \sigma(W)/\sigma(Z)$).

CDF has recently presented [70] a direct measurement of the W width Γ_W based on the measured shape, and especially the high-mass tail, of the transverse mass distribution of $W \to e\nu$ events. In the region far above M_W the effects of the largely gaussian resolution of the detector die off and the distribution is determined by the Breit-Wigner distribution of M_W. From the data of Run 1a, CDF obtains $\Gamma_W = 2.11 \pm 0.28(stat) \pm 0.16(syst)$ GeV. The systematic error comes from measured detector resolutions, and consequently both errors should improve with statistics approximately as $\sqrt{N}$. After 1 fb^{-1} of data (Run 2), an error in Γ_W of about 30 MeV is expected [70] from CDF and DØ combined; this precision approaches the level of radiative corrections to the width.

Furthermore, the direct measurement of the W width may be combined with the leptonic branching ratio measurement to provide a measurement of $\Gamma(W \to l\nu)$. This leptonic partial width is predicted to be $\Gamma(W \to l\nu) = g^2 M_W/48\pi$, and thus its measurement may be used to determine g. In conjunction with the measured leptonic branching ratio, this would give a measurement of $\Gamma(W \to l\nu)$ at the 1.8% level, and consequently an error on the W-$l\nu$ coupling g of 0.9%. The extraction of g from $\Gamma(W \to l\nu)$ is more reliable because it is not sensitive to QCD corrections that are present in the $\Gamma(W \to q\bar{q})$ widths.

7 The Top Quark

The most exciting recent result in high-energy physics has been the first evidence [2] for the top quark, presented in April 1994 by the CDF Collaboration and based on the data from Run 1a. A measured cross section of $\sigma_{t\bar{t}} = 13.9^{+6.1}_{-4.8}$ pb from dilepton and lepton+jet events was reported, with an estimated mass for the top quark of $174 \pm 10(stat) \pm 12(syst)$ GeV based on the reconstruction of lepton+jet events. The DØ experiment reports no significant signal [71,72]. In terms of cross section, the DØ result is $\sigma_{t\bar{t}} = 8.2 \pm 5.1$ pb if a mass of 180 GeV is assumed. The two reported cross sections are compatible both with each other and with the expectations of the Standard Model [73], as shown in Figure 1.

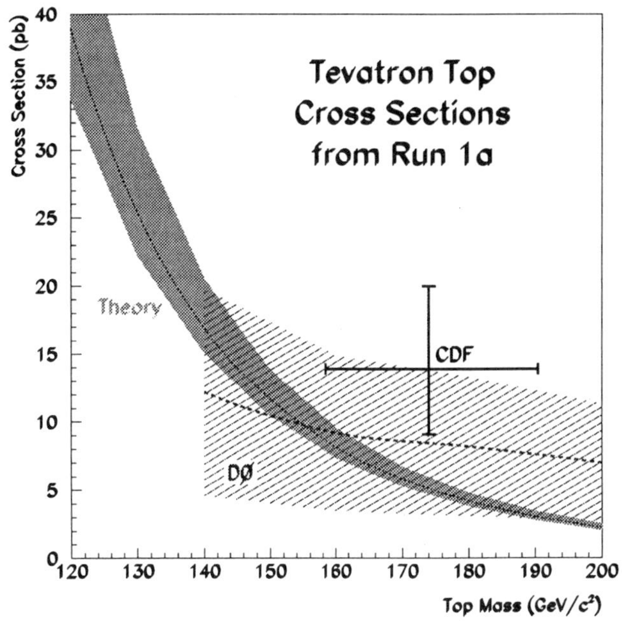

Figure 1: Top cross section versus M_t. The solid band is the theoretical NNLO prediction, the cross-hatched region is the 1-σ range from DØ and the point is the measurement from CDF. The horizontal error bar on the CDF point comes from the mass estimation from lepton+jet events. DØ does not report a measurement of the mass.

Although the statistical significance of the data, from Run 1a only ($\sim$ 20 pb^{-1}), is too limited at present to unambiguously establish the existence of the top quark, the Fermilab data sample has already doubled in Run 1b. It will double again within the next year, and with a continued high level of accelerator performance it is expected to reach a total of 160–200 pb^{-1} for Runs 1a and 1b combined. If $\sigma_{t\bar{t}} \geq 5$ pb (less than the central value reported by either experiment, and corresponding to the theoretical prediction for $M_t \approx 180$ GeV), then each experiment would expect to see a signal of at least 2-3 σ with 100 pb^{-1} of luminosity. A Standard Model top with a mass even as high as 200 GeV, which is at the edge of consistency with present data from Fermilab and LEP, should be definitively measured with 200 pb^{-1}.

It is clear that the accumulation of high statistics at the Tevatron deserves the highest priority. The importance of the discovery of the top quark is three-fold. First, the top is an essential element of the basic SU(2) structure of the Standard Model. Its discovery is crucial for us to claim to have fully established the basic elements of the theory; if the top were not found the Standard Model would have to be rejected or drastically modified.

Secondly, the top quark has effects on other measurable parameters of the Standard Model through higher-order loop diagrams. These effects are calculable as a function of M_t, and their measurement provides a major test of the renormalization structure of the Standard Model and of the quantum effects predicted by the theory. The essential ingredient for these tests is a precise measurement of the mass of the top quark, discussed below.

Finally, the only property of the top quark for which we now have direct evidence is its mass, and this might be regarded as anomalously high (more than an order of magnitude above the mass of any other quark or lepton); it may well be that other properties of the top quark hold additional surprises. Moreover, the top mass is on the scale of electroweak symmetry-breaking; this makes the top a potential laboratory for studying symmetry-breaking effects and for finding possible new particles or new effects in top decays that might take us beyond the Standard Model. The essential ingredient for these investigations is a thorough study of all accessible parameters of the top quark, including its V and A coupling strengths, and its width, decay modes, branching ratios, and production mechanisms.

7.1 Current Status of the Direct Measurement of M_t

The cross section $\sigma_{t\bar{t}}$ is an imprecise indicator of M_t, and using it would in any case introduce a theoretical bias which should be avoided. Consequently the determination of the top mass must be based on the reconstruction of $t\bar{t}$ events, and is sensitive to event reconstruction, detector resolution, particle identification, and backgrounds.

Each $t\bar{t}$ final state is expected to decay into $b\bar{b}W^+W^-$, with each W decaying into either $l\nu$ ($BR = \frac{1}{3}$) or qq' ($BR = \frac{2}{3}$). Events with two leptonic W decays are quite distinctive, but at the present level of statistics they do not allow an accurate reconstruction of M_t due to the missing neutrinos. Events with two hadronic decays might allow mass reconstruction, but at present they cannot be separated from the large backgrounds. Events with one leptonic decay and one hadronic decay have a distinctive signature from the high-p_t lepton and sufficient information from the hadronic decay to allow a reconstruction of the parent top mass. The CDF measurement of the top mass is based solely on events of this lepton+4-jet topology.

The CDF analysis starts with a sample of 52 $W+3$-jet events, and selects the subsample of 10 events which show a b-quark tagged by either a displaced vertex or a final-state lepton. The jet identification criteria are then relaxed for the 4th jet to obtain higher acceptance, giving a subsample of 7 events with 4 jets and a b-tag. The background in this sample, obtained by using Monte Carlo calculations to extrapolate from 3-jet data, is estimated to be $1.4^{+2.0}_{-1.1}$ events. Each jet is then identified with one of the final-state quarks, and a constrained fit is carried out allowing the measured energies and angles to vary within experimental errors and under the constraints that the b-tagged jet be assigned to one of the b quarks, that the two jets assigned to the W reconstruct to M_W, and that the reconstructed masses of the t and $\bar{t}$ be equal. The assignment of jets giving the best χ^2 is used to calculate the value of M_t for that event. The distribution of reconstructed masses is compared to Monte-Carlo expectations in Figure 2.

The number of possible jet permutations for each event is large and the situation is exacerbated by the presence of initial and final state gluon radiation; the requirement of the b-tag reduces the number of viable combinations by a factor of 2. Nevertheless, from Monte-Carlo studies the CDF group finds that the assignment of jets giving the best χ^2 gives a wrong solution about two-thirds of the time. This produces a smearing of the mass resolution relative to that for correct solutions, while apparently leaving the central value unshifted. This is demonstrated for Monte Carlo events in Figure 3. The width of the Monte Carlo distribution suggests that the effective σ is in the range 20-25 GeV for a single event; this agrees with the quoted statistical error of ± 10 GeV on M_t. The central value and the errors on M_t are obtained from maximum likelihood fits to the data, using admixtures of the postulated top signal and estimated backgrounds, as shown in the inset in Figure 2.

The major systematic errors [2,74] are due to uncertainties in (a) the energy scale of the detector (1.8%), (b) the effects of gluon radiation on measured jet energy (4.4%), and (c) the shape assumed for the background ($^{+5.3}_{-4.4}$%).

Both (a) and (b) affect the assignment of momenta to the final state jets, so systematic errors here have a direct effect on the reconstructed top mass. The question of gluon radiation from jets is particularly important because parton quantities are required to calculate the mass of the top. A check of the systematics, and a possible correction for any bias, is provided by the hadronic W decay in the top event sample and also by other data samples such as γ + 1-jet events. In general, systematic uncertainties in the reconstructed jet energy are obtained from the data and should scale as $1/\sqrt{N}$ for the foreseeable future.

The other significant source of systematic error (c) is due to errors in the assumed shape of the background from QCD events and from W+jets production. When

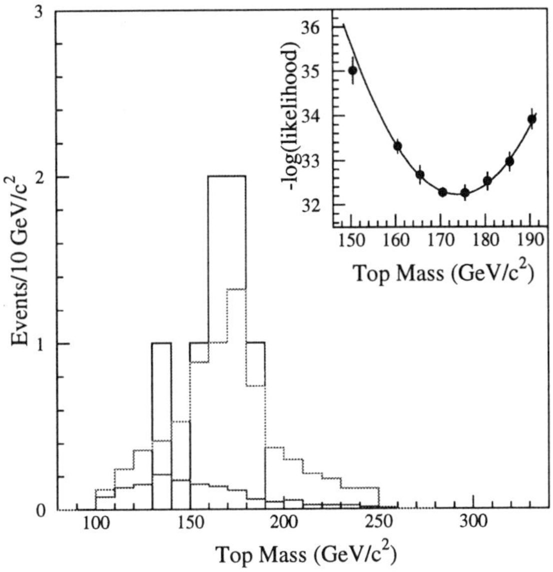

Figure 2: Top mass distribution for CDF data. The solid histogram is the data, the dotted histogram indicates the estimated W+jets background. The dashed histogram is the sum of the appropriate numbers of Monte Carlo signal and W+jets background.

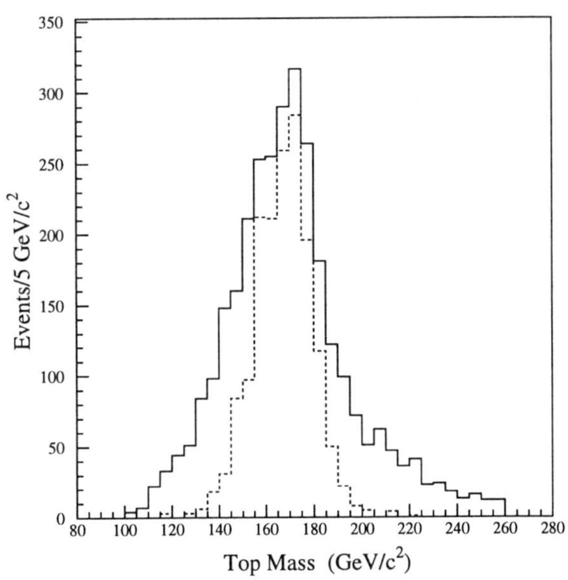

Figure 3: Monte Carlo showing the best χ^2 solutions as a function of mass when one of the b jets is correctly identified. The dotted line indicates the distribution of correct solutions.

fitted to the top hypothesis, the background tends to peak at masses near 140 GeV, as shown in Figure 2. The higher the top mass, the less this is important and, as sample sizes increase, more sophisticated discriminators can be used. Studies of $W+n$-jet events will clearly help in understanding backgrounds and reducing systematic errors; these studies will improve with higher statistics.

7.2 Future Measurements of M_t

7.2.1 M_t at the Tevatron

As discussed above, the systematic errors in M_t are expected to decrease with increasing statistics, roughly as $1/\sqrt{N}$ in the near future. Consequently, one might naively extrapolate the future error on M_t by scaling from the present CDF error of ± 16 GeV. This would give an expected top resolution of $\sim$8 GeV after 100 pb^{-1} and, allowing for a slower reduction in systematic error during Run 2, perhaps $\sim$4 GeV after 1000 pb^{-1}. This extrapolation procedure predicts a satisfying factor of two reduction after each subsequent run, and it has the overwhelming advantage of simplicity; but it neglects a number of important factors.

At present, the only measurement of M_t comes from the CDF data of Run 1a. As discussed above, we expect a clear top signal to emerge in both detectors in Run 1b if $M_t < 200$ GeV, and the total Tevatron data sample will consequently be enhanced over that of CDF alone. Improvements in acceptance and the inclusion of additional channels and new analysis techniques will lead to additional improvements. On the other hand, there are still large statistical limitations in our understanding of both $\sigma_{t\bar{t}}$ and M_t, so projections of future precision are necessarily uncertain.

Consequently we consider the scenario of $\sigma_{t\bar{t}}$=5 pb, which would correspond to the theoretical cross section for M_t=180 GeV. We assume slightly improved efficiencies for b-tagging and for acceptance and include a 30% loss in efficiency from the 4-jet requirement. We use an efficiency×branching-ratio of $\sim$5% summed over CDF and DØ. If M_t is determined only from events with 4 jets plus a b-tag, then with 100 pb^{-1} of data we find approximately 25 events from both detectors combined. Assuming a resolution of 25 GeV from each event gives a statistical error of 5 GeV on M_t. With the additional assumption that systematic errors in jet energy and in background subtraction will scale as $1/\sqrt{N}$ during Run 1b, we obtain a systematic error of $\sim$6 GeV, giving a final error on M_t of $\sim$ 8 GeV.[c] These estimates have

[c] Note added in proof: In March 1995, CDF and DØ each presented definitive confirmation ($> 4\sigma$) of the existence of a massive top quark. The CDF mass was $176\pm8\pm10$ GeV, and the DØ result was $199^{+19}_{-21} \pm 22$ GeV. Both systematic errors are expected to improve with additional analysis work. These results are completely in accord with the expectations discussed in this report.

been based on events with lepton+jets(with b tag). Additional information will come from non-tagged events, selected through kinematic analyses with backgrounds suppressed with event shape cuts, which appear to have only a slightly broader mass distribution than the tagged events [75]. The di-leptons have also been shown to have potential for mass determination [72,76] which we have not included. We conclude that a mass resolution of 8 GeV on M_t is a reasonable projection for the end of Run 1b. It may well be that this estimate will be surpassed with new analysis techniques or a larger event sample.

It is likely that systematic errors will improve more slowly than statistical ones as the data sample increases beyond 100 pb^{-1}. But before the beginning of Run 2, there will be significant upgrades to both detectors in addition to the critical upgrade in luminosity provided by the Tevatron Main Injector. DØ will install a solenoid and a silicon vertex detector before running with the Tevatron Main Injector, and CDF has a program for upgrading its muon and calorimeter systems as well as for installing an upgraded silicon vertex detector. Both systematic and statistical errors should be improved by these upgrades. We believe a resolution of 4 GeV on M_t by the end of Run 2 is a reasonable extrapolation from present knowledge.

As is the case with W mass measurements, there are no "brick walls" to absolutely limit improvement in the resolution of M_t. Additional running should lead to improvements in systematic as well as statistical error (although perhaps not as fast an $1/\sqrt{N}$), and so an extended high-luminosity run of 10 fb^{-1}, using an upgraded Tevatron such as TeV*, will further improve δM_t. But the uncertainties in our projections are already large, and there seems little point in trying to estimate the size of further improvements until we know more about the properties of the top and have more experience with the application of different analysis techniques to a larger data sample. This is another important reason for trying to maximize statistics in Run 1b.

7.2.2 M_t at the LHC

Top production at the LHC proceeds through gluon fusion 90% of the time and is complementary to the Tevatron where quark antiquark annihilation dominates. The $t\bar{t}$ production cross section at LHC will be about 300 times greater than that at the Fermilab Tevatron; at equal luminosities of 10^{32} $\text{cm}^{-2}\text{sec}^{-1}$ (1.0 fb^{-1}/year), one day of LHC will produce as many $t\bar{t}$ events as one year of Tevatron running. With LHC running at 10 fb^{-1}/year (one tenth design luminosity), one year should produce $> 10^7$ $t\bar{t}$ events (for $M_t \leq 200$ GeV). This will make it possible to study decay modes with very small branching ratios. We describe two methods for measuring the top mass that have been examined by the ATLAS [77] and CMS [78] collaborations.

The first method uses lepton+jet events, requiring an isolated lepton with $p_t >$40 GeV from $t \to bW, W \to l\nu$, with three jets in the opposite hemisphere from $\bar{t} \to \bar{b}W$, $W \to 2$ jets. The top mass is reconstructed from the three jets, requiring that two of them reconstruct to a W and that the third be tagged as a b-jet using its displaced vertex. Backgrounds from W+jets are large, but can be greatly reduced by requiring a tag for the second b-jet. Poor statistics are not at issue: 1.0 fb^{-1} should give a statistical error of less than 1 GeV, so the statistical error is small and one can apply additional cuts to obtain a cleaner sample of events. The LOI's [77,78] for ATLAS and CMS estimated a precision of ≤ 5 GeV. More recent work [79] indicates that an error of < 3 GeV can be attained for $M_t = 175$ GeV. As in the case of the Tevatron, the jet reconstruction scale is expected to be established by the hadronic W decays in the top event sample.

The second method requires both W's to decay leptonically, and for one of the $b's$ to decay semileptonically. The mass distribution of the $l\bar{l}$ in the same hemisphere (from the same parent t) can be used to determine the t mass. This method should work even at high luminosities, when jet energy measurement is worsened by the high event rate. This method requires integrated luminosities of 10 fb^{-1} or more, but provides an important alternative procedure which is independent of the energy scale of jets. The method is calibrated by Monte Carlo simulation of the leptonic decays. Recent studies [80] indicate that a resolution of ≤ 2.5 GeV can be attained for $M_t = 175$ GeV.

Since there will be two experiments with comparable resolutions, a precision on M_t of 2 GeV seems achievable after about two years of LHC running.

7.2.3 M_t at the NLC

The most precise measurements of the top mass which have been considered are those done at a proposed linear e^+e^- collider (NLC), with a center of mass energy which can be varied from below $t\bar{t}$ threshold to 500 GeV and with an integrated luminosity of $\sim$50 fb^{-1} per year. Two methods for measuring M_t have been discussed [81].

A nine-point scan over the threshold region, using a total integrated luminosity of 10 fb^{-1}, is expected to yield resolutions of 300 MeV for M_t=150 GeV and 520 MeV for M_t=180 GeV [81]. The determination of M_t uses both the measured excitation curve and the momentum distributions of the top at each scan point. Cuts selecting lepton+4-jet events were used for the momentum measurements, giving about 15% efficiency and negligible backgrounds.

At energies of 100 GeV above $t\bar{t}$ threshold, M_t can be reconstructed from both 6-jet and 4-jet+lepton final states. Selection cuts yield efficiencies of $\sim$30% with background levels of ~10%. A statistical precision of ~150 MeV can be obtained with 10 fb^{-1}. There are systematic shifts of ~500 Mev from various sources which must be taken out, and residual systematic uncertainties of also ~500 MeV; it is thought that both of these can be reduced in the future. Although the threshold scan probably yields the better precision, the high-energy data give very different systematics and provides an important check of M_t (as well as additional information as described below).

7.3 Other Aspects of Top Physics

The mass of the top quark is anomalously high, being on the order of the electroweak symmetry-breaking scale rather than on the mass scale of other quarks. This makes it especially important to measure other properties of the top with the best precision possible. These properties include the decay modes, branching ratios, width, couplings, and production cross sections.

A detailed study of the top quark is envisaged as a major part of the program for an e^+e^- linear collider [81,82]. However, a subset of such studies may also be accessible to higher luminosity running at the Tevatron and at LHC. Studies of these programs are underway for both LHC [77,78] and the Tevatron [83,84].

7.3.1 Top Physics at Hadron Colliders: LHC and TeV*

At both the Tevatron and the LHC, a number of important quantities can be determined from the final states of $t\bar{t}$ events, with precision mainly limited by statistics. The measurements needed include partial widths which may rely heavily on double B-tagging capability and multiple lepton detection and momentum determination. Techniques [85,86] exist for determining the ratio of $V-A$ to $V+A$ decay couplings using for example the correlations between the b and the lepton (from the same t) in lepton+3-jet events. These require accurate determination of the b quark direction and a reliable algorithm for reconstruction of the W four-momentum. Such studies are especially interesting for the top since it decays primarily into $b-W_{long}$, rather than through the W_L polarization state which dominates all weak decays of lighter particles. From the statistics expected in Tevatron Run 2, it is expected that the branching ratio to W_{long} can be measured to a precision of better than 10% [86]. A CP violation test is also possible in top pair production by determining the spin component perpendicular to the plane formed by the b and lepton [85,87].

The single-top "W-gluon-fusion" production mechanism ($W^+g \to t\bar{b}X$) is a source of complementary top physics [88,89]. This process occurs through a strictly electroweak mechanism, and since the t is produced in association with a light $\bar{b}$, each particle is produced in a pure negative-helicity state; measurement of the energy and angular distributions of the b resulting from t-decay therefore determines the relative $V-A$ and $V+A$ charged-current couplings of the top (expected to be pure $V-A$). Moreover, since this process occurs through the $W-t\bar{b}$ vertex it provides a measurement of the magnitude of this coupling and consequently gives the total width of the top quark. The cross section for $t\bar{b}+\bar{t}b$ is roughly half that of $t\bar{t}$; detection efficiencies using conventional cuts are estimated to be ~20% those of $t\bar{t}$. Initial studies [89,90] have indicated that backgrounds from W+2-jet events in this channel are severe, which may limit its importance. Additional studies are under way [84].

At equal luminosities of 10 fb^{-1}, the number of reconstructed $t\bar{t}$ events produced should be about 5,000 for TeV* and on the order of 500,000 for the LHC. Although the LHC has a clear statistical advantage, one should note that the relative mix of $q\bar{q}$ and gg production mechanisms is quite different for $p\bar{p}$ (Tevatron) versus pp (LHC) colliders. At the LHC, gg production of $t\bar{t}$ will dominate while at the Tevatron, $q\bar{q}$ production will dominate. The two machines are consequently different both in backgrounds and in the physics of top production.

7.3.2 Top Physics at a Linear e^+e^- Collider (NLC)

The NLC would provide a number of advantages in top studies: a clean environment, control of E_{cm} (in the $t\bar{t}$ frame), and the ability to run with polarized beams. Although statistics are small compared to those of the LHC, systematics and backgrounds are much better. As described above, a threshold scan can determine M_t to ±500 MeV; the same scan could provide a measurement of the width of the t to a precision of 5-10% [91].

In a run at energies ~100 GeV above $t\bar{t}$ threshold, the t and $\bar{t}$ quarks produced will be highly polarized due to the interference between the electromagnetic and weak currents, and the fast decay of the top via the weak charged current will preserve the helicity information [81,82]. Semileptonic decays can be reconstructed to determine the helicity structure of *both* the production (neutral current) and decay (charged current) couplings simultaneously. Control of the polarization of the beams would increase the sensitivity of these measurements.

A run of 50 fb^{-1} at E_{cm}=400 GeV is expected to produce about $10-15 \times 10^3$ reconstructed events, and to allow the measurement of each of the SM couplings to the level of a few percent. The anomalous magnetic moment could be measured to ~0.02 Bohr magnetons, compared to the expected moment of α_s/π=0.04 from QCD [81].

7.3.3 Beyond the Standard Model

Since the top is the heaviest of the known elementary particles, its decays may produce states which have not been observed previously. At any of these potential top factories, a number of non-standard model searches are possible with a sufficient top sample. Within supersymmetric models, additional light Higgs-like particles are required which would naturally couple to the heaviest particles: decay from the top, $t \to bH^+$ and subsequent decay to tau leptons. This could conceivably occur at a rate of 10% of the conventional one. SUSY also requires the existence of a scalar top SUSY partner which could be produced at sizable fractions of the rate of conventional top and would have a similar signature. These and other effects are discussed in the "Beyond the Standard Model" section of this review.

8 Global Analysis of Electroweak Data

In presenting the status of combined electroweak fits, we have borrowed from work of the LEP Electroweak Working Group in Sections 8.1-8.4 . A more complete discussion of the data and of their fitting procedures can be found in their 1994 paper [38]. A more detailed review of LEP electroweak physics, including future prospects, is given in reference [92].

8.1 Lepton Universality

The most precise test of universality for the charged-current interaction comes from the ratio $R_{e/\mu} = \Gamma(\pi \to e\bar{\nu}_e)/\Gamma(\pi \to \mu\bar{\nu}_\mu)$. This ratio is not affected by the strong interactions to zeroth order in α; the $O(\alpha)$ corrections have been calculated to be $\approx 3.7\%$, and general arguments given in Refs. [93,94] indicate that the calculation is very precise. The ratio of experimental results [95] to theory [94] gives

$$\frac{R^{exp}_{e/\mu}}{R^{th}_{e/\mu}} = 0.9966 \pm 0.0030 \pm 0.0004, \tag{34}$$

where the errors are experimental and theoretical, respectively. Alternatively, introducing effective $SU(2)_L$ couplings associated with the $(e\nu_e)$ and $(\mu\nu_\mu)$ currents, one obtains

$$\frac{g_e}{g_\mu} = 0.9983 \pm 0.0015 \pm 0.0002, \tag{35}$$

which is an impressive verification of e-μ universality. As the amplitudes involved are proportional to the W four-momentum q^μ, this test probes the longitudinal W couplings, and constrains heavy neutrino mixing, possible contributions from additional charged Higgs scalars, leptoquarks, and compositeness [96].

The tau coupling ratio can be extracted from $Br(\tau \to e\bar{\nu}_e\nu_\tau(\gamma)) = \tau_\tau\Gamma(\tau \to e\bar{\nu}_e\nu_\tau(\gamma))$, giving

$$\left(\frac{g_\tau}{g_\mu}\right)^2 = Br(\tau \to e\bar{\nu}_e\nu_\tau(\gamma))\frac{\tau_\mu}{\tau_\tau}\left(\frac{m_\mu}{m_\tau}\right)^5 \times 0.9996 \tag{36}$$

where g_τ and g_μ are again effective couplings associated with the $(\nu_\tau\tau)$ and $(\nu_\mu\mu)$ currents. Tau decays also provide a measurement of e-μ universality using

$$\left(\frac{g_\mu}{g_e}\right)^2 = \frac{Br(\tau \to \mu\bar{\nu}_\mu\nu_\tau(\gamma))}{Br(\tau \to e\bar{\nu}_e\nu_\tau(\gamma))} \times 1.02821, \tag{37}$$

where the numerical factors in both equations reflect phase-space differences, radiative corrections, and propagator effects. Inserting the experimental values [96] gives

$$\frac{g_\tau}{g_\mu} = 0.9949 \pm 0.0064 \tag{38}$$

$$\frac{g_\mu}{g_e} = 0.9983 \pm 0.0060 \tag{39}$$

providing a very nice test of e-μ-τ universality for charged current couplings.

Universality of the neutral current couplings has been tested at LEP from the partial widths of the Z to each of the three lepton flavors (which primarily determines g_{A_ℓ} for each lepton l), and from the forward-backward asymmetries for each lepton flavor (giving g_{V_ℓ}), as presented in Section 4.1.2; tau polarization measurements, described in Section 4.1.3, have also been used to extract $\mathcal{A}_e$ and $\mathcal{A}_\tau$. A combined fit to the LEP line-shape, forward-backward asymmetries, and tau polarization data give the following ratios of leptonic coupling constants:

$$\begin{aligned} g_{A_\mu}/g_{A_e} &= 1.0014 \pm 0.0021 , \\ g_{A_\tau}/g_{A_e} &= 1.0034 \pm 0.0023 , \\ g_{V_\mu}/g_{V_e} &= 0.83 \pm 0.16 , \\ g_{V_\tau}/g_{V_e} &= 1.044 \pm 0.091 . \end{aligned} \tag{40}$$

The best-fit values of the couplings are presented and compared graphically in Figure 4.

8.2 The Invisible Width of the Z

About 20% of the Z decays are expected to be into $\nu\bar{\nu}$ final states, which cannot be directly observed. The invisible width of the Z can be experimentally measured using

$$\begin{aligned} \frac{\Gamma_{inv}}{\Gamma_{\mu\mu}} &= \frac{\Gamma_Z - \Gamma_{had} - 3 \times \Gamma_{ll}}{\Gamma_{\mu\mu}} \\ &= \sqrt{\frac{12\pi R_l}{M_Z^2\, \sigma^0_{had}}} - R_l - 3. \end{aligned} \tag{41}$$

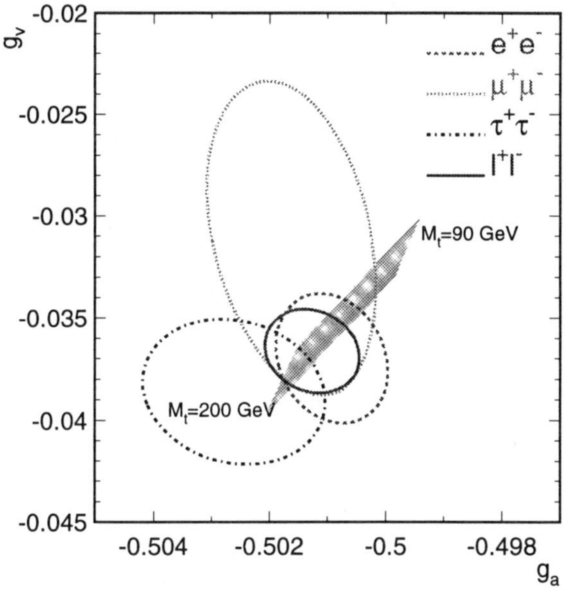

Figure 4: 68% probability contours in the g_{V_ℓ}-g_{A_ℓ} plane. The solid contour results from a fit assuming lepton universality. The shaded band represents the Standard Model prediction.

From this ratio and the expected SM couplings, one can calculate the number of light neutrinos to be [38]

$$N_\nu = 2.988 \pm 0.023 \qquad (42)$$

to be compared to an expected value of 3 (for 3 leptonic doublets). The ratio Γ_{inv}/Γ_{ll} is insensitive to α_s and to loop corrections (and therefore insensitive to M_t and M_H). This measurement is consequently an important test of the $SU(2)\times U(1)$ structure of the couplings of the leptonic doublets.

8.3 The Effective Electroweak Mixing Angle $\sin^2\theta_{\rm eff}$

If the universality of leptonic couplings, and the analogous universality of couplings for each of the three doublets of quarks is assumed, then the SM predicts the relative strengths of all of these couplings as a function of the single parameter $\sin^2\theta_{\rm W}^{eff}$. Therefore $\sin^2\theta_{\rm W}^{eff}$ can be measured independently in a number of different ways, as shown in Table 2. The comparison of these measurements provides a major test of the standard model, and in particular of the g_V and g_A structure of the couplings.

The first 6 lines of Table 2 are from the LEP experiments [38] and give a χ^2 of 2.8 for 5 degrees of freedom. The values of $\sin^2\theta_{\rm W}^{eff}$ obtained from lepton asymmetries, tau polarization ($\mathcal{A}_\tau$ and $\mathcal{A}_{\rm e}$ combined), and $A_{\rm FB}^{0,\,\rm b}$ all give independent and consistent results, with similar precision.

The last line gives the SLC value of $\sin^2\theta_{\rm W}^{eff}$ from the measurement of A_{LR}, and differs by 2.4 σ from the average of the others. The ratio of the LEP and SLC quantities is not sensitive to higher-order loop corrections, and it is difficult to find any natural theoretical reason for a difference. SLC measurements should improve by a factor of two within the next few months. We assume in this report that the present small discrepancy will be resolved with better measurements; if this is not the case, understanding the reason for the difference will be crucial.

8.4 Standard Model Fits to Electroweak Data (Not Including M_t)

The Standard Model predicts the masses of the gauge bosons W and Z as a function of $\sin^2\theta_W$, as well as the magnitudes of the couplings. This allows precise comparisons of a number of independent measurements; their precision makes them sensitive to α_s, and to the masses M_t (top quark) and M_H (Higgs boson) through loop corrections. The leading top quark dependence is quadratic and allows a determination of M_t. The dependence on M_H is logarithmic, and consequently M_H cannot be reliably determined without a direct measurement of M_t.

The LEP Electroweak Group has combined LEP measurements using the full correlation matrix between the different measurements, and has simultaneously fitted M_t and α_s to the data under Standard Model assumptions. Table 3 shows the constraints obtained on M_t and $\alpha_s(M_Z^2)$ when fitting all electroweak measurements to Standard Model calculations. The fits have been repeated for $M_H = 60$, 300 and 1000 GeV and the difference in the fitted parameters is quoted as the second uncertainty in each parameter. The second column gives the results from fitting LEP data alone, the third column includes M_W measurements and neutrino data, and the fourth column includes A_{LR} from SLC. The χ^2 for all these fits is acceptable.

The LEP measurements of the partial widths and of the asymmetries are sensitive in varying degrees to α_s and M_t. The individual measurements (LEP averages) and their errors are summarized in Table 4(a), which also gives the final errors expected from LEP 1 (discussed in Section 4.1 above and used in Section 8.6 below), the Standard Model fit assuming M_H=300 GeV, and the pull of each measurement on the fit. The sensitivities of each of these values to α_s and M_t is shown in Figure 5.

The only measurements that show any significant deviation from the SM predictions are $R_b = \Gamma_b/\Gamma_{had}$ from LEP and $\sin^2\theta_{eff}$ from the SLC measurement of A_{LR}. R_b is about 2.2 σ high (favoring a lower top mass), and A_{LR} is about 2.4 σ high (favoring a higher top mass), as further discussed in Section 8.5. These deviations do not seem remarkable in view of the large number of measurements that are included in the fit. Moreover, there is a correlation between the measurements of R_b and R_c;

Table 2: (from Ref. [38]) Comparison of several determinations of $\sin^2\theta_{eff}^{\rm lept}$ from asymmetries. Averages are obtained as weighted averages assuming no correlations.

	$\sin^2\theta_{eff}^{\rm lept}$	average by group of observations	$\chi^2/(d.o.f.)$	cumulative average	$\chi^2/(d.o.f.)$
$A_{FB}^{0,\ell}$	0.2311 ± 0.0009				
$\mathcal{A}_\tau$	0.2320 ± 0.0013				
$\mathcal{A}_e$	0.2330 ± 0.0014	0.2317 ± 0.0007	1.4/2	0.2317 ± 0.0007	1.4/2
$A_{FB}^{0,b}$	0.2327 ± 0.0007				
$A_{FB}^{0,c}$	0.2310 ± 0.0021	0.2325 ± 0.0006	0.6/1	0.2321 ± 0.0005	2.8/4
$\langle Q_{FB} \rangle$	0.2320 ± 0.0016	0.2320 ± 0.0016	–	0.2321 ± 0.0004	2.8/5
A_{LR} (SLC)	0.2294 ± 0.0010	0.2294 ± 0.0010	–	0.2317 ± 0.0004	9.0/6

Table 3: (from Ref. [38]) Results of fits to LEP and other data for M_t and $\alpha_s(M_Z^2)$, assuming the Standard Model with a single Higgs boson. No external constraint on $\alpha_s(M_Z^2)$ has been imposed. In the third column the combined data from $p\bar{p}$ and neutrino experiments is also included, and the fourth column gives the result when the SLD measurement of A_{LR} is added. The central values and the first errors quoted assume $M_H = 300$ GeV (not the best fit to M_H). The second errors correspond to the variation of the central value when varying M_H in the interval $60 \leq M_H$ [GeV] ≤ 1000.

	LEP	LEP + M_W and ν data	LEP + M_W and ν data + A_{LR} from SLC
M_t (GeV)	$173^{+12}_{-13}\,^{+18}_{-20}$	$171^{+11}_{-12}\,^{+18}_{-19}$	$178^{+11}_{-11}\,^{+18}_{-19}$
$\alpha_s(M_Z^2)$	$0.126 \pm 0.005 \pm 0.002$	$0.126 \pm 0.005 \pm 0.002$	$0.125 \pm 0.005 \pm 0.002$
$\chi^2/(d.o.f.)$	7.6/9	7.7/11	15/12
$\sin^2\theta_{eff}^{\rm lept}$	$0.2322 \pm 0.0004\,^{+0.0001}_{-0.0002}$	$0.2323 \pm 0.0003\,^{+0.0001}_{-0.0002}$	$0.2320 \pm 0.0003\,^{+0.}_{-0.0002}$
$1 - M_W^2/M_Z^2$	$0.2249 \pm 0.0013\,^{+0.0003}_{-0.0002}$	$0.2250 \pm 0.0013\,^{+0.0003}_{-0.0002}$	$0.2242 \pm 0.0012\,^{+0.0003}_{-0.0002}$
M_W (GeV)	$80.28 \pm 0.07\,^{+0.01}_{-0.02}$	$80.27 \pm 0.06\,^{+0.01}_{-0.01}$	$80.32 \pm 0.06\,^{+0.01}_{-0.01}$

Table 4: Summary of measurements included in the combined analysis of Standard Model parameters. Section a) summarizes LEP averages, Section b) gives $\sin^2\theta_{eff}^{lept}$ from the measurement of the left-right polarization asymmetry at SLC, c) gives $\sin^2_\theta$ from νN-scattering, d) gives M_W from hadron colliders, and e) gives the recent M_t measurement from CDF. The third column gives the anticipated future errors. These are the final results for LEP 1 (1996), SLC (1998), νN (1998), and Tevatron Run 1b (1996). LEP 2 results (1999) are not included. The Standard Model fit result in column 4 and the pulls in column 5 are derived from the fit to all data *except* M_t for a fixed value of $M_H = 300$ GeV.

	measurement and error in 1994	anticipated error in 1998	Standard Model fit	pull (1994)
a) LEP 1				
line-shape and lepton asymmetries:				
M_Z [GeV]	91.1888 ± 0.0044	± 0.0025	91.1887	0.0
Γ_Z [GeV]	2.4974 ± 0.0038	± 0.0025	2.4973	0.0
σ_h^0 [nb]	41.49 ± 0.12	± 0.06	41.437	0.4
R_ℓ	20.795 ± 0.040	± 0.025	20.786	0.2
$A_{FB}^{0,\ell}$	0.0170 ± 0.0016	± 0.0010	0.0153	1.0
τ polarization:				
$\mathcal{A}_\tau$	0.143 ± 0.010	± 0.005	0.143	0.0
$\mathcal{A}_e$	0.135 ± 0.011	± 0.006	0.143	−0.7
b and c quark results:				
$R_b = \Gamma_{b\bar{b}}/\Gamma_{had}$	0.2202 ± 0.0020	± 0.0014	0.2158	2.2
$R_c = \Gamma_{c\bar{c}}/\Gamma_{had}$	0.1583 ± 0.0098	± 0.0080	0.172	−1.4
$A_{FB}^{0,b}$	0.0967 ± 0.0038	± 0.0022	0.1002	−0.9
$A_{FB}^{0,c}$	0.0760 ± 0.0091	± 0.0060	0.0714	0.5
$q\bar{q}$ charge asymmetry:				
$\sin^2\theta_{eff}^{lept}$ from $\langle Q_{FB}\rangle$	0.2320 ± 0.0016	± 0.0016	0.2320	0.0
b) SLC				
$\sin^2\theta_{eff}^{lept}$ from $\mathcal{A}_e$	0.2294 ± 0.0010	± 0.00025	0.2320	−2.6
c) νN scattering				
$1 - M_W^2/M_Z^2(\nu N)$	0.2256 ± 0.0047	± 0.0032	0.2242	0.2
d) M_W from $p\bar{p}$				
M_W [GeV] (CDF, DØ, UA2)	80.23 ± 0.18	± 0.110	80.32	−0.5
e) M_t from Fermilab				
M_t from lepton+jet events	$174. \pm 16.$	$\pm 8.$	178.	+0.2

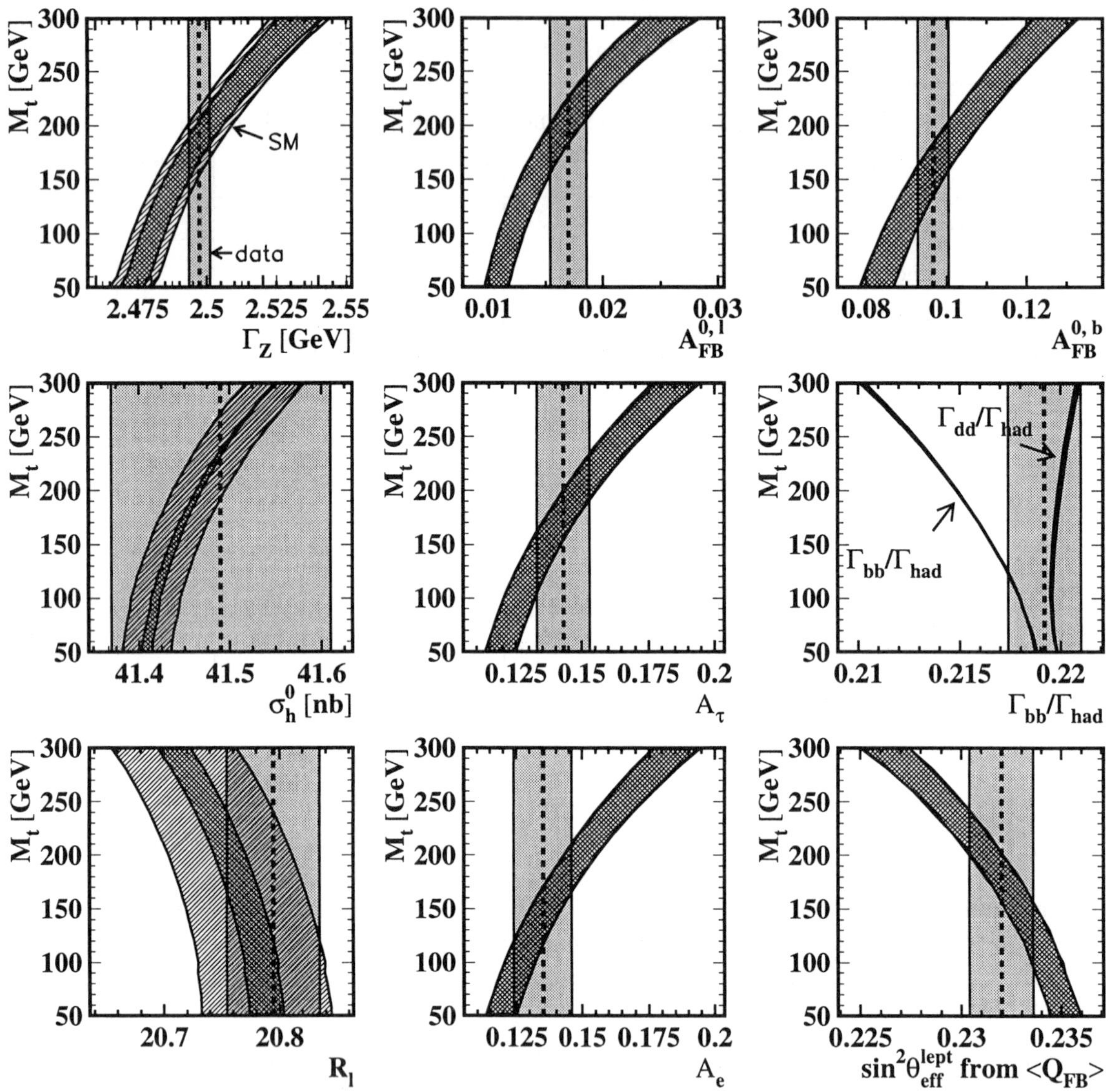

Figure 5: Comparison of LEP measurements with the Standard Model prediction as a function of M_t. The cross-hatched area shows the variation of the Standard Model prediction with M_H spanning the interval $60 < M_H\,(\mathrm{GeV}) < 1000$ and the singly-hatched area corresponds to a variation of $\alpha_s(M_Z^2)$ within the interval $\alpha_s(M_Z^2) = 0.123 \pm 0.006$. The total width of the band corresponds to the linear sum of both uncertainties. For the ratios of partial widths, $\Gamma_{b\bar{b}}/\Gamma_{\mathrm{had}}$ and $\Gamma_{d\bar{d}}/\Gamma_{\mathrm{had}}$, this variation nearly cancels. For the comparison of R_{b} with the Standard Model the value of R_{c} has been fixed to the Standard Model. The experimental errors on the parameters are indicated as vertical bands.

if R_c is fixed at its Standard Model value of 0.171, then the deviation of R_b is reduced to 1.9 σ. It is also worth noting that the electroweak corrections to R_b are different from all other electroweak corrections, since R_b is sensitive to the $W - t\bar{b}$ vertex. This quantity is therefore more sensitive to some kinds of deviations from the Standard Model. On the other hand, it is difficult to theoretically account for a difference between the values of $\sin^2\theta_{\rm eff}$ measured at LEP and SLC.

The internal consistency of all of these measurements, including R_b and A_{LR}, is very good. Moreover, the values of α_s and M_t obtained from these fits are consistent with external measurements of these parameters: event shape measurements at LEP give ($\alpha_s(M_Z^2) = 0.123 \pm 0.006$ [97]), and the top mass extracted by CDF from lepton+jet events is $M_t = 174 \pm 10^{+13}_{-12}$ GeV. This gives strong support to the idea that we are indeed seeing electroweak quantum corrections in the measured electroweak parameters, and that a massive top quark is a major source of these corrections. However, the remarkably close agreement between the predicted and measured values of M_t is of course fortuitous. One should note that the prediction given in Table 3 has assumed the fixed value $M_H = 300$ GeV; if M_H is left as a free parameter, a somewhat lower value of M_t is found as described below.

8.5 Comparison to Direct Measurements of M_W and M_t

As discussed in Section 2, the Standard Model is completely specified at tree level by the 3 parameters $\alpha(M_Z)$, G_F, and M_Z. When loop-level corrections are included, two additional parameters (*e.g.*, M_t and M_H) are required. It is convenient to choose the 2 parameters to be M_t and M_W, since the direct measurements of these masses are expected to improve very dramatically over the next decade.

Both M_t and M_W can be predicted from the e^+e^- and νN measurements listed in parts *a*), *b*), and *c*) of Table 4. The $M_t - M_W$ contours [98] from a Standard Model fit to this data (referred to below as "Z data", even though νN is also included) are compared to the direct measurements in Figure 6. The mass M_H of the Higgs (which can be calculated as a function of M_t, M_W, and the other SM parameters) is unconstrained in these fits. The shaded bands correspond to lines of constant M_H, with the width of the bands being due mainly to the uncertainty in $\alpha(M_Z)$. The best fit and the limits of the 1-σ contour, from the Z and νN data alone, correspond to the values $M_W = 80.315^{+0.066}_{-0.068}$ GeV and $M_t = 159.^{+18.}_{-15.}$ GeV. These values agree well with the direct measurements of M_W and M_t listed in Table 4 and displayed as the direct M_W-M_t point in Figure 6.

The shapes of the constraints from different measurements can be understood on an $M_t - M_W$ plot in terms of the leading order M_t and M_H dependence of the loop corrections. Since electroweak corrections involve two unknown parameters, any given measurement will produce a band in the $M_t - M_W$ plane; for example, the asymmetry measurements of $\sin^2\theta_{\rm eff}$ give a band which slopes upward slightly with increasing M_t. Both the size and the shape of the contours resulting from combining many measurements depend on the central values of those measurements, as well as on their errors. In particular, the contours of Figure 6 are significantly influenced by the two measurements which deviate from the best-fit SM values at the 2-σ level: these are R_b from LEP and A_{LR} from SLC. Figure 7 shows the 1-σ contour (shaded region) from the fit to all e^+e^- data *except* these two measurements, as well as the 1- and 2-σ contours from each of these measurements alone (the experimental constraints on M_Z and α_s are included in all contours).

The solid contour in the center of Figure 7 is determined mainly by the measurements of $\sin^2\theta_{eff}$ (giving a band with a slight upward slope to the right) and Γ_Z (giving a wider roughly horizontal band). The intersection of these bands defines the solid contour. The SLC measurement of A_{LR} gives a band similar to the LEP $\sin^2\theta_{eff}$ band, but displaced upward due to the corresponding difference in $\sin^2\theta_{eff}$. The band from R_b has a very different dependence because of the strong $b - t$ coupling and the fact that the M_H dependence cancels in the ratio of $\Gamma_{b\bar{b}}$ to Γ_Z. Consequently R_b is virtually independent of M_H and gives an approximately vertical band in the figure between the limits imposed by the constraints on M_H.

As can be seen from Figure 7, the R_b data tend to pull M_t strongly to the left. The data A_{LR} contours, on the other hand, tend to pull almost vertically upward in M_W. The result of these two pulls is that M_t and M_H are both significantly pulled to lower values. The effects from these measurements, due to their deviation from SM predictions, contribute to the relatively small contours of Figure 6 and the relatively tight constraints on M_t, and especially on M_H.

If all measurements agreed precisely with the SM predictions, the contours would be somewhat larger. This is shown in Figure 8, in which the central values of all measurements have been shifted to the best fit using all the data of Table 4 (including M_W and M_t). Although Figure 6 unquestionably gives the more accurate picture of our present knowledge, it is impossible to predict how the contours will evolve in time since we do not know how the central values of individual measurements may change. Consequently we use the contours of Figure 8 as a reference point to gauge future improvements. When the measurements of LEP 1, LEP 2, SLC, νN, and Tevatron Run 2 are completed, the considerable reduction in errors should result in the contours shown in Figure 9

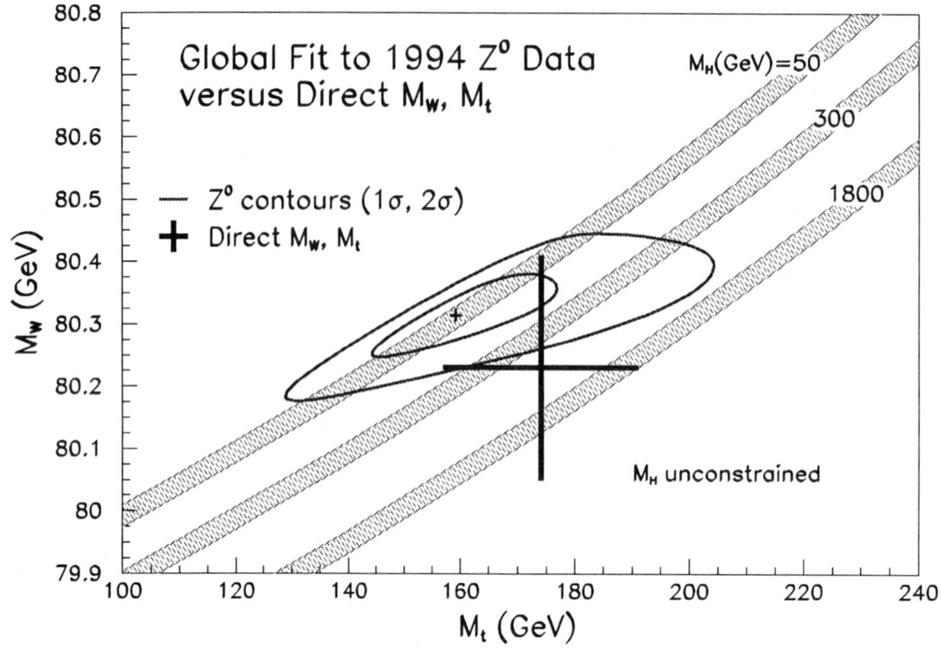

Figure 6: The $M_t - M_W$ contours from a global SM fit to all e^+e^- and νN data (parts a, b, and c of Table 4) are compared to the direct measurements. In the global fit, M_H is a free parameter; it ranges from 22 to 180 GeV on the 1-σ contour. The cross-hatched bands correspond to the indicated Higgs masses; the widths of the bands are due primarily to the uncertainty in $\alpha(M_Z)$.

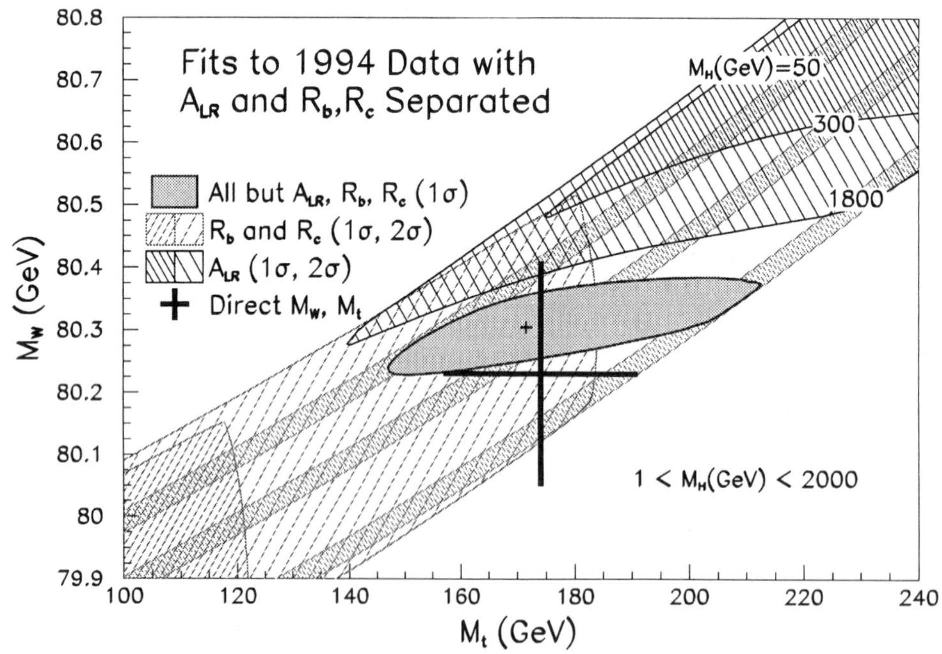

Figure 7: The 1- and 2-σ contours from A_{LR} measurements (upper right) and from R_b, R_c measurements (lower left) are separately displayed. All contours include the constraints from M_Z, $\alpha_s = 0.123 \pm 0.006$, and $1 < M_H(GeV) < 2000$. The solid contour in the center is the 1-σ limit from all other Z and νN data; M_H ranges on this contour from 37 GeV to the upper cutoff value of 2000 GeV.

(the specific resolutions assumed here are given in the following section). However, one should bear in mind that the actual contours in 2002 might look quite different, just as the "expected" contours of Figure 8 differ from the actual contours of Figure 6.

8.6 Future Expectations and Constraints on M_H

At the present time, electroweak measurements are in excellent agreement with the predictions of the Standard Model. Moreover, significant restrictions on the allowed Higgs mass are now just beginning to emerge from the data. Fortunately, major improvements in electroweak measurements are expected over the next decade, accompanied by major improvements in direct searches for the Higgs during the next two decades. The confrontation of the indirect determinations with the results of the direct searches will perhaps provide the ultimate test of the Standard Model. The precision of the predictions of M_H from electroweak data provides a useful benchmark of electroweak measurements, and is one indicator of which measurements are most important.

The sensitivity of the current Z_0 data alone to M_H is limited, since loop corrections are usually proportional to linear combinations of M_t^2/M_Z^2 and $\ln(M_H^2/M_Z^2)$, which are dominated by M_t. For this reason, an external measurement of M_t greatly improves the sensitivity to M_H. If the CDF determination of M_t from lepton+jet events is included in the full electroweak fit, then the sensitivity of the data matches the expectations from Monte-Carlo calculations and weakly favors a relatively low value of M_H. Reference [99] presents several ways of fitting all of the data, and obtains a best-fit Higgs mass in the range $50 < M_H < 130$ GeV in various fits, with an uncertainty on the order of a factor of two or three in each fit. The results vary by about a factor of two depending on the details of data treatment, as discussed in the paper. A similar fit by the LEP Electroweak Group [38] gives similar results.

However, we note that the resolution on the Higgs mass, in particular, is sensitive to the details of the fitting program as well as to the selection of input data. Moreover, the change in the value of $\alpha(M_Z)$ suggested by reference [33] would increase the central value of M_H by a factor of 3 or 4. The experimental evidence that the Higgs is light is not compelling at this time.

We defer to other work in progress [38,100] for more complete discussions of current data and analysis; in this report our emphasis is on the relative improvements to be expected from future data. Figure 10 shows how we expect the χ^2 curve for the Higgs mass to evolve in time as measurements improve. In producing this plot,[d] we have adjusted all measurements to agree with the Standard Model predictions for $M_H = 100$ GeV, and have then adjusted the errors of each measurement as a function of time in accordance with expectations, assuming that final results will lag behind data-taking by at least a year. At each step, we have examined the sensitivity of δM_H to a factor-of-two reduction in 5 sources of error: δM_t, δM_W, $\delta\alpha(M_Z)$, $\delta sin^2\theta_{eff}$ from all LEP and SLC measurements, and all other LEP measurements (M_Z, Γ_Z, σ_h^0, R, $\Gamma_{b\bar{b}}$, and $\Gamma_{c\bar{c}}$).

The major advances between now and 1996 come from the improvements in δM_t and δM_W from the Tevatron, in final LEP 1 measurements, and in $\mathcal{A}_e$ from SLC. The anticipated errors for both 1994 and 1996 are given in Table 4; we also assume a reduction in $\delta\alpha_s$ from 0.006 to 0.003 from LEP data (*e.g.*, event shapes, τ decays, etc.). The SLC error has been reduced to $\delta\sin^2\theta_{\rm eff}$=0.0005, and the Tevatron error on M_W has become 110 MeV. The biggest effect comes from the reduction of δM_t from 16 GeV to 8 GeV. The resulting uncertainty in M_H is reduced from $f_\sigma \approx$3.5 to $f_\sigma \approx 2.3$. At that point in time, δM_H will be most sensitive to improvements in δM_t and $\delta\alpha(M_Z)$.

Between 1996 and 1999, we expect a new measurement of $\sin^2\theta_W$ from the Fermilab neutrino experiment, and a reduction in the SLC error in $\sin^2\theta_{\rm eff}$ from 0.00040 to 0.00025, and a measurement of M_W from LEP 2 at a sensitivity of $\pm$50 MeV (preliminary LEP 2 errors). Since M_W will be predicted to a precision of 42 MeV from the LEP 1/SLC/νN data (31 MeV if M_t is included in the fit), this last measurement provides an impressive test of the Standard Model. There will be no new Tevatron collider results since Run 2 will only begin in 1998-9, and consequently there will be no reduction in δM_t. Since the uncertainty on M_H is still dominated by δM_t, the 1999 improvements o not give a major immediate improvement in δM_H although they will be important when δM_t is further reduced.

Between 1999 and 2002, we expect results from 1 fb^{-1} from Tevatron Run 2 and final results from LEP 2, giving errors of 40 MeV on M_W and 4 GeV on M_t; the resulting error on M_H is reduced to $f_\sigma \approx 2.0$.

During the first two or three years of LHC running, we assume a reduction on δM_t from 4 GeV to 2 GeV. No other improvements are assumed.

At this point in time (taken to be 2006) the error on M_H has been reduced to $f_\sigma \approx$1.9, and is most sensitive to improvements in M_W and $\alpha(M_Z)$; improvements in other measurements are relatively unimportant. Both M_W and

[d]Note that the horizontal scale in Figure 10 is logarithmic, and that the χ^2 curve appears to be parabolic in shape as a function of $\ln(M_H)$. This is because the loop diagrams contribute terms proportion to $\ln(M_H^2)$. As a consequence, the 1-σ deviations from M_H scale approximately with M_H, and it is better to quote the errors on M_H as a factor f_σ rather than as an absolute number. For example, in 1996 the 1-σ limits for 1996 are 40 and 225 GeV for M_H^0=100 GeV ($f_\sigma = 2.3$); these would be 85 and 409 GeV for M_H^0=200 GeV ($f_\sigma = 2.2$).

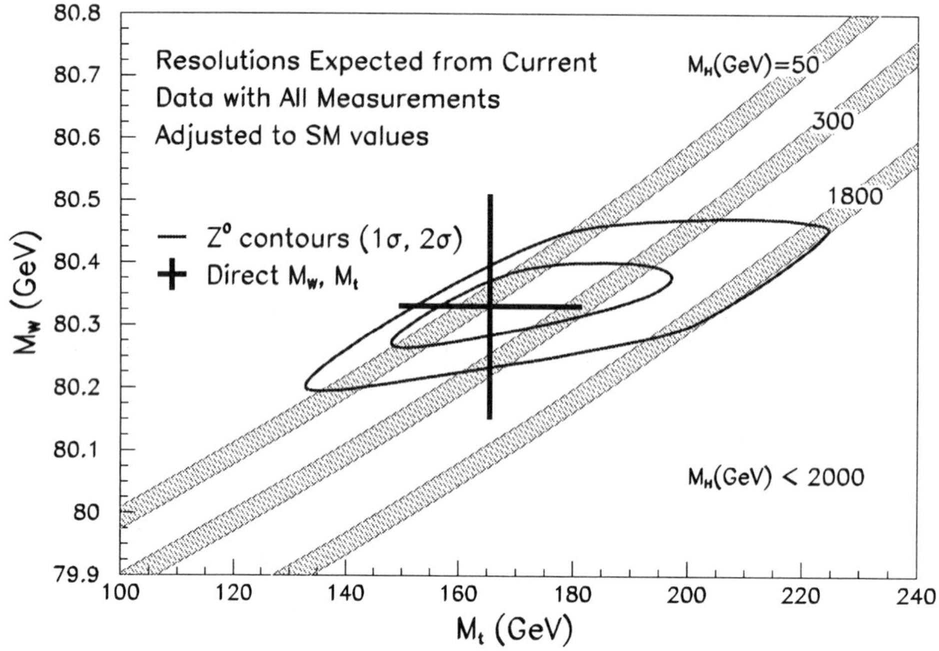

Figure 8: If the central values of all current measurements are shifted to agree with the predictions of the best-fit Standard Model, but with the correct current uncertainties as given in the left-most column of Table 4, then the contours of Figure 6 become the contours shown above.

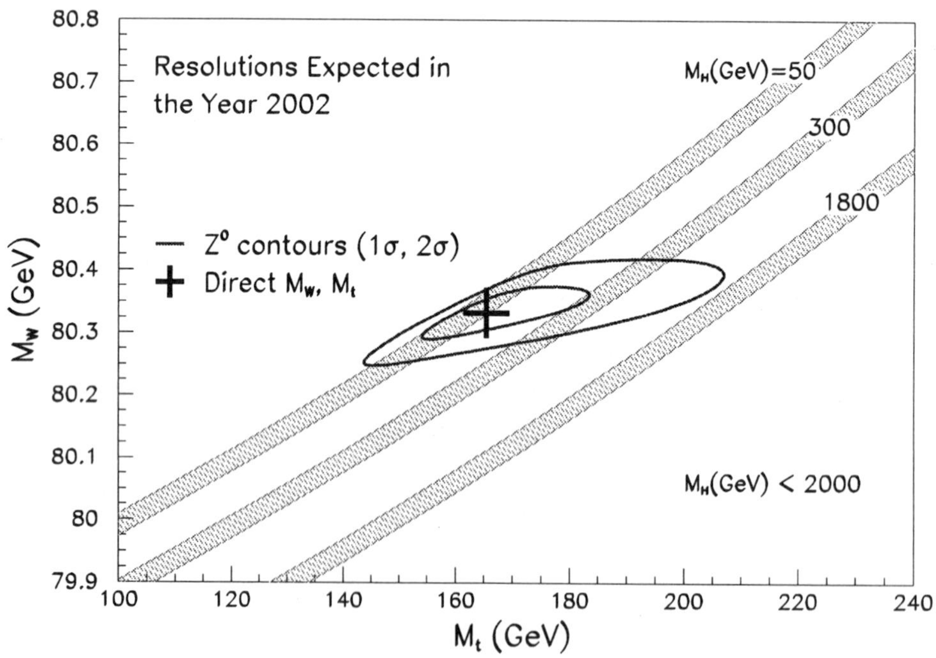

Figure 9: In calculating the EW contours from measurements other than M_t and M_W, the errors of all measurements have been reduced to be those that we anticipate after completion of the SLC, LEP 1, and Fermilab νN programs (~1998). The uncertainties in the direct measurements are taken to be δM_W = 40 MeV and δM_t = 4 GeV (~2002).

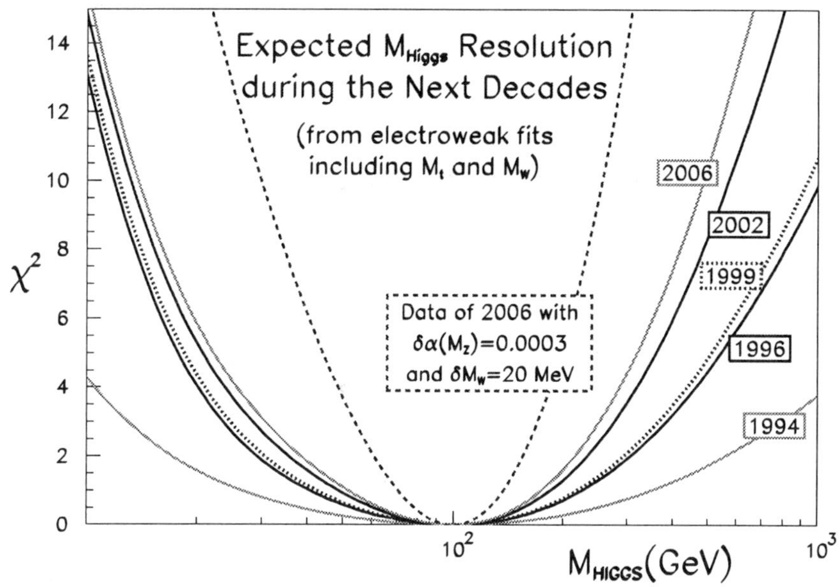

Figure 10: Anticipated sensitivity of the χ^2 curve from SM fits to the Higgs mass assuming $M_H = 100$ GeV, given as a function of time during the next decade. The inner-most (dashed) curve shows the effect of improving the measurements of both $\alpha(M_Z)$ and M_W beyond current expectations.

$\alpha(M_Z)$ are accessible to improve measurements, and we believe these should be a focus for additional work during the period 1994–2006.

Improvements in M_W might be possible at LEP 2 before LHC turns on, and at either LHC or the Tevatron after that. A reduction in δM_W from 40 to 20 MeV is certainly not easy to accomplish, but it might be attainable; indeed, the δM_W error of 40 MeV we have assumed for 2002 is probably somewhat pessimistic.

Similarly, during the latter half of the present decade, measurements at Novosibirsk, Daphne, Beijing, possibly Cornell, and perhaps other accelerators, combined with theoretical work, could lead to a reduction in the error on $\alpha(M_Z)$ of a factor of 2 or more; we believe this effort needs increased emphasis from both experimenters and theorists. As the final step in the evolution of Higgs resolution, we have assumed errors of $\delta M_W = 20$ MeV and $\delta\alpha(M_Z) = 0.0003$; these are to be taken as proposed goals rather than as projections.

If these improvements in both $\alpha(M_Z)$ and M_W are achieved, the resulting Higgs sensitivity is shown in the inner-most curve of Figure 10, and corresponds to an uncertainty in the Higgs mass of $f_\sigma \approx 1.4$. δM_H will not be dominated by any one measurement, and in fact its sensitivity to factor-of-two improvements in each of the five quantities considered is about the same. If M_H differs from 100 GeV, then the minimum of the χ^2 curves is shifted and the details of the shapes also change, but the general features of the curves are the same. In all of this analysis, we have ignored possible theoretical uncertainties in higher-order diagrams, including those in the $t-b$ vertex, in the expectation that theory will keep pace with experiments during this time.

The comparison between Z contours and direct measurements of M_t and M_W after these improvements is shown in Figure 11. The comparison of this figure to Figure 9 shows the effects of $\delta\alpha(M_Z)$ on both the contours and the widths of the Higgs bands, and also shows the factor-of-two improvements in the direct measurements of both M_t (from LHC) and M_W.

9 Direct Searches for the Higgs Boson

Future searches for the Higgs boson for other electroweak symmetry-breaking phenomena are discussed in more detail by the "Beyond the Standard Model" section of this report. But because of the importance of the comparison of the M_H predictions from precision experiments with the direct searches for the Higgs, we will briefly review here the expectations for Higgs searches over the next 15 years.

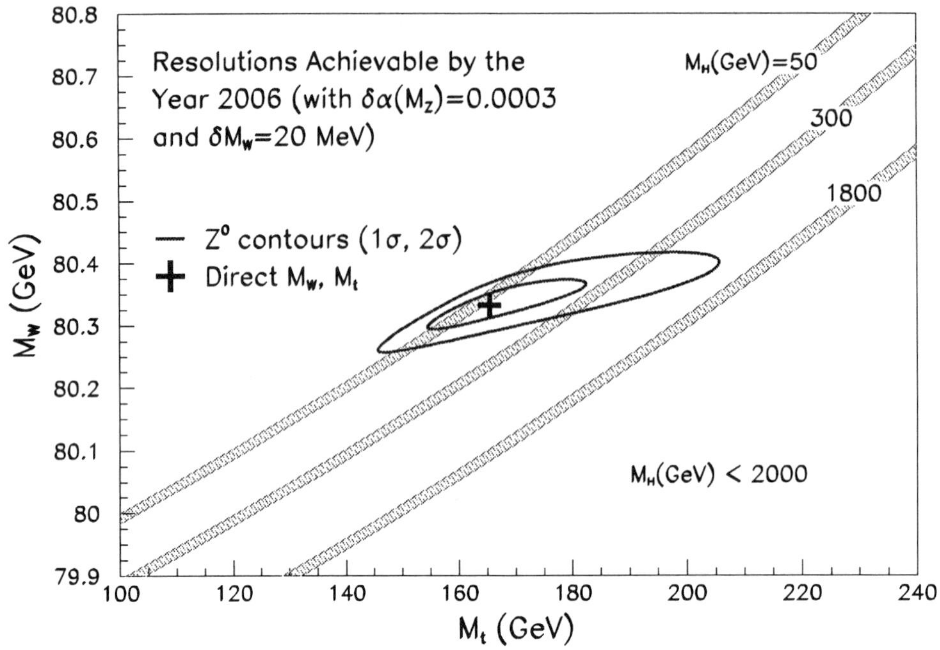

Figure 11: The electroweak contours here are calculated from the same data as in Figure 9, but with the error in $\alpha(M_Z)$ reduced by a factor of 3 to ± 0.0003. The uncertainties in the direct measurements are taken to be $\delta M_W = 20$ MeV and $\delta M_t = 2$ GeV (from LHC). Each of these 3 errors is felt to be perhaps optimistic, but nevertheless feasible.

9.1 Low Mass Region: $M_H < 100$ GeV

The best limits on the Higgs mass now come from the LEP experiments. The Higgs boson would be produced near the Z resonance through the reaction

$$e^+e^- \to Z \to Z^* + H, \tag{43}$$

where Z^* denotes an off-mass-shell Z. Such events can be detected if the Z^* decays into l^+l^- or $\nu\bar{\nu}$. In the former mode, events are identified by the acoplanar l^+l^- (where $l = e$ or μ), and in the latter by the significant missing energy and transverse momentum of the event. In both modes, searches can be sensitive to inclusive hadronic decays of the Higgs.

All four of the LEP experiments have searched for such events in both of these modes (and others). No signal has been observed, and present lower limits from each experiment lie in the range of 55 to 60 GeV. Combining all of the four LEP searches gives a LEP limit of $M_H > 62.5$ GeV [101]. The experiments are seeing some events, consistent with the expectations from background, so it is unlikely that there will be a significant improvement in 1995-6.

The predominant production diagram for a Higgs at LEP 2 would be

$$e^+e^- \to Z^* \to Z + H \tag{44}$$

similar to the LEP 1 production diagram except now it is the intermediate Z which is off-mass-shell.

The threshold for this production mechanism is $M_H + M_Z$, and the cross section rises fairly steeply above threshold. For a luminosity of 400 pb^{-1}, the 5-σ discovery limit for a single experiment for $E_{cm} = 175$ GeV is 82 GeV, and for $E_{cm} = 190$ GeV it is 92 GeV [102]. At the present time, it appears that the LEP 2 energy will be approximately $E_{cm} = 180$ GeV, corresponding to a Higgs limit of approximately 86 GeV. $E_{cm} = 196$ GeV could be attained with construction of additional RF cavities, extending the Higgs range up to 95-100 GeV. An integrated luminosity of 400 pb^{-1} or more appears feasible under either scenario.

9.2 High Mass Region: $M_H > 130$ GeV

It appears that this region will be covered adequately by the LHC. The decays $H \to ZZ$ and $H \to ZZ^*$, with both Z's decaying into lepton pairs (e^+e^- or $\mu^+\mu^-$), provide a clear signature and sufficient rate for Higgs masses up to 800 GeV. Both ATLAS and CMS calculate good sensitivities over the region $130 < M_H < 800$ GeV, and conclude that a Higgs with $130 < M_H(\text{GeV}) < 500$ can be discovered with 10 fb^{-1}, and that 100 fb^{-1} (1 year at design luminosity) would extend this range up to 800 GeV [77,78]. Other decay modes such as $ll\nu\nu$ and $lljj$

could extend this range up to 1 TeV with additional running.

The lower limit of this range is set by the branching ratio into ZZ^*, which falls off rapidly due to the Z^* propagator as M_H falls below about 130 GeV.

9.3 Intermediate Mass Region: $85 < M_H < 130$ *GeV*

This region, between the LEP 2 limit and the region of significant branching ratio for $H \to ZZ^*$, is experimentally the most difficult one to cover. Unfortunately, there are both theoretical and experimental reasons to suspect that this may be one of the most interesting regions.

LEP 2 will cover the lower part of this region, up to a limit of perhaps 80-90 GeV. The LHC, and possibly the Fermilab collider after luminosity upgrades, will have some sensitivity. But the dominant decay mode $H \to b\bar{b}$ does not allow precise mass reconstruction and may be swamped by hadronic backgrounds in hadron colliders. The two final states which appear to offer the best opportunity for a discovery of the Higgs in this mass region are discussed below:

9.3.1 $H \to \gamma\gamma$

In the intermediate mass range ($80 < M_H < 130$ GeV) the decay $H \to \gamma\gamma$ gives the best signature for Higgs detection, in spite of a low branching ratio ($\sim 0.1\%$). Good mass resolution on the Higgs requires measurement of the γ angles as well as their energies, and backgrounds are severe.

Both ATLAS and CMS have calculated their expected sensitivity to a Higgs in this mass region [77,78], and each estimates a sensitivity $S/\sqrt{B} \geq 5$ for $90 \leq M_H \leq 150$ GeV (S=signal, B=background). However, the sensitivities depend critically on calorimeter performance and pile-up effects. Better estimates should be available soon in the Technical Design Reports of each detector, which will include improved calorimeter design parameters.

However, even if the LHC detectors can adequately cover this decay mode, the Higgs sensitivity would still be quite sensitive to the $\gamma\gamma$ branching ratio, which would be reduced in some modest extensions to the Standard Model, including the minimal supersymmetric model. This could make detection of the Higgs by its $\gamma\gamma$ decay impossible at the LHC. In any case, one would certainly put a high priority on the detection of the dominant decay modes of the Higgs, and on the measurement of its branching ratios.

9.3.2 WH Production, with $W \to l\nu$ and $H \to b\bar{b}$

Several groups [103,104,105] have recently investigated the possibility of detecting the $H \to b\bar{b}$ decay, with the Higgs produced via WH or ZH production and using the leptonic decays of the gauge bosons as an additional tag to reduce backgrounds. Tagging of each of the b's is essential, as are kinematic cuts to suppress $t\bar{t}$ backgrounds. Estimates of $S/\sqrt{B}$ range from about 4.8 to 6.6 for a 100-GeV Higgs with 30 fb^{-1} of data, with improved sensitivity for a lower mass Higgs.

Both references [103,104] have found that, for equal luminosities of 30 fb^{-1}, Fermilab and LHC have similar sensitivities to this mode; the signal is larger at the LHC, but so are the backgrounds. The required luminosity would correspond to one or two years of low-luminosity running at the LHC, and probably a longer period of running at full luminosity at an upgraded Tevatron (TeV*). It is not clear that experiments at either accelerator could carry out the b-tagging and reconstruction necessary to identify these events, since rates and multiple-interaction problems will be severe. Because backgrounds are quite different between these two machines, further studies of both possibilities are needed.

9.4 Higgs Physics at the NLC

At a 500 GeV linear collider, the Standard Model Higgs boson would be produced by WW fusion and Z bremsstrahlung ($e^+e^- \to \nu\bar{\nu}H, ZH$). The primary sources of background are the processes: $e^+e^- \to W^+W^-$, $e^+e^- \to q\bar{q}$, and $e^+e^- \to e\nu W$. A data sample corresponding to one year of operation at design luminosity (50 fb^{-1}) would contain about 200 reconstructed ZH events in the 4-q, 2-q2-ℓ, and 4-ℓ channels if the Higgs mass is 150 GeV. Beam energy and Z mass constraints enhance the resolution of the kinematic event reconstruction and permit the isolation of reasonably pure event samples (the background to the 4-q sample is about 20%). A 50 fb^{-1} sample would permit the observation of a significant signal (5σ) up to a Higgs mass of 300 GeV [106].

Within the mass reach of the machine (a 1 TeV collider would have about twice the mass reach), a number of Higgs properties can be determined. The large signal-to-noise ratio enables the precise measurement of the Higgs mass (a resolution of 180 MeV is possible at a Higgs mass of 110 GeV) and the determination of the Higgs spin and CP parity. The measurement of the cross section for $e^+e^- \to ZH \to \ell^+\ell^- H$ with a 50 fb^{-1} sample would determine the ZHH couplings with a statistical precision of about 8% at $M_H = 110$ GeV. In the region $M_H < 2M_W$, the Higgs branching ratios to $b\bar{b}$, $c\bar{c} + gg$, $\tau^+\tau^-$, and WW^* can be measured (the precisions for $M_H = 110$ GeV are: 7%, 40%, 15%, and 50%, respectively [107]).

Since the Standard Model Higgs boson is not necessarily the most likely source of spontaneous symmetry breaking, we note that non-standard Higgs states (including composite ones and those that decay into invisible particles) can also be observed and studied at a linear collider (see the Spontaneous Symmetry Breaking Working Group section).

10 Couplings of the Gauge Bosons

The standard electroweak model is based on non-abelian gauge groups and as such makes clear predictions about the mutual couplings of the gauge bosons, γ, W and Z. These couplings can be related to the static or transition moments of those particles. The measurement of these couplings represents a program not unlike that of measuring g_A and g_V for the fermions. Initially one tests the basic structure. Eventually, with sufficient precision in the measurements, there is sensitivity to other sectors of the theory, for example the Higgs structure.

While there are some low energy experiments with indirect sensitivity [108] to the values of these couplings, much more direct measures are afforded in high-energy experiments when pairs of bosons, $W\gamma$, WW, WZ, $Z\gamma$ and ZZ are kinematically allowed. Several new results have been recently presented [109] from the Tevatron experiments. As an example, the current limits on the parameter $\Delta\kappa_\gamma$ related to the anomalous magnetic moment of the W can be expressed as $-2.3 \leq \Delta\kappa_\gamma \leq 2.2$ from either the CDF or DØ experiments [110]: $\Delta\kappa_\gamma = 0$ is the standard model value.

With 1 fb^{-1}, the limits are expected to reduce by an order of magnitude. Similar sensitivity is expected at LEP 2; the correlations between the couplings are different between e^+e^- and $p\bar{p}$ processes and this affects the sensitivities to different couplings.

At LHC with much higher energy and with an integrated luminosity of 100 fb^{-1}, approximately two orders of magnitude reduction in the bounds would be realised, assuming standard behavior. This achieves the level of sensitivity at which different extensions of the standard model suggest observable effects. Similarly, at an e^+e^- linear collider with 500 GeV in the center of mass, the limits offered are $-0.0024 \leq \Delta\kappa_\gamma \leq 0.0024$.

The reader is referred to the section prepared by the Electroweak Symmetry Breaking subgroup for more details.

11 Conclusions

Precision electroweak measurements from LEP, SLC, and Fermilab now provide stringent tests of the Standard Model. A dramatic test of the SM is given by the comparison of the mass of the top quark predicted from precision measurements of M_W and electroweak couplings with a direct measurement from hadron collider experiments. Evidence for a top quark of the expected mass has been presented by CDF this year. During the next decade, there will be a number of other major improvements in measurements of M_W, M_t, and electroweak couplings. These measurements provide important tests of the Standard Model and will give significant constraints on the mass of the Higgs boson. The confrontation of these indirect predictions of M_H with the results of direct searches for the Higgs will be perhaps the most exciting development of the next two decades in the field of particle physics.

In this report, we have reviewed the status of all the major electroweak measurements, including precision measurements of electroweak couplings and of the masses M_W and M_t, and have attempted to project the resolutions that will be attained in each measurement as a function of time over the next 15 years. We have examined how these improved measurements will affect tests of the Standard Model, and how the resolution on the predicted mass of the Higgs boson is expected to evolve during this time. We have also briefly reviewed the status and expectations of direct searches for the Higgs at LEP and LHC.

In general, we believe the presently planned program of electroweak measurements is an excellent one. It requires completion of the current programs at LEP 1, SLC, and Fermilab (Run 1b), followed by major new measurements at LEP 2, Fermilab (Run 2), and the LHC.

We believe there are three ways in which the present program might be improved:

- The accurate measurement of M_t is a key element in electroweak tests, and consequently the accumulation of high statistics during the present Tevatron Run 1b is of prime importance; after this run, there will be no new information on M_t for about 5 years. An accurate knowledge of M_t and the consequent improvement in predictions of M_H may be important for future planning during that time. We believe it is crucial to establish the observation of the top quark beyond any doubt, and highly desirable to improve the mass resolution to the level of ~8 GeV during Run 1b. This might require up to 200 pb^{-1} of integrated luminosity.
- As measurements at LEP and SLC continue to improve, the uncertainty in $\alpha(M_Z)$ will become an increasing limitation in the precision of electroweak tests. The determination of this quantity is limited by the knowledge of the $\gamma-$hadron coupling at low q^2, needed for the evaluation of radiative corrections to $\alpha(M_Z)$. We believe there needs to be an increased emphasis on overcoming the experimental and theoretical limitations in determining $\alpha(M_Z)$, and reducing the error on this quantity by a factor of three (to ± 0.0003).
- The other parameter which will have the most significant effect on the ultimate precision of electroweak tests, and on the predictions of the mass of the Higgs

boson, is M_W. It appears that a resolution of 40 MeV is achievable under the the current program, – but an improved resolution of 20 MeV is desirable, and would give a better match to other resolutions. This improvement is difficult to attain, but it can be addressed at a number of different accelerators: LEP 2, the Tevatron, the LHC, and the NLC. The possibilities for improvements in these measurements should be more thoroughly explored.

With regard to direct searches for the Higgs boson and for the origin of electroweak symmetry breaking, we believe the situation is not altogether satisfactory. Data taken at LEP 2 and LHC will extend the region of sensitivity enormously, and are likely to lead to the discovery of the Higgs if the Standard Model is indeed correct. However, coverage of the low-mass region at the LHC (from 80 to 130 GeV) is predominantly from the $\gamma\gamma$ decay mode, and complete coverage is a significant challenge. Every effort must be made to avoid a hole in Higgs coverage between LEP 2 and LHC. In addition, detection of other (dominant) decay modes should be a priority. The LHC and LEP 2 physicists are addressing these problems, and we will have a better understanding of the experimental limitations at both facilities within the next year.

Some studies have indicated that the Fermilab Tevatron, with major upgrades in luminosity and perhaps energy, could be sensitive to a Higgs in this mass region. More detailed studies of the physics potential of an upgraded Tevatron, and of anticipated detector resolutions and overall performance, should be emphasized during the next two years. In conjunction with this, studies of necessary upgrades to both the experiments and the accelerator should be carried out.

Acknowledgements

While the conveners of the Working Group take responsibility for the contents of this report, they were helped in its production by many others. These contributed either through private communications, verbal or written, or comments on some draft. In addition some groups met to discuss specific issues such as the projection of the top mass measurement error, the W mass measurement error, and the tri-boson couplings. The LEP Electroweak Working Group at CERN was especially helpful, and made many of their results available to us ahead of publication. We are grateful for all the help received, and want to particularly acknowledge the following contributors:

H. Aihara, D. Amidei, D. Baden, T. Barklow, U. Baur, R. Bernstein, E. Blucher, R. Brock, D. Burke, W. Carithers, P. Clarke, M. Della Negra, M. Demarteau, S. Errede, T. Ferbel, H. E. Fisk, L. Foa, A. Galtieri, H. Frisch, N. Hadley, U. Heintz, C. Hill, I. Hinchliffe, P.Grannis, G. Kane, Young Kee Kim, B. Klima, S. Kopp, A. Liss, G. Mikenberg, M. Narain, L. Nodulman, L. S. Orozco, M. Pang, S. Parke, M. Peskin, C. Quigg, R. Raja, J. Rosner, D. Saltzberg, D. Schaile, D. Schlatter, M. Shochet, M. Strovink, J. Thompson, C. Wieman, C. Wendt, B. Winer, D. Wood, and C. Wyss.

References

[1] A. Sirlin, "The Standard Electroweak Model circa 1994: a Brief Overview," Comments Nucl. Part. Phys. 21, no. 5, 287 (1994).

[2] CDF Collaboration: F.Abe *et al.*, Phys. Rev. Lett. **73** (1994) 225;
F.Abe *et al.*, Fermilab PUB-94/097-E(1994), PRD 50, 2966 (1994).

[3] D. Schaile and P.M. Zerwas, Phys. Rev. D45, 3262 (1992).

[4] C. Bouchiat, J. Iliopoulos, and P. Meyer, Phys. Lett. B38, 519 (1972).

[5] W.J. Marciano, *Radiative Corrections to Neutral Current Processes,* to appear in "Precision Tests of the Standard Electroweak Model," Advanced Series in High Energy Physics, World Scientific Publishing Co., P. Langacker, editor.

[6] M. Sher, Phys. Lett. B317, 159 (1993) and Addendum; G. Altarelli, Electroweak Summary Talk at the Tennessee International Symposium on "Radiative Corrections: Status and Outlook", Gatlinburg, Tennessee, June 27-July 1, 1994.

[7] E. Farhi and L. Susskind, Phys. Rep. 74, 277 (1981); M. Peskin and T. Takeuchi, Phys. Rev. D46, 381(1992).

[8] W. Marciano, Phys. Rev. D41, 219(1990); W. Bardeen, C. Hill, and M. Lindner, Phys. Rev. D41, 1647 (1990).

[9] For a review, see J.F. Gunion *et al.*, "The Higgs Hunters Guide", Addison-Wesley, Redwood City, 1990.

[10] P. Langacker and M. Luo, Phys. Rev. D44, 817 (1991); U. Amaldi, W. de Boer, and H. Furstenau, Phys. Lett. B249, 441 (1990); A. Sirlin, Phys. Lett. B232, 123 (1989).

[11] *Review of Particle Properties* by the Particle Data Group, Phys. Rev. D50, No. 3, p.1173, August 1994.

[12] A. Sirlin, Phys. Rev. D22, 971 (1980); ibid. 29, 89 (1984).

[13] A. Sirlin, Phys. Lett. B232, 123 (1989); G. Degrassi, S. Fanchiotti, and A. Sirlin, Nucl. Phys. B351, 49 (1991).

[14] M. Veltman, Nucl. Phys. B123, 89 (1977).

[15] W. Marciano and A. Sirlin, Phys. Rev. D22, 2695 (1980); G. Degrassi and A. Sirlin, Nucl. Phys. B352, 342 (1991).

[16] R. Barbieri *et al.*, Phys. Lett. B**288**, 95 (1992); J. Fleischer, O.V. Tarasov, and F. Jegerlehner, BI-TP-93/24, PSI-PR-93-14.

[17] A. Djouadi and C. Verzegnassi, Phys. Lett. B**195**, 265 (1987); A. Djouadi, Nuovo Cimento **100A**, 357.

[18] J. Fleischer *et al.*, Phys. Lett. B**293**, 437 (1992); G. Degrassi, Nucl. Phys. B**407**, 271 (1993); G. Buchalla and A.J. Buras, MPI-PTh111-92, TUM-T31-36/92 (1992).

[19] P. Gambino and A. Sirlin, Phys. Rev. D49, R1160 (1994).

[20] A. Sirlin, "Universality of the Weak Interactions," to appear in Precision Tests of the Standard Electroweak Model, Advanced Series in high-energy Physics, ed. P. Langacker (World Scientific); A. Sirlin, Nucl. Phys. B71, 29 (1974); A. Sirlin, Rev. Mod. Phys. 50, 573 (1978).

[21] P. Gambino and A. Sirlin, Phys. Rev. Lett. 73, 621 (1994)..

[22] M. Peskin and T. Takeuchi, Phys. Rev. Lett. 65, 964 (1990); M. Golden and L. Randall, Nucl. Phys. B361, 3 (1991); W. Marciano and J. Rosner, Phys. Rev. Lett. 65, 2963 (1990); D. Kennedy and P. Langacker, Phys. Rev. Lett. 65, 2967 (1990); B. Holdom and J. Terning, Phys. Lett. B247, 88 (1990).

[23] See, for example, the second paper in Ref. [6].

[24] P. Langacker, talk presented at the Tennessee International Symposium on "Radiative Corrections: Status and Outlook," Gatlinburg, Tennessee, June 27 - July 1, 1994.

[25] G. Altarelli, R. Barbieri, and S. Jadach, Nucl. Phys. B369, 3 (1992); G. Altarelli, R. Barbieri, and F. Caravaglios, Nucl. Phys. B405, 3 (1993).

[26] W.J. Marciano, Lectures given at the 1993 SLAC Summer Institute, BNL-60177.

[27] G.L. Kane, in Proceedings of the International Workshop on Supersymmetry and Unification of Fundamental Interactions, ed. P. Nath (World Scientific, Singapore, 1993) p. 209.

[28] We would like to thank I. Hinchliffe, F. Jegerlehner, M. Peskin and C.Quigg for helpful discussions on this issue.

[29] G. Degrassi, S. Fanchiotti and A. Sirlin, Nucl. Phys. **B351** (1991) 49.

[30] F. Jegerlehner, Z. Phys. **C32** 195, 1986.

[31] H. Burkhardt, F. Jegerlehner, G. Penso and C. Verzegnassi, Z. Phys. **C43**, 497, 1989.

[32] J. Thompson, Private Communication.

[33] Morris L. Swartz, *Reevaluation of the Hadronic Contri-*
butions to $\alpha(M_Z^2)$, SLAC-PUB-6710, November 1994.

[34] S. Eidelman and F. Jergerlehner, *Hadronic Contribution to g-2 of the Leptons and to the Effective Fine Structure Constant* $\alpha(M_Z^2)$, PSI-PR-95-1, January 1995.

[35] ICFA Workshop on Linear Colliders; Saariselka, Finland, September 1991; ed. R. Orava, P. Eerola, nd M. Nordberg. World Scientific, 1993.

[36] Physics and Experiments with Linear e^+e^- Colliders; Hawaii, 1993; pp 141-165, ed. F.A.Harris, S.L.Olsen, S.Pakvasa, X. Tata; World Scientific, 1993.

[37] L. Arnaudon *et al.*, (LEP Energy Working Group), *Accurate Determination of the LEP Beam Energy by Resonant Depolarization*, CERN-SL/94-71 (BI), August 1994.

[38] *Combined Preliminary Data on Z Parameters from the LEP Experiments and Constraints on the Standard Model*, by the LEP Collaborations and the LEP Electroweak Working Group; CERN/PPE/94-187, 25 November 1994 (prepared from contributions of the LEP experiments to the *27th International Conference on High Energy Physics*, Glasgow, Scotland, 20-27 July 1994).

[39] LEP Electroweak Working Group, *LEP Electroweak Heavy Flavour Results for Summer 1994 Conferences*, LEPHF/94-03, ALEPH Note 94-119 PHYSIC 94-103, DELPHI 94-108 PHYS 425, L3 Note 1630, OPAL Technical Note TN242.

[40] SLD Collaboration, K. Abe *et al.*, SLAC-PUB-6456, March 1994, submitted to Physical Review Letters.

[41] P. Langacker, M. Luo, A. Mann, Rev. Mod. Phys. 64, 87 (1992).

[42] J.V. Allaby *et al.*, Z. Phys. C36 (1987) 611.

[43] A. Blondel *et al.*, (CDHS Collaboration), Z. Phys. C45 (1990) 361.

[44] CCFR Collaboration, Phys. Rev. Lett., Vol. 72, p. 3452 (1994).

[45] CCFR Collaboration, Phys. Rev. Lett., Vol 70, p. 134 (1993).

[46] P. Vilain *et al.*, *Precision Measurement of Electroweak Parameters from the Scattering of Muon-Neutrinos on Electrons*, CERN-PPE/94-124, June 30, 1994.

[47] W.J. Marciano and J.L. Rosner, *Phys. Rev. Lett.* **65**, 2963 (1990) and P.G.H. Sandars, *J. Phys.* **B23**, L655 (1990).

[48] M. Swartz, private communication.

[49] M.C. Noecker, B.P. Masterson, and C.E. Wieman, *Phys. Rev. Lett.* **61**, 310 (1988).

[50] M.A. Blundell, W.R. Johnson, and J. Sapirstein *Phys. Rev. Lett.* **65**, 1411 (1988).

[51] B.W. Lynn and P.H. Sandars, SSCL-Preprint-183, June 1993.

[52] D.M. Meekof, *et al.*, *Phys. Rev. Lett.* **71**, 3442 (1994); D.M. Meekof, *et al.*, Washington APS Meeting, April 1994; P. Baird, NATO Summer Institute July 18-30 (1994).

[53] W.R. Johnson, private communication.

[54] C.E. Wieman, private communication.

[55] L.A. Orosco, private communication.

[56] D. Budker, D. DeMille, E.D. Commins, and M.S. Zolotorev, *Phys. Rev. Lett.* **70**, 3019 (1993).

[57] CDF Collaboration: Young-Kee Kim, Presented at the Electroweak Session of the XXIXth Rencontres de Moriond, Meribel, France, March 12-19, 1994. FERMILAB-Conf-94/169-E, 1994.

[58] DØ Collaboration: Chang-Kee Jung, Presented at *The 27th Inter. Conf. on High Energy Physics*, Glasgow, July, 1994.

[59] J. Alitti *et al.*, (UA2 Collaboration), Phys. Lett. B276, 354 (1992).

[60] F. Abe *et al.*, (CDF Collaboration), Phys. Rev. Lett. 65, 2243 (1990).

[61] M. Demarteau *et al.*, *Combining W Mass Measurements*, CDF/PHYS/CDF/PUBLIC/2552 and D0NOTE 2115.

[62] Young-Kee Kim, CDF Collaboration, Private Communication.

[63] J. Bijnens et al, LEP 200 Study, Aachen, 1986.

[64] C. Wyss, private communication.

[65] Akiya Miyamoto, Physics and Experiments with Linear e^+e^- Colliders, Hawaii, 1993, pp 141-165, Ed. F.A.Harris,S.L.Olsen, S.Pakvasa, X. Tata, World Scientific, 1993.

[66] D. Wood, presented at *Physics in Collision 1994*, Tallahassee, FL, June 15-17, 1994.

[67] F. Abe *et al.*, Phys. Rev. Lett. **73**, 220 (1994); F. Abe *et al.*, Phys. Rev. Lett. **69**, 28 (1992); F. Abe *et al.*, Phys. Rev. Lett. **64**, 152 (1990); J. Alitti *et al.*, Phys. Lett. **276B**, 365 (1992); C. Albajar *et al.*, Phys. Lett. **253B**, 503 (1991).

[68] CDF Collaboration, paper submitted by W. Badgett to IX Topical conference on $p\bar{p}$ Physics, Tsukuba, Japan, October 1993; DØ Collaboration, presented by N.A. Graf at the same conference.

[69] J. Rosner, M.P. Worah, and T. Takeuchi, Phys. Rev. **D 49**, 1363 (1994).

[70] F. Abe, *et al.*, *A Direct Measurement of the W Boson Width $\Gamma(W)$)*, CDF/ANAL/ ELECTROWEAK/CDFR/2642, September 1994; submitted to Phys. Rev. Lett.

[71] DØ Collaboration: P. Grannis, presented at *The 27th Int. Conf. on High Energy Physics, Glasgow*, July,1994.

[72] DØ Collaboration: S.Abachi *et al.*,, Phys. Rev. Lett. **72** (1994) 2138.

[73] E.Laenen, J.Smith and W. van Neerven, Phys. Lett. **321 B**(1994) 254.

[74] CDF Collaboration: H.H. Williams, presented at *The 27th Int. Conf. on High Energy Physics, Glasgow*, July,1994.

[75] DØ Top Mass Fitting Group:Private Communication.

[76] DØ Collaboration: M. Strovink,European Physical Society Conference, Marseille, 1994.

[77] ATLAS Letter of Intent, CERN/LHCC/92-4, October 1992.

[78] CMS Letter of intent, CERN/LHCC/92-3, October 1992.

[79] ATLAS Internal Note, A.Mekki and L.Fayard, PHYS-NO-028,1993

[80] ATLAS Internal Note, Nicholas Andell, PHYS-NO-037, 1994.

[81] P. Igo-Kemenes, Physics and Experiments with Linear e^+e^- Colliders, Hawaii, 1993, pp 95-121, Ed. F.A.Harris,S.L.Olsen, S.Pakvasa, X. Tata, World Scientific, 1993.

[82] Johann H. Kuhn, Physics and Experiments with Linear e^+e^- Colliders, Hawaii, 1993, pp 95-121, Ed. F.A.Harris,S.L.Olsen, S.Pakvasa, X. Tata, World Scientific, 1993.

[83] D.Amidei, D.Baden, C. Hill, B.Klima, S.Parke, Top Quark Factory, 1994, unpublished.

[84] A workshop concerning top physics and Tevatron upgrades was held at the University of Michigan in November 1994. Working groups were formed, and a report is expected in summer 1995.

[85] G. Kane, C.P. Yuan, G. Ladinsky, Phys. Rev. D45 (1992) 124.

[86] D. Amidei and D.Winn, CDF Internal Note 2914, Sept 9, 1994.

[87] C.P. Yuan, *Strategies for Probing CP Properties in the Top Quark System*, MSUHEP-50123, January 1995; G. Grzadkowski and J. Gunion, Phys. Lett. B287 (1992) 237.

[88] C.P. Yuan, Phys. Rev. **D 41** (1990) 42; D.O. Carlson and C.P. Yuan, Phys. Lett. B306 (1993) 386.

[89] R.K. Ellis and S. Parke, Phys. Rev. D46 (1992), 3785.

[90] D. Gerdes, CDF Internal Note 2865.

[91] K. Fujii, T. Matsui, Y. Sumino, Phys. Rev. D50, 4341 (1994).

[92] D. Schaile, "Tests of the Electroweak Theory at LEP," CERN-PPE/93-213, December 1993.

[93] M.V. Terent'ev, Yad. Fiz. **18**, 870 (1973) [translation: Sov. J. Nucl. Phys. **18**, 449 (1974)].

[94] W.J. Marciano and A. Sirlin, Phys. Rev. Lett. **36**, 1425 (1976); *ibid.* **71**, 3629 (1993).

[95] D.I. Britton, *et al.*, Phys. Rev. Lett. **68**, 3000 (1992); C. Czapek *et al.*, Phys. Rev. Lett. **70**, 17 (1993).

[96] D. Bryman, Comments Nucl. Part. Phys. 21, 101 (1993); W.J. Marciano, "e-μ-τ Universality", Cornell University Seminar, November 1993.

[97] S.Bethke, Proceedings of the Linear Collider Workshop in Waikoloa/Hawaii, April 1993;
S. Catani, presented at the International Europhysics Conference, Marseille, France, 22nd - 28th July 1993;
S. Banerjee, *α_s Measurements from Event Shape Variables using Z Data*, presented at the EPS Conference on High Energy Physics, Marseille, July 22-28, 1993, L3 Note 1460.

[98] The program EW-BHM (G. Burgers, W. Hollik, M. Martinez) was used for the contours in Figures 5, 6, 7, 8, 9, and 10 of this report. The program is described in CERN PPE 93-157.

[99] J. Ellis, G.L. Fogli, and E. Lisi, CERN TH-7261/94.

[100] D. Bardin *et al.*, LEP Electroweak Working Group Report, document in preparation.

[101] F. Richard, LAL-94-50, September 1994 (talk presented at the 27th International Conference on High Energy Physics, Glasgow, Scotland, 10-27 July 1994.

[102] P. Janot, Proceedings of the XXVIIth Rencontre de Moriond (*1992 Electroweak Interactions and Unified Theories*, edited by J. Tran Thanh Van), 1992, p. 317.

[103] A.Stange, W.Marciano, and S.Willenbrock, BNL-60340, ILL-(TH)-94-8, 1994.

[104] S.Mrenna and G.L.Kane, UM-TH-94-24, 1994.

[105] D. Chao, presented at Albuquerque DPF.

[106] For recent studies, see P. Grosse-Wiesemann, D. Haidt, and H.J. Schreiber, in *Workshop on e^+e^- Collisions at 500 GeV: the Physics Potential*, Munich, February; Annecy, June; Hamburg, September 1991, ed. P.M. Zerwas, DESY 92-123A (1992), p. 37; P. Janot, *ibid.*, p. 107; A. Djouadi, D. Haidt, B.A. Kniehl, B. Mele, and P.M. Zerwas, *ibid.*, p. 11, and references therein; J.F. Gunion, in *Physics and Experiments with Linear e^+e^- Colliders*, Hawaii, 1993, p. 166; P. Janot, *ibid.*, p. 192; and references therein.

[107] M.D. Hildreth, T.L. Barklow, and D.L. Burke, Phys. Rev. **D49**, 3441 (1994).

[108] Dieter Zeppenfeld, Physics and Experiments with Linear e^+e^- Colliders, Hawaii, 1993, pp 695-703, Ed. F.A.Harris,S.L.Olsen, S.Pakvasa, X. Tata, World Scientific, 1993.

[109] CDF: F.Abe *et al.*, FERMILAB-Conf-94/158-E, 1994; DØ:A.Spadafora, FERMILAB-Conf-94/016-E, 1994.

[110] F. Abe *et al.*, (CDF Collaboration), contributed paper to the 27th International Cponverence on High Energy Physics, Glasgow, Scotland, July 20-27, 1994, FERMILAB-Conf-94/158-E (June 1994); A. Spadafora (DØ Collaboration), FERMILAB-Conf-94/016-E (January 1994).

Quantum Chromodynamics

Alfred H. Mueller

Columbia University, New York, NY 10027

Berndt Müller

Duke University, Durham, NC 22708

Claudio Rebbi

Boston University, Boston, MA 02215

Wesley H. Smith

University of Wisconsin, Madison, WI 53705

Working Group Members

M.G. Albrow	*Fermilab*
G.C. Blazey	*Fermilab*
A. Brandt	*Fermilab*
G. Bunce	*Brookhaven National Laboratory*
A.C. Caldwell	*Columbia University*
N.M. Cason	*University of Notre Dame*
N. Christ	*Columbia University*
J. Collins	*Pennsylvania State*
J.M. Conrad	*Columbia University*
L. Dixon	*Stanford Linear Accelerator Center*
S. Ellis	*University of Washington*
G.E. Forden	*University of Arizona*
S.H. Geer	*Fermilab*
W. Guryn	*Brookhaven National Laboratory*
R.T. Hamilton	*Harvard University*
E. Hughes	*Stanford Linear Accelerator Center*
Z. Jin	*Fermilab*
R. Kauffman	*Brookhaven National Laboratory*
G. Kilcup	*Ohio State University*
A.S. Kronfeld	*Fermilab*
G. Levin	*Fermilab*
M.-P. Lombardo	*Univ. of Illinois Urbana-Champaign*
J.P. Mansour	*Michigan State University*
R.L. McCarthy	*SUNY at Stony Brook*
S.A. Mendoza	*SUNY at Stony Brook*
E.J. Mirkes	*University of Wisconsin*
A.H. Mueller	*Columbia University*
D. Muller	*Stanford Linear Accelerator Center*
F.I. Olness	*Southern Methodist University*
J.-W. Qiu	*Iowa State University*
A.A. Bhatti	*Rockefeller University*
F.O. Borcherding	*Fermilab*
R. Brower	*Boston University*
P. Burrows	*Massachusetts Institute of Technology*
J.B. Carroll	*University of California-Los Angeles*
N.-P. Chang	*City College of CUNY*
E. Church	*University of Washington*
J.M. Conrad	*Columbia University*
S.R. Dasu	*University of Wisconsin*
A. Dzierba	*Indiana University*
B.F. Flaugher	*Fermilab*
J.W. Gary	*University of California-Riverside*
W.T. Giele	*Fermilab*
T.J. Hallman	*University of California-Los Angeles*
B. Harris	*SUNY Stony Brook*
J.W. Huston	*Michigan State University*
J.I. Kapusta	*University of Minnesota*
S. Keller	*Florida State University*
E. Kovacs	*Fermilab*
J.F. Lagae	*University of Kentucky*
S.J. Lindenbaum	*Brookhaven National Laboratory*
J.E. Mandula	*U.S. Department of Energy*
R. Mawhinney	*Columbia University*
H. Melanson	*Fermilab*
R. Milner	*Massachusetts Institute of Technology*
J.M. Moss	*Los Alamos National Laboratory*
B. Muller	*Duke University*
J. Negele	*Massachusetts Institute of Technology*
V. Papavassilio	*Argonne National Laboratory*
K. Rajagopal	*Harvard University*

C. Rebbi	*Boston University*
W. Smith	*University of Wisconsin*
G. Sterman	*SUNY at Stony Brook*
D. Toussaint	*University of Arizona*
W.-K. Tung	*Michigan State University*
A.R. White	*Argonne National Laboratory*
B. Wyslouch	*Massachusetts Institute of Technology*
S. Sanghera	*Cornell University*
H.M. Spinka	*Argonne National Laboratory*
P. Stevenson	*Rice University*
L. Trueman	*Brookhaven National Laboratory*
H. Weerts	*Michigan State University*
J. Womersley	*FermilabDØ*
Z. Zhang	*SUNY at Stony Brook*

1 Preface

This report of the American Physical Society Division of Particles and Fields Long Range Planning Working Group on Quantum Chromodynamics (QCD) is in large part the result of contributions made in preparation for and during a two day workshop held in Madison, WI on April 9 & 10, 1994. A number of experts in various areas of QCD experiment and theory contributed reports on specific areas and presented talks that covered the present activities in the area, the expected activities that are planned to take place in the next year or so, and the options for future activities in the more long range time scale. These presentations were accompanied with discussions on the important issues for the future of QCD physics. The agenda for the the workshop is given in Appendix A.

2 Introduction and Executive Summary

2.1 The Beginnings

With the discovery of asymptotic freedom in 1973 the development of Quantum Chromodynamics (QCD) as a quantitative theory of the strong interactions began in earnest. The approximate scaling behavior which had been found in deep inelastic electron scattering experiments at SLAC found a natural explanation. The parton model, developed in response to the experimental discovery of scaling, followed naturally from QCD with quarks and gluons being the previously mysterious partons.

However, quarks and gluons were not produced as free particles. The growth of the QCD coupling at long distances suggested that free quarks would have infinite energy and so could only appear in color singlet combinations with other quarks and gluons. That is, quarks are confined. It was quickly realized that quark confinement could only emerge from essentially nonperturbative aspects of QCD. In 1974 lattice QCD was born with the suggestion that one could perform nonperturbative calculations in QCD by taking the continuum limit of a discretized, but gauge invariant,theory on a finite lattice. This lattice gauge theory contained quark confinement in a way which was manifest, so long as the lattice spacing was not too small. That confinement remains as the lattice spacing goes to zero was later verified numerically.

Although quarks and gluons were not supposed to be present as physical particles in QCD they could manifest themselves in a rather direct way as jets of hadrons having nearly the same energy and direction as the parent quarks and gluons. Evidence for quark jets appeared in e^+e^- annihilation experiments at SPEAR while jets, including the gluon, become clearly visible at PETRA. The successful description of three-jet events in e^+e^- annihilation and of scaling violations in inelastic lepton scattering using perturbative calculations established QCD as a successful quantitative theory of the strong interactions.

2.2 The Major Achievements

From the late 1970's to the present there has been an intense activity, both theoretical and experimental, toward developing QCD into a mature and all encompassing theory of the strong interactions. QCD factorization has been established at a semirigorous level and tested, for example, in the comparison of inelastic lepton scattering, μ-pair production and W and Z production. Determinations of the single coupling parameter in QCD, α_S, have been made from experiments using a large variety of independent methods, and a nontrivial consistency has emerged with $\alpha_S(\mathrm{M}_Z)$ now known to about 5 percent. Fixed target deep inelastic muon and neutrino scattering experiments at Fermilab and CERN have measured consistent distributions for the nucleon valence quark, sea quark and gluon constituents and verified their interactions within the QCD framework. The results from *ep* scattering at HERA are now extending these measurements to much higher Q^2 and lower x, providing new understanding of the gluons. Another new Q^2 and x region is also explored by the Tevatron collider photon measurements.

The accurate measurements of jet cross sections for single jet transverse energies as large as 400 GeV in proton-antiproton collisions at Fermilab has tested QCD predictions to very short distances and ruled out quark substructure down to distances of 1.4×10^{-17}cm. Data on jets at LEP have allowed accurate determinations of α_S, have stimulated important progress in resumming perturbation theory, and have checked predictions of color coherence phenomena in QCD. Striking progress in calculating hadron masses and weak matrix elements as well as much information and understanding about the finite temperature QCD phase transition have been obtained from numerical studies of lattice gauge theory.

2.3 The Future

QCD is a successful theory of the strong interactions which has been tested by confronting theory and experiment in many different ways in both perturbative and nonperturbative regimes. QCD is unique within the standard model of particle physics in that it is a theory which must exist and be internally consistent over all possible energy scales, both in perturbative and nonperturbative domains, at zero temperature and at high temperature. Thus QCD is a theory which deserves a most intense scrutiny in a diverse set of circumstances, and such a scrutiny will certainly be rewarded by exciting surprises and a deeper and more satisfying understanding of this profound theory of matter. Indeed, it is important to make precise tests of QCD in order to check the very framework within which particle physics is viewed. QCD is the only theory where relativistic quantum field theory, as a basis for particle physics, can be tested in a way which goes beyond a few orders of perturbation theory.

While there is much evidence that QCD is the correct theory of the strong interactions there is still a very incomplete understanding of quark confinement, of the high temperature and high density phases of QCD, of the absence or unnatural smallness of strong CP violation, and of how a perturbative (partonic) picture of QCD is connected with nonperturbative regimes, including bound states.

In the next ten years much new progress can be expected in lattice gauge theory calculations as more computing power becomes available and further progress is made in the development of algorithms. The dream of a precise, convincing and reliable calculation of the hadron mass spectrum and of low energy weak matrix elements is perhaps not too far away. The latter will provide important information for the interpretation of data gathered at the B Factories at the end of this decade. Lattice techniques for evaluating α_S, which are already competitive with perturbative derivations, will probably provide the most precise calculation of the strong coupling constant. The derivation of ever more accurate results on high temperature QCD and on the properties of the quark-gluon plasma will complement experimental information to be obtained at heavy ion colliders. New methods of doing nonperturbative QCD calculations, such as discrete light-cone quantization, are providing complementary insights to the field.

New and increasingly more precise jet data are being gathered in e^+e^-, hadron, and ep collisions. Hadron collider data, photoproduction data, and deep inelastic scattering data continue to provide a wealth of precise information on gluon and other parton densities. Comparisons of observations depending on the same underlying mechanisms, such as gluon distributions or colorless exchanges, provided by different measurements, represent valuable opportunities to enhance our understanding of QCD. The high precision emerging in these measurements warrants higher order calculations well beyond those that presently exist. Experimentalists should be encouraged to continue to find ingenious techniques and new areas in which consequences of QCD may be explored. Theorists must refine present calculational techniques and develop new and deeper ways of doing perturbation theory to make use of the new experimental information which is becoming available.

Heavy ion reactions at colliders offer the possibility of studying a new phase of QCD at high temperatures and densities such as existed in the early universe. A major challenge here is to understand the transition at early times in the collision from hadronic degrees of freedom through a pre-thermalized system of quarks and gluons to an equilibrated quark-gluon plasma. This is a field where nuclear physics and particle physics merge, and where perturbative physics, nonperturbative physics and phenomenology are essential in order to extract the desired information on high density and high temperature QCD.

Small-x physics involves high-energy scattering in a regime of weak coupling, but where very high parton densities may be achieved leading to a new nonperturbative domain of QCD. Important information has recently been produced by experiments at HERA and relevant data should come soon from the Fermilab Tevatron collider. The rapid increase in parton density with decreasing x observed in ep scattering must eventually be constrained by unitarity. Therefore, future running at HERA should observe effects from the saturation of the parton densities. Low-x physics results would also emerge from LHC. Theoretical progress in this field may come from techniques in mathematical physics and two-dimensional field theories as well as from traditional perturbative and soft physics approaches.

As the US nuclear physics community is moving to explore higher energies at facilities such as CEBAF and RHIC, its interaction with the particle physics community in the field of QCD will intensify. In the future, CEBAF will play an increasingly important role in the study of hadron spectroscopy. Interactions between the nuclear and particle physics communities, *e.g.*, through the shared use of experimental facilities, as already occur at a less extensive scale at Brookhaven, Fermilab, and SLAC, are beneficial to both communities and should be encouraged to expand. In particular, the fixed target program at Fermilab has been very successful and offers an important area for collaboration between nuclear and particle physicists in the future.

While the past decade has been one of much progress in QCD and the same can be expected for the decade to follow, it should be noted that in comparison to the European effort at HERA and LEP the present experimental program in the US suffers from a shortage of experimenters and theorists dedicated to QCD studies as well as organized collaboration between them. The Fermilab Tevatron collider is and will remain a leading facility for QCD studies, and the CDF and DØ collaborations have produced many important QCD results. Excellent theoretical support has been provided in producing these results, but more extensive collaboration between experimenters and theorists will be needed for the great potential of this vital program to be fully realized. DESY is one example of such a collaboration. Through an active series of workshops and with the help of theorists at DESY, experimenters have found broad and effective support from theorists in Europe and in the US. To foster activity on a similar scale within the US program, a renewed commitment to QCD studies should be made at the same level. Elsewhere in the US, the QCD programs at both SLC/SLD and CLEO/CESR have produced significant results and promise to continue to do so. A QCD program at the NLC would offer an opportunity to make further precision tests. Finally, important US efforts exist that participate in the European QCD programs at HERA and LEP. In addition, the new window on QCD physics to be opened up by the energy regime at the LHC should produce a rich program of QCD measurements. US involvement in these European programs is vital to keeping US physicists engaged at this QCD frontier. Of primary importance, however, remains the need for leadership in the maintenance and expansion of QCD experiment and theory in the US.

3 Experimental Studies of QCD

3.1 Introduction

The experimental study of QCD has evolved considerably since the last decade when the principal goal was to "test QCD" to the present activity of measuring QCD elements such as α_S, fragmentation, and parton distribution functions. Present tests of QCD explore the frontiers of the theory, such as the boundary between perturbative and non-perturbative QCD of small-x and diffractive phenomena. The experiments of the future will provide precise measurements of QCD that will significantly enhance our knowledge of strong interactions and should aid in the discovery of the Higgs and the detection of deviations from the standard model such as composite quarks. Future experiments to explore the quark-gluon plasma will also open a new area of QCD exploration.

3.2 Measurements of α_S

The measurement of α_S for a variety of processes and Q^2 scales provides a fundamental test of QCD. The most precise direct measurement of α_S to date has been made by the LEP experiments for $Q^2 = M_Z^2$[1]. SLD has also produced comprehensive results on measurements of α_S from jet rates, energy-energy correlations and event shapes[2]. A summary of these results is shown in Figure 1. SLD and LEP results also provide evidence for the flavor independence of α_S. Ultimately, it is hoped to obtain a precision approaching a couple of percent for the α_S value measured for each quark flavor. Measurements of α_S will also be pursued with high priority at LEP 2, using jet rates, energy correlations and event shape variables such as thrust. The large lever arm in c.m. energy provided by an NLC and the reduced uncertainty due to hadronization effects should allow determination of α_S to about 4% and tests its running with energy[3].

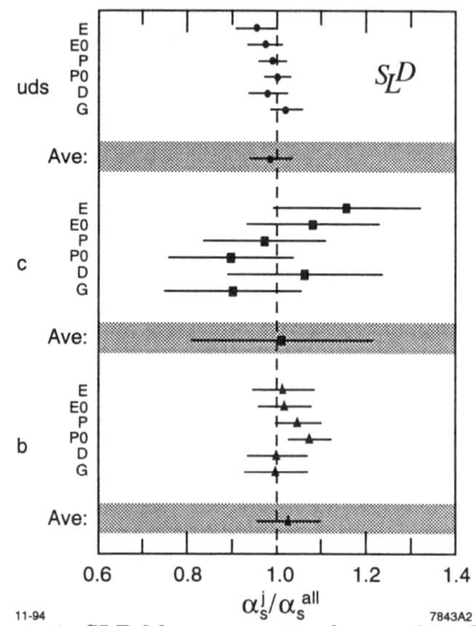

Figure 1: SLD Measurements of α_S and its flavor dependence.

The CCFR (FNAL E770)[4] νN scattering value of α_S measured from xF_3 and F_2 at high x is in good agreement with results from E-665 as well as the CERN NMC and EMC μN scattering experiments and other deep inelastic and low energy measurements. However, the low energy measurements are systematically slightly different from the LEP results. This difference may be due to the QCD scale value and definition in terms of renormalization scheme as well as higher order corrections. The forthcoming νN experiment, NuTeV (FNAL 815), expects a factor of 2 improvement in the statistical error on α_S, along with improvement on systematics, which could increase the significance of this difference.

Hadron collider measurements of α_S can be obtained by comparing ratios of cross sections $\sigma(X + (n + 1)\,\text{jets})/\sigma(X + n\,\text{jets})$ where $X = \gamma$, W, Z, or jet. For example, DØ[5] has performed a measurement of α_S from $W + 1$ *jet*/$W + 0$ *jet*. Hadron collider measurements will complement the LEP results, test the process independence and flavor independence of α_S, and provide measurements of the running of α_S over a large Q^2 range.

HERA measures α_S from deep inelastic scattering and the evolution of the structure functions. The experiments ZEUS and H1 measure deep inelastic scattering over a large range of Q^2 from a fraction of a GeV^2 to 10^5 and x below 10^{-4}. HERA also measures the relative rate for 2 jet events (+ the proton remnant) to 1 jet events which is sensitive to α_S. At HERA, this measurement can be performed for different values of Q^2, so that it may be possible to see the evolution of α_S within one experiment over a large range of Q^2.

The running of α_S over a large range of Q^2 is also shown by the dijet angular distributions from CDF[6] and DØ[7]. This range will be greatly extended at the LHC.

3.3 Parton Distributions and Structure Functions

The HERA ep collider has greatly extended the range of x and Q^2 accessible for experimental measurement beyond that of the previous fixed target experiments. This is shown in figure 2. The HERA measurements[8] show a rapid rise in F_2 as x decreases at low x (x values down to $\approx 10^{-4}$ were probed.) The conclusion is that the data prefer a rising gluon density leading to a rising sea quark density as x decreases. New data from H1 and ZEUS indicate that the rise at fixed Q^2 at low x appears to be softer for lower Q^2. The x region accessed by HERA is important since existing structure function parametrizations are checked with data only down to $x \approx 0.04$. HERA is now providing knowledge about the gluon distribution at low x. The HERA measurements will be considerably refined in the next few years as luminosity increases and the detectors are better understood. Already, preliminary measurements down to $Q^2 = 4.5$ GeV^2 have been presented. Upgrades and special running conditions (shifted vertex, lower energies) should provide the measurements to extend to Q^2 below 1 GeV^2. This will allow detailed tests of the scaling behavior of QCD via $dF_2/d\ln Q^2$, as well as the refined extraction of the gluon density. Data from electron and positron beams in principle allows a measurement of the non-singlet valence distribution, $F_3(x, Q^2)$, limited only by HERA statistics. An extraction of F_L will be possible, which provides information about the gluon density, along with F_2. The gluon structure function can also be measured via the boson-gluon fusion rate and other independent ways at HERA.

So far, the main tools in studying small-x behavior have been the Balitsky, Fadin, Kuraev, Lipatov (BFKL)[9] and Dokshitzer, Gribov, Lipatov, Altarelli, Parisi (DGLAP)[10] equations. These equations generate the leading growths at small x in different kinematical regimes. At present there is an intense activity to decide which of these mechanisms is responsible for the small-x behavior seen at HERA.

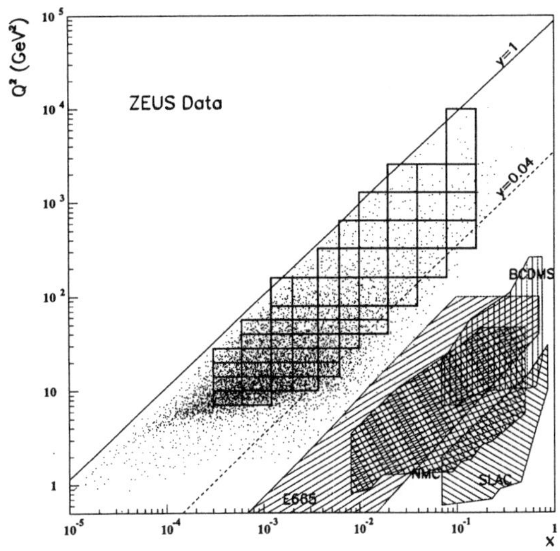

Figure 2: ZEUS 1993 data with the bins used for structure function analysis compared with the range of coverage of fixed target experiments.

Hadron collider measurement involve the convolution of two parton distribution functions. As the accuracy of the perturbative QCD predictions improves, these measurements can also reliably be incorporated into the global fitting programs used to extract parton distributions. Consistency of the extracted distributions amongst the various experiments is in itself a QCD test, and checks the factorization hypothesis. The kinematic range of the Tevatron collider is $2 \times 10^{-3} \lesssim x \lesssim 1$ and 10^2 GeV$^2 \lesssim Q^2 \lesssim 10^5$ GeV^2. CDF data provide measurements of the gluon, sea-quark and heavy flavor parton distributions in the proton, and yield constraints on flavor asymmetries in the quark distributions. Jet and γ production are sensitive to the gluon distribution. Drell-Yan production can be used to constrain the sea-quark distributions. $W + c$ and $\gamma + c$ production provide direct measurements of the strange and charm content of the proton. In addition, measurements of the asymmetry in W production are sensitive to the u/d ratio.

CDF[11] and DØ[12] are studying the triple differential cross section $d^3\sigma/dE_t d\eta_1 d\eta_2$. The result from DØ is shown in Figure 3. This can be directly mapped to the x_1 and x_2 of scattering partons and measuring this truly triple differential gives information of parton distributions between: $x = 0.003$ and 0.7 or higher. It is hoped that the new hadron collider data will extend the results to x as low as .001. Data from the LHC will probe to much lower x. The triple differential cross section is also a direct measure of the gluon content of the proton since at low p_t all scattering occurs through gluon-gluon scattering. However, understanding of the results awaits further NLO predictions.

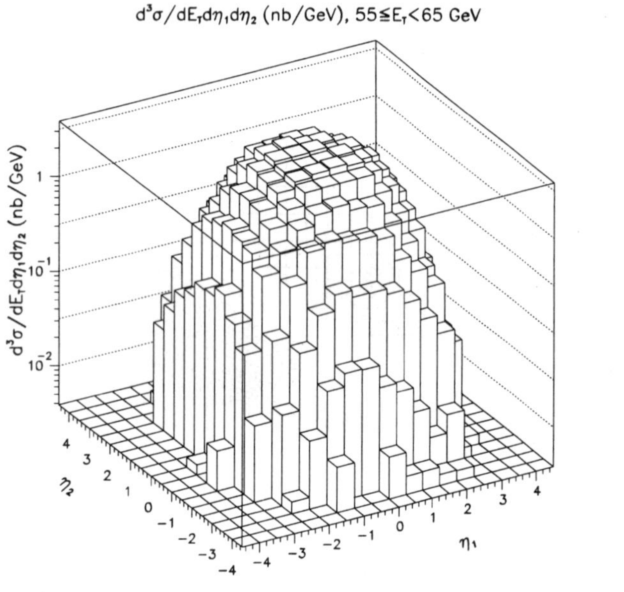

Figure 3: The triple differential cross section $d^3\sigma/dE_t d\eta_1 d\eta_2$ from DØ

In fixed target μN scattering, FNAL E665 has new measurements of the structure functions F_2^p and F_2^N in new regions of x and Q^2 as well as a measurement of σ_n/σ_p down to x of 10^{-3}. Comparison of F_2 in μ and ν scattering shows disagreement at low x[13,14]. The discrepancy is in x bins measured by only single experiments - CCFR (ν) and NMC (μ). New, higher precision results in both ν and μ will help address this problem. Through measurement of opposite sign dimuons, νN scattering also determines the strange sea and NuTeV should improve these measurements.

3.4 Spin Structure Functions

3.4.1 Background

In 1988, the EMC μN scattering experiment[15] presented a measurement of the nucleon spin structure function that violated a sum rule developed by Ellis and Jaffe[16]. The interpretation of the sum rule violation was that the strange sea in the proton is highly polarized. Another interpretation of the measurement was that the valence quarks carry little spin, the remainder carried either by the gluons or by orbital angular momentum. Experimental programs at CERN, at SLAC, and more recently at DESY were launched to explore this. The SLAC experimental program consists of four experiments (E142, E143, E154, E155) aimed at studying the proton and neutron spin structure in detail. SLAC experiment E142 ran in 1992 and found small negative asymmetries in the scattering of polarized electrons off polarized neutrons in the ^{3}He nucleus[17]. In parallel, the SMC program at CERN looked at the scattering from polarized deuteron targets[18]. These experiments address the significant questions concerning the fraction of the nucleon spin carried by the quarks and the strange sea polarization in the neutron spin structure function[19]. The combined neutron and proton results allow for a test of a fundamental QCD sum rule developed by Bjorken[20].

3.4.2 Recent Developments & Future Outlook

Recently, SLAC Experiment E143 used polarized ammonia and deuterated ammonia targets to remeasure both the proton and deuteron spin structure functions to high precision. The electron beam polarization and energy were improved to 80% and 29 GeV. Most recently, in 1994, E143 at SLAC[21] and SMC at CERN[18] reported new data on polarized protons. Latest indications are that between 30% to 40% of the proton's spin is carried by the bare valence quarks. This still seems somewhat less than might have been expected from predictions based on constituent quark models of the nucleon. It has been suggested that this result signals a breakdown of the quark model and possibly an indication that some aspects of the nucleon are better described by a chiral (Skyrme) model[22]. It has also been noted that the flavor singlet axial anomaly complicates the traditional interpretation of the first moment of the g_1 structure function in terms of the quark spin in the proton[23]. A good measurement of the distribution of gluon spin in the nucleon would be likely to give strong support to one or the other of these suggestions. The issues here are very important. The reliability of the quark model, and the viability of Skyrme-type models will rest on determining just which partons carry how much of the nucleon's spin.

In the future we can expect data from two more experiments at SLAC which have been approved to use a higher energy 50 GeV beam. With this higher energy, lower x and higher average Q^2 measurements of the spin structure functions will be measured. The total combined data set from these SLAC experiments alone should be adequate to test the Bjorken sum rule to better than 10% of its value. The forthcoming HERMES experiment at DESY will also provide a precise determination of the g_1 sum rules for the proton and neutron. The resulting uncertainty in the Bjorken sum rule is expected to be 5-8%. HERMES will also determine the g_2 structure function (associated with transverse target polarization) for both the proton and neutron. This structure function allows the first direct study of quark- gluon correlation functions without contamination by leading twist-2 structure functions. HERMES will carry out measurements of coincident hadrons produced in spin-dependent deep inelastic scattering. These measurements will provide information on the flavor dependence of the polarized structure functions as well as the contribution of orbital angular momentum. In addition, these measurements are expected to be sensitive to $h_1(x)$, the chiral-odd leading-twist structure function.

3.4.3 Polarized Proton Physics

The polarized beam capability proposed for RHIC offers a unique array of spin measurements[24]. This allows the measurement of $g(x)$ for nuclei, $\Delta g(x)$ for pN, $h_1(x)$ (which counts the valence quark polarization), parity violating terms in W and Z production and the sea quark helicity difference in W production. The Drell-Yan process can also be used to measure spin observables with longitudinal and transverse polarization to yield new information about antiquark and valence quark polarization. Gluon polarization can be measured by using direct photons from the dominant quark-gluon Compton graph and through medium p_t jets (20 - 50 GeV/c) which are predominantly quark-gluon produced. Parity violation in W^+ production measures the difference between u and $\bar{d}$ polarization, and for W^- production it measures the difference between d and $\bar{u}$ polarization. h_1 will be measured with transverse spin using both γ^* and Z production. Experiments at RHIC should also be able to search for parity violation in other processes (jets, γ).

3.5 Tests of Perturbative QCD

3.5.1 Jets and Final States

CDF and DØ have produced a wealth of data on jets and final states. DØ has compared the overall inclusive cross section, $d^2\sigma/dE_t d\eta$, jet and dijet cross sections, and direct predictions[5]. They have uniform rapidity coverage for $-4.0 < \eta < 4.0$ and can extend the coverage for jets down to transverse energies (E_t) of 15 GeV. They observe that jet cross sections measured out to $\eta = 3.0$ agree within errors with NLO predictions over more than 10 orders of magnitude. DØ has measured the dijet angular distribution over an extended range, which tests the correctness of LO and NLO hard parton cross sections. DØ has compared the location of a third jet in an event with respect to the second leading jet ordered by E_t and see an enhancement predicted by color coherence.

CDF also observes color coherence effects in hadronic angular ordering which provides evidence for the interference between the initial and final state gluon emission from color connected partons. CDF has compared the inclusive jet E_t spectrum, the two-jet mass spectrum and angular distribution, and the two-jet differential cross section with the NLO predictions[6,25]. There is good agreement between the measured and predicted differential cross sections. These NLO comparisons agree at the ± 20% level, which is about a factor of two better than the precision of the corresponding comparisons at LO. In the future, a combination of improved theoretical calculations and data from luminosities between 100 pb^{-1} and 1 fb^{-1} should result in QCD measurements with a precision of a few percent.

Not all the current CDF measurements show good agreement with QCD predictions. There are significant discrepancies between the data and NLO theory for the ratio of scaled inclusive jet cross sections measured at two values of $\sqrt{s}$ (1800 GeV and 546 GeV)[26] and the inclusive photon spectrum at low p_t[27]. While the newer measurement of the b production cross section by CDF[28] appears closer to the QCD prediction than previous measurements[29], the b cross section measurement by DØ[30] agrees with QCD expectations. At LO there are order of magnitude discrepancies with the measured prompt ψ and ψ' production cross sections. The resolution of these discrepancies is clearly important to establish the reliability of the QCD predictions and may lead to some new insights.

The new kinematic regime available at the Tevatron should provide a test of QCD calculations in the semi-hard region. As dijet separations reach large rapidities a second scale, $\ln(s/p_t^2)$, enters the perturbative expansion, leading naturally to a resummation in terms of the BFKL Pomeron[31]. It will be one of the exciting challenges, over the next five to ten years, to make precise comparisons between theory and experiment and to measure the BFKL Pomeron. Already some data on rapidity gaps at FNAL energies had appeared[32,33].

Numerous tests of perturbative calculations have been performed at SLD and LEP 1. These include tests of second order formulae (for quantities like thrust, oblateness and other event shape variables) and resummed calculations[34]. Most of these studies will be repeated at LEP 2, where the reduced theoretical and hadronization uncertainties will make the comparisons of theory and data even more unambiguous. There are still relatively few variables for which resummed calculations have been performed. New calculations utilizing perturbation theory beyond second order will be tested at SLD, LEP 1 and LEP 2, as they become available. Tests of perturbative calculations relevant for coherence phenomena, the interference effects associated with the coherent emission of soft gluons, have been performed at LEP 1 and will be continued. Studies of coherence should benefit from the smaller hadronization uncertainties at LEP 2.

Tests of matrix elements at LEP comprise measurements of the absolute and relative energies and angles of jets in three-, four- and five-jet events and a comparison to matrix element formulae[35]. Investigations of the three-jet matrix element at LEP have yielded much more rigorous tests than were possible at PETRA and PEP, because the jet resolution parameter (y_{cut}) can be made much smaller: this means that the pole structure arising from soft gluon radiation can be observed and compared to theory. Also, it is possible to study the matrix elements at much smaller scaled jet energy values, x.

The properties of quark and gluon jets have been studied at LEP, using tagged quark jets and rare, two-fold symmetric three-jet events to obtain unambiguous results on quark and gluon jet differences[35]. These investigations, which require large data statistics, will continue to the end of the LEP 1 program.

CLEO is performing a QCD analysis of jet rates in e^+e^- 4-flavor continuum around 10 GeV to study if there is a difference between the 4-flavor and 5-flavor cross sections due to flavor dependence [36].

Studies are underway to determine whether perturbative QCD is able to describe deep inelastic scattering jet data, both in the FNAL E665 [37] kinematic regime and in the high-energy regime of HERA[38,39]. QCD models of final state radiation have been well tested at other accelerators, such as LEP. However, the initial state radiation predictions are not well tested. HERA offers the opportunity to perform such tests.

Initial data from HERA indicates that deep inelastic scattering two-jet data are dominated by QCD Compton and boson-gluon-fusion processes. The overall energy flow in deep inelastic scattering events should manifest the filling of the phase space between the current jet and photon remnant with particles materializing from the emission of additional gluons and quarks created by color transfer between the struck quark and proton remnant. Effects such as color coherence are being studied. Several QCD models tested by these data are found to agree. As the increased statistics from HERA become available, there will be an iterative process as QCD models are refined to agree with more precise data. A measurement of the gluon density from jets should be available within about a year.

Jet production in deep inelastic scattering with $Q^2 > 4$ GeV^2 has been investigated at HERA for events with center-of-mass energies W between 30 and 280 GeV[38,39]. The characteristics of the multijet events are not consistent with a bare quark-parton model, but are well described by the first order QCD matrix elements plus higher order corrections implemented via parton showers. Multijet production has also been studied in events with a large rapidity gap. The properties of these events with $W > 140$ GeV are consistent with a leading twist diffractive production mechanism. These analyses indicate that considerable information may be gleaned from inelastic and diffractive jet production at HERA. Studies at HERA of multijet production and differential cross sections in quasi-real photoproduction at center-of-mass energies in the range 130 - 275 GeV have been compared with leading logarithm and parton shower predictions, including resolved and direct processes. Future data should allow a determination of the photon structure function via photoproduction of jets.

Jets in lepton-hadron interactions have also been observed in FNAL experiment E665[37]. They have measured 2 + 1 jet rates corrected to the partonic level, and compared these with LO and NLO calculations, which they find underestimate the data. They are also analyzing the propagation of multijets in nuclear matter using nuclear targets. They expect to analyze a sample 30 times larger in the next 2 years. Increases in fixed target energy, however, cannot match the kinematic range at HERA, which is the equivalent of a 52 TeV electron beam on a fixed proton target.

3.5.2 Photoproduction

The next few years should see detailed measurements of the photon structure function at HERA from photoproduction events ($Q^2 \approx 0$). Besides $p\bar{p}$, γp scattering is the only other hadronic interaction which can be currently measured at c.m. energies above 100 GeV. As is the case for purely hadronic interactions, photoproduction has a soft component due to peripheral processes and a hard component due to the scattering of a parton in the proton with a parton in the vector meson from the photon. However, the photon also has a direct pointlike coupling to quarks that leads to additional hard scattering processes that are not present in hadron-hadron interactions. "Direct" processes include the photon-gluon fusion into a quark-antiquark pair and the photon scattering off a quark in the proton with the emission of a gluon (QCD Compton scattering). These processes are also measured in direct photon production (see Section 3.5.4). "Resolved" processes include the hard scattering of the hadronic (vector meson) photon, and the hard scattering due to the quark and gluon content of the photon. Studies of these processes at HERA over the next few years will result in a detailed measurements of the hadronic structure of the photon and the gluon distribution of the proton.

Other interesting processes that can be studied at HERA with photoproduction events include boson gluon fusion producing heavy quarks, such as in J/ψ production. These processes will also help measure the gluon distribution of the proton. In general, the hadronic properties of the photon will be studied in great detail. Already, the direct and resolved scattering of the photon on the proton have been observed. In addition, the observation of hard scattering photoproduction events with rapidity gaps offers the opportunity to measure the elastic and inelastic diffractive contribution. Two-jet production in hard-scattering photoproduction events will allow determination of the photon structure function.

Measurements of the photon structure function, $F_2^\gamma(x,Q^2)$, in the two-photon reaction $e^+e^- \rightarrow e^+e^-$ + *hadrons* have been performed at LEP 1[40]. This topic will be pursued actively at LEP 2 and will allow an investigation of the scaling behavior of $F_2^\gamma(x,Q^2)$ over a large Q^2 range.

3.5.3 Heavy Quarks

Heavy quark experimental results have some unresolved QCD issues. As mentioned above, there is a disparity between the NLO QCD calculation of the b cross section and experimental data from UA1 and CDF. However, the new DØ cross section results are in good agreement with theory. These results have been compared to heavy quark predictions and are sensitive to gluon density in their normalization and rapidity dependence. The transverse momentum (p_t) distribution of B mesons rises faster than predicted by NLO calculations. In Tevatron fixed target experiments, there is a deviation between the angular distributions of back to back peaking of charm and beauty particles and NLO QCD predictions, although the b-particle statistics are limited[41]. Higher energy fixed target experiments would permit sufficiently precise measurements for b-particles.

The measured Feynman x-distributions of J/ψ and other charmed hadrons present an important challenge for QCD theory and phenomenology. Such distributions, especially at large values of x, are very sensitive to higher twist effects. Understanding the nuclear A-dependence and large-x dependence of J/ψ production may also have important consequences for the use of J/ψ production as a signal for the quark-gluon plasma formation in heavy ion collisions. At the moment there is no general consensus as to what physics determines the large-x behavior of heavy particle production. However, progress is being made by both nuclear and particle theorists, which should hopefully lead to a satisfactory understanding of this area in the future.

3.5.4 Single Photons

DØ has measured the direct photon cross section in the central region[42], and will extend this in the next year to the $\eta > 1.0$ region. This is used to measure the gluon density of the proton. CDF finds a 20% discrepancy with QCD predictions for the single photons with $p_t \lesssim 30$ GeV[43]. This discrepancy could be due to inadequate knowledge of the gluon distribution in the x range probed or to photon fragmentation. Better measurements of these phenomena should help to resolve this.

LEP 1 can study photon radiation from quarks since the Z^0 pole suppresses initial-state photon radiation[44]. So far, QCD studies of final-state photon radiation have concentrated on isolated, high-energy photons in order to test the parton shower mechanism, since photons are emitted according to a matrix element similar to that for gluons, without the complications of hadronization. These studies will continue to the end of the LEP 1 program. Studies of final-state photon radiation *within* jets have now begun, with the purpose of measuring the quark to photon fragmentation function.

3.5.5 Other Tests of Perturbative QCD

The SLD experiment plans to exploit particle identification, secondary vertex identification and polarization to explore QCD quark/antiquark and flavor dependence. They have compared multiplicities of events with b- and uds- tags[45]. They will focus on the charm sector where tests of perturbative QCD can be made at the c-mass, close to the confinement scale. SLD also plans to compare gluon radiation patterns from heavy and light quarks. They are studying parton polarization in hadronic jets.

CLEO is comparing semihard reactions in two-photon collisions, *e.g.*, $\gamma\gamma \rightarrow$ V+X, with calculations in perturbative QCD. They will also continue tests of QCD predictions for $\gamma\gamma \rightarrow \pi^+\pi^-, p\bar{p}$ as well as $\Upsilon \rightarrow gg\gamma$, and the two-photon decay widths of the η_c and χ_{c2}[46]. CLEO also intends to test some of the predictions of the modified leading log approximation and local parton hadron duality for multiplicity differences between charm quark and light quark events.

Resummed QCD predictions for the W and Z p_t spectrum have been tested with CDF data and found to be in good agreement[47]. The Drell-Yan cross section makes an ideal measurement for a systematic comparison between NLO, NNLO and resummed QCD. CDF expects to measure this within 5-10% from their present data. DØ also uses the p_t spectrum of the Z to test QCD techniques employed in calculations, including, resummation and NLO high p_t predictions[48]. DØ has examined the energy flow in W and jet events and shown that there is a difference between the energy flow around a colored versus a colorless object, as predicted by QCD.

3.6 Boundary between Perturbative and Non-Perturbative QCD

There are many examples where collider data give information on phenomena at the boundary between perturbative and nonperturbative QCD. Examples include the onset of small x effects, and the study of the hard diffractive process and rapidity gap events.

3.6.1 Small-x

As smaller values of x are probed for a fixed Q^2, the density of partons is predicted to increase. The approximations made in deriving the perturbative evolution equations (leading $\ln Q^2$) are no longer valid. The opportunity to explore the small-x region at HERA has generated much interest. For example, for $Q^2 = 10$ GeV^2, x values as small as 10^{-4} can be reached at HERA. Early data from HERA indicate that the parton distributions are rising quickly, and are close to the predicted $x^{-0.5}$ behavior[49]. This rapid increase in the parton density would eventually lead to a violation of unitarity. This implies that a saturation of the parton densities will occur below some value of x. Searches for this saturation are an important part of the tests of QCD at HERA. The data from H1 and ZEUS will also indicate how large the gluon densities become, if there is gluon recombination, if partons overlap and whether there are "hot spots"[50].

While CDF and DØ have begun to study small-x effects, their analysis would be enhanced by increasing kinematic coverage of jet and photon measurements and improving their detector understanding. However, more theoretical input is also needed to understand the types and magnitude of phenomena to be expected. For instance, it is possible to look for small-x effects in jet events with large rapidity gaps. Both DØ and CDF have investigated these effects[32,33], although the theoretical framework here also needs better understanding.

3.6.2 Diffractive Scattering & Rapidity Gaps

The observation at HERA of deep inelastic scattering events with a large rapidity gap in the final state between the proton direction and the first energy deposit in the detector is an indication of diffractive scattering[51]. The flatness of the rapidity gap distribution, as well as other properties of the events such as independence of the cross section on W, are consistent with photon diffractive dissociation off a Pomeron. A detailed study of these events may yield insight into the transition from perturbative to non-perturbative scattering. Exclusive reactions, such as elastic vector meson production, should provide stringent tests of calculations in perturbative QCD, as well as new methods for extracting gluon distributions. Finally, if the Pomeron has small dimensions compared to the proton, then new low-x effects such as saturation would set in even earlier. The two HERA experiments will upgrade their detectors in order to measure the scattered particle in these events and allow a precise measurement of the kinematics of the final state. This will enable measurements of the Pomeron structure function as well as measurements of the structure function of virtual pions.

Rapidity gaps have been found at the Tevatron[32,33] and are now a subject of considerable interest. In principle, such events are an excellent place to study high-energy semi-hard physics including the BFKL Pomeron. However, the analysis is complicated by the presence of hadrons coming from soft interactions involving the spectator quarks in the colliding hadrons. More experimental work, such as a comprehensive search for rapidity gaps in various channels, is essential. More theoretical work is necessary in order to make sharp contact between QCD theory and experiment. This promises to be an interesting subject in the coming years.

Diffractive physics at the LHC promises to be a rich source of information since certain topologies will be cleaner. due to cleaner events. It will be possible to look for diffractive Higgs events with reduced hadronic activity in the rapidity region near the Higgs particle[52]. Both single and double pomeron exchange can be observed, particularly in $H \to \gamma\gamma$ events. In addition, single diffraction at the LHC can be studied in $b\bar{b}$ production.

3.6.3 Jet Fragmentation

CDF has made comparisons of the average jet shape (energy flow inside jets) with NLO two-jet predictions that show good agreement[25], indicating that there is a perturbative piece of jet fragmentation that can be calculated. In the future, the scope of the QCD measurements will widen to include more detail on jet shapes, fragmentation, energy flow in the "underlying event" accompanying hard processes and the energy flow in rapidity gap events. CDF also has the capability of measuring the fragmentation of heavy flavor jets, the fragmentation of generic jets to γ's and π^0's, and differences between quark jets and gluon jets. DØ has measured the E_t profile of jets as a function of the distance from the jet axis for different E_t's and rapidities[5]. Clear differences are observed and in general the behavior is described well by the parton shower model.

The CDF and DØ studies complement the information from the LEP and SLC experiments and provide checks of the universality of jet fragmentation. The large statistics of the LEP 1 experiments permit a substantial number of heavy resonance states to be reconstructed and their production characteristics and correlations to be studied[44]. These studies will include such topics as baryon creation from the vacuum. SLD plans to measure heavy quark fragmentation functions and decay topologies and study particle fragmentation properties separately in quark and antiquark jets.

3.6.4 pp and $\bar{p}p$ Elastic Scattering

At Fermilab there is a program of measuring $\bar{p}p$ elastic scattering up to $\sqrt{s}$ range 0.6 – 2 TeV. At RHIC, there will be a systematic study of the proton-proton elastic scattering in the $\sqrt{s}$ range 60 – 500 GeV and covering the four-momentum transferred $|t|$ up to 8 (GeV/c)^2. The small $|t|$, or Coulomb nuclear interference region, determines the total cross section, the ratio of real to imaginary part of the scattering amplitude, and the slope parameter. These measurements should clarify some of the present puzzles connected with elastic scattering[53].

At large $|t|$, where data can be taken with the standard lattice, $|t|$ values beyond the expected second dip can be reached. At these $|t|$ values it is possible to make a comparison with perturbative QCD calculations. The same setup can be used to study elastic scattering using polarized protons at RHIC, to measure the difference in the total cross sections as function of initial transverse spin, the analyzing power, and the transverse spin correlation parameter. At lower energies, the spin dependence in polarized proton-proton elastic scattering has a very strong angular dependence which has never received satisfactory explanation. It will be interesting to see whether the strong t-dependence of A_{NN} persists at higher energies.

3.7 Quark Substructure

Quark substructure is investigated by measurement of the inclusive jet cross section. Composite quarks would manifest themselves through deviations from the standard QCD expectations at high transverse momenta where valence quark scattering dominates. The present CDF limit on the compositeness scale is $\Lambda_c > 1.4$ TeV (95% C.L.). By the end of the decade the expected two orders of magnitude in statistics and increase in energy from 1.8 TeV to 2.0 TeV should improve this limit to 1.8 TeV. LHC exposures of 10^4 and 10^5 pb^{-1} would result in sufficient statistics to observe a compositeness scale Λ_c as large as 15 and 20 TeV, respectively. The largest systematic error is expected to be that caused by a nonlinear performance of the calorimeter. The anticipated performance of the ATLAS and CMS detector calorimeters should be adequate to explore the compositeness scale out to the 15 - 20 TeV range.

3.8 Light Hadron Spectroscopy

3.8.1 Light Quark Spectroscopy

A thorough understanding of the conventional $q\bar{q}$ nonets is an important area in which to look for extraneous states, or states with peculiar production or decay characteristics. Exotic J^{PC} combinations can also be looked for. Some of the processes used in these studies include peripheral production, peripheral production which emphasize OZI-disallowed processes, central production, $p\bar{p}$ annihilation, J/ψ radiative decays and photon-photon scattering. Brookhaven experiment E852 will collect a substantial sample of final states with charged particles and neutrals. This should address the issue of the only reported exotic to date, the $J^{PC} = 1^{-+}$ M(1405)[54], since E852 will be sensitive to many more modes and able to search for the exotic hybrid which should decay into b1(1235) π, as well as investigating $\eta - \eta$, $\eta - \eta'$ and various vector-vector states.

3.8.2 Glueballs

Glueballs are a direct consequence of the non-abelian nature of color $SU(3)$ and so their observation would be an important demonstration of non-perturbative QCD in hadron spectroscopy[55]. They are the only unobserved hadrons predicted by $SU(3)$, but there is no consensus on whether they exist. Lattice gauge theory (see Section 6.2.2) predicts the lowest lying glueball states to be a 0^{++} state with mass $\approx 1740 \pm 71$ MeV and a 2^{++} with mass $\approx 2359 \pm 128$ MeV. In the case of the 0^{++} the number of isoscalars found so far are 4 [$f_0(975)$, $f_0(1400)$, $f_0(1590)$, $f_0(1720)$], only one of which is expected to be a $q\bar{q}$ state. Thus, at least two must be eliminated to have a possible candidate for a 0^{++} glueball. Multiquark, quark molecule, or hybrid states are all candidates for these other states[56], which adds to the uncertainty.

In the 2^{++} channel the reaction $\pi^- p \to \phi\phi n$ at BNL revealed a breakdown of the OZI suppression and later the existence of three candidate $(\phi\phi)$ glueball states, the $f_2(2010)$, $f_2(2300)$ and $f_2(2340)$, with classic Argand diagrams. The many Godfrey-Isgur $q\bar{q}$, and multiquark states expected in this region were not observed, suggesting that these 2^{++} glueball candidates break the OZI suppression. In the next year or two, runs are expected in Brookhaven experiment E-881 to enhance non pion exchange processes. By studying $K^- p \to \phi\phi\binom{\Lambda}{\Sigma}$ (OZI allowed) and $\bar{p}p \to \phi\phi\pi^0$ (OZI forbidden) one hopes to further clarify the situation.

3.9 Quark-Gluon Plasma

3.9.1 RHIC Program

The Relativistic Heavy Ion Collider (RHIC) and the PHENIX, STAR and PHOBOS detectors are designed to study the reorganization of matter predicted to occur in high-energy nucleus-nucleus collisions. These collisions range from Au + Au at $\sqrt{s}$ of 60 – 200 GeV per NN pair at a design luminosity of 2×10^{26} to pp at $\sqrt{s}$ of 50 – 500 GeV at a design luminosity of 10^{31}, with the option to provide longitudinal and transverse polarized beams. The large systems of highly excited, strongly interacting matter formed in these collisions will be probed to search for signatures of the existence of the quark-gluon plasma - a state of matter in which quarks and gluons are 'deconfined' and where chiral symmetry is restored. QCD issues that are addressed by this program are to establish the existence of a quark-gluon phase of matter and measure the properties of matter in the region of the phase change. Also to be studied are nuclear shadowing and color transparency. The program is scheduled to start in 1999.

The detectors at RHIC should address almost all aspects of quark-gluon plasma formation and chiral symmetry restoration in nuclear collisions. The early stages of the collision, expected to be a gluon-dominated QCD plasma, can be studied in PHENIX using electromagnetic probes (lepton-pairs and photons). Alternatively, one can look for open charm through the decays $D \to$ leptons in PHENIX and potentially $D \to$ charged hadrons in STAR. The quark-gluon plasma phase will be investigated by measurements of predicted signatures, such as suppression of heavy vector mesons ($J/\Psi, \Psi, \Upsilon$) through μ and e-pairs measurements in PHENIX. Similarly, it is planned to search for the enhancement of strangeness production using strange hadron measurements in STAR and PHENIX and multiple-strange baryons and antibaryons in STAR.

If the QGP phase transition is first-order, a dramatic increase in the differential source size is expected due to the time-delay associated with particle emission accompanying a first-order phase transition. Such behavior can be inferred from two-particle interferometry measurements in PHENIX, PHOBOS and STAR. The characteristic phase transition behavior of the temperature as a function of entropy density in this case can be measured in STAR and PHENIX; and local density fluctuations in particle number, energy, and entropy could be observed in STAR and PHOBOS. Direct photons from the mixed phase will be measured by PHENIX. Measurements of the mass, width and yield of ϕ-mesons via $\phi \to e^+e^-$ in PHENIX and $\phi \to K^+K^-$ in PHENIX and STAR could detect possible modifications of particle masses and widths in the medium due to partial chiral symmetry restoration.

Localized regions of disoriented chiral condensate, formed in the chiral phase transition, expected to be second-order, could be measured in STAR, via abnormal ratios of the various charges of pions in localized regions of momentum space event by event.

To be able to understand RHIC collisions and perform reliable perturbative QCD calculations, the initial conditions of the collision must be well understood. This requires measurements of the gluon structure function and nuclear shadowing of quarks and gluons in the x region $0.02 < x < 0.1$. These measurements can be done directly at RHIC. The process gluon + quark $\to \gamma$ + jet will be measured in STAR to determine the gluon structure function. Both STAR and PHENIX, through γ + jet coincidences and direct γ measurements, respectively, can determine the degree of nuclear shadowing.

3.9.2 LHC Program

The LHC heavy ion program will involve a dedicated detector, ALICE, as well as operation of the two general purpose pp collider detectors, ATLAS and CMS. These experiments will study the production of heavy quarkonium states ($c\bar{c}$ and $b\bar{b}$) to look for signs of quark-gluon plasma. These studies will concentrate on bottomonium production rather than charmonium since J/ψ production from B-meson decay dominates at the LHC, except at very low p_t. These detectors will study the global event characteristics as a function of energy density and also examine the suppression of Υ' and Υ'' with respect to Υ production. At the LHC with 10^5 Υ, corresponding to an integrated luminosity of 15 nb^{-1}, the estimated precision of the measurement of the suppression is 7 (20)% for the Υ' (Υ'')[57].

3.10 Conclusions

Many areas of QCD experimentation are producing new and important results. There is new information on parton distribution functions, QCD parameters and tests of consistency between perturbative QCD and data. Quantitative collider measurements are yielding new insights in processes such as Drell-Yan production, jets, direct photon production and heavy quark production. The frontier of our knowledge of QCD is being explored in studies of large transverse momentum processes, small-x scattering, diffractive processes and rapidity gaps. Finally, we are just beginning to explore areas of exclusive reactions, spin physics and QCD in nuclei.

The wealth of new and more precise QCD measurements presents an important opportunity to enhance our understanding of the strong interaction and the structure of the nucleon. However, the sophistication of QCD theory beyond the parton model and the complexity of modern experiments requires much more close collaboration between theorists and experimentalists. Efforts such as that of CTEQ (Collaboration between Theorists and Experimentalists on QCD) are particularly important to provide this and should be broadened.

There are many open questions to be explored in QCD. These include the gluon distribution, the flavor-dependence of the sea, features of direct photon production and jet differential cross sections, heavy quark production, the small x region, diffractive phenomena, spin physics and the quark-gluon plasma. QCD represents an extraordinarily rich area for continued theoretical and experimental exploration. The confrontation of QCD theory with improved experiments will produce insight on the structure of hadrons, understanding of the underlying quantum field theory and clues that may point to new physics.

4 Theoretical Studies of QCD

4.1 Factorization in Processes with a Hard Scattering

There are a few processes, such as $\sigma(e^+e^- \to hadrons)$ at high energies and $\Gamma(Z \to hadrons)$, where perturbative QCD gives a complete description of an observable. However, in the general situation where hadrons make up one or both of the initial particles in a scattering, or when a hadron with some definite momentum is measured in the final state, nonperturbative effects are important. Factorization in a high-energy hadronic reaction, which at the same time has a hard scattering, gives the cross section for that process as a product of a perturbative hard scattering cross section of quarks and gluons times parton distributions in the initial hadrons[58,59]. These parton distributions give the flux of quarks and gluons available for the hard scattering. They are universal, depending only on the hadron from which they come but not on the particular hard process. Thus the quark and gluon distributions in the proton can be determined, say, from deep inelastic lepton-proton scattering and used to make absolute predictions for μ-pair, W, Z and jet cross sections in proton-antiproton collisions.

The determination of deep inelastic lepton-proton scattering in terms of deep inelastic lepton-quark scattering times the distribution of quarks in the proton is completely equivalent to the use of the short distance operator product expansion. Such a partonic description is thus as solid and well founded as any result obtained in a four-dimensional field theory.

When two hadrons initiate a hard scattering process the operator product expansion is not effective and the partonic description is the only available formalism. Here factorization is nontrivial. For example, in Z production in proton-antiproton collisions the proton interacts with the antiproton to produce many low transverse momentum pions in addition to the Z. At first glance, it would appear that such interactions are non-factorizable since they involve different interactions of the proton than occur in lepton-proton scattering. It was one of the important achievements of the 1980's to argue convincingly that hard scatterings initiated by hadron-hadron collisions factorize.

Since factorization is the foundation on which so many predictions of high energy hard interaction are based (including predictions for the production rates of top quarks, Higgs particles and various supersymmetric particles) it is important to have rigorous arguments for factorization at the leading twist level, to better understand the nature of higher power corrections and to determine to what extent they can be described in terms of partons and/or higher twist operators.

4.2 Parton Distributions

The quark distribution $q_f(x, Q^2)$ is the average number of quarks of flavor f and longitudinal momentum fraction x in the infinite momentum frame wavefunction of the proton[59]. The quarks are bare at a transverse coordinate scale $\Delta x_\perp \sim 1/Q$. The spin averaged quark and gluon distributions of the proton have been reasonably well determined over a fairly wide region of x and Q^2 from deep inelastic lepton-nucleon scattering experiments. This gives an average picture of an unpolarized proton in terms of its bare quark and gluon constituents. It would be very interesting to have a similarly detailed picture of a polarized proton in terms of quark and gluon spins.

Polarized quark distributions give much information on chiral symmetry-breaking. For example, the Bjorken sum rule gives g_A, the coupling of the axial vector current to the proton, as

$$\Gamma_1^P - \Gamma_1^N = \frac{g_A}{6}\left[1 - \frac{\alpha}{\pi} - 3.58\left(\frac{\alpha}{\pi}\right)^2 - 20.2\left(\frac{\alpha}{\pi}\right)^3 + \cdots\right] \tag{1}$$

in the $\bar{MS}$ scheme with $\alpha = \alpha(Q)$ and where Γ_1^P and Γ_1^N are the first moments of the $g_1(x, Q^2)$ structure functions of the proton and neutron respectively. $g_1(x, Q^2)$ gives the number density of longitudinally polarized quarks in a longitudinally polarized proton. The difference of g_A from one is a direct consequence of chiral symmetry breaking.

An even more interesting link to chiral symmetry comes from determining the total amount of the proton's spin carried by bare quarks from polarized parton distributions. The constituent quark model predicts that most of the proton's spin should be carried by bare quarks, but with little of the spin carried by strange quarks. On the other hand, Skyrme models suggest that strange quarks carry a significant amount of the proton's spin, though the total spin carried by all flavors of quarks is predicted to be small. The experimental situation is summarized in Section 3.4. The issues here are very important for a good understanding of the wavefunction of the proton and, in particular, whether the constituent quark model can adequately describe chiral symmetry breaking or whether that symmetry breaking must be introduced more explicitly through a low energy effective theory like the Skyrme model. Our theoretical understanding of these issues has grown considerably over the past five to six years, and continued progress can be expected in the future.

4.3 α_S-Determinations

Since the early days of QCD analyses a great deal of attention has been given to determining α_S, or equivalently the Λ parameter of QCD[60]. It has become conventional to give α_S in terms of its value at the mass of the Z. At present, an average over the various ways of determining α_S from experiment gives $\alpha_S(M_z) = 0.118$ with an error between $\pm$ 0.007. It is remarkable that such diverse methods of determining α_S as event shapes in Z decay, hadronic decays of the τ and the analysis of scaling violation in deep inelastic scattering give compatible results. α_S-values obtained from some recent analyses are shown in Figure 4. Present data on jets from Z-decay and on deep inelastic scattering are now of such precision that complete next-to-next-to-leading order calculations are of considerable interest for both of these processes. Such calculations are now in progress. The accuracy of present data and the large number of processes where next-to-leading order (in some cases even next-to-next-to-leading order) calculations have been done and resummation procedures applied has renewed interest in understanding if there are preferred ways of fixing the scale of the running coupling in terms of the kinematic scale of the hard process. The time may now be ripe for a deeper understanding of this thorny issue.

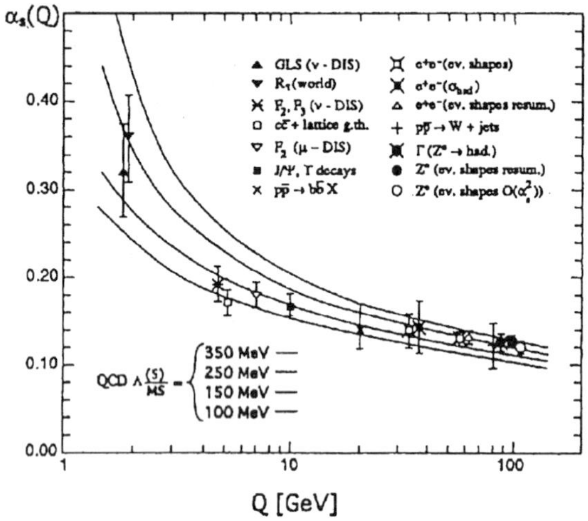

Figure 4: Summary of the Q^2 dependence of α_S(Q). The lines indicate the QCD predictions in NNLO for four different values of $\Lambda^{(5)}_{\overline{MS}}$. (From Ref. [61]).

4.4 Higher Order Calculations

Higher order calculations are at the heart of comparing QCD theory and experiment[59]. Next-to-leading order corrections are required in order to keep factorization scale dependence down to an acceptable level and to obtain reliable normalizations. There are now a relatively large number of processes, such as the cross section for electron-positron annihilation into hadrons, the τ-lepton hadronic decay rate and the Bjorken sum rule, for which higher order corrections have been done through order ${\alpha_S}^3$. For two-jet production in hadronic collisions and for three-jet production in electron-positron annihilation the one-loop corrections have been calculated while for deep inelastic lepton scattering and the Drell-Yan process parts of next-to-next-to-leading order corrections have been done. There are now vigorous programs underway to calculate the next-to-leading order terms for four-jet production in electron-positron annihilation and for three-jet production at hadronic colliders. Recently, resummation techniques have been developed and calculations performed which allow application of QCD predictions near the edge of phase space for jet production in electron-positron annihilation and for the Drell-Yan process.

The progress made in performing and understanding higher order QCD corrections has been impressive. However, the complexity of calculation grows exceedingly rapid with the order of the higher order correction when a straightforward calculation of Feynman diagrams is performed. In the past few years an approach inspired by string theory has been developed which may effectively organize the large number of perturbative diagrams into simpler structures.

It is important to continue to develop new techniques which allow higher order calculations to be applied to more complicated processes, for example, multi-jet production, and which extend present calculations to ever higher order. It would be very interesting to have corrections at order ${\alpha_S}^4$, one order beyond what is presently available, for a number of different processes. At this order the approach of the perturbation series toward its asymptotic form should be emerging. In addition, calculations at this order are important in matching perturbative and nonperturbative contributions and for understanding scale fixing.

4.5 Jets

Because of confinement in QCD, individual quarks and gluons cannot be isolated. However, the short distance and short time part of a high-energy hard reaction is described in terms of quarks and gluons. At later times and longer distances these quarks and gluons evolve into well-collimated sets of high-energy hadrons called jets. Thus, a jet is the directly measurable quantity which comes closest to a quark or gluon. Measurement of jet cross sections are, effectively, measurements of quark and gluon scattering cross sections. Indeed, the early jet observations made at SPEAR, and a little later but more clearly at PETRA and PEP, were crucial in establishing quarks and gluons as "real" entities.

Understanding jet production and identification at a very refined level is important for many diverse reasons:

- Jets can be a background masking new physics.
- Jets may be useful in discovering new particles, such as the Higgs Particle.
- A discrepancy between theory and experiment in extremely high p_t jet production would give the earliest signal for quark substructure. Present limits from Fermilab indicate that quarks are point-like down to 1.4×10^{-17}cm.

The QCD theory of jet production and jet evolution into hadrons has developed over the last 20 years into a sophisticated and precise description of most aspects of jet physics. The early stages of jet evolution are well described by coherent QCD branching. General properties of this branching such as the momentum dependence of single particle distributions, the average multiplicity of produced hadrons and multiplicity fluctuations are known in analytic form. Detailed properties of jet evolution can be obtained from Monte Carlo implementations of coherent QCD branching. The very final stages of jet evolution, when quarks and gluons turn into hadrons, is not well understood in a fundamental way. However, there are good phenomenological descriptions of this conversion of partons to hadrons which once fixed can be used for jets at all energies independently of how the jets are produced.

Comparison between theory and experiment in jet physics has reached a level of accuracy of about 10 percent, a remarkable achievement in strong interaction physics. Major questions, however, need to be better understood. The identification of jets with quarks and gluons is clearly not perfect since jets have no color. It is important to understand the fundamental limitations to jet physics. For example, in present jet algorithms there are $1/Q$ (with Q the energy of the jet) corrections to perturbative predictions which depend on the details of the final fragmentation of quarks and gluons into hadrons. Is it possible to find algorithms which eliminate these corrections? Perhaps there are observables, unrelated to our present view of jets, which more efficiently express the physics of hard parton-parton scattering.

4.6 Exclusive Processes

Exclusive processes such as wide angle elastic hadron-hadron scattering and elastic lepton nucleon scattering do not yet furnish precise tests of QCD but they have some unique and unusual features which illustrate essential features of gauge theories[62].

It is generally agreed that perturbative QCD determines the hard part of an exclusive reaction at sufficiently high energy. There is not general agreement as to what energy is necessary for perturbation theory to apply, however. One part of the problem is knowing the exclusive wavefunctions of hadrons while another part of the problem is to better understand Sudakov effects which help an early application of perturbative QCD by forcing the valence quarks to be in a more point-like configuration than otherwise might be expected. There has recently been considerable progress in understanding Sudakov effects. There is not yet consensus, however, on the shape of exclusive wavefunctions. This is a difficult nonperturbative problem but one for which considerable progress should be possible using lattice gauge theory and light-front quantization techniques.

Whether or not short distance effects dominate exclusive processes at moderate momentum transfers can be tested experimentally. A hadron in a point-like configuration of its wavefunction should interact weakly, through its small color dipole moment, with nuclear matter. This is the phenomenon of "color transparency." By measuring quasi-elastic reactions in nuclei the effective sizes of hadrons partaking in a hard exclusive reaction can be determined. Early experiments at Brookhaven and SLAC have given intriguing but inconclusive results[63]. New results are expected soon from Brookhaven and Fermilab. These experiments and the physics issues which they shed light on lie at the interface of particle and nuclear physics. With vigorous experimental programs at Brookhaven, Fermilab, and with a somewhat upgraded energy at CEBAF, a good understanding of the space-time regions involved in hard exclusive processes should emerge in the next five to ten years.

4.7 The QCD Perturbation Series

The perturbative expansion, for measurable quantities, is certainly not a convergent expansion in QCD[64]. This is signalled by the presence of factorially growing coefficients. At present three known sources of $n!$ terms, multiplying ${\alpha_S}^{n+1}$ in the perturbation series, are known. They are ultraviolet renormalons, infrared renormalons and instanton-anti-instanton pairs. Instanton-anti-instanton pairs reflect the factorial behavior due to graph counting, while ultraviolet and infrared renormalons reflect ultraviolet and infrared renormalization logarithms in internal Feynman integrals. Ultraviolet renormalons do not hinder a reconstruction of observables from perturbation theory by using a Borel transform; however, infrared renormalons and instantons indicate that QCD perturbation theory is missing essential information.

For certain quantities like the total electron-positron annihilation cross section into hadrons, the τ-lepton hadronic width and the Bjorken sum rule the important $n!$'s, due to the leading infrared renormalons, can be removed and the missing nonperturbative information inserted using higher twist terms called QCD condensates[65]. However, the questions of scheme dependence and uniqueness have not yet been answered. Also, there are quantities like the Drell-Yan cross section where infrared renormalons are especially important but do not seem to be related to normal QCD condensates.This same phenomena occurs in heavy quark effective theories. These are far from academic problems. In many cases higher twist terms are important for an accurate description of data, and it is now apparent that the higher twist terms and high orders of perturbation theory are inextricably related.

Partly stimulated by work on possible baryon number violating effects in the electroweak theory at high energy, there has been renewed interest in instantons in intermediate energy interactions. It is clearly important to clarify whether or not knowledge of specific instanton-anti-instanton configurations provides sufficient information to remove the $n!$'s due to graph counting in QCD. There has been some early work in this direction.

Work on the nature of the QCD perturbation series and connections to higher twist terms is exciting and it can be expected that much progress will be made in the next five-ten years. Such work concerns the very nature of asymptotically free field theories. At the same time practical applications are at hand.

4.8 Small-x Physics

Small-x physics has recently become one of the most active areas of QCD research both for theorists[66] and for experimenters. This field has been greatly stimulated by recent data at HERA showing a rapid rise of the proton structure function, F_2, with decreasing x.

Perhaps the most important goal in small-x physics is to reach high enough energies (small enough x) so that parton densities in the proton become sufficiently large that the parton picture breaks down and a new strong field regime of QCD is reached. This is often referred to as the parton saturation regime of QCD. This limit of QCD has many features in common with the early stages of formation of the quark-gluon plasma which is expected in high-energy heavy ion collisions. However, in the small-x case the focus is on partons having transverse momentum in the multi GeV range while in heavy ion collisions dense partonic systems with such high transverse momentum partons are unlikely.

A second goal is to study the approach to the dense partonic regime. On theoretical grounds it has long been expected that parton distributions would exhibit the rapid growth at small values of x which has now been seen at HERA. Such a rapid growth of a cross section cannot be sustained over too large a range in $\ln(1/x)$ before unitarity corrections become important. Such corrections may be important in the energy regimes available at HERA and Fermilab. These corrections are expected to dampen the small-x growth.

In many respects the small-x problem, including unitarity corrections, resembles that of solving a two-dimensional field theory. There are now vigorous attempts in the particle physics and mathematical physics communities to reformulate small-x behavior in terms of a definite two-dimensional field theory.

Small-x behavior also has much in common with the high-energy behavior of the total cross section for hadron-hadron scattering, a process clearly dominated by nonperturbative QCD. Indeed, as Q^2 becomes small the small-x behavior of F_2 is expected to weaken and closely resemble the slow growth in energy of total hadronic cross sections. Thus, progress in small-x behavior is likely to lead to a renewed interest in trying to understand high-energy hadronic scattering at small momentum transfer, a process for which there is a highly developed, and successful, phenomenology but not yet a real connection to the fundamental degrees of freedom in QCD.

Other processes closely related to small-x behavior are diffractive scattering in deep inelastic lepton-proton interactions, recently measured at HERA, and hadronic collider events having large rapidity gaps bordered by jets, recently seen at Fermilab.

Small-x physics is still in its infancy. Its relations to heavy ion physics, mathematical physics and soft hadron physics along with a rich variety of possible experimental signatures make it central for QCD studies over the next decade.

5 QCD and Dense Matter

5.1 Introduction

Dense states of hadronic matter are believed to occur in nature in three places: in atomic nuclei, in the interior of neutron stars, and within the first millisecond of the evolution of our universe. The first of these, atomic nuclei, are generally the subject of the discipline of nuclear physics, but they are also of interest to high-energy physicists because, as we shall discuss below, there exist certain aspects of perturbative QCD that can be studied more advantageously in interactions between hadrons and nuclei than in hadron-hadron collisions. Hadronic matter in neutron stars and in the early universe can be characterized by two bulk thermodynamic quantities: the temperature T and the baryon chemical potential μ_B. In the early universe hadronic matter was essentially baryon symmetric ($\mu_B/T \approx 10^{-9}$), whereas in the inner core of neutron stars thermal effects are negligible ($T/\mu_B < 10^{-5}$).

Quite general arguments indicate that the internal structure of hadronic matter changes drastically when $\mu_B, T \gg \Lambda_{QCD}$, because the interactions between quarks and gluons become weak at high densities. The high density phase, usually called the *quark-gluon plasma*, is characterized by a lack of spontaneous chiral symmetry breaking and of local quark confinement, thus its properties are very different from those of low density hadronic matter, which is well described as a gas of weakly interacting hadrons.

The quark-gluon plasma is of more than academic interest, because conditions required for its existence can, most likely, be produced in highly energetic collisions between heavy nuclei that will be studied before the end of the present decade at the Brookhaven Relativistic Heavy

Ion Collider (RHIC), and later at even higher energy at the CERN-Large Hadron Collider (LHC). The physics of nonabelian gauge fields at high temperature, which is of central importance in the evolution of the early universe, is experimentally accessible only in the context of QCD. Methods developed theoretically and results obtained experimentally at RHIC will be seminal for other areas of particle cosmology. This aspect, together with the potential for the discovery of new forms of hadronic matter, underpins the importance of RHIC for particle physics.

Our discussion of the current status of the field, and of the immediate and long range goals will focus on three conceptually different, but interrelated issues: Hadron-nucleus interactions at high energy, QCD at finite temperature, and QCD aspects of relativistic heavy ion collisions.

5.2 QCD and Hadron-Nucleus Interactions

5.2.1 Anomalous A-dependence

Hadron-nucleus collisions have played a key role in testing QCD as a theory of the strong interaction. Although first discovered almost twenty years ago, anomalous nuclear dependence in hadron-nucleus collisions [67,68] received little attention until very recently. It is now recognized that it represents a unique physical observable that cannot only provide new tests of QCD dynamics, such as multi-parton correlations which are normally difficult to measure in hadron-hadron collisions, but also probe the nuclear structure at high energy, which is crucial for understanding and extracting quark-gluon plasma signals from relativistic heavy ion collisions.

The A-dependence in jet and single particle production is also known as the "Cronin effect". The invariant cross ection for single particle production in hadron-nucleus collisions is proportional to A^{α} where α is a function of the momentum of the produced particle and can be as large as 1.3 [69,70]. Theoretically, it is agreed that multiple scattering is responsible for the observed anomalous nuclear dependence [71]. Until recently, calculations for the multiple scattering were based on addition of a number of single Born cross sections, and consequently, the results were sensitive to the infrared cutoff [72]. It was recently shown that double scattering can be consistently treated in perturbative QCD in terms of the factorization where the hard cross section is infrared safe and non-perturbative matrix elements are well-defined [73].

As already discussed in Sections 3.5 and 3.9, J/ψ suppression is considered as a possible signal for the formation of a quark-gluon plasma in relativistic heavy ion collisions. Experiments at Fermilab and CERN have revealed a strong J/ψ suppression in hadron-nucleus collisions [74,75], possibly invalidating the usefulness of J/ψ suppression as a quark-gluon plasma signal. Several models have been proposed to explain the nuclear suppression, which are based on completely different physical pictures, but seem to explain the data equally well [76,77,78]. This makes it difficult to extrapolate the observed suppression to nucleus-nucleus collisions. More work is needed before the prospects of J/ψ production as a quark-gluon plasma signature can be reliably assessed [79].

5.2.2 Multiparton Correlations

The predictive power of perturbative QCD as applied to hadrons relies on the factorization theorem which states that a physical observable can be factorized into a series of perturbative QCD cross sections and nonperturbative, but universal parton distributions. The well-known twist-2 parton distributions, which can be interpreted as the probability densities for finding a parton within a hadron, provide the leading term in this expansion. The next-to-leading term is given by the twist-4 matrix elements, which are known as four-parton correlation functions [73,80]. Knowledge of these correlation functions is crucial for understanding the non-perturbative aspects of QCD beyond the probabilistic parton distributions.

The actual size of the non-leading terms is very important for precision tests of QCD, in particular, for tests of QCD beyond the simple collisions between two partons. Because of the dominance of the leading term, it has been difficult to extract information on the non-leading terms from reactions involving isolated hadrons. It is here where nuclear targets can be of help [81].

A strong nuclear dependence in the two-jet momentum imbalance, known as acoplanarity, has been found in hadron-nucleus as well as in photon-nucleus collisions [82,83]. Since the acoplanarity due to radiation of additional partons should be the same in hadron-hadron and hadron-nucleus collisions, it is not expected to exhibit a strong nuclear dependence. The observed anomalous effect must therefore be caused by multiple scattering of final state partons while they travel through nuclear matter. This fact can be utilized not only to derive parton-parton correlations in the nuclear structure functions, but also to provide information on interaction between a colored parton with nuclear matter [84].

5.2.3 Color Transparency

In hadron-hadron elastic scattering at large angles, only small-size configurations contribute. The nucleus can act as a filter permitting only these small-size components of the incident hadron wavefunction to penetrate deep into the nucleus and scatter elastically off nucleons in the core of the nucleus [85,86]. This phenomenon, commonly known as "color transparency", was already discussed in Section 4.6. After the hard collision, the scattered hadron will initially also be in a small-size configuration which eventually expands into the full sized final state hadron. The nuclear dependence of elastic scattering can therefore provide information on the formation time of the final state hadron [87].

5.2.4 Nuclear Shadowing

At small x, the parton densities in a nucleus become so large that the sea quarks and gluons overlap spatially and the nucleus cannot be viewed as a collection of uncorrelated nucleons. This occurs when the longitudinal size of a parton, in the infinite momentum frame of the nucleon, becomes larger than the size of the nucleon. Partons from different nucleons then interact and annihilate, effectively reducing the parton density in the nucleus. This shadowing effect can be described by nonlinear corrections to the parton evolution equations that describe the parton fusion processes. Whereas a considerable amount of information about the shadowing of sea quarks exists today [88,89], the experimental and theoretical understanding of gluon shadowing is still quite limited.

Parton distribution functions at small x in very large nuclei may be systematically calculable within the framework of perturbative QCD, because there exists a new large scale parameter in the area density of valence quarks. This allows for a systematic exploration of the onset of screening of partons and the consequent shadowing of parton distributions observed in deep inelastic scattering on large nuclei [90]. This new approach could also prove to be useful as basis for a perturbative description of particle production in the central rapidity region in collisions between heavy nuclei.

5.3 QCD at Finite Temperature

5.3.1 The QCD Phase Transition

The case of baryon-symmetric hadronic matter ($T \neq 0$, $\mu_B = 0$), which is relevant to the early universe and to heavy ion collisions at collider energies, is much better understood than the general case of dense hadronic matter (T, $\mu_B \neq 0$), because computer simulations of lattice QCD have been feasible for this case. The numerical results (see Figure 5) show that bulk thermodynamic quantities, such as energy and entropy density, exhibit a strong structural change – probably a true phase transition – around a critical temperature $T_c \approx 150$ MeV for QCD with two light quark flavors [91]. The transition is accompanied by a rapid change in the strength of the light quark condensate, which plays the role of the chiral order parameter.

The order of this chiral phase transition is of considerable phenomenological importance. Universality arguments predict that there is a second-order transition

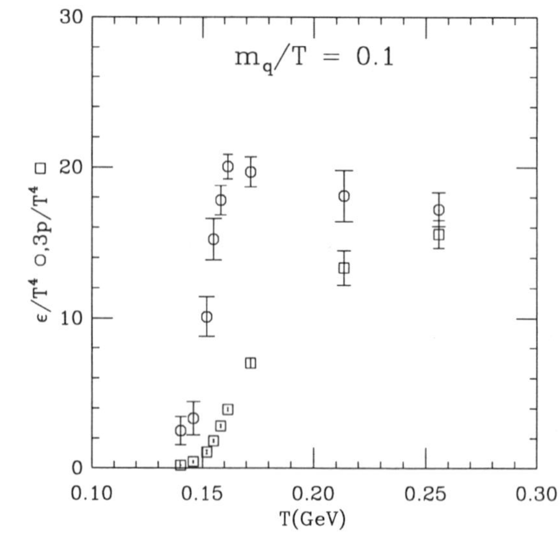

Figure 5: Energy density and pressure for two-flavor QCD on lattices with temporal extent $N_\tau = 4$. (From Ref. [91]).

for QCD with two massless quark flavors [92,93]. As discussed in more detail in Section 6.2.3, lattice results for two very light quark flavors and one medium heavy quark flavor confirm this by not revealing hysteresis effects, at least for presently accessible lattice sizes [94,95]. On the other hand, bulk thermodynamic quantities exhibit a very rapid change in the vicinity of T_c, indicating the possibility of a weak first-order transition or of a rapid cross-over between regimes with different structure but without singular behavior of thermodynamic quantities. Simulations on much larger lattices, of order 100 lattice spacings in each of the spatial directions, are required in view of the large mismatch between the light quark masses and the QCD scale parameter Λ.

The development of numerical techniques for the study of lattice-QCD at nonvanishing baryon chemical potential remains a long term goal which, if it could be achieved, would significantly improve our understanding of the physics of neutron star cores and of relativistic nuclear collisions at energies accessible in fixed target experiments.

5.3.2 Hadrons in Medium

The structure of the low-temperature phase appears to be well described as a weakly interacting gas of mesons, in good agreement with results obtained by chiral perturbation theory and by methods based on QCD sum rules. The main potentially observable effects are shifts in hadron masses, with the general tendency to restore degeneracy of parity doublets, and changes in the width of unstable mesons. The medium modifications appear to be especially sensitive to an increase in the net baryon density. It is possible that dense cold baryon matter contains a strong component of strange hadrons, maybe in form of a kaon condensate. Thermal excitations are predicted to be less effective in inducing medium modifications of hadrons, except in the immediate vicinity of the phase transition [96].

5.3.3 Quark Gluon Plasma

Most numerical results of lattice gauge theory for the high temperature phase are in good agreement with the picture of a weakly interacting quark-gluon plasma, and are well described by thermal QCD perturbation theory, if the nonperturbative changes in the quark and gluon condensate are taken into account. The energy of an isolated quark at $T > T_c$ is finite and its color charge is screened at short distance by plasma polarization.

However, some nonperturbative features survive even at very high temperature. In particular, space-like Wilson loops continue to obey an area law, in agreement with arguments based on dimensional reduction to three-dimensional QCD [97,98,99]. As a result, space-like correlators of hadron currents remain strongly localized although color charges are deconfined [100]. This puzzling phenomenon occurs at the energy scale g^2T (g is the QCD coupling constant), where the thermal perturbative expansion diverges due to the absence of screening of static color-magnetic fields in perturbation theory [101]. However, the survival of spatial correlations appears to have a negligible effect on thermodynamic quantities and on the propagation of isolated quarks according to the lattice simulations. The phenomenological implications of the survival of space-like hadronic correlations in the high temperature phase require further study.

Several other phenomenologically relevant transport properties of the quark-gluon plasma are also known to be sensitive to the physics at the scale g^2T, in particular, various dissipation rates such as color conductivity [102], the damping rate of collective excitations [103], and the entropy growth rate [104]. A better theoretical understanding of the infrared limit of thermal gauge theories and of the physical mechanisms operating at the inverse length scale of order g^2T is needed [105].

5.4 QCD and Relativistic Heavy Ion Collisions

Relativistic heavy ion collisions afford the unique possibility to study QCD at high density in the laboratory. The basic questions are whether a quark-gluon plasma can be formed in these reactions, and if so, for what range of energy and nuclear size, and how it can be detected experimentally. The collision between two nuclei is an enormously complicated problem, since one has to deal with the evolution of a complex system of many particles from an almost pure initial quantum state into a very complicated final state best described by a statistical ensemble. This is a fundamental problem in statistical physics, here with the added difficulty that relativity is a crucial ingredient. The solution of the conceptual and technical difficulties posed by relativistic heavy ion collisions will bring many new insights and should help us better understand some basic aspects of statistical physics.

5.4.1 Thermalization

There has been a great deal of progress in the past few years in our understanding of the processes leading to the formation of a thermalized QCD plasma in a nuclear collision, and it now appears that perturbative QCD can be applied to describe the formation process [106,107], at least at high collision energies where perturbative interactions among partons, often called mini-jets, constitute an important contribution to the energy deposition mechanism. The evolution from the initial parton distributions of the colliding nuclei into a locally thermalized phase-space distribution of quarks and gluons in the central rapidity region can be visualized as a multiple parton cascade [108]. An example of the particle multiplicity distributions predicted by theoretical models (here the HIJING model [109]) based on these concepts is shown in Figure 6.

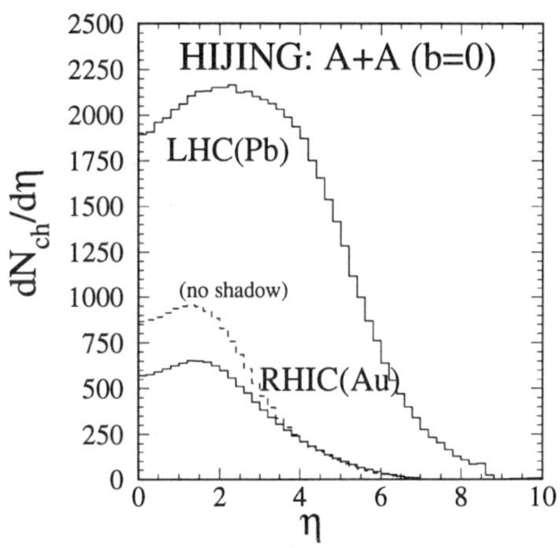

Figure 6: Predictions of the HIJING model [109] for charged particle multiplicity distributions versus pseudorapidity for central Au+Au collisions at RHIC and Pb+Pb collisions at the LHC.

The parton cascade model presently relies on two parameters that artificially cut off the infrared singularities of perturbative QCD cross sections: the minimal momentum transfer in parton-parton interactions, $p_T^{min} \approx 2$ GeV/c, and the minimal virtuality in radiative processes, $\mu_0 \approx 0.7$ GeV, which are determined from hadron-hadron collision phenomenology. These parameters may have to be modified before they can be applied to nuclear collisions. Recent theoretical studies [110,111] indicate that the infrared singularities can be regulated dynamically by medium effects in relativistic heavy ion collisions, due to color screening and collision-induced radiation suppression. If this can be implemented into the parton cascade approach, parameter independent predictions for the initial phase of high-energy nuclear collisions would be possible.

5.4.2 Quark-Gluon Plasma Signatures

As mentioned in Section 3.9.1, the experimental observation of the formation of a quark-gluon plasma relies on a variety of probes carrying information about the structure of the deconfined, chirally restored phase. Quantitative predictions for these signatures are quite sensitive to the dynamics of the thermalization and consecutive evolution of the plasma.

Numerical studies of partonic cascades have shown that the phase space distribution of gluons becomes quasi-isotropic on a timescale of 0.5 fm/c, through a combination of two-body collisions, radiative processes, and free-streaming separation of partons with different rapidities. Chemical equilibration of parton flavors takes considerably longer, of the order of 2 fm/c. The short time-scale for the thermalization of gluons implies the formation of a very hot ($T \geq 500$ MeV) initially gluon-dominated QCD plasma. This "hot glue" scenario has important phenomenological consequences: the spectra of electromagnetic signals (lepton-pairs and photons) are predicted to be much harder than previously thought, prevailing over backgrounds up to several GeV/c in transverse momentum or invariant mass, and there may be substantial thermal and epithermal contributions to charmed meson yields [112].

These signals constitute experimental probes of the earliest stages of the nuclear collision. Other proposed signatures for the formation of a quark-gluon plasma are the suppression of heavy vector mesons ($J/\psi, \psi', \Upsilon$), the enhanced production of strange hadrons (especially strange antibaryons), and kinematic consequences of the existence of a mixed phase. A comprehensive assessment of these signatures in nuclear collisions is needed, in order to find the best combination of signals that can establish the formation of a locally color-deconfined phase with restored chiral symmetry [113].

5.4.3 Hadronization and chiral transition

As the quark-gluon plasma expands and cools, it eventually reaches the hadronization threshold. Present wisdom is that the hadronization (chiral) phase transition is of second order and hence allows for long range fluctuations of the chiral order parameter. Numerical studies have recently shown that, if this transition proceeds out of thermal equilibrium, fluctuations in temporarily unstable infrared modes can develop into extended domains characterized by a misaligned chiral order parameter [114,115,116]. Such a "disorientated chiral condensate" would eventually decay into a collective low-energy excitation of the pion field resembling an isospin singlet pion laser. This scenario would have dramatic experimental signatures, most notably a highly unusual charge ratio of the emitted pions, with a large probability for the strong depletion of neutral pions (see Section 3.9.1).

The dynamics of the chiral phase transition out of equilibrium is still poorly understood. Considerable progress has been recently made through a combination of lattice simulations and modelling of instanton effects on the origin of chiral symmetry breaking in QCD [117]. More work is required to extend these studies to dynamical properties of the quark condensate in the vicinity of the phase transition.

5.4.4 Strange quark matter

The baryon-rich quark gluon plasma, which may be formed in nuclear collisions at presently accessible energies, is characterized by a high content of strange quarks. The strange quark concentration can be further increased by distillation mechanisms during hadronization, opening exciting prospects for the formation of multiple hypernuclei and, quite possibly, of multi-strange quark matter droplets that might have a long life-time. Experimental searches for such new metastable states of nuclear matter ("strangelets") are under way [118,119,120]. Improved models for the description of hadronic matter with a high strangeness content, possibly based on ideas of chiral perturbation theory, need to be developed.

5.4.5 Nonperturbative effects

Semiclassical studies indicate that there exist efficient nonperturbative mechanisms that transfer energy from high-energy to low-energy modes in gauge theories. A better understanding of the meaning of (semi-) classical solutions in the framework of quantum field theory could shed light on this issue, which has potential implications, besides QCD, for nonperturbative electroweak processes, such as multiparticle production at ultrahigh energies and baryogenesis. A long term goal is to find a practicable scheme for the study of nonequilibrium processes in the framework of lattice QCD. This would also facilitate investigations of the nonequilibrium aspects of the hadronization process, which hold considerable promise of providing unambiguous signatures for the occurrence of a phase transition in relativistic heavy ion collisions.

5.5 *Effective Theories of QCD*

5.5.1 Chiral perturbation theory

Low energy effective theories of QCD are a powerful tool in strong interaction physics. They have been successfully applied to electroweak decays of hadrons [121,122], to low energy hadron scattering [123,124], and most recently, to few nucleon systems [125,126,127]. Many of these theories make use of the approximate chiral invariance of the QCD Lagrangian caused by the near masslessness of the light quark flavors u, d, and to a lesser extent, s. The gradient expansion of the chiral field allows for a systematic perturbation theory approach, at energies much lower than the chiral symmetry breaking scale $\Lambda_\chi \approx 1$ GeV, which conserves all symmetries of full QCD.

There are a number of interesting new results in QCD at low energy. The properties of the pseudoscalar mesons (π, K, η) have been studied in great detail using chiral perturbation theory to two loop order. There are a large number of free parameters which must be fit, but there are a far greater number of experimental measurements which can be predicted. The results are in excellent agreement with the experimental data. Previous discrepancies, such as in $\gamma\gamma \to \pi^0\pi^0$ have all been resolved using the new two-loop results [128].

Chiral perturbation theory has also been successfully applied to the baryon sector. It was thought that chiral corrections to the baryons were large, and that chiral perturbation theory might not be that useful for the baryons. However, it has been shown how one can consistently apply chiral perturbation theory to the baryons, with small corrections [129]. The basic idea is that one must include both the nucleon and the Δ resonance in the chiral Lagrangian, since there is a large $\Delta N\pi$ coupling. The loop effects of intermediate $\Delta's$ and nucleons cancel, so that the radiative corrections are under control. The baryon calculations are still in their infancy, compared with the sophisticated results obtained for the meson sector.

5.5.2 Large N_c expansion

The $1/N_c$ expansion (N_c is the number of colors) provides a means of computing hadronic quantities in a systematic expansion in QCD [130,131]. In the mesonic sector, the $1/N_c$ expansion can be used to obtain relations between chiral Lagrangian parameters that are well satisfied experimentally [132]. The $1/N_c$ expansion has also been used to justify certain features of QCD inspired models of hadrons, such as the Nambu-Jona-Lasinio model. Recently, the $1/N_c$ expansion has been used to obtain systematic predictions for baryons in QCD [133]. Some results, which are proved up to order $1/N_c^2$, are in excellent agreement (15%) with experiment.

5.5.3 Applications

QCD is known to provide an excellent description of strong interaction dynamics. The computations of low energy hadron properties at the 10% level are not accurate enough to provide a precision test of QCD. Nevertheless, these calculations are very important, and should be pursued in the future. Firstly, tests of the electroweak theory invariably require knowledge of hadronic matrix elements, and these can be calculated using the methods discussed above. For example, the prediction of rare K decay rates or ϵ'/ϵ in the standard model requires knowing the matrix elements of 4-quark operators between K and π states. The sensitivity to deviations from the standard model is limited by how accurately one can calculate these hadronic matrix elements. Secondly, QCD provides us with a testing ground for the theoretical methods used to study non-perturbative effects in gauge theories. Non-perturbative effects may well be responsible for electroweak symmetry breaking (either through technicolor, or by dynamical supersymmetry breaking). It is important to test the theoretical ideas on QCD to see how well they work, before applying them to new theories.

6 Lattice QCD

6.1 *Introduction*

Many important features of QCD lie outside the reach of perturbation theory. In order to study them one must resort to non-perturbative techniques. In particular, one must be able to regularize and renormalize the theory in a non-perturbative manner. The formulation of QCD on a lattice in Euclidean space-time, introduced by Wilson in 1974, provides such a regularization and establishes a powerful framework for studying the non-perturbative properties of QCD and other quantum field theories. Since its introduction, lattice QCD has formed the basis for a very large number of investigations of hadron properties. Of special importance has been the fact that the lattice regularization permits the application of numerical simulation techniques to the analysis of the quantum fluctuations. These have been used successfully to derive several quantitative predictions from the first principles of QCD[134].

Lattice QCD does not hold the unique key to the study of the non-perturbative properties of hadrons, even if we add the challenging constraint of a meaningful regularization of the ultraviolet divergences. For instance, a calculation based on the expansion into quantum fluctuations around a semiclassical solution of $O(1/g)$, even if the expansion takes a perturbative form, would embody non-perturbative effects. However, because of the very special role that the lattice formulation has played in the study of non-perturbative QCD phenomena and because of the many results that have been obtained through its application, this entire section will be dedicated to it.

6.2 Lattice QCD in Euclidean Space-Time

6.2.1 Methodology

The most important aspect of lattice QCD is that it provides a gauge regularization of the ultraviolet divergences which does not require a gauge fixing. This is accomplished by taking finite elements of the gauge group, rather than the gauge potentials $A^i_\mu(x)$ which are elements of the gauge algebra, as dynamical variables. These finite elements of the gauge group $U^{c,c'}_\mu(x)$ are color $SU(3)$ matrices and are defined over the oriented links of a lattice in Euclidean space-time[135]. In most applications this is a hypercubical lattice with lattice spacing a. Matter fields (the quark fields ψ, $\bar\psi$ in QCD) are defined over the sites of the lattice. The gauge field variables $U_\mu(x)$ and the quark fields are combined into gauge invariant expressions, which form the building blocks of the discretized space-time action. This consists of a pure gauge term $S_g(U)$, which reduces to $\int \frac{1}{4}F_{\mu\nu}F^{\mu\nu}d^4x$ in the continuum limit, and of a matter field term $S_q(\psi,\bar\psi,U)$, which discretizes the Dirac term of the continuum action. In terms of these variables the quantum expectation value of any observable is given by

$$\langle O\rangle = Z^{-1}\int dU\, d\bar\psi\, d\psi\, O(U,\bar\psi,\psi)\, e^{-S_g-S_q} \qquad (2)$$

with

$$Z = \int dU\, d\bar\psi\, d\psi\, e^{-S_g-S_q} \qquad (3)$$

If one considers a system of finite space-time volume V at first, letting $V\to\infty$ at the end of the calculations, the integrals in the two equations above are integrals over a finite, albeit very large, number of variables. These integrals are either over a compact domain (for the group elements U) or over Grassman variables ($\bar\psi$, ψ), and thus they represent mathematically well-defined, finite quantities. Since O can be any observable, Eqs. (2) and (3) provide in principle the description of all QCD phenomena. Of course, in principle is the keyword. Although the integrals are well defined, they are quite complex and calculating them, even in an approximate manner, is a formidable task. Moreover, at the end of the calculation the regulator given by the finite lattice spacing a must be removed in order to obtain continuum results. This is done by readjusting the coupling constant g which appears in S_g and determines the strength of quantum fluctuations of the gauge field. g plays the role of a bare coupling constant. In the process of renormalization g and a are sent simultaneously to zero, with a functional relation $a = a(g)$ determined in its leading orders by asymptotic freedom, in such a way that all physical observables tend to a finite limit[136].

The regularization of QCD given by Eqs. (2) and (3) is non-perturbative and permits the implementation of many calculational techniques, frequently similar to techniques used in statistical mechanics, which are not available in the more conventional perturbative schemes of renormalization. Of particular importance is the possibility of applying powerful computational methods to an approximate calculation of the quantum expectation values.

The computational analysis of lattice QCD proceeds first through the integration over the quark fields $\bar\psi,\psi$, which can be done explicitly because the matter part of the action S_q is bilinear in the quark fields. This leads to integrals over the gauge variables only

$$\langle O\rangle = Z^{-1}\int dU\, \langle O\rangle_U\, e^{-S_{eff}} \qquad (4)$$

where $\langle O\rangle_U$ stands for the average of O over the quark field fluctuations alone, in the background provided by the gauge field U, and

$$S_{eff} = S_g - \log\det[D(U)] \qquad (5)$$

$D(U)$ being the lattice Dirac operator that appears in S_q.

Because of γ_5 invariance, $\det[D(U)]$ is a positive, semidefinite quantity. $e^{-S_{eff}}$ can therefore be taken as a measure factor in the space of the gauge variables $U_\mu(x)$ and the integrals giving $\langle O\rangle$ can be approximately calculated by numerical simulation techniques. This is the essence of the computational methods underlying the majority of the numerical studies of QCD performed in the past, or envisioned for the future. There are, however, some important remarks which must be appended even to the most concise description of the methodology of lattice QCD.

• Numerical simulations techniques proceed by averaging over a very large number of "configurations" of the system (in our case the collection of all $U_\mu(x)$) distributed according to the desired measure. These are obtained through repeated "upgrades" of the dynamical variables $U_\mu(x)$, in which these are either individually or collectively replaced by new values, according to some definite stochastic or deterministic algorithm. Since the number of dynamical variables is huge and the number of upgrades required for reasonably accurate averages can also be very large, it is crucial that the upgrades be done by the computer as rapidly as possible. With a measure factor that involves only couplings between neighboring variables, such as the exponential of the pure gauge part of the action e^{-S_g}, an individual upgrade requires a small number of arithmetic operations (these can range in the thousands, but this is still a small number with respect to the typical number of dynamical variables and to the overall scale of the computation) and, in any case, independent of the volume of the system. But this is no longer the case when the non-local $\det[D(U)]$ is incorporated in the measure. Algorithms to account for the effects of the fermionic determinant, either in an approximate manner or exactly, have been introduced and are routinely applied. They require a few orders of magnitude ($10^2 - 10^4$) more arithmetic operations than are needed with a local measure alone. All of this has prompted the use of an approximation, called the quenched or valence approximation, whereby the gauge field configurations are generated according to the pure gauge measure factor e^{-S_g}. Since in field theoretic terms the fermionic determinant accounts for the creation and annihilation of virtual quark-antiquark pairs, the quenched approximation consists in neglecting $q - \bar{q}$ vacuum polarization effects. Various arguments can be given to support the validity of such approximation. Also, there is a considerable effort in lattice QCD investigations to go beyond the quenched approximation. It is a fact, however, that many computational analyses of QCD, especially those aiming at the largest lattices or smallest quark masses, have been or are currently based on the quenched approximation.

• There are some notorious problems in the lattice discretization of the continuum Dirac operator. It is not possible to define a lattice Dirac operator with the formal chiral properties of the continuum one[137]. There are formulations of the lattice Dirac operator which permit meaningful simulations of QCD, but one pays with either an explicit breaking of chiral symmetry (Wilson formulation)[135], which must be recovered through the careful tuning of a mass counterterm, or with a breaking of flavor symmetry, which is restored only in the continuum limit, and limitations on the possible number of flavors (Kogut-Susskind or staggered formulation)[138,139].

• The γ_5 invariance which guarantees the reality of $\log\det[D(U)]$ and makes it possible to incorporate the fermionic determinant in the measure is no longer true in presence of a quark chemical potential. Thus numerical studies of QCD at finite baryon number density, while not impossible, are computationally much more demanding and the results much more approximate.

• Perturbative techniques can also be applied to the lattice formulation of QCD. The perturbation expansion is more complicated on the lattice than in the continuum because of the loss of Lorentz invariance, but can still be carried out. Lattice perturbative calculations are important and have been done to determine crucial renormalization parameters and to establish a bridge to the more conventional perturbative results[140,141].

6.2.2 Hadron spectroscopy

Among the non-perturbative observables of QCD hadron masses occupy a very prominent role. Accordingly, through the years many lattice calculations have been devoted to the calculation of the hadron spectrum. One considers an observable $O(t)$ with non-vanishing matrix elements between the vacuum and the states with the quantum numbers of the hadron whose mass is being sought. O can therefore act as source for creation of the hadron, $\bar{O}$ as a sink for its annihilation. Typically O will consist of a quark-antiquark bilinear for the calculation of meson masses, a product of three quark fields for baryon masses or some expression involving the gauge fields for the study of glueballs. Also, it is convenient to project over states of zero spatial momentum by including into the definition of O a sum over spatial sites (a projection over definite, non-zero spatial momentum can also be easily implemented). One uses then simulation techniques to evaluate the Euclidean correlation function (or Green function) $\langle \bar{O}(t)O(t')\rangle$. On general grounds this is given by

$$\langle \bar{O}(t)O(t')\rangle = \sum |\langle \phi|O|0\rangle|^2 e^{-E(\phi)|t-t'|} \tag{6}$$

where the sum ranges over all physical states with the quantum numbers of O and the exponential fall-off is due to the fact that one is considering Green functions in Euclidean space-time. From a numerical determination of the leading exponential behavior(s) one can then derive the energy (mass, if one has performed a projection over zero space momentum) of the lowest state(s).

The basis for such calculations was established in the early eighties[142,143]. The intervening years have brought, however, very important refinements in the construction of the source (sink) operators O ($\bar{O}$), by which crucial enhancements of the matrix elements between the vacuum and the desired hadron states have been obtained, as well as constant improvements in the scope and accuracy of the calculations[144]. The actual precision is limited by the statistical nature of the calculations as well as by a variety of systematic errors. The latter are due to the finite volume of the lattice, the finite lattice spacing, practical limitations on the quark mass (the rate of convergence of the algorithms for calculating quark propagators becomes prohibitively slow for small quark masses) and to the finite extent in Euclidean time over which one can calculate the Green function with sufficient accuracy (this limits the precision in the calculation of the masses due to mixing with higher states). In regard to this last point, it is to be noticed that the calculation of Green functions for operators built out of quark fields proceeds through an initial calculation of the quark propagators, which are then combined as appropriate and averaged over several gauge field configurations. Of the overall fall off of the Green functions, a large fraction is then due to the fall off of the quark propagators themselves, and only part to the averaging procedure. This is to be contrasted to the case of the Green functions for gluonic operators, where the entire fall off comes from cancellations among quantities which are typically of order one. As a consequence, the masses of states whose Green functions are given by connected quark lines can be determined with better accuracy than the masses of purely gluonic states (glueballs). Even worse is the situations for states that would involve disconnected quark lines, such as admixtures of q-$\bar{q}$ states and glueballs, for which up to now it has been possible to do very little with lattice techniques.

Currently, large scale calculations of the hadron spectrum in the quenched approximation involve lattices with a spatial extent ranging up to 32^3 sites, time extent up to 64 sites, ultraviolet cutoffs ranging up to $a^{-1} \approx 3GeV$ and spatial volumes ranging up to $(3-4\,\mathrm{f})^3$ [144]. The lowest values of the quark mass is best characterized by the corresponding value of the pseudoscalar mass, as it emerges from the calculation. One obtains ratios m_π/m_ρ down to ≈ 0.3 as opposed to the experimental value 0.175. Because of the current algebra relation $m_{ps}^2 \sim m_q$ the square of the pseudoscalar mass is a better indicator of the quark mass and this gives a current value $(m_{ps}/m_{vect})^2 = 0.09$ versus a target 0.03. A lot of attention is being paid to the effects of the finite lattice spacing, finite volume and other sources of systematic effects. Extrapolations based on analyses done for several values of these parameters are used to reach the physical domain. One recent study[145], based on a very careful set of extrapolations over volume, lattice spacing and quark mass, has produced a set of results in excellent agreement with the experimental data (*e.g.*, $m_N/m_\rho = 1.216(104)$ (experimental value 1.222); $m_\Delta/m_\rho = 1.565(122)$ (exp. 1.604); $m_\phi/m_\rho = 1.333(32)$ (exp. 1.327); $f_\pi/m_\rho = 0.106(14)$ (exp. 0.121). In this study m_π, m_K are used to determine the bare quark masses, while m_ρ sets the scale for the lattice spacing. The same study found a value $1740 \pm 71MeV$ for the 0^{++} glueball mass. A slightly lower but not inconsistent value of $1625 \pm 92MeV$ has recently been found by another group[146]. These values pertain to the pure gauge system and do not account for possible mixing with $q\,\bar{q}$ states which, as mentioned above, are much harder to calculate. Figures 7 and 8 illustrate the results obtained for hadron masses and for the scalar glueball decay constant in the study mentioned above[145,147].

There are also many investigations which do not rely on the quenched approximation (full QCD simulations or simulations with dynamical quarks). Since including $\det[D(U)]$ (and thus the effects of virtual q-$\bar{q}$ pairs) in the measure is algorithmically very costly, such calculations are typically limited to lattice sizes about one half of the corresponding quenched calculations. Even more important is the fact that the simulation algorithms can be effectively implemented only with rather large quark masses leading to $(m_{ps}/m_{vect})^2 \gtrsim 0.25$, versus 0.09 for the quenched approximation and the experimental value of 0.03 for light quarks. With such large quark masses

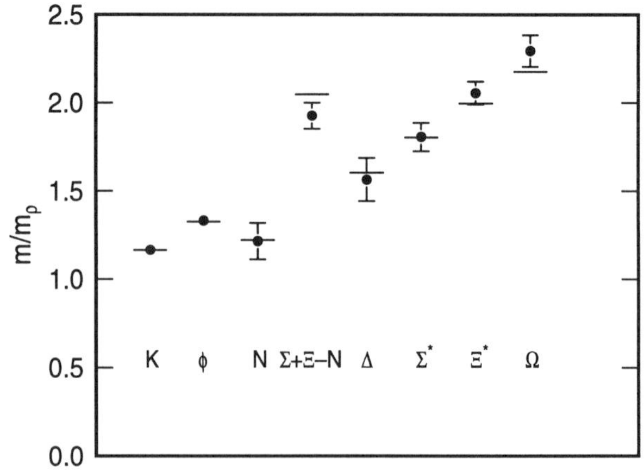

Figure 7: Masses of several hadronic states in units of m_ρ as obtained in a recent large scale quenched calculation. The horizontal lines represent the experimental values.

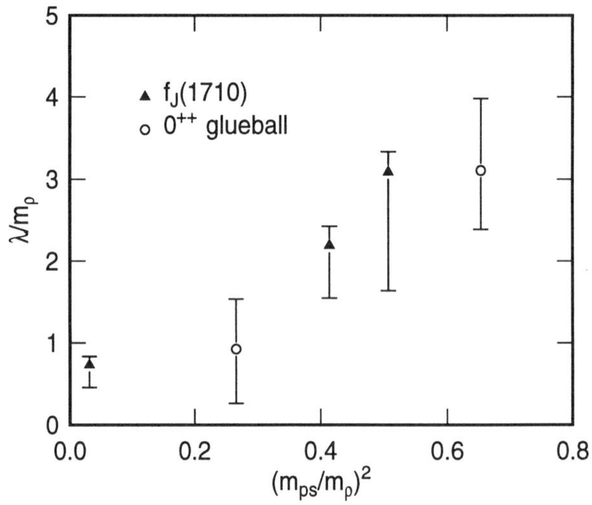

Figure 8: Decay constants of the f_0 (1720) in comparison to lattice results for the scalar glueball.

the whole effect of the fermionic determinant appears to be limited to the renormalization of the bare coupling constant and, moreover, the ratio between the masses of the nucleon and the vector meson stays very close to the heavy quark limit of 3/2. This situation is reminiscent of the earlier quenched calculations, where because of the more modest computer resources and in absence of the recent algorithmic improvements, one was similarly limited to large quark masses. It is very likely that in the near future the progress in non-quenched calculations ill parallel the advances achieved y quenched spectrum calculations during the last few years.

Another set of spectral data of great interest in QCD are the masses of states containing heavy quarks. These are too large for a direct lattice calculation based on the formalism outlined above, but can still be calculated with good accuracy either by using the lattice to evaluate the potential binding the heavy quarks (and the spin dependent potentials) or by developing an effective theory to describe the degrees of freedom of the heavy quarks in a non-relativistic approximation. These approaches give origin to interesting issues of renormalization, where substantial progress has recently been made[148]. The splittings among different states in the heavy quark families can be calculated with precision, and these results can in turn be used to determine the value of the coupling constant α_S. Recently values for $\alpha^{(5)}_{\overline{MS}}$ at M_Z clustering around 0.110, with errors ± 0.008, have thus been found, with the major element of uncertainty coming from the corrections one has to make to the quenched approximation to include short-distance quark polarization effects[149]. It is to be noticed that these calculations of the strong coupling constant are already competitive with those based on perturbative QCD and that the lattice may soon provide the way to produce the most precise determinations of α_S. A compilation of values of α_S, obtained by perturbative and by lattice methods, is presented in Figure 9 [149].

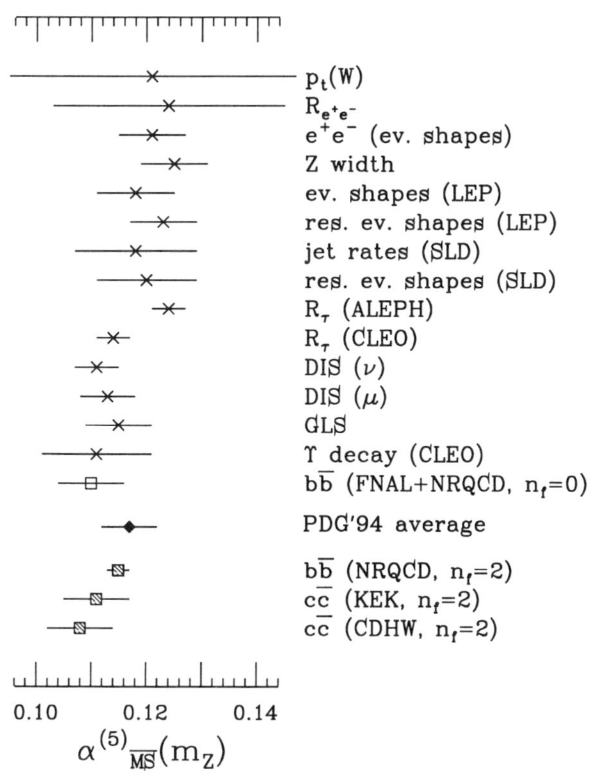

Figure 9: Compilation of results for α_S. The crosses denote results based on perturbative QCD, the boxes results based on lattice QCD. The shaded boxes represent very recent results which are not part of the PDG'94 average.

6.2.3 High temperature QCD

The demonstration that quenched QCD undergoes a deconfining transition at a temperature of approximately $200 MeV$ was one of major successes of lattice QCD. The study of QCD at a finite temperature T proceeds through a path integral formulation of the thermal average

$$\langle O \rangle_T = Tr(Oe^{-H/T}) \tag{7}$$

The exponential is interpreted as a propagation factor for an Euclidean time $t = 1/T$ and one is thus lead to consider a system quantized in an Euclidean domain of infinite extent in the spatial directions but finite extent $1/T$ in the temporal direction, where periodic (antiperiodic) boundary conditions are imposed on bosonic (fermionic) fields to implement the trace in Eq. (7). In practice this system is simulated on a lattice of N_s sites in the space directions and $N_t \ll N_s$ sites in the time direction. The temperature is related to lattice spacing and temporal extent by $T = 1/(N_t a)$.

The properties of the deconfining transition in quenched QCD have been by now rather well established[150]. The critical temperature, the order of the transition (weakly first order) and other observables, such as the surface tension of nucleating hadrons, have been determined. Many of the current efforts are being devoted to simulating hot QCD with dynamical quarks[151,152,153]. As one would expect, the creation and annihilation of virtual q-$\bar{q}$ pairs has strong effects on the dynamics of the thermal fluctuations. These are felt even for moderately large quark masses and lattice simulations have shown that they can alter the nature of the transition. For large quark masses, of course, one expects little departure from the results of quenched QCD. For intermediate quark masses, full QCD simulations indicate a weakening of the transition, which appears to change to a rapid cross over from the hadronic medium to a quark-gluon plasma with no discontinuities. Also, the temperature of the transition is lowered with respect to the quenched case. At exactly zero quark mass there are theoretical arguments for a transition driven by the restoration of chiral symmetry, which is of the second order for two quark flavors and of the first order with $N_f > 2$. The interesting case is, of course, the one of light, but not vanishing quark masses and current investigations have been focusing on this situation. There is good numerical evidence that a first order transition persists into the domain of finite (non-zero) quark mass with four flavors of light quarks. For the more realistic case of two light quarks and one quark of intermediate mass the results are still inconclusive, but it is realistic to expect that substantial progress will soon be made.

6.2.4 Weak matrix elements

Hadronic matrix elements of weak operators are very important for extracting the parameters of the electroweak theory from the experimental data and, more generally, for testing the predictions of any fundamental theory of weak interactions against experiment. Lattice techniques can be used to calculate many of these matrix elements[154,155,156].

The difficulty in calculating these observables, and consequently the precision which can be achieved, depends a lot on the type of matrix element under consideration. If we recall that individual hadronic states are isolated on the lattice by the "filtering ability" of the propagation in Euclidean time, which enhances the contribution from the lightest states, it will be clear that matrix elements between the vacuum and a single particle state, such as those encountered in the calculation of pseudoscalar decay constants, are much easier to calculate that those involving two or three external particle states. Additional difficulties lattice calculations have to contend with, beyond the general need of correcting for finite lattice spacing, finite volume etc., come from the fact that the weak interactions frequently involve scales much larger than the lattice momentum cutoff. This can be taken care of by an operator product expansion of the interaction followed by renormalization down to the scale of lattice momenta. Problems of renormalization thus play a very important role in the lattice determination of weak matrix elements and much progress has been made in this field. Perturbative and non-perturbative methods of renormalization have been developed and are routinely used to relate the quantities calculated on the lattice with their continuum counterparts[154,155,156].

Pseudoscalar decay constants for the light mesons (f_π, f_K) are calculated together with the masses in the hadronic spectrum and the results are in good agreement with experiment. Recently much attention has been paid to the calculation of the decay constants for heavy-light mesons, f_B and f_D. Since the mass of the B meson is larger that the lattice cutoffs which can be reached in present calculations, the determination of f_B can proceed either through the use of a static approximation for the heavy meson or via an extrapolation of results obtained for lighter mesons. Earlier calculations showed a marked discrepancy between the results obtained by the two methods, but a better understanding of the static approximation and of various renormalization factors has brought the two sets of results in much better agreement. The values found for f_B thus tend to cluster around $200MeV$, with quoted errors of the order of $10-15\%$ and variances between the results obtained by different groups of about as much. Values $f_D \approx 210MeV$ and ratios $f_{Bs}/f_B \approx 1.1$, $f_{Ds}/f_D \approx 1.1$ are also found[155].

Quantitatively meaningful results have begun to appear for semileptonic form factors in the decays $D \to K$, $D \to K^*$, $B \to K^*\gamma$ and for the Isgur-Wise function. Another quantity for which very substantial progress has been made is B_K, the $K^0 - \bar{K}^0$ mixing parameter. It is possible to quote today a lattice value $B_K(2GeV) = 0.616 \pm 0.020 \pm 0.017$ ($\hat{B}_K = 0.825 \pm 0.027 \pm 0.023$)[157].

Many other matrix elements have been considered in the literature, including those governing the non-leptonic decays $K \to \pi\pi$, trying in particular to find a computational explanation for the $\Delta I = 1/2$ rule. As implied above, these are very challenging and for the moment cannot be calculated with confidence, but with the expected improvements in algorithms and computational resources they should also become calculable within the next few years.

6.2.5 Hadron structure and other observables

There are many more QCD observables which can be calculated by lattice techniques. Reasons of space prevent us to go at any depth into their list. Many of these observables are discussed in detail in [134]. First steps have been taken toward the calculation of structure functions. Charge density correlations within hadrons have been determined. A very interesting recent calculation has shown that these are left almost unchanged if one uses so-called cooling techniques in the simulation to suppress short range quantum fluctuations, leaving only long range instanton excitations. This points to an intriguing role played by topologically non-trivial structures in the dynamics of hadrons.

The properties of the QCD vacuum have been the subject of many investigations. Lattice techniques have been used to evaluate observables such as the magnitude of the fluctuations of the topological charge and the gluon condensate. They have helped clarify the effects of monopoles in the maximally Abelian gauge and investigate the gauge fixing ambiguities encountered for large fields. Altogether, the lattice formulation of QCD is much more than a tool for the numerical determination of experimental observables. Suitably used, it can provide valuable insights into the whole dynamics of strong interactions.

6.2.6 Discussion of the errors

Since lattice QCD calculations are based on sampling techniques, the results are affected by statistical errors. In general it is rather straightforward to estimate the magnitude of the statistical errors (exceptions are the cases where metastabilities make it difficult to reach statistical equilibrium) and these are universally quoted together with the results. Somehow more difficult is the estimate of the systematic errors coming from finite lattice spacing, finite volume, the quenched approximation (if used) and all other approximations required to implement the numerical simulations. A lot of attention is generally paid to these sources of error and various procedures, such as repeating the calculations with different lattice sizes and different values of the bare coupling constant (which, through the renormalization relation $a = a(g)$ implies different lattice spacings), are used to estimate the magnitude of the systematic effects and, if possible, to correct for them.

Nevertheless, although these (statistical and systematic) errors can be quantified, there are other elements of uncertainty which depend to a large extent on the questions which are being asked and on what one is willing to assume. This is what makes the often heard question "when will lattice calculation produce a result accurate to (say) 5% for the ratio m_ρ/m_N?" difficult to answer. Depending on what theoretical assumptions one is willing to accept, such a rate of precision has already been achieved or may be still several years far away. The recent calculation of the spectrum considered above is a case in point. Large samples of configurations have been used to reduce the statistical errors and very careful extrapolations in lattice volume and lattice spacing have been made. Still the calculation could only be performed for quark masses larger or equal to approximately one half the strange quark mass and in the quenched approximation. Recent theoretical studies, based on chiral perturbation theory, of the quenched approximation indicate that the limit of zero quark mass is singular. Taken per se this would seem to invalidate completely the extrapolation in quark mass that was used to derive the masses of hadrons made of u and d quarks: on theoretical grounds one would not trust a linear extrapolation for the quenched approximation. At the very least, one would want to see the values it produces with much lighter quark masses. But then the effects of $q - \bar{q}$ vacuum polarization effects are expected to become important and one would not trust the quenched approximation anyway. This road leads to the conclusion that the only reliable results would be those of full QCD simulations done with light dynamical quarks. Such simulations are certainly several years away.

But one can look at things from a different perspective. One can give theoretical arguments in support of the fact that hadron masses should exhibit a smooth behavior as function of the masses of the quarks. From this point of view, one can then assume the legitimacy of a linear extrapolation in m_q (using squared masses, on current algebra arguments, for the lightest pseudoscalars), which finds confirmation in the experimental data. Notice that even with this assumption, the slopes and intercepts of the linear fits remain as important, and quite non-trivial, non-perturbative observables of QCD. The lattice calculation of the spectrum, done within a range of values for m_q where the quenched approximation is expected to be valid, provides then a quantitative determination of these observables. This is a major accomplishment, for which one would have held little hope prior to the advent of lattice QCD.

6.3 Alternative Discretization Techniques

6.3.1 Null-plane quantization

The null-plane quantization is a well established, alternative method of defining a quantum field theory where one of the light cone coordinates, *e.g.*, $x^+ = (x^0 + x^3)/\sqrt{2}$, replaces the time coordinate x^0 as the evolution variable. It offers some important advantages over the more conventional $x^0 = const$ quantization, such as better properties of the vacuum state, explicit invariance under Lorentz boosts in the x^3 direction and a more direct relationship between deep inelastic structure functions and the wave-functions of quarks within hadrons. Computational techniques based on the null plane quantization have been introduced and studied during the past few years[158]. In this approach one focuses directly on the wave-functions of the hadronic components. In the restricted two dimensional space spanned by the $x^+, x^- = (x^0 - x^3)/\sqrt{2}$ coordinates the gauge field interaction produces a linear potential, which gives origin to confinement. The extension to four dimension can be accomplished by discretizing the space of transverse coordinates x^1, x^2. The challenge is then to show that confinement survives this extension of the degrees of freedom and to incorporate all appropriate renormalization effects. As a computational technique, the null-plane quantization of QCD has not been as widely studied as the Euclidean lattice formulation, but it constitutes a quite different approach with the potential of producing valuable complementary results.

6.3.2 Hamiltonian QCD and other approaches

In the Hamiltonian approach to lattice QCD one discretizes the space coordinates, but maintains a continuous time variable. The gauge dynamical variables are finite group elements associated with the oriented links of the spatial lattice (very much like in the Euclidean formulation) and their conjugate momenta, which are the components of the chromoelectric field. The evolution is in real time and is generated by a well-defined Hamiltonian operator[138]. Indeed, if one considers a system of finite volume, this is a many-body Hamiltonian with a finite number of degrees of freedom. One tries then to find good approximations for the wave function of the vacuum and of the hadrons, and for the energy levels of these states, typically by using variational techniques. The major difficulty in this approach is the need of incorporating a very large number of components in the wave functions, a problem which is bypassed in Euclidean lattice QCD by simulating directly the quantum fluctuations. Thus, unless one succeeds in producing extremely good *Ansätze* for the wave functions, there are serious limitations to the accuracy which can be achieved.

Several other computational techniques, for example methods based on the derivation of equations relating the expectation values of transport factors (Wilson loops), have been proposed and studied. In addition, one should mention the large body of analytical work that has been and is been done in the context of lattice QCD. Research in this field is indeed far from being exhausted by the numerical simulations. Perturbative calculations, done by analytic expansion techniques, play a crucial role in defining various renormalizations that must be made to bridge the gap to the continuum. Analytic methods have been used to study finite size effects, to study gauge fixing ambiguities and their implications, to calculate the spectrum of QCD in a small box, to perform large N_c and strong coupling expansions etc. Very much like what is happening in other fields of physics, in lattice QCD one is also finding that analytical and computational methods complement each other and together provide a very powerful tool for deriving quantitative predictions.

6.4 Expected Progress

6.4.1 Computational resources

Lattice QCD calculations are very demanding computer applications. The size of the lattices one can consider, as well as other important parameters such as the values of the quark mass, depend in a crucial manner on the number of variables one can store and on total number of arithmetic operations one can perform. For a computation of a reasonably limited duration, the latter converts in number of floating point operations per second (flops). Indeed, scope and accuracy of lattice QCD calculations have steadily increased over the years as computers have gained in memory capacity and speed. The pace of progress in computer technology is forecast to continue for years to come and thus one can correspondingly foresee very substantial, hardware driven improvements in lattice QCD calculations.

While the advance in computer technology is obviously quite independent of lattice QCD applications, which can thus ride the wave of commercial development, the very special computational features of such applications has stimulated the design and construction of dedicated computers, to be used exclusively (or mostly) as a laboratory for the numerical simulation of QCD[159]. The rationale behind such developments is that the highly organized structure of data and communications in QCD applications permits an optimal utilization of parallelism, so that one can gain in economy and efficiency by designing and building a supercomputer targeted to these calculations. Dedicated machines capable of sustained speeds of several Gigaflops have been built and used successfully in the US and abroad[159].

At present there are two projects within the US for dedicated QCD supercomputers capable of reaching into the Teraflops domain:

• A project pursued by a group at the Columbia University in collaboration with researchers from several other institutions plans to use digital signal processors and a rather streamlined communications architecture to achieve a peak speed of 0.8 Teraflops and a sustained speed of 0.5 Teraflops[160]. The total cost of this project is estimated at $3M.

• The QCD Teraflops project plans to enhance a commercially available machine with special multiprocessor boards carefully designed to take advantage of the locality features exhibited by QCD calculations (and of many other large scale applications as well)[161]. This supercomputer, with a peak speed of 1.6 Teraflops and an estimated sustained speed in excess of 1 Teraflops, would anticipate the pace of commercial development by a few years and at a fraction of the cost (estimated cost $10M development, $25M construction).

The two projects are quite different and, to a large extent, complementary. The Columbia project is for a rather rigid machine, designed to implement the currently available algorithms in an outstandingly efficient and economical manner. The QCD Teraflops project is for a much more general purpose and easier to program supercomputer, which could be fruitfully used also for a wide range of non QCD applications. The importance that the development and implementation of new algorithms are likely to play for the progress of lattice QCD speaks of course in favor of the flexibility of the QCD Teraflops machine, but the Columbia project has on its side its substantially lower cost.

Since either project would require a substantial allocation of funds, issues of access become important. To formulate these in terms familiar to particle physicists, the question is whether a dedicated machine should be considered more like an accelerator, *i.e.*, a facility to serve several groups of experimenters, or like a detector, where the group who built it is entitled to take and analyze the data in an exclusive manner. Given the expectation that the funds allocated to a QCD machine may, directly or indirectly, reduce the total amount of computer resources otherwise available, many researchers within the lattice community have expressed a strong sentiment that any such machine should be operated as a facility. However, the physicists who design and build a special purpose computer can legitimately expect to see their efforts rewarded by some kind of priority in the use of machine. These are important issues, which will require careful consideration.

6.4.2 Algorithms

Advances in supercomputer technology alone are not sufficient for the progress of lattice QCD. The development of Teraflops supercomputers will bring an increase in computer speed and memory of one to two orders of magnitude with respect to what is available today. The number of operations required by a lattice simulation obviously contains the volume of the lattice as a factor. This implies that a sheer increase of computer power, even into the Teraflop domain, can produce little more than a doubling of the size of the largest lattices that can be studied. As a matter of fact, the situation is worse than that. A major motivation for considering larger lattices is to reduce the lattice spacing, coming closer to the continuum limit. One also wants to be able to consider smaller quark masses. But smaller lattice spacings and smaller quark masses both imply a decrease in computational efficiency, through the phenomenon of critical slowing down. The algorithms for calculating quark propagators (a crucial component of almost all QCD simulations) are based on iterative procedures, whose rate of convergence decreases dramatically as the quark mass (or, better, its value in lattice units $m_q a$) is reduced. Thus, in absence of progress leading to more efficient computational procedures, one cannot expect, from hardware developments alone, even the gains that a naive scaling of the number of degrees of freedom would suggest.

The lattice community has always been aware of the importance of algorithm development and the progress in accuracy of lattice QCD calculations has been accompanied, indeed made possible, by crucial advances in computational techniques. Examples of this progress are all the techniques that have been developed to incorporate fermionic degrees of freedom in the simulations, with the discovery of the "hybrid Monte Carlo" algorithm topping the list of the most important breakthroughs[162]. Another example is given by the refinement in the source and sink operators used for spectroscopy and matrix element calculations. Here, indeed, the line between what should be considered algorithm development and what ought to be considered theoretical progress becomes blurred, but correctly so, because the development of better computational techniques typically finds its roots in a better understanding of the physics of the phenomena under investigation.

Current areas of algorithmic research include the development of better methods for calculating quark propagators, which may overcome, or at least moderate, critical slowing down. Multigrid methods[163] as well as other techniques are being studied. Some progress has been made, but more progress will require a better understanding of the properties of the lattice Dirac operator in presence of fluctuating gauge fields.

Another promising direction of progress consists in the computational use of improved actions[164]. Renormalization group ideas have recently been applied to the definition of a "perfect" action, an action which remains unaltered in the renormalization leading to the continuum limit. Actions approximating the properties of perfect actions may permit to recover the features of the continuum limit working with coarser lattices and therefore with smaller number of dynamical variables. The increased computational power (of the computers of the next generation) could then be applied to an improvement of the accuracy of the simulations and to an expansion of their scope.

6.4.3 Observables

The increase in computer power and the progress in algorithms which are anticipated to occur in a time of three to four years will permit substantial improvements in the accuracy of QCD lattice calculations and a widening of their scope.

One expects that it will be possible to perform quenched calculations of the spectrum for light quark masses close to the experimental value (more properly, for values of m_{ps}/m_{vect} close to m_π/m_ρ). For full QCD, barring unanticipated progress in the algorithms for simulating dynamical fermions, one will probably be able to perform calculations with quark masses half way between m_s and m_u, m_d. This should be sufficient to see the $q-\bar{q}$ vacuum polarization effects go beyond a mere renormalization of the bare coupling constant. One should also be able to see genuine departures from the quenched approximation. Precise determinations of α_S from heavy quark spectroscopy can be expected.

The nature of the transition to a quark-gluon plasma will probably be resolved and progress will be made toward a precise calculation of several thermodynamic observables.

One may expect a rather accurate determination of weak matrix elements (perhaps with errors of $\pm 5\%$) for which one is beginning to get quantitative results, as well as an extension of the calculations to matrix elements which cannot be evaluated today.

More observables will become calculable. These may range from phenomenological parameters, such as the coupling constants in effective chiral Lagrangians, to scattering lengths, to moments of structure functions, to quantities relevant to the interface between perturbative and non-perturbative lattice QCD. While this list of observables is potentially very rich and interesting, it would futile to try to define it too exactly now: its overall span will depend on the detailed progress of the methods of lattice QCD as well as on the ingenuity of the scientists who will apply them.

6.5 Long Range Outlook

Looking farther ahead into the future, given the rapid progress of computer technology and the theoretical and algorithmic developments which are likely to occur, it is to be expected that, five to ten years from now, lattice QCD will be able to produce an accurate determination of a wide range of observables, which will provide stringent tests for the underlying theory of QCD and valuable input for theories describing non-strong interactions. However, the actual rate of progress of lattice QCD will depend on decisions which reflect the policies of the entire particle physics community. This leads us to the following conclusions.

Computer resources - The progress of lattice QCD is heavily dependent on the availability of ever more powerful means of computation. Fortunately the development of supercomputer technology is receiving a lot of support at the policy making level, and thus lattice QCD automatically benefits from the expansion of computer resources that this support generates. Lattice QCD applications, because of the huge volumes of data that they manipulate and the extremely large number crunching capability they require, have been acknowledged as one of the driving forces of supercomputer development. Thus there is the possibility of obtaining funding for QCD applications (either as direct funds for the development of dedicated machines, or in the form of increased allocations of supercomputer time) from non-HEP sources because of the impact that QCD applications may have on supercomputer technology. However, this also requires a strong endorsement from the particle physics community of the value of this mode of research, including the willingness of allocating funds to support the necessary hardware and algorithm developments. While any such investment can be leveraged by appreciable funding from non-HEP sources, it would be difficult to expect the latter to occur if particle physicists, first, do not recognize the value of QCD applications.

Exchange of information - Lattice QCD calculations are not a black box which turns out numbers affected by smaller or larger errors. The details of the calculations are frequently as significant as the final results. On the other hand, lattice QCD investigations are of marginal value if pursued in isolation, without exposure to the whole problematic of strong interactions. Thus, it would be valuable if there were better contacts between scientists working on lattice QCD and other areas.

Experiments - Lattice QCD has the potential of producing one day very accurate results, derived entirely from first principles, on many hadronic observables. It is conceivable, for instance, that it will be possible to predict spectroscopic data with a precision sufficient to put stringent tests to the validity of QCD. However, this would be of little value in absence of experimental data to compare with. As the experimental frontier moves to ever higher energies and smaller distances, it is important not to neglect those energy domains, which do not lie at the boundary of technology, but where a lot of very valuable information can still be collected.

References

[1] B.R. Webber, "QCD and jet physics", in *Proc. of the XXVII International Conference on High Energy Physics*, Glasgow Scotland, July 1994; J. W. Gary, "QCD at high energy colliders", in *Proc. of the Fifth Conference on the Intersections of Particle and Nuclear Physics*, St. Petersburg Florida, Jun 1994.

[2] P.N. Burrows, "QCD at the Z^0", *Nucl. Phys. B* **23** (Proc. Suppl.) 30 (1991); S. Bethke, Aachen preprint PITHA 94/29 (1994) (Talk at the Tennessee International Symposium on Radiative Corrections, Gatlinburg, USA, June 27 - July 1 1994.)

[3] Workshop on physics and experiments with Linear e+e- colliders, Hawaii, 26-30 April 1993, edited F.A. Harris *et al.*, World Scientific, Vol II p. 670.

[4] P.Z. Quintas *et al*, *Phys. Rev. Lett.* **71** 1307 (1993).

[5] V. Elvira, "Inclusive jet Cross Sections at the DØ Detector", FERMILAB- CONF-94-323C-E, to be published in the proceedings of DPF 1994.

[6] F. Abe *et al.*, *Phys. Rev. Lett.* **69**, 2896 (1992).

[7] F. Nang, "Measurement of the Triple Differential Inclusive DiJet Cross Section, $d^3\sigma/dE_t d\eta_1 d\eta_2$ in $p\bar{p}$ collisions at $\sqrt{s} = 1.8TeV$, FERMILAB-CONF-94-323E-E, to be published in the proceedings of DPF 1994.

[8] G. Wolf, "HERA Physics", DESY 94-022, from the proceedings of the 42nd Scottish Universities Summer School in Physics, 1993; F. Sciulli, "Summary of the Eilat Workshop", DESY 94-147, Eilat, Israel, 1994.

[9] A.P. Bukhvostov, G.V. Frolov, L.N. Lipatov, E.A. Kuraev, *Nucl. Phys. B* **258**, 601 (1985).

[10] V.N. Gribov and L.N. Lipatov, *Sov. J. Nucl. Phys.* **15**, 438 (1972); L. N. Lipatov, *Sov. J. Nucl. Phys.* **20**, 94 (1975); G. Altarelli and G. Parisi, *Nucl. Phys. B* **126**, 298 (1977).

[11] E. Kovacs, "Testing QCD with Jet Physics at CDF", FERMILAB-CONF-94-215-E, to be published in the proceedings of DPF 1994; E. Buckley-Geer, Studies of Prompt Photon Production and Multijet Production at the Tevatron Collider, FERMILAB-CONF-94-336-E to be published in the proceedings of ICHEP, Glasgow, 1994.

[12] "DØ Quantum Chromodynamics: A Compilation of Results Presented at DPF 1994", FERMILAB-Conf-94/323-E DØ; H.Weerts (DØ Collaboration), "Studies of Jet Production with the DØ Detector" in the Proceedings of the 9th Topical Workshop in $p\bar{p}$ Collider Physics, Tsukuba (1993), edited by K.Kondo.

[13] D. Allasia *et al.*, *Phys. Lett. B* **249**, 366 (1990).

[14] "Measurements of Nucleon Structure Functions, F_2 & xF_3, and Precision Tests of PQCD at the FNAL Tevatron", S. R. Mishra *et al.* in *Proceedings of the International Conference on High Energy Physics*, Dallas, TX, August, 1992.

[15] J. Ashman *et al.*, *Nucl. Phys. B* **328**, 1 (1989).

[16] J. Ellis and R. Jaffe, *Phys. Rev. D* **9**, 1444 (1974), erratum-ibid. **10**, 1669 (1974).

[17] P.L. Anthony *et al.*, *Phys. Rev. Lett.* **71**, 959 (1993).

[18] B. Adeva *et al.*, *Phys. Lett. B* **329**, 399 (1994).

[19] B. Adeva *et al.*, *Phys. Lett. B* **320**, 400 (1994).

[20] J. D. Bjorken, *Phys. Rev. D* **1**, 1376 (1970).

[21] K. Abe *et al.*, *Phys. Rev. Lett.* **74**, 346 (1995).

[22] M. Kutschera, *Phys. Lett. B* **325**, 271 (1994).

[23] G.T. Gabadadze, "An Estimate of the Proton Singlet Axial Constant" JINR-E2-94-518, Jan 1995.

[24] "Polarized Protons at RHIC", G. Bunce, J. Collins, S. Heppelmann, R. Jaffe, S.Y. Lee, Y. Makdisi, R.W. Robinett, J. Soffer, M. Tannenbaum, D. Underwood, A. Yokosawa, Particle World Volume 3, No. 1, p. 1-12, 1992.

[25] A. Bhatti, "Inclusive Jet and Two-Jet Differential Cross-Sections at CDF", FERMILAB-CONF-94-159-E, contributed to *27th International Conference on High Energy Physics (ICHEP)*, Glasgow, Scotland, 20-27 Jul 1994.

[26] S. Behrends, "Scaling Behavior of Jet Production in CDF", *Proceedings of Particles & Fields 92: 7th Meeting of the Division of Particles Fields of the APS (DPF 92)*, Batavia, IL, Nov 1992, DPF conf.1992:999-1001 (QCD161:A6:1992).

[27] F. Abe *et al.*, *Phys. Rev. D* **48**, 2998 (1993); *Phys. Rev. Lett.* **73**, 2662 (1994); *Phys. Rev. Lett.* **74**, 1891 (1995).

[28] D. Crane, "*B* Production at CDF", FERMILAB-CONF-93-362-E, Jan 1994, presented at 9th Topical Workshop on Proton - Antiproton Collider Physics, Tsukuba, Japan, 18-22 Oct 1993.

[29] F. Abe *et al.*, *Phys. Rev. Lett.* **71**, 2537 (1993).

[30] S. Abachi *et al.*, "Inclusive μ and B-quark Production Cross-Sections in $\bar{p}p$ Collisions at $\sqrt{s}$ = 1.8 TeV", FERMILAB-PUB-94-409-E, Dec 1994, Submitted to *Phys. Rev. Lett.* .

[31] W. J. Stirling, *Nucl. Phys. B* **423**, 56 (1994).

[32] B. May, "Rapidity Gaps Between Jets at DØ", FERMILAB-CONF-94-323K-E, Sep 1994, presented at *1994 Meeting of the American Physical Society, Division of Particles and Fields (DPF 94)*, Albuquerque, NM, 2-6 Aug 1994.

[33] F. Abe *et al.*, *Phys. Rev. Lett.* **74**, 855 (1995).

[34] S. Catani, L. Trentadue, G. Turnock, B.R. Webber, *Nucl. Phys. B* **407**, 3 (1993).

[35] M. Acciarri *et al.*, *Phys. Lett. B* **345**, 74 (1995); R. Akers *et al.*, *Z. Phys. C* **63**, 197 (1994).

[36] R. Balest *et al.*, "A Study of Jet Production Rates in the Four Flavor Continuum and a Test of QCD", CLEO-CONF-94-28, submitted to Int. Conf . on High Energy Physics, Glasgow, Scotland, Jul 20-27, 1994.

[37] M. R. Adams *et al.*, *Phys. Rev. Lett.* **69**, 1026 (1992).

[38] M. Derrick *et al.*, "Jet Production in high Q^2 Deep-inelastic *ep* Scattering at HERA", DESY-95-016, submitted to *Z. Phys. C* , 1995.

[39] I. Ahmed *et al.*, *Phys. Lett. B* **346**, 415 (1995).

[40] D.J. Miller, "Gamma-gamma Topics and a look at Mesonic Goo", Lepton-Photon Symposium, Cornell, August 1993.

[41] J. Appel, "New Preliminary Results on the Physics of Charm Hadroproduction Subprocesses", FERMILAB-CONF-94-347-E, contrib-

uted to 27th International Conference on High Energy Physics (ICHEP), Glasgow, Scotland, 20-27 Jul 1994.

[42] S. T. Fahey, "Direct Photon Production at DØ", FERMILAB-CONF-94-323A-E, Presented at *1994 Meeting of the American Physical Society, Division of Particles and Fields (DPF 94)*, Albuquerque, NM, Aug 1994.

[43] F. Abe *et al.*, "Direct Photon Results from CDF" FERMILAB-CONF-93-202-E, Submitted to *16th International Symposium on Lepton and Photon Interactions*, Ithaca, NY, Aug 1993.

[44] R. J. Hemingway, "Inclusive particle production in Z0 decays", Proc. of the XXVII International Conference on High Energy Physics, Glasgow Scotland, July 1994

[45] P. N. Burrows, "Measurement of the Charged Multiplicity of $Z^0 \to B\bar{B}$ Events", SLAC-PUB-6602, Aug 1994. Presented at *1994 Meeting of the American Physical Society, Division of Particles and Fields (DPF 94)*, Albuquerque, NM, Aug 1994.

[46] J. Dominick *et al.*, *Phys. Rev. D* **50**, 4265 (1994).

[47] F. Abe *et al.*, *Phys. Rev. Lett.* **67**, 2937 (1991); *Phys. Rev. Lett.* **66**, 2951 (1991).

[48] J.Zhi-Yu Jiang, "Measurement of the W/Z P_t Distributions at DØ", FERMILAB-CONF-94-323I-E, Sep 1994. 3pp. Presented at *1994 Meeting of the American Physical Society, Division of Particles and Fields (DPF 94)*, Albuquerque, NM, Aug 1994.

[49] M. Derrick *et al., Z. Phys. C* **65**, 379 (1995); I. Abt *et al., Phys. Lett. B* **321**, 161 (1994).

[50] I.M. Dremin, B.B. Levchenko, *Phys. Lett. B* **292**, 155 (1992).

[51] M. Derrick *et al., Phys. Lett. B* **338**, 483 (1994; T. Ahmed *et al., Nucl. Phys. B* **435**, 3 (1995).

[52] G. Ingleman, LHCC Workshop, CERN, Nov., 1994

[53] S.V. Goloskokov, S.P. Kuleshov, O.V. Selyugin, *Mod. Phys. Lett.* A **9**, 1207 (1994).

[54] D. Alde *et al., Phys. Lett. B* **205**, 397 (1988).

[55] S.J. Lindenbaum, "Glueballs in 2++ Final States, Glueballs, Hybrids, and Exotic Hadrons", AIP conf. Proc. no. 185 pp 68-87(1988); S.J. Lindenbaum, "The Status of Glueballs" in *Future Directions in Particle and Nuclear Physics at Multi-Gev Hadron Beam Facilities*, proceedings of the workshop at BNL 4-6 March 1993, BNL-52389, pp. 384-389.

[56] D. Morgan, *Phys. Lett. B* **258**, 444 (1991), erratum-ibid. **269**, 477 (1991).

[57] ATLAS Technical Proposal, CERN/LHCC/94-43, 1994;
CMS Technical Proposal, CERN/LHCC/94-38, 1994.

[58] J. C. Collins, D. E. Soper and G. Sterman, "Factorization of Hard Processes in QCD", in *Perturbative Quantum Chromodynamics*, ed. A. H. Mueller, World Scientific (1989).

[59] The CTEQ Collaboration, "Handbook of Perturbative QCD", to be published in Reviews of Modern Physics, 1995.

[60] B. R. Webber, "QCD and Jet Physics" in *Proceedings of the XXVII International Conference on High Energy Physics*, Glasgow, Scotland (1994).

[61] S. Catani, *Proceedings of the International Europhysics Conference on High Energy Physics*, Marseille, France, 22-28 July, 1993.

[62] S. J. Brodsky and G. P. Lepage, "Exclusive Processes in Quantumchromodynamics", in *Perturbative Quantum Chromodynamics, op. cit.*

[63] A. S. Carroll *et al., Phys. Rev. Lett.* **61**, 1698 (1988).

[64] A. H. Mueller, "The QCD Perturbation Series" in *QCD 20 Years Later*, eds. P. M. Zerwas and H. A. Kastrup, World Scientific, 1993.

[65] M. A. Shifman, "QCD Sum Rules: The Second Decade" in *QCD 20 Years Later, op. cit.*

[66] E. M. Levin "Parton Density at Small x_{bj}" in *QCD 20 Years Later, op. cit.*

[67] J. W. Cronin *et al., Phys. Rev. D* **11**, 3105 (1975).

[68] L. Kluberg *et al., Phys. Rev. Lett.* **38**, 670 (1977).

[69] W. Busza and R. Ledoux, *Ann. Rev. Nucl. Part. Sci.* **38**, 119 (1988).

[70] W. Geist, *Nucl. Phys.* A **525**, 149c (1991).

[71] P. M. Fishbane and J. S. Trefil, *Phys. Rev. D* **12**, 2113 (1975).

[72] K. Kastella, *Phys. Rev. D* **36**, 2734 (1987).

[73] K. Kastella, G. Sterman, and J. Milana, *Phys. Rev. D* **39**, 2586 (1989).

[74] D. M. Alde *et al.*, *Phys. Rev. Lett.* **66**, 153 (1991).

[75] P. Amandruz *et al.*, CERN-PPE-91-198 (1991).

[76] R. Vogt, S. J. Brodsky, and P. Hoyer, *Nucl. Phys. B* **360**, 67 (1991).

[77] S. Gavin and J. Milana, *Phys. Rev. Lett.* **68**, 1834 (1992).

[78] C. Benesh, J. Qiu, and J. Vary, preprint ISU-NP-93-15 (1994) ⟨hep-ph/9403265⟩.

[79] S. Gavin, H. Satz, R. L. Thews, and R. Vogt, *Z. Phys. C* **61**, 351 (1994).

[80] J. Qiu and G. Sterman, *Nucl. Phys. B* **353**, 105 and 137 (1991).

[81] M. Luo, J. Qiu, and G. Sterman, *Phys. Rev. D* **49**, 4493 (1994).

[82] M. D. Corcoran *et al.*, *Phys. Lett. B* **259**, 209 (1991).

[83] D. Naples *et al.*, *Phys. Rev. Lett.* **72**, 2341 (1994).

[84] M. Gyulassy, M. Plümer, M. H. Thoma, and X. N. Wang, *Nucl. Phys. A* **538**, 37c (1992).

[85] G. Bertsch, S. J. Brodsky, A. S. Goldhaber, and J. F. Gunion, *Phys. Rev. Lett.* **47**, 297 (1981).

[86] S. J. Brodsky and P. Hoyer, *Phys. Rev. Lett.* **63**, 1566 (1989).

[87] L. L. Frankfurt, G. A. Miller, and M. Strikman, Ann. Rev. Nucl. Part. Sci. **45**, 501 (1994).

[88] M. R. Adams *et al.*, *Phys. Rev. Lett.* **68**, 3266 (1992); *Phys. Lett. B* **287**, 375 (1992).

[89] M. R. Adams *et al.*, *Z. Phys. C* **65**, 225 (1995).

[90] L. McLerran and R. Venugopalan, *Phys. Rev. D* **49**, 2233 and 3352 (1994).

[91] T. Blum, S. Gottlieb, L. Kärkkäinen, and D. Toussaint, preprint AZPH-TH/94-22, ⟨hep-lat/9410014⟩.

[92] R. Pisarski and F. Wilczek, *Phys. Rev. D* **29**, 338 (1984).

[93] F. Wilczek, *Int. J. Mod. Phys. A* **7**, 3911 (1992).

[94] F. R. Brown *et al.*, *Phys. Rev. Lett.* **65**, 2491 (1990).

[95] F. Karsch, preprint BI-TP-95-11 (1995), ⟨hep-lat/9503010⟩.

[96] B. Boyd *et al.*, preprint BI-TP-94-42 (1995), ⟨hep-lat/9501029⟩.

[97] J. Borg, *Nucl. Phys. B* **261**, 455 (1985).

[98] E. Manousakis and J. Polonyi, *Phys. Rev. Lett.* **58**, 847 (1987).

[99] F. Karsch, E. Laermann, and M. Lütgemeier, *Phys. Lett. B* **346**, 94 (1995).

[100] S. Schramm and M. C. Chu, *Phys. Rev. D* **48**, 2279 (1993).

[101] A. D. Linde, Phys. Lett. **96B**, 289 (1980).

[102] A. V. Selikhov and M. Gyulassy, *Phys. Lett. B* . **316**, 373 (1993).

[103] E. Braaten and R. D. Pisarski, *Phys. Rev. D* **42**, 2156 (1990).

[104] C. Gong, *Phys. Lett. B* **298**, 257 (1993); Phys. Rev. D **49**, 2642 (1994).

[105] E. Braaten, *Phys. Rev. Lett.* **74**, 2164 (1995).

[106] J. P. Blaizot and A. H. Mueller, *Nucl. Phys. B* **289**, 847 (1987).

[107] K. J. Eskola, K. Kajantie, and J. Lindfors, Phys. Lett. B **214**, 613 (1988).

[108] K. Geiger and B. Müller, *Nucl. Phys. B* **369**, 600 (1992).

[109] X. N. Wang and M. Gyulassy, *Phys. Rev. D* **44**, 3501 (1991).

[110] T. S. Biró, B. Müller, and X. N. Wang, Phys. Lett. B **283**, 171 (1992).

[111] T. S. Biró *et al.*, Phys. Rev. C **48**, 1275 (1993).

[112] K. Geiger, *QCD based space-time description of heavy ion physics*, Phys. Rept. (in print, 1995).

[113] B. Müller, Rep. Prog. Phys. **28**, 1 (1995).

[114] K. Rajagopal and F. Wilczek, *Nucl. Phys. B* **404**, 577 (1993).

[115] S. Gavin, A. Gocksch, and R. D. Pisarski, *Phys. Rev. Lett.* **72**, 2143 (1994).

[116] M. Asakawa, Z. Huang, and X. N. Wang, *Phys. Rev. Lett.* **74**, 3126 (1995).

[117] M. C. Chu, J. M. Grandy, S. Huang, and J. W. Negele, *Phys. Rev. D* **48**, 3340 (1993); **49**, 6039 (1994).

[118] Experiments 864, 878, 882, 886, and 891 at the AGS at Brookhaven, and NA-52 at the SPS at CERN.

[119] K. P. Pretzl, Europhys. News **24**, 167 (1993).

[120] H. J. Crawford, Sci. Am. **270**, 58 (1994).

[121] J. Gasser and H. Leutwyler, *Nucl. Phys. B* **250**, 465 (1985).

[122] J. Gasser, *Nucl. Phys. B* **279**, 65 (1987).

[123] J. Gasser and H. Leutwyler, Ann. Phys. **158**, 142 (1984).

[124] V. Bernard *et al.*, Phys. Rept. **246**, 315 (1994).

[125] S. Weinberg, *Nucl. Phys. B* **363**, 3 (1991); *Phys. Lett. B* **295**, 114 (1992).

[126] C. Ordonez and U. van Kolck, *Phys. Lett. B* **291**, 459 (1991).

[127] C. Ordonez, L. Ray, and U. van Kolck, *Phys. Rev. Lett.* **72**, 1982 (1994).

[128] M. Knecht, B. Moussallam, and J. Stern, *Nucl. Phys. B* **429**, 125 (1994).

[129] E. Jenkins and A. V. Manohar, Phys. Lett B **255**, 558 (1991); **259**, 353 (1991).

[130] G. 't Hooft, *Nucl. Phys. B* **72**, 461 (1974); **75**, 461 (1974).

[131] E. Witten, *Nucl. Phys. B* **160**, 57 (1979).

[132] M. P. Mattis and M. E. Peskin, *Phys. Rev. D* **32**, 58 (1985).

[133] R. Dashen and A. V. Manohar, *Phys. Lett. B* **315**, 425 and 438 (1993).

[134] *Proceedings of Lattice 90-94*, *Nucl. Phys. B* (Proc. Suppl.) 20 (1991); 26 (1992); 30 (1993); 34 (1994); 42 (1995).

[135] K. Wilson, *Phys. Rev. D* **10**, 2445 (1974).

[136] M. Creutz, L. Jacobs and C. Rebbi, Phys. Rep. **95**, 201 (1983).

[137] H. B. Nielsen and M. Ninomiya, *Nucl. Phys. B* **185**, 20 (1981); *Nucl. Phys. B* **195**, 541 (1982).

[138] J. Kogut and L. Susskind, *Phys. Rev. D* **11**, 395 (1975).

[139] L. Susskind, *Phys. Rev. D* **16**, 3031 (1977).

[140] A. Hasenfratz and P. Hasenfratz, Phys. Lett. **93B**, 165 (1980).

[141] R. Dashen and G. Gross, *Phys. Rev. D* **23**, 2340 (1981).

[142] H. Hamber and G. Parisi, *Phys. Rev. Lett.* **47**, 1792 (1981).

[143] D. Weingarten, *Phys. Lett. B* **109**, 57 (1982).

[144] C. Michael, *Proceedings of Lattice 94 Nucl. Phys. B* **42**, (Proc. Suppl.): 147 (1995).

[145] F. Butler, H. Chen, J. Sexton, A. Vaccarino and D. Weingarten, *Nucl. Phys. B* **430**, 179 (1994).

[146] G. Bali *et al.*, *Phys. Lett. B* **309**, 378 (1993).

[147] J. Sexton, A. Vaccarino and D. Weingarten, *Proceedings of Lattice 94*, *Nucl. Phys. B* **42** (Proc. Suppl.): 279 (1995).

[148] P. MacKenzie, *Proceedings of Lattice 92*, *Nucl. Phys. B* **30** (Proc. Suppl.): 35 (1993).

[149] A. El-Khadra, *Proceedings of Lattice 93*, *Nucl. Phys. B* **34** (Proc. Suppl.): 141 (1994).

[150] A. Ukawa, *Proceedings of Lattice 89*, *Nucl. Phys. B* **17** (Proc. Suppl.): 118 (1990).

[151] B. Peterson, *Proceedings of Lattice 92*, *Nucl. Phys. B* **30** (Proc. Suppl.): 66 (1993).

[152] F. Karsch, *Proceedings of Lattice 93*, *Nucl. Phys. B* **34** (Proc. Suppl.): 63 (1994).

[153] C. De Tar, *Proceedings of Lattice 94*, *Nucl. Phys. B* **42** (Proc. Suppl.): 73 (1995).

[154] C. Sachrajda, *Proceedings of Lattice 92*, *Nucl. Phys. B* **30** (Proc. Suppl.): 20 (1993).

[155] C. Bernard, *Proceedings of Lattice 93*, *Nucl. Phys. B* **34** (Proc. Suppl.): 47 (1994).

[156] G. Martinelli, *Proceedings of Lattice 94*, *Nucl. Phys. B* **42** (Proc. Suppl.): 127 (1995).

[157] S. Sharpe, *Proceedings of Lattice 93*, *Nucl. Phys. B* **34** (Proc. Suppl.): 403 (1994).

[158] K. Wilson, T. Walhout, A. Harindranath, W-M. Zhang, R. Perry and S. Glazek, *Phys. Rev. D* **49**, 6720 (1994).

[159] Y. Iwasaki, *Proceedings of Lattice 93*, *Nucl. Phys. B* **34** (Proc. Suppl.): 78 (1994).

[160] R. Mawhinney, *Proceedings of Lattice 94*, *Nucl. Phys. B* **42** (Proc. Suppl.): 140 (1995).

[161] J. Negele, *Proceedings of Lattice 92*, *Nucl. Phys. B* **30** (Proc. Suppl.): 295 (1993).

[162] S. Duane, A. Kennedy, B. Pendleton, D. Roweth, *Phys. Lett. B* **195**, 216 (1987).

[163] R. Brower, R. Edwards, C. Rebbi and E. Vicari, *Nucl. Phys. B* **366**, 689 (1991).

[164] T. DeGrand, A. Hasenfratz, P. Hasenfratz, F. Niedermayer and U. Wiese, *Proceedings of Lattice 94*, *Nucl. Phys. B* **42** (Proc. Suppl.): 67 (1995).

Appendix: DPF QCD Workshop Agenda

Saturday, April 9, 1994 University of Wisconsin, Madison, WI

QCD Experimental Session - Organizer: Wesley Smith, University of Wisconsin

8:30 - 8:40	Wesley Smith	U of Wisconsin	Introduction
8:40 - 9:00	Bill Gary	UC Riverside	LEP
9:00 - 9:20	Phil Burrows	MIT	SLD
9:20 - 9:40	Sukhpal Sanghera	Cornell	CLEO
9:40 - 10:00	Steve Geer	FNAL	CDF
10:00 - 10:20	Harry Weerts	Michigan State	DØ
10:20 - 10:40	Allen Caldwell	Columbia U.	HERA
11:00 - 11:20	Harry Melanson	FNAL	FNAL-E665
11:20 - 11:40	Janet Conrad	Columbia U.	FNAL-E815
11:40 - 12:00	Emlyn Hughes	SLAC	SLAC-E142
12:00 - 12:20	Richard Milner	MIT	HERA-HERMES
12:20 - 12:40	Alex Dzierba	Indiana U.	Light Quark Spectroscopy
12:40 - 13:00	Wu-Ki Tung	Michigan State	CTEQ
14:00 - 14:20	Jim Carroll	LBL	RHIC-STAR
14:20 - 14:40	Joel Moss	Los Alamos	RHIC-PHENIX
14:40 - 14:50	Gerry Bunce	BNL	RHIC-SPIN
14:50 - 15:00	Wlodek Guryn	BNL	RHIC-pp elastic
15:00 - 15:10	Bolek Wyslouch	MIT	RHIC-PHOBOS

Lattice Gauge Session - Organizer: Claudio Rebbi, Boston University

15:30 - 16:00	Norman Christ	Columbia U.	Solving QCD: Why, How, When?
16:00 - 16:30	Greg Kilcup	Ohio State	Weak Matrix Elements
16:30 - 17:00	Andreas Kronfeld	FNAL	SM Phenomenology & Lattice QCD
17:00 - 17:30	Jeff Mandula	D.O.E.	Resources & Impact of L.G.T.
17:30 - 18:00	John Negele	MIT	Teraflops Project
18:00 - 18:30	Doug Toussaint	U. of Arizona	Numerical Calculations in QCD

Sunday, April 10, 1994

QCD Theory Session - Organizer: Berndt Muller, Duke University

8:00 - 8:30	Registration/Coffee & Pastries		
8:30 - 9:00	Joe Kapusta	U. of Minnesota	Finite T. QCD
9:00 - 9:30	Berndt Muller	Duke	RHIC - QCD Theory
9:30 - 10:00	K. Rajagopal	Harvard	Chiral Phase Transition
10:00 - 10:30	J. Qiu	Iowa State	QCD & A-Dep in Had-Nucleus
10:30 - 10:35	N. Chang	CCNY	Chiral Restoration at High-T

QCD Theory Session - Organizer: Al Mueller, Columbia University

11:00 - 11:30	George Sterman	Stony Brook	In face of Pert. & Non-Pert. QCD
11:30 - 12:00	John Collins	Penn State	Polarization & QCD
12:00 - 12:30	Steve Ellis	U. Washington	JETS: Past, Present, Future
13:30 - 14:00	Genya Levin	FNAL	small-x and Diffractive Phys.
14:00 - 14:30	Lance Dixon	SLAC	H0 Multi-parton Calculations
14:30 - 14:45	Paul Stevenson	Rice University	Scheme dep., opt. & Freezing
14:45 - 15:00	Sam Lindenbaum	BNL	Glueball Investigations

Heavy Flavor Physics and CP Violation

John P. Cumalat

University of Colorado, Boulder, CO 80309

Estia J. Eichten

Fermi National Accelerator Laboratory, Batavia, IL 60510

Helen R. Quinn

Stanford Linear Accelerator Center, Stanford, CA 94309

Michael P. Schmidt

Yale University, New Haven, CT 06511

Edward H. Thorndike

University of Rochester, Rochester, NY 14627

Yau W. Wah

University of Chicago, Chicago, IL 60637

1 Overview

Assuredly one of the deep mysteries of particle physics is why there are families — apparently blind, unnecessary repetitions of e^-, ν_e, u, d. The existence of families gives rise to many of the free parameters of the Standard Model, in particular the four parameters that describe the Cabibbo-Kobayashi-Maskawa (CKM)[1] matrix, the masses of the c, s, t, and b quarks, and the masses of the μ and τ leptons. A determination of all these parameters is needed if for no other reason than to fully define the Standard Model. Further, patterns evident among the masses and matrix elements may provide clues to guide the development of a theory to explain families, and their values certainly provide constraints on attempts at such a theory.

Many quantities which can be measured in several quite different experiments are related to the same parameter in the Standard Model. Physics beyond the Standard Model could be revealed by inconsistencies in the results obtained in different experiments. In addition physics beyond the Standard Model can lead to signals much larger than those predicted by the Standard Model, especially for channels that are forbidden or highly suppressed in the Standard Model. The search for, and eventually the measurement of, partial rates for very rare decays offers the opportunity to test predictions of the Standard Model at the quantum loop level; significant deviations would point to the presence of new physics.

Searches for both these types of discrepancies, inconsistencies between parameters measured in different channels and larger than predicted signals for forbidden or loop-suppressed decays, provide important low-energy windows on the high-energy-scale physics beyond the Standard Model and should be energetically pursued. To date the searches for rare decays have been pursued to the most stringent limits in K decays, where the extremely small Standard Model predictions leave many orders of magnitude to search for non-Standard effects. However searches in D, B and tau decays probe possibly different coupling sectors and thus interesting constraints on models also come from these searches, despite the fact that at first glance the numbers appear less stringent.

The existence of three families of quarks is by now well established, and with that the existence of a nontrivial phase in the most general matrix of weak coupling constants (the CKM matrix). Consequently, unless there is some symmetry which forces this phase to be identically 0° or 180°, there will be CP violation. However, while the occurrence of CP violation should not be viewed as unusual in the context of the Standard Model, it remains to be seen whether this phase is indeed the source of CP violation in nature.

The CKM matrix can be written as

$$V = \begin{pmatrix} V_{ud} & V_{us} & V_{ub} \\ V_{cd} & V_{cs} & V_{cb} \\ V_{td} & V_{ts} & V_{tb} \end{pmatrix} \quad (1)$$

$$\simeq \begin{pmatrix} 1-\lambda^2/2 & \lambda & A\lambda^3(\rho - i\eta) \\ -\lambda & 1-\lambda^2/2 & A\lambda^2 \\ A\lambda^3(1-\rho-i\eta) & -A\lambda^2 & 1 \end{pmatrix} + O(\lambda^4) \quad (2)$$

In the above expression $\lambda = \sin(\theta_{Cabibbo})$, and the three remaining parameters A, ρ, and η encode the remaining two weak mixing angles and the phase that introduces CP violation. This parametrization, due to Wolfenstein [2], is frequently used in discussing CP violating effects. It is given here up to terms of order λ^3; higher order terms in λ can be calculated from the Unitarity requirements. The CP violation in the Standard Model all arises from the parameter η in this matrix and vanishes if $\eta = 0$. We introduce this notation here because experimental results, particularly those on CP violation, are often presented in terms of the Wolfenstein parameters. Determination of the magnitudes of CKM matrix elements, obtained independently of CP violation measurements, are a necessary preliminary to the test of the hypothesis that CP violation is "explained", (*i.e.*, parametrized) by the phase in the CKM matrix. Further, they are an essential component of the general program to overdetermine Standard Model parameters and thereby search for effects due to physics beyond the Standard Model.

CP violation, while readily accommodated in the three–generation Standard Model with the appearance of η in the CKM matrix, remains one of the least understood aspects of particle physics. There is as yet no evidence that η is in fact the only or even the principal source of the observed CP violation in K decays. Cosmologists attempting to understand the dominance of matter over antimatter in the universe generally agree that it is unlikely that the Standard Model with a single Higgs multiplet can provide sufficient CP violation to drive a mechanism for baryon number generation at the weak scale that can explain the observed baryon to entropy ratio in the universe [3]. Furthermore baryon generation at higher temperatures, for example at the grand unification scale must be protected by special symmetries to avoid dilution by baryon-number-violating effects that continue to occur in most models at all temperatures above the weak scale. Thus it is extremely important and interesting to study all possible sources of CP violation, in order to find what mechanisms are at work in this physics.

As with the quark sector, so in the lepton sector the family structure and conservation laws of lepton flavor are little understood, and may in fact be only approximate in models beyond the Standard Model. Once again it behooves us to study the parameters of this sector with as much redundancy as possible to seek pointers to non-Standard effects. The search for forbidden processes, exemplified by the search for decays violating lepton number, holds forth the possibility of revealing a "smoking gun" of evidence that *there is* physics beyond the Standard Model.

In the Standard Model the tau is a heavy electron-like particle, which comes with its partner, the left-handed tau neutrino. From the measurements of the Z width at SLC and LEP [4] we know that the tau lepton is the heaviest sequential lepton, *i.e.*, the last charged lepton with a light neutrino. Many extensions of the Standard Model contain mass–dependent couplings, and therefore, predict much larger deviations from the standard theory in tau data than in the muon sector, so again measurements that appear redundant in the Standard Model need to be pursued.

In addition to the extraction of the fundamental Standard Model parameters and searches for non-Standard effects related to heavy quark and lepton sectors there is also considerable physics to be learned from studies of spectra and branching ratios to particular decay channels of heavy flavor mesons and baryons, and also from hadronic decays of the tau. Recent developments in the theory of these decays include heavy quark effective theory which provides a systematic expansion in $1/m_q$ about a rigorous infinite quark mass limit. Improved lattice calculational techniques are also likely to shed new light on hadronic parameters relevant to these measurements such as hadronic matrix elements and decay form factors. Much remains to be learned in this area, both on the theoretical side in improved calculations and on the experimental side in new or improved branching ratio and spectrum determinations.

In what follows Section 2 discusses the determination of Standard Model parameters, Section 3 is devoted to CP violation, Section 4 to rare and forbidden decays and Section 5 to hadronic physics effects in heavy flavor decays. Finally in Section 6 we discuss the present and proposed facilities for furthering these investigations.

2 Standard Model Parameters and Tests

Below we discuss the determination of the CKM matrix, the determination of heavy quark masses, and detailed studies within the tau–lepton sector.

2.1 CKM Matrix Determination

Given the constraint of unitarity, a 3-family CKM matrix is described by 4 parameters, three angles and a phase. The Standard Model provides no guidance as to their values. Tree-level processes involving the first two families have been used to determine one of these four parameters, the Cabibbo angle, which essentially determines V_{ud}, V_{us}, V_{cd}, and V_{cs}. Tree-level processes involving b decay can determine two more, $|V_{cb}|$ and $|V_{ub}|$, although obtaining reasonable precision on $|V_{ub}|$ is not easy. The fourth parameter is somewhat of a dilemma. It can be obtained from *CP* violation measurements, as discussed in Section 3. However, one would like to determine the CKM matrix *without* using such measurements, to obtain a sharper test of the origin of *CP* violation. From Eqs. (1) and (2) it is clear that measurements of $|V_{ud}|$ and $|V_{td}|$ determine ρ and η and hence the phase in the CKM matrix. In principle, this is measurable from tree-level decays of top, but in practice such an approach seems very far off. So, one resorts to loop processes involving b and s quarks. Below we discuss determination of $|V_{cb}|$, $|V_{ub}|$ and $|V_{td}|$. ($|V_{tb}|$ and $|V_{ts}|$ are obtained from unitarity, to much better accuracy than is possible from their direct measurement.) The present knowledge of the CKM matrix is summarized in the Particle Data Group report [1].

2.1.1 $|V_{cb}|$

Two methods have been used to determine $|V_{cb}|$. Both work. At this stage it is a matter of improving accuracy.

- From average lifetime and inclusive semileptonic decay branching ratio.

The semileptonic width of the b quark, in the naive parton model, is

$$\Gamma(b \to c\ell\nu) = \frac{G_F^2 m_b^5}{192\pi^3}|V_{cb}|^2 f_1\left(\frac{m_c}{m_b}\right) \tag{3}$$

Inclusive lifetime measurements now have an accuracy of $\pm$ 4%, and inclusive semileptonic decay branching ratio measurements an accuracy of $\pm$ 3%, which implies a determination of $|V_{cb}|$ to $\pm\ 2\frac{1}{2}$%. There are perturbative QCD corrections to Eq. (3), which are not a problem. The problem is the factor m_b^5, because what to use for m_b is not clear. An uncertainty of 200 MeV in m_b translates into a $\pm$ 10% uncertainty in $|V_{cb}|$. If the difference $m_b - m_c$ is assumed known, this uncertainty drops to $\pm$4%. Theoretical work is needed here, to determine m_b and m_c, and to account for the kinetic energy of the b quark inside the hadron [5,6].

- From $B \to D^*\ell\nu$, extrapolating to zero-recoil point.

At the zero-recoil point, the value of the form factors can be determined, from Heavy Quark Effective Theory (HQET, see Section 5.1), to an accuracy of $\pm$ 4%, implying a theoretical limitation on the determination of $|V_{cb}|$ of $\pm$ 4%, with opportunities for improvements [6,7]. The problem with this method is that the zero-recoil point is at one end of the physically allowed range. Consequently, the rate cannot be measured by interpolating from both sides of the point of interest, but only by extrapolating from one side. Improvements to a determination of $|V_{cb}|$ by this method will come both from experimental work (higher statistical accuracy, so the extrapolation can be done more accurately), and from theoretical work (understanding the shape of the form factor near the zero-recoil point, to guide the extrapolation, and determining the form factor at zero recoil more accurately).

2.1.2 $|V_{ub}|$

Four methods for determining $|V_{ub}|$ will be discussed. So far only the first has given useful results.

- Lepton spectrum near the endpoint.

By measuring the yield of leptons beyond the endpoint for $b \to c\ell\nu$, both CLEO [8] and ARGUS [9] have established that the process $b \to u\ell\nu$ occurs, and therefore $|V_{ub}|$ is non-zero. Unfortunately, there are major theoretical problems in determining what fraction of the spectrum from $b \to u\ell\nu$ lies in this endpoint region [10]. Current theoretical understanding limits the accuracy of a determination of $|V_{ub}|$ by this method to about $\pm$ 25%, but recent work with HQET shows promise of improving the situation [11].

- Exclusive Semileptonic Decays, with Tagged B's.

If one of the B's in a $B\bar{B}$ event at the Υ(4S) is fully reconstructed, then the four-momentum of the second B is known, and its exclusive semileptonic decays can be reconstructed in spite of the neutrino. Given a sufficiently large sample of $B\bar{B}$ events with one B fully reconstructed (a "tagged B" sample), one could observe the charmless exclusive decays $B \to \pi\ell\nu$ and $B \to \rho\ell\nu$. HQET relates these decays to $D \to \pi\ell\nu$ and $D \to \rho\ell\nu$. If these D decays have also been measured, one can extract $|V_{ub}/V_{cd}|$. (Since the D decay measurements are far easier than the B decay measurements, they will likely be done far sooner.)

The major stumbling block to this method is obtaining a sufficiently large tagged B sample. The technique requires that the $B\bar{B}$ pair be produced at the Υ(4S), and so rates will be those of e^+e^- facilities. CLEO obtains $\sim$ 1000 tagged B's/fb^{-1}. The branching ratios for $B \to \pi\ell\nu$ and $B \to \rho\ell\nu$ are $\sim 10^{-4}$. So a reasonable measurement (100 events) requires 10^6 tagged B events (either $\sim$ 1000 fb^{-1} or substantial improvements in B reconstruction efficiency). For scale, CLEO has logged 1 fb^{-1}/yr for the past few years, and a B Factory running at design luminosity would log 30 fb^{-1}/yr. Said more directly, in its simplest form this approach won't work.

- "Neutrino Detection".

Given a sufficiently hermetic detector operating at the Υ(4S), one can use the missing 4-momentum of the event to select events for which the only undetected particle is a single neutrino. Then, taking the missing 3-momentum of the event to be that of the neutrino, one can study $b \to u\ell\nu$ decays in two ways.

(a) Exclusive Semileptonic Decays. Given the neutrino, $B \to \pi \ell \nu$ and $B \to \rho \ell \nu$ can be reconstructed using the standard approach that works for hadronic decays. Sample sizes of 20 fb^{-1} might be adequate, a considerable reduction over that required with method 2. (At an asymmetric collider, one may be able to use the vertexing of the two B's to reconstruct these exclusive modes even without a fully hermetic detector.)

(b) Inclusive Semileptonic Decays. Given the neutrino and the lepton, one can determine, to reasonable accuracy, the mass M_X in a decay $B \to X \ell \nu$. (If the B were at rest, M_X could be precisely determined. The B momentum of 300 MeV/c, with unknown direction, smears the determination of M_X by a few hundred MeV.) In this way, one can separate $B \to X_u \ell \nu$ from $B \to X_c \ell \nu$, for $M_{X_u} \stackrel{\sim}{<} 1.6$ GeV. This makes accessible a much larger fraction of the phase space for $b \to u \ell \nu$ than is accessible with method 1.
Preliminary work with "neutrino detection" by CLEO shows promise.

- Exclusive Hadronic $b \to u$ Decays.

CLEO, with 2 fb^{-1}, has a hint of $\bar{B}^0 \to \pi^+\pi^-$ [12]. With a sample of 20 fb^{-1}, several charmless hadronic decays will be measured. However, it is questionable whether theory will be adequate to the task of obtaining $|V_{ub}|$, at the sub-25% level, from measurements of hadronic decays.

2.1.3 $|V_{td}|$

Five methods for determining $|V_{td}|$ will be discussed. So far, only the first has given useful results.

- $B^0 - \bar{B}^0$ mixing, m_t, and f_B from theory.

The $B^0 - \bar{B}^0$ mixing parameter $x_d = \Delta M/\Gamma$, measured to ± 15% [13,14], is related to $|V_{td}|$ via the expression [15]:

$$x_d = \frac{G_F^2}{6\pi^2} m_t^2 \tau_B B_B f_B^2 M_B |V_{td} V_{tb}^*|^2 \frac{A(z_t)}{z_t} \eta_{QCD}, \qquad (4)$$

with $z_t = m_t^2/M_W^2$, and $A(z_t)$ a known function. The QCD correction factor η_{QCD} is known to ± 5%, posing no problem. The problems are the top quark mass m_t and $B_B f_B^2$, where B_B is the "bag parameter" arising from the vacuum insertion approximation, and f_B is the B meson decay constant. At present m_t is known to ± 10%, and it is reasonable to expect improvements to ± 5% soon. Then the only problem will be $B_B f_B^2$.

There are calculations of B_B and of f_B^2, by lattice gauge techniques and other methods [16]. Currently these determine $\sqrt{B_B} f_B$ to perhaps 20 – 25 %, and this uncertainty dominates the error on $|V_{td}|$ extracted in this way. Significant improvements in this method depend on improvements in lattice gauge calculations (in particular better understanding of the quenched approximation), and also on improved measurements of f_D and/or f_{D_s}, which can provide tests of the theoretical calculations.

- Ratio of $B_d - \bar{B}_d$ and $B_s - \bar{B}_s$ Mixing.

An expression similar to Eq. (4) applies to $B_s - \bar{B}_s$ mixing. Taking ratios of equations, one has

$$\frac{x_d}{x_s} = \frac{\tau_{B_d}}{\tau_{B_s}} \frac{B_{B_d}}{B_{B_s}} \frac{f_{B_d}^2}{f_{B_s}^2} \left| \frac{V_{td}}{V_{ts}} \right|^2. \qquad (5)$$

Now, theory is only required to calculate the ratio $(B_{B_d} f_{B_d}^2)/(B_{B_s} f_{B_s}^2)$, which should pose much less of a problem, since many common uncertainties cancel in the ratio. However, measuring $B_s - \bar{B}_s$ mixing is a major challenge. Because x_s is expected to be large, it is necessary to measure the time evolution of the mixing process. This might be done with SLD, at LEP, at an asymmetric B Factory or with a hadron collider. These experiments may be able to determine x_s if it is smaller than 20. Beyond that value, a measurement will be extremely difficult.

- $b \to d\gamma$.

Just as the QCD-uncorrected process $b \to s\gamma$ has a width proportional to $|V_{ts} V_{tb}^*|^2$, so the QCD-uncorrected process $b \to d\gamma$ has a width proportional to $|V_{td} V_{tb}^*|^2$ (see Section 4.3.2). Thus, one might expect, for example, that $Br(B \to \rho\gamma)/Br(B \to K^*\gamma)$ should equal $|V_{td}/V_{ts}|^2$, times a known kinematic correction factor. One needs to worry about the large QCD correction handled by renormalization group techniques, and about long distance effects. The former are believed to be under control [17]. The size of long distance effects is currently a matter of intense debate [18], [19], but there is agreement that $b \to d\gamma$ contains information on $|V_{td}|$, and that the theoretical work needed to extract this information is quite feasible. (Measurement of $D \to \rho\gamma$ would provide an important test of the theoretical understanding.) This would mean that a measurement of $b \to d\gamma$ exclusive decays will give V_{td}. With 3 fb^{-1}, CLEO has an upper limit [20], and with 20 times more data (2 years of design-level B Factory running), would have a useful measurement.

The processes $b \to s\ell^+\ell^-$ and $b \to d\ell^+\ell^-$ (*i.e.*, the same processes but with virtual photons) have smaller QCD corrections. Should the QCD corrections prove to be a major stumbling block for $b \to d\gamma$, then these might be more suitable. In particular, an eventual measurement of $(B^- \to \pi^-\mu^+\mu^-)/(B^- \to K^-\mu^+\mu^-)$, at a hadron collider, seems feasible.

- Determining f_B from $B^- \to \tau^-\nu$ or $B^- \to \mu^-\nu$.

These decays have widths proportional to $f_B^2|V_{ub}|^2$, and thus a successful measurement, combined with $B^0 - \bar{B}^0$ mixing (and theory for B_B) gives one $|V_{ub}/V_{td}|^2$, which would be very nice. The expected branching ratio for $B \to \tau\nu$ is $\sim 10^{-4} - 10^{-5}$, while that for $B \to \mu\nu$ is smaller by a factor of 230. The present experimental upper limit on $B \to \tau\nu$ is $\sim 1/4\%$ [21]. This limit was obtained with a 3 fb^{-1} data sample, so it is clear that improvements in technique are required in addition to increases in luminosity. The separate vertexing provided by an asymmetric machine may help to reduce backgrounds and so improve this situation, but hermiticity will still be a major problem.

- $K^+ \to \pi^+ \nu\bar{\nu}$.

This process, proportional to $|V_{td}|^2$, has an expected branching ratio in the range of $(0.5 - 3)\times 10^{-10}$. Unlike some other kaon decays, long distance corrections are very small and theoretical uncertainties are at the 10% level [22]. The current upper limit on the branching ratio is 3×10^{-9}. If the actual value is near 2×10^{-10}, then experiment BNL787 expects a measurement to $\pm$ 30%, giving $|V_{td}|$ to $\pm$ 15%. Upgrades in the detector and increases in beam intensity could improve the accuracy a factor of two.

Summary:

- Two methods have been used to give $|V_{cb}|$. Both work, and give determinations in the $\pm 5\%$ range. Accuracy will improve with more data and with theoretical refinements.
- For $|V_{ub}|$, "neutrino detection" is the most promising approach, with "proof of concept" by CLEO imminent.
- There are 5 methods which can give $|V_{td}|$. None is easy, and none is direct. All should be pursued. Should two methods give different answers, that could indicate additional particles in the loops, and would strongly suggest new physics.

2.2 Heavy Quark Masses

The problem of definition of quark masses has many implications for theoretical predictions. For example, meson decay rates for free quarks depend on the fifth power of the quark mass and thus predictions are sensitive to small shifts in this value. However, since quarks cannot be isolated there is no direct connection between the Lagrangian mass parameters that appear in such perturbative calculations of quark processes and any measurable quantity. Even in the context of perturbation theory there is a plethora of definitions in the literature, since the quark mass parameter is convention and renormalization scale dependent. Further it is difficult to relate these parameters to measurable quantities, especially in the

Current algebra relationships between meson masses and Lagrangian parameters (which lead to the old term "current quark masses" for these quantities) also leave some uncertainties because of higher order corrections, both from weak interaction perturbation theory and from corrections to the chiral limit. Lattice calculations of spectra can determine mass parameters on the lattice, and recently there has been substantial progress in relating these quantities to perturbatively-defined and to experimentally-measured quantities. Recent progress in heavy quark effective theory has introduced yet another definition of the quark mass in terms of the equation

$$M_Q = m_q + \bar{\Lambda} - \frac{\lambda_1 + 3\lambda_2}{2m_q} + \mathcal{O}(1/m_q^2) \qquad (6)$$

where M_Q is the pseudoscalar meson mass, m_q is the quark mass, $\bar{\Lambda}$ is a parameter which may be interpreted as the energy of the light degrees of freedom, λ_1 parametrizes the kinetic energy of the heavy quark inside the hadron, and λ_2 parametrizes the effects of the chromomagnetic moment operator. One readily determines λ_2 from the vector-pseudoscalar mass difference, since for vector mesons the above equation is modified by the replacement of $3\lambda_2$ with $-\lambda_2$. $\bar{\Lambda}$ and λ_1 cannot be so readily determined, and are the objects of current theoretical activity, as are the $1/m_q^2$ corrections. Systematic treatment of this type, and other similar work starting from lattice calculations of meson properties, offer perhaps the best approach yet to the problem of the relationship of theoretically defined quark mass parameters to experimentally measured quantities. This problem is one where further theoretical work is needed, to determine the most precise way to extract these important parameters from experiment.

The experimental inputs used so far are the masses of the $q\bar{q}$ bound states and the masses of the heavy-light mesons. A measurement of the photon energy spectrum in $b \to s\gamma$ would provide additional information on the b quark mass; at the naive spectator model level $\langle E_\gamma \rangle = m_b/2$. HQET calculations bear this result out, giving small corrections [23,24]. Extrapolating from studies with CLEO [25], a 30 fb^{-1} data sample would determine m_b to $\pm$ 100 MeV (statistical error only).

For the top quark, which decays before bound states form, the relationship between the Lagrangian parameter in perturbation theory and predicted top-quark production cross sections is more direct and hence there is less ambiguity in the definition of the top quark mass parameter. (Determination of the top quark mass is considered by the Electroweak Working Group.)

2.3 Tau parameters and Tests

Thanks to its mass, the tau is the only lepton that can decay to hadrons (*i.e.*, $\bar{u}d$ and $\bar{u}s$ quark pairs). This provides an opportunity to study strong interaction phenomena in weak charged–current decays in particularly clean experimental and theoretical environments. When approached in an inclusive way, semi-hadronic decays of the tau provide interesting tests of perturbative QCD and give a determination of the scale of strong interactions. The phenomenology of low–mass hadronic resonances finds a rich, and often unique, source of information in individual semi-hadronic tau decays modes.

The tau lepton also plays a special role at LEP and SLC experiments performed at the Z peak, since it is the only lepton that copiously decays inside the apparatus, offering an independent method of measuring the Lorentz structure of weak neutral currents via lepton–polarization measurements.

We discuss lepton universality tests and measurements of Lorentz structure of charged and neutral weak currents involving τ's below. Searches for flavor violating and other rare decays are discussed in Section 4. Measurements of the tau neutrino mass and searches for neutrino mixing are described by the Neutrino Working Group. QCD tests in semi-hadronic decays of tau are covered in Section 5.

2.3.1 Lepton Universality Tests

- Purely leptonic decays

The Standard Model predicts that couplings of leptons to charged and neutral currents are flavor independent. Test of this assumption can be obtained by comparison of the electronic decay of tau, $\tau \to \nu_\tau e \bar{\nu}_e$, to muon decay, $\mu \to \nu_\mu e \bar{\nu}_e$. These two processes differ only by different phase-space (dependence on m_τ and m_μ) and possibly different couplings of τ and μ (g_τ and g_μ). In comparison of the measured decay branching fractions, one also has to include different lifetimes (τ_τ and τ_μ) and radiative corrections (R_c):

$$\left(\frac{g_\tau}{g_\mu}\right)^2 = \frac{\tau_\mu}{\tau_\tau}\left(\frac{m_\mu}{m_\tau}\right)^5 \frac{B(\tau \to \nu_\tau e \bar{\nu}_e)}{B(\mu \to \nu_\mu e \bar{\nu}_e)} (R_c) \qquad (7)$$

Since each of the μ parameters is known with accuracy roughly three orders of magnitude better than the corresponding τ parameters the above test is totally dominated by experimental errors for τ lepton. Theoretical uncertainty is only of the order of 0.01%. Using the world average values, $\tau - \mu$ universality is now tested to about 1%: $g_\tau/g_\mu = 0.996 \pm 0.010$ [26]. There has been significant progress in accuracy of the measurements of τ parameters over last few years, and we are now close to the accuracy reached in $\mu - e$ universality tests: $g_\mu/g_e = 1.003 \pm 0.003$. (Further improvement on this ratio can be expected from a new branching ratio measurement at TRIUMF). At present, the contribution to the error on g_τ/g_μ are as follows: $\Delta\tau/\tau = 0.8\%$, $5\Delta m_\tau/m_\tau = 0.08\%$, and $\Delta B_e/B_e = 0.6\%$. The tau mass has been precisely measured by the BES experiment at the Beijing collider (BEPC), which operates at the τ–production threshold [27]. Limited further improvement is possible by devoting more beam time for this measurement at BEPC, in combination with improved beam-energy determination, or by construction of a τ–charm factory.

There are two complementary paths to improve tau lifetime measurements: experiments at the Z peak which are statistics limited; measurements at intermediate e^+e^- collision energies (CESR or future B Factories), which are also systematics limited. Increased statistics at LEP, and upgrade of the CLEO-II detector to include a silicon vertex detector and upgrade of CESR to higher luminosity will yield improvements in the lifetime measurements in next few years. Asymmetric B Factories are also expected to play an important role here. The most precise measurement of B_e has been obtained by CLEO [28]. Even though, there is some room for improvement, CLEO measurements will eventually be limited to $\sim 0.5\%$ by the error on theoretical cross section for $\tau^+\tau^-$ and Bhabha events (which are used to determine the luminosity). Even if these theoretical limitations are overcome by more precise calculations, improved measurement would require a dedicated effort to control experimental systematics at this level. High statistics measurement at the Z peak offers perhaps more attractive route to improve B_e measurement in the long run, because theoretical normalization errors are already smaller and because of possibility of directly counting number of produced tau leptons (better separation from backgrounds).

- Semi-hadronic decays

An independent test of $\tau - \mu$ universality can be obtained from comparison of $\tau \to \pi\nu_\tau$ with $\pi \to \mu\bar{\nu}_\mu$ (or $\pi \leftrightarrow K$). At present this test is limited to precision of about 2% due to the error on $B(\tau \to \nu_\tau\pi, K)$. Improvements are expected from both LEP experiments and CLEO. A τ–charm factory which would operate at the τ-pair production threshold, where two body decays can be isolated from other τ decays by kinematic constraints, may be able to achieve a somewhat better measurement, perhaps down to a fraction of a percent.

- Tests at hadron machines

Universality of $\tau - \mu - e$ couplings can be also tested in decays of $W \to l\bar{\nu}_l$. So far, these measurements have rather large errors: 8% on g_τ/g_e (UA1+UA2). CDF and DØ are expected to improve the accuracy of these measurements to about 4% (See the Electroweak Working Group report).

2.3.2 Lorentz Structure of the Weak Currents

- Purely leptonic decays

The Lorentz structure of four-fermion interactions in τ decays can be studied in full analogy with μ decays. The charged lepton momentum spectrum and angular distribution in the τ rest frame can be written in the most general form which allows for scalar, vector and tensor interactions with help of the Michel parameters

$(\rho, \eta, \xi, \delta)$ [29]:

$$\frac{d\Gamma}{dx\, d\cos\vartheta} \propto x^2 \left[12(1-x) + \rho\tfrac{4}{3}(8x-6) + 24\eta\tfrac{m_l}{m_\tau}\tfrac{1-x}{x}\right] \\ -\xi\mathcal{P}_\tau \cos\vartheta\, x^2 \left[4(1-x) + \delta\tfrac{4}{3}(8x-6)\right] \tag{8}$$

where $\mathcal{P}_\tau$ is τ lepton polarization, $x = P_l/P_{max}$, and ϑ is an angle between τ spin and $\vec{P_l}$. The Standard Model ("V-A") predicts: $\rho = \frac{3}{4}$, $\eta = 0$, $\xi = 1$, and $\delta = \frac{3}{4}$. Values of these parameters are sensitive to small admixtures of scalar or right-handed intermediate bosons, and thus their measurement provides a search for new physics.

The spectral shape parameters, ρ and η, are measured from fits to the observed electron and muon spectra in τ decays. Sensitivity to η parameter is better in $\tau \to \nu_\tau \mu \bar{\nu_\mu}$ because this parameter enters multiplied by the m_l/m_τ ratio. Since the η parameter changes also the overall rate, precision measurements of B_e and B_μ allow determination of η. The decay asymmetry parameters, ξ and δ, are more difficult to measure because they depend on tau polarization which averages out to zero when integrated over all angles in $\gamma \to \tau^+\tau^-$ production. Thus, measurements at lower energy e^+e^- colliders must study correlations in leptonic decays of the two τ's in the event. Measurements using $Z \to \tau^+\tau^-$ have an advantage of average $\mathcal{P}_\tau \neq 0$. However, usually the value of $\mathcal{P}_\tau$ and not the Michel parameters are the subject of measurements at LEP and SLC.

The measured Michel parameters in τ decays [29], $\rho = 0.733 \pm 0.026$, $\eta = -0.010 \pm 0.052$ and $|\xi| = 0.90 \pm 0.18$, are consistent with the Standard Model but far less precise than corresponding measurements in muon decays: $\rho = 0.752 \pm 0.003$, $\eta = -0.007 \pm 0.013$, $\xi = 1.003 \pm 0.008$ and $\delta = 0.745 \pm 0.004$. The MEGA experiment at LAMPF will further improve measurements of ρ parameter in muon decays. The best measurements in τ decays come so far from the ARGUS experiment at DORIS. CLEO and LEP should soon improve these measurements by analysis of already recorded data. Further improvements are likely to be systematics limited and will require a dedicated effort. A tau-charm factory would be the best suited for precision measurements of the Michel parameters because sensitivity to the Michel parameters is best in the tau rest frame. Also cross-feed from the other τ decays can be better controlled at the τ-pair production threshold. Precision of a fraction of a percent is achievable at a τ-charm factory.

- Semi-hadronic decays

Semi-hadronic decays offer independent opportunity to test V-A coupling at the $\tau - W - \nu_\tau$ vertex. Studies of energy and angle correlations among final state hadrons from decays of both τ's in the event allow one to measure tau neutrino helicity (-1 in the Standard Model). Measurements with accuracy of a few percent are now available from ARGUS, CLEO-II and ALEPH experiments.

- Neutral currents

At LEP and SLC, tau decays act as polarimeter for measurement of lepton polarization in $e^+e^- \to Z \to \tau^+\tau^-$. This allows the determination of V and A couplings to Z of τ's and e's independently (test of $\tau - e$ universality). Furthermore, determination of the relative sign of the V and A couplings is possible (not available from other measurements at Z). The Standard Model predictions for these couplings depend on top quark and Higgs masses (See the Electroweak Working Group report). These measurements are part of the core program at LEP and SLC. LEP experiments have larger data samples. The SLD experiment at SLC offers measurements with a different set of systematics. The measurements at the SLC also benefit considerably from the electron beam polarization.

3 *CP* Violation and *T* Violation

For neutral pseudoscalar mesons with non-trivial flavor content the mass eigenstates are linear combinations of the flavor eigenstates

$$X_{1,2} = (pX^0 \pm q\overline{X}^0)/\sqrt{pp^* + qq^*} \tag{9}$$

where X^0 may be K^0, D^0, B^0 etc. If *CP* were an exact symmetry then these states would have $p = q = 1$, *i.e.*, the mass eigenstates would also be *CP*-eigenstates. However, as is well known in the kaon system, *CP*-violating phases in the mixing amplitudes can introduce deviations from this form. In principle this is true for vector as well as pseudoscalar mesons, but since the former decay strongly there is never a chance to observe the effects of mixing in such states. Similarly, in the Standard Model, mixing in the D^0 states is negligible, *i.e.*, they decay before any appreciable mixing effects can be observed.

In general one can distinguish three major types of *CP* violation [30]. Firstly there are two distinct categories of indirect *CP* violation. Indirect *CP* violation effects can arise because of interference between the amplitudes for the decay of a neutral meson to a final state of definite *CP* and the decay to that same state after mixing of the neutral meson with its antiparticle. This effect has been observed in neutral K decays, for example to two pions. Similar effects could also occur in the neutral D and B systems, although in the Standard Model the prediction is that such effects are much smaller in the D system, because of the smaller mixing in that system. The second type of indirect *CP* violation occurs if mixing effects in the neutral K, D, or B sectors lead to a deviation from unity for the absolute value of the ratio of the weights (p and q) of the two *CP*-conjugate mesons in the mass eigenstates of the system ($|q/p| \neq 1$). This then leads to *CP*-violating rate asymmetries in the decay of neutral mesons to final states of definite flavor. Effects of this latter type are also observed in the K system, for example the asymmetries in the semileptonic decay channels are of this type. However, such effects are expected to be very small in the B system in the Standard Model.

The third type of CP violation is direct CP violation, *i.e.*, that due to a difference between the magnitude of the decay amplitude A for a particular process and the CP-conjugate amplitude $\overline{A}$, $|A/\overline{A}| \neq 1$. This requires that there be at least two independent contributions to the amplitude that contribute with different strong phases and different weak phases [31]. In the Standard Model the tree diagrams for a given decay all contribute with the same weak phase and thus direct CP violation effects occur only when there are loop (penguin) contributions. Direct CP violation can occur in charged decays, in neutral decays to flavor tagged eigenstates or in neutral decays to CP eigenstates. No convincing evidence for direct CP violation has yet been established.

Discussions of CP violation in the K system in general do not make a strong distinction between the two types of indirect CP violation, since both occur in that system. However note that the first type of indirect process – interference of mixed and unmixed decays to CP eigenstates in the neutral meson sector – can occur even when both $|A/\overline{A}| = 1$ and $|q/p| = 1$. It requires only that the weak phase of the mixing be different from the weak phase of the decay amplitude. This is the predominant effect predicted by the Standard Model for B decays.

Since different models can give quite different predictions for the each of the three types of CP violation, and for the different strengths of expected effects in K, D and B systems, as well as in baryonic decays, there are a plethora of channels that need to be explored. Only by exploiting as many opportunities as possible to explore CP violating effects can we hope to unravel the mystery of its source and perhaps to understand thereby the baryon domination of the universe. This is a field where much remains to be learned.

3.1 CPT and T Violation

In addition it is interesting to ask what experimental limits can be set on the possibility of violations of CPT symmetry. CPT symmetry is fundamental and is an exact symmetry in any field theory [32], so any experimental evidence that it is broken would require a serious re-evaluation of our understanding of all particle processes. Thus searches for CPT non-conservation are in the interesting situation where there is a very small probability of success, but where it would certainly have very large effect on our understanding of particle physics if a convincing CPT-violating result were to be obtained.

CPT violating effects could be observed via small differences in particle-antiparticle masses and lifetimes. The proton-antiproton and electron-positron fractional mass difference is is less than approximately 4×10^{-8}, at 90% confidence level, which is not sufficiently accurate to provide a sensitive probe of CPT symmetry. The experimental situation is considerably better, however, in the K^0 sector. Defining

$$\eta_{+-} = \frac{A(K_L^0 \to \pi^+\pi^-)}{A(K_S^0 \to \pi^+\pi^-)} = \epsilon + \epsilon' \quad (10)$$

$$\eta_{00} = \frac{A(K_L^0 \to \pi^0\pi^0)}{A(K_S^0 \to \pi^0\pi^0)} = \epsilon - 2\epsilon' \quad (11)$$

$$\Phi_{+-} = Arg(\eta_{+-}),\ \Phi_{00} = Arg(\eta_{00}),\ \Delta\Phi = \Phi_{00} - \Phi_{+-}$$

it can be shown that [33]

$$\frac{m_{K^0} - m_{\bar{K}^0}}{m_K} \simeq (\frac{\Delta m}{m_K})\sqrt{2}|\eta_{+-}|\tan(\Phi_{+-} - \Phi_\epsilon + \Delta\Phi/3). \quad (12)$$

Here $\Phi_\epsilon = Arg(\epsilon)$ and $\Delta m = 3.479 \times 10^{-6}$ eV is the $K_L^0 - K_S^0$ mass difference. This small value for Δm provides considerable leverage in testing CPT since this gives $\Delta m/m_K = 6.99 \times 10^{-15}$. Measurement of $(\Phi_{+-} - \Phi_\epsilon)$ and $\Delta\Phi$ therefore can provide a direct test of CPT violation. Current best measurements from FNAL which are at present the most sensitive to this effect are consistent with CPT symmetry, giving

$$\Delta\Phi = 0.67^\circ \pm 1.0^\circ, \quad (13)$$

$$\Phi_{+-} - \Phi_\epsilon = -0.38^\circ \pm 1.1^\circ \quad (14)$$

There is no particularly natural scale at which CPT symmetry might be violated. Theorists view renormalizable field theory as the low-energy limit of some more general theory which might not itself be a field theory, and which could in principle contain the seeds of CPT violation. Continued efforts on measurements of Φ_{+-}, Φ_ϵ, and $\Delta\Phi$ could yield a surprise of surely fundamental importance.

Given CPT invariance, T violation is equivalent to CP violation. One of the strongest available constraints on model building for beyond-Standard-Model CP-violating physics is the upper limit on the value of the T-violating neutron electric dipole moment $d_n < 1.2 \times 10^{-25}$ e cm [34]. In the Standard Model, despite the known CP violation, the prediction for this quantity is very small, $\leq 10^{-31}$ e cm [35]. However the introduction of other sources of CP violation into the theory can dramatically change this prediction [36]. Many models have been ruled out by their failure to keep this quantity smaller than the measured upper limit. Experiments searching for an effect due to such a moment should continue.

3.2 CP Violation in K Decays

So far, the only established laboratory manifestations of *CP* violation are the indirect effects in the decays of neutral K mesons, such as the observation of $K_L^0 \rightarrow \pi\pi$ decays. The Standard Model predicts also a direct *CP* violation due to the interference between tree and penguin amplitudes for weak decays. In the Standard Model direct *CP* violation would imply a non-zero phase in the CKM matrix, ($\eta \neq 0$ in the Wolfenstein parametrization), whereas indirect *CP* violation effects, of the types already observed in K decays, could occur in a model where all weak coupling decay amplitudes are real, ($\eta = 0$), and the only source of *CP* violation is a phase in the mixing amplitude coming from some non-Standard-Model contribution [37].

In the kaon system, the traditional way to look for direct *CP* violation is to measure the parameter ratio ϵ'/ϵ for the 2π decays of the neutral kaon. This amplitude is directly proportional to η but with a factor that is difficult to calculate; thus a non-zero value of ϵ'/ϵ will establish direct *CP* violation but to obtain a precise value for η will require improvement in calculational techniques, such as lattice calculations, for hadronic matrix elements. Progress continues to be made in this area. However, the value of ϵ' is suppressed for high values of m_t because of a cancellation between different contributions; estimates now are in the range of about $3 \times 10^{-3}\eta$ with about a factor of 2 uncertainty. For a recent review of the search for direct *CP* violation effects see [38]. The current experimental situation is ambiguous. One experiment (CERN NA31) reports a non-zero result [39]

$$\epsilon'/\epsilon = (23 \pm 6.5) \times 10^{-4}, \tag{15}$$

while a second experiment (FNAL E731) sees a result consistent with zero [40]

$$\epsilon'/\epsilon = (7.4 \pm 5.9) \times 10^{-4}. \tag{16}$$

Each of these experiments is now entering a second round of measurement with improved apparatus (FNAL E832 and CERN NA48). Clearly it is important to resolve this issue.

The other decays in the kaon system that hold promise for the detection of direct *CP* violation are the modes $K_L^0 \rightarrow \pi^0 l\bar{l}$ where l can be either e, μ, or ν [41]. For the charged lepton states, the branching ratio (proportional to $|\eta|^2$) are in the range of $10^{-11} - 10^{-12}$. It is most interesting to note that these decay modes have both direct and indirect *CP*-violating contributions. Their ratio is not known but may be of order of unity. The direct piece is primarily a short-distance effect that can be fairly reliably calculated. The indirect *CP*-violating amplitude can be determined most reliably by a measurement of $K_S^0 \rightarrow \pi^0 e^+ e^-$. In the absence of this a measurement of the rate and form-factor in $K^+ \rightarrow \pi^+ e^+ e^-$ and $K^+ \rightarrow \pi^+ \mu^+ \mu^-$, plus Chiral Perturbation theory can determine this amplitude. There is also the *CP*-conserving amplitude that is potentially of the same size as the *CP*-violating one, coming from a $\pi^0\gamma\gamma$ intermediate state. This can be determined by measurements of $K_L^0 \rightarrow \pi^0\gamma\gamma$. Both the indirect and the *CP*-conserving contributions must be known so that the indirect *CP* effect, if observed, can be interpreted correctly. The latest Brookhaven, FNAL and CERN kaon results (*i.e.*, the M_{ee} spectrum of the $K^+ \rightarrow \pi^+ e^+ e^-$ and the $M_{\gamma\gamma}$ spectrum of the $K_L^0 \rightarrow \pi^0\gamma\gamma$ seem to indicate that these two amplitudes are small. An observation of $K_L^0 \rightarrow \pi^0 e^+ e^-$ or $K_L^0 \rightarrow \pi^0 \mu^+ \mu^-$ will thus be a very exciting opportunity to look into direct *CP* violation effect.

Perhaps the most theoretically attractive probe for direct *CP* violation in the K system is the decay $K_L^0 \rightarrow \pi^0 \nu\bar{\nu}$. Unlike virtually all other decays in the kaon system, the decay is essentially pure direct *CP* violating; its branching ratio is given approximately by

$$\begin{aligned} BR(K_L^0 \rightarrow \pi^0 \nu\bar{\nu}) &= 8 \times 10^{-11} (m_t/m_W)^{2.2} A^4 \eta^2 \\ &= 2 \times 10^{-10} \eta^2 \end{aligned} \tag{17}$$

where A is the parameter that appears in V_{cb} in the Wolfenstein parametrization of the CKM matrix. Given the lack of theoretical ambiguity the measurement of this branching ratio would be one of the cleanest possible ways to extract η. Table 1 shows the current best limits from FNAL E799 [42,43].

Table 1: Limits on $K_L^0 \rightarrow \pi l\bar{l}$

Process	Limit
$K_L^0 \rightarrow \pi^0 e^+ e^-$	4.3×10^{-9}
$K_L^0 \rightarrow \pi^0 \mu^+ \mu^-$	5.1×10^{-9}
$K_L^0 \rightarrow \pi^0 \nu\bar{\nu}$	5.8×10^{-5}

It is important that these rare branching ratios be measured with whatever precision is possible. They afford one of the best ways of determining the parameter η directly and they will complement the eventual measurements of the angles of the unitarity triangle from neutral B decays. Ways to search for direct *CP* violation in strange quark decays other than neutral kaon decays include:

- a comparison of decay asymmetry parameters from hyperon and its *CP* conjugate antihyperon decays (A new search for this effect is planned at FNAL.) [44];
- a comparison of Dalitz slope difference of $K^+ \rightarrow \pi^+\pi^+\pi^-$ and $K^- \rightarrow \pi^-\pi^-\pi^+$ (The current limit is about 1%.); and
- a non-zero value of the transverse muon polarization in $K_{\mu 3}$ decay, which is a triple-vector correlation and

odd under time-reversal (thus *CP*-violating if one assumes *CPT* invariance). The current limit is of order 0.03 [45]. A new search for this effect is planned at KEK [46]. The Standard Model predictions for these asymmetries are of order 10^{-5}.

3.3 CP violation in the Charm Sector

Because the GIM cancellations in the d, s, b sector are more exact than in the u, c, t sector both mixing and penguin effects are expected to be small in the charm mesons. Thus Standard Model predictions for mixing of the neutral D mesons are generally very low, 10^{-8} or less. Predictions for direct *CP* violation in D decays are somewhat more encouraging. Particle and antiparticle partial decay rates to charge conjugate final states can differ only if (at least) two amplitudes with different weak phases and different strong phases lead to the same final state. In the Standard Model this occurs only for Cabibbo-suppressed decays. Furthermore decay modes for which one anticipates that at least two amplitudes contribute to the same final state are generally quite difficult to pick out experimentally. For decay modes with branching ratios on the order of 0.1% the expected asymmetries are at most a few times 0.1%, assuming the Kobayashi-Maskawa phase is fairly large ($\eta > 0.3$). Thus neither mixing nor direct *CP* violations in D decays are expected to be observable, at the level predicted by the Standard Model, in foreseeable experiments. However new physics from beyond the Standard Model may give much larger effects and hence these searches should be undertaken in the same spirit as searches for rare or forbidden decays [47].

A 5 σ signal for direct *CP* violation at the level of the Standard Model predictions would require a parent meson sample of at least 10^{10} events. While this is nominally within the reach of a dedicated fixed target experiment at Fermilab, triggering and reconstruction inefficiencies make it unlikely that measurements can really be made with this precision. A dedicated tau-charm factory would produce of order 10^8 charmed mesons per year at design luminosity, of which 15% might be reconstructed. A *B* Factory is likely to produce a similar number of reconstructed D's. Thus, as stated above, it will be difficult to observe an effect at the level of the Standard Model prediction, but searches for unexpectedly large effects should be pursued.

Physics beyond the Standard Model in general does not predict direct *CP* violation greater than a fraction of a percent (with the possible exception of models with with a fourth non-sequential generation of quarks). However some models predict significant changes in the mixing mechanism and hence greatly enhanced indirect *CP* violation in D^0 decays. For example a 5% asymmetry in a doubly Cabibbo-suppressed decay rate would correspond to a new amplitude almost a factor of 100 less than

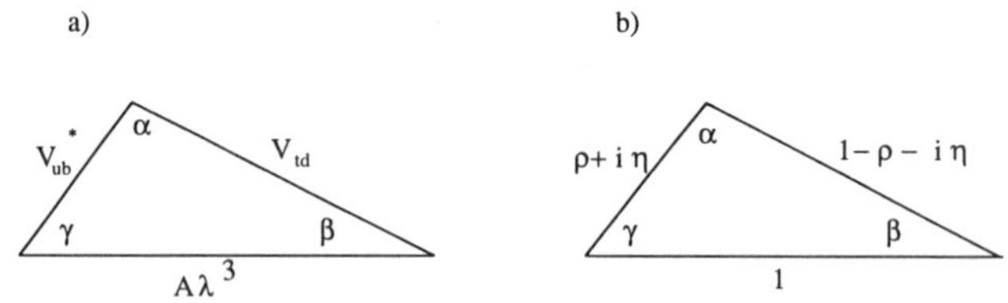

Figure 1: The unitarity triangle, (a) the relation obeyed by the CKM elements; (b) the relation obeyed by (CKM elements)/$A\lambda^3$

the dominant spectator decay amplitudes of the Standard Model. Assuming that new physics produces amplitudes $\propto (g'/M_X)^2$, where g' is a coupling strength and M_X is the mass scale of the new physics, such a search for direct *CP* violation would be probing conventional coupling strength physics at the TeV scale.

3.4 CP Violation in B decays

The B system provides a rich laboratory for the study of *CP* violation [48]. The neutral B mesons are expected to exhibit indirect *CP* violation, like the neutral K's. However in the B case there are a number of possible *CP* eigenstates that can be studied and the relationships between the measured asymmetries in these different channels will provide an excellent way to discover whether the pattern of *CP* violation is as expected in the Standard Model or quite different. For a review of the theoretical predictions for *CP* violation in B decays in the Standard Model and beyond see [49] and references therein.

Unitarity of the CKM matrix gives, among others, the relationship

$$V_{tb}^* V_{td} + V_{cb}^* V_{cd} + V_{ub}^* V_{ud} = 0, \tag{18}$$

which can be displayed as a triangle in the complex plane, as shown in Figure 1a. The base of this triangle can be rescaled to be of unit length, as shown in Figure 2b.

Constraints on the unitarity triangle come from the value of ϵ in K_L^0 decays, from the measured rate for $B_d^0 - \bar{B}_d^0$ mixing, and from the determination of $|V_{ub}|/|V_{cb}|$. At present there are significant theoretical and experimental uncertainties in the relationship between these measured quantities and the parameters of the underlying theory. However, the trend of steadily improving lattice calculations of quantities such as the f_B and B_B parameters should help this situation. The values of the parameters used here are [1,50,51,52]

$$\begin{aligned} &x_d = 0.65 \pm 0.08\,, && f_B\sqrt{B_B} = 200 \pm 40\,\mathrm{MeV}\,,\\ &|\epsilon| = (2.259 \pm 0.018) \times 10^{-3}\,, && B_K = 0.75 \pm 0.05\,,\\ &|V_{ub}|/|V_{cb}| = 0.076 \pm 0.026\,, && |V_{cb}| = 0.040 \pm 0.003\,,\\ &\tau_{B^0} = 1.50 \pm 0.11\,\mathrm{ps}\,, && \end{aligned} \tag{19}$$

and with the value of the running top-quark mass, $m_t(m_t)$ corresponding to the physical range claimed [53] by CDF, $m_{t_{phys}} = 174 \pm 16$ GeV. The status of the constraints on the unitarity triangle either from existing measurements or theoretical estimates of these parameters, is shown in Figure 2. This result translates into allowed ranges for the three angles α, β and γ of the unitarity triangle of

$$\begin{aligned} -0.89 \leq & \sin 2\alpha \leq 1.00\,, \\ 0.18 \leq & \sin 2\beta \leq 0.81\,, \\ -1.00 \leq & \sin 2\gamma \leq 1.00\,. \end{aligned} \tag{20}$$

To learn more about the nature of *CP* violation from the study of B decays, a systematic study of many channels is required. Simply observing a single *CP*-violating effect in the most accessible channel, while interesting, will not do enough to determine the source of *CP*-violation in particle theories. One must make a number of measurements with sufficient precision, and carry out the theoretical calculations with sufficient accuracy, so that one would be sensitive to sources of *CP* violation other than the CKM phase. In practice, what this means is overdetermining the CKM matrix, measuring the magnitudes of the elements, and the phases between elements, via *CP* violation measurements in a variety of B-decay channels, and seeing if it all fits together. Fortunately nature offers us the opportunity to do just that.

CP violation measurements are thus an essential part of a larger effort to look for new physics by overdetermining the CKM matrix. If the overdetermined CKM matrix elements were found to be inconsistent, with different values obtained from different ways of fixing the quantities, that would imply physics from beyond the Standard Model. This additional physics could in itself be *CP*-conserving, for example an additional *CP*-conserving contribution to $B^0 - \bar{B}^0$ mixing would destroy the predicted relationship between the phase of the mixing amplitude and the phases of CKM matrix and hence lead to contradictory results for the value of such elements when treated as if the mixing has only Standard Model sources. Alternately there could be additional types of *CP*-violating phase introduced from additional couplings in the extended model.

Tables 2 and 3 list the channels for B decay for which *CP*-violating asymmetries arising from the interference of mixed and unmixed decays can occur, and the Standard Model prediction for the relationship of the tree decay amplitude weak phase, ϕ_D $(\overline{A}/A = e^{-2i\phi_D})$ and the relative phase of the two terms in the heavy neutral B mass eigenstate, ϕ_M $(q/p = e^{-2i\phi_M})$. These govern the expected asymmetries in the Standard Model.

In compiling these tables various simplifying approximations have been made[54], for example that a single weak decay amplitude dominates the decay. The contribution of penguin processes could invalidate this assumption in some channels, and further analysis and cross-checks are needed, to determine whether such contributions are large. Isospin analysis of related channels can be used to remove possible QCD penguin contributions [55], but Z-penguin contributions cannot be removed in this way and can introduce significant effects in certain cases [56].

Table 2: *CP* Modes for B_d Decays and Standard Model Asymmetry Predictions.

Quark process	B_d Mode	$\phi_M - \phi_D$
$b \to c\bar{c}s$	$J/\psi K^0_S$	β
$b \to c\bar{c}s$	$J/\psi K^{0*}$	β
$b \to c\bar{c}d$	D^+D^-	β
$b \to c\bar{c}d$	$D^{+*}D^{-*}$	β
$b \to u\bar{u}d, d\bar{d}d$	$\pi^+\pi^-$	α
$b \to u\bar{u}d, d\bar{d}d$	$\rho^0\pi^0$	α

Table 3: *CP* Modes for B_s Decays and Standard Model Asymmetry Predictions.

Quark process	B_s Mode	$\phi_M - \phi_D$
$b \to c\bar{c}s$	$J/\psi\phi$	0
$b \to c\bar{c}d$	$J/\psi K^0_S$	0
$b \to s\bar{s}s$	$\phi\phi$	0
$b \to u\bar{u}d, d\bar{d}d$	$\rho^0 K^0_S$	γ
$b \to s\bar{s}d$	ϕK^0_S	β

The channels for B_s which the Standard Model predicts zero asymmetry because of a cancellation of the decay weak phase with the mixing amplitude weak phase are interesting. In models with very different mixing mechanisms these channels would be expected to have asymmetries comparable to those in other channels.

All of the above effects are of the same type as the already observed *CP* violation in K decays, that is they occur because of interference between decay with and without mixing in B^0 decays to flavor-neutral channels. In addition one expects direct *CP* violation at the few percent level in B decays in the Standard Model, larger in channels with Cabibbo-suppressed tree contributions. These effects can most readily be searched for in the decays of charged B's. Searches for asymmetries in any charged B decay channels are thus of interest and should be pursued. The largest asymmetries are expected in those channels where tree and penguin diagrams give comparable strength contributions because of the Cabibbo suppression of the tree amplitude ($b \to u\bar{u}s$ *e.g,* $B^+ \to K^{*+}\rho^0$) [57]. While the magnitude of the expected asymmetry depends on strong phase shifts and

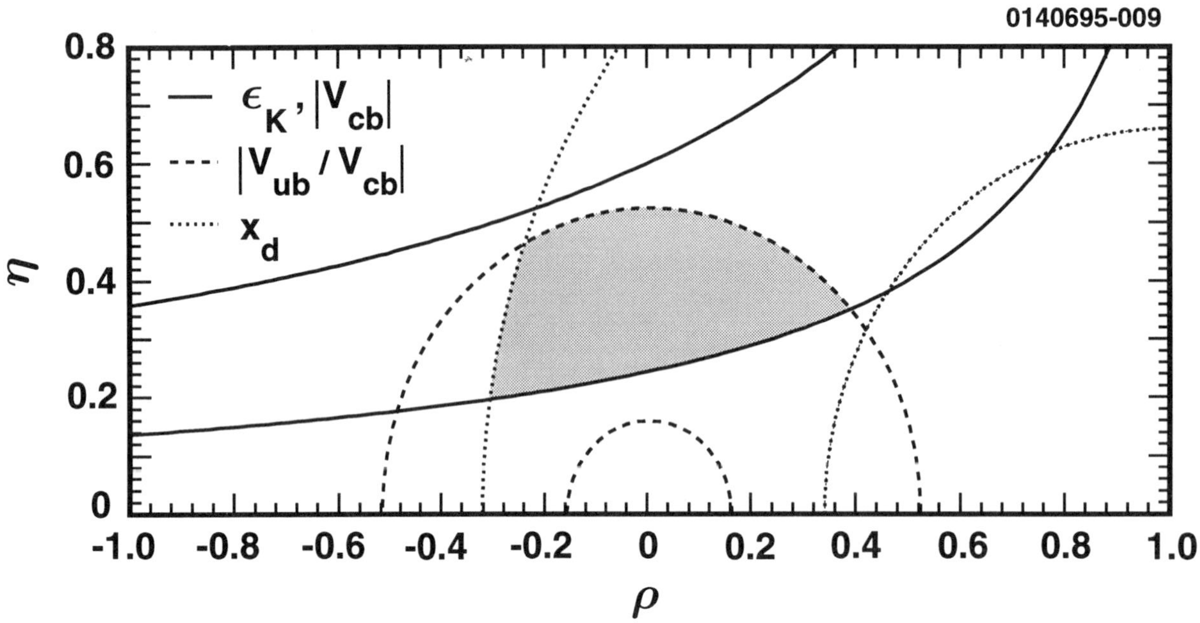

Figure 2: Constraints on the unitarity triangle.

cannot be reliably predicted, the observation of such an asymmetry would be a clear demonstration of the existence of direct *CP* violation.

Experiments which can study many of the channels above will be a feature of the next generation of *B* physics. These are discussed further in the section on facilities.

3.5 Tests of CP Invariance in Tau Decays

CP violation is not expected in the leptonic sector, in the context of the Standard Model. If, however, neutrinos have mass and mix in analogy with the CKM mechanism in the quark sector, *CP* violation can occur in tau decays. Since we know that any neutrino masses are much smaller than quark masses, this source of *CP* violation is already well below our ability to detect in foreseeable experiments. Thus, if *CP* violation is indeed observed in tau decays, it would be an exciting signal for an unanticipated source beyond the Standard Model.

Both the total cross section and differential rate for $e^+e^- \to \tau^+\tau^-$ are sensitive to *CP*-violating form-factors. For unpolarized beams, only two form factors contribute to *CP*-odd correlation functions in tau-pair production: the electric dipole moment of the τ, $d_\tau(q^2)$, and the weak dipole moment $d_\tau^w(q^2)$. The LEP experiments collect data near the peak of the Z and are sensitive to a weak dipole moment, at $q^2 = m_Z^2$. A non-zero value of $d_\tau^w(q^2)$ would contribute to the width $\Gamma(Z \to \tau^+\tau^-)$ and would introduce *CP*-odd correlations of the momenta of the τ^+ and τ^- decay products [58]. These correlations have been searched for by the ALEPH and OPAL collaborations by forming *CP*-odd triple products from the momenta of the decay products of the two taus in the event. They obtain results consistent with zero *CP*-violating correlations. Combined with the measurements of $\Gamma(Z \to \tau^+\tau^-)$ in comparison with Standard Model predictions a limit of $|d_\tau^w| < 2.9 \times 10^{-17}\ e \cdot \mathrm{cm}$ at 95% C.L. is obtained. Improvements in these limits, as well as tests below the Z where $d_\tau(q^2)$ is probed, can be expected in the next decade. The remote possibility that a non-zero dipole moment can be seen remains an enticing possibility, well worth the effort.

3.6 Summary

There are clearly many areas of *CP* violation physics to study. Standard Model predictions relate heavy flavor *CP* violating effects to the magnitudes of CKM parameters through the unitarity of the three-generation theory. Models beyond the Standard Model can violate these relationships and at the same time may introduce significant contributions in channels that are forbidden or loop-suppressed in the Standard Model. Thus the multiple avenues of precision rare decay searches in *K* physics, multiple CKM parameter determinations via different channels, and systematic study of *CP* Violations, both direct and indirect, in as many channels as possible should all be explored as possible low-energy windows to physics beyond the Standard Model.

4 Rare and Forbidden Decays

4.1 Introduction

The search for processes that are either forbidden or very rare in the Standard Model is a powerful technique for testing the model as well as for obtaining important constraints on physics beyond the Standard Model. The search for forbidden processes, exemplified by the search for decays violating lepton number, holds forth the possibility of revealing a "smoking gun" of evidence that *there is* physics beyond the Standard Model. The search, and finally, the measurement of partial rates for very rare decays offers the opportunity to test predictions of the Standard Model at the quantum loop level where significant deviations can again point to the presence of new physics [59,60,61].

4.2 Lepton Flavor Violation

The search for decays involving the violation of lepton flavor conservation gets to the heart of the matter in that these searches can hopefully shed light on the *generation* puzzle posed by the Standard Model. The replication of the fundamental fermions to include three *families* of quarks and leptons is accommodated but not explained within the Standard Model. Although mixing among the quark flavors in the charged–current weak interactions occurs, governed by the unitary Cabibbo–Kobayashi–Maskawa matrix, mixing of the lepton families does not. The conservation of lepton family number, separately for e, μ and τ, must be invoked to explain the absence of decays such as $\mu \to e\gamma$, etc. At face value, without any extension to the Standard Model, the conservation of lepton family number would require an exact global symmetry, or an unbroken local gauge symmetry with unobserved massless gauge bosons mediating long–range forces.

Flavor–changing transitions resulting in lepton number violating decays are present at some level in many extensions of the Standard Model. In some models, for example those involving a (broken) horizontal gauge symmetry, lepton flavor violation is quite natural. Some of the branching ratio limits that have been obtained experimentally on lepton flavor violating processes are given in Table 4.

Table 4: Limits on Lepton Number Violating Processes

Process	Upper Limit 90% CL	Laboratory
$\mu \to e\gamma$	4.9×10^{-11}	LAMPF
$\mu \to eee$	1.0×10^{-12}	PSI
$\mu N \to eN$	4.3×10^{-12}†	PSI
$\pi^0 \to \mu e$	1.7×10^{-8}	FNAL
$K_L^0 \to \mu e$	3.3×10^{-11}	BNL
$K^+ \to \pi^+\mu^+e^-$	2.1×10^{-10}	BNL
$K_L^0 \to \pi^0\mu^+e^-$	3.5×10^{-9}	FNAL
$\tau \to \mu\gamma$	4.2×10^{-6}	CESR
$\tau \to eee$	3.3×10^{-6}	CESR
$\tau \to \mu ee$	3.4×10^{-6}	CESR
$\tau \to \mu\mu e$	3.6×10^{-6}	CESR
$\tau \to \mu\mu\mu$	4.3×10^{-6}	CESR
$D^0 \to \mu e$	1.0×10^{-4}	DESY
$B^0 \to \mu e$	5.9×10^{-6}	CESR
$B^0 \to \tau\mu$	8.3×10^{-4}	CESR
$B^0 \to \tau e$	5.3×10^{-4}	CESR
$B^0 \to K^{*0}\mu e$	2.7×10^{-5}	CESR
$B^+ \to K^+\mu e$	1.2×10^{-5}	CESR

†This is a limit on conversions/captures

Many of these limits (especially from muons and kaons) are stringent enough to place important constraints on model building. The relative importance of these constraints, and those that can be obtained in the future from continued searches depends on the nature of the new physics. And the interplay of limits obtained in several processes is often required to confront the plethora of conceivable extensions to the Standard Model (*e.g.,* technicolor and its extensions, left–right symmetric models, models with composite quarks and/or leptons an extended Higgs sector, supersymmetry, leptoquarks, etc.) [62,63].

The decay $K_L^0 \to \mu e$ serves as a good illustration of the importance of conducting the search for new physics over a broad range of processes, as well as a chance to point out the sensitivity of the search for rare decays to physics at high mass scales. The required $s\bar{d} \to \mu^- e^+$ transition could proceed via a tree–level exchange of a leptoquark that coupled s quarks to muons and d quarks to electrons; such a interaction would not contribute to purely leptonic processes ($\mu \to e\gamma$) or to the $K_L^0 - K_S^0$ mass difference ($\Delta m_K = 3.5 \times 10^{-15}$ GeV, otherwise the most important constraint on flavor–changing neutral current interactions.) Assuming the new interaction has a $V - A$ form, the $K_L^0 \to \mu e$ decay rate can be compared directly with the familar decay $K^+ \to \mu^+\nu_\mu$. If the leptoquark–fermion vertices carry the weak interaction coupling, this comparison leads to the result [64]:

$$BR(K_L^0 \to \mu e) = 10^{-12} \times (\frac{220\text{TeV}}{M_{LQ}})^4 \tag{21}$$

where M_{LQ} is the mass of the leptoquark. With these assumptions, the current experimental limit translates into a lower bound on M_{LQ} of 90 TeV, a mass scale well beyond the reach of direct production with present or planned accelerators. However, since the branching fraction depends on the fourth power of M_{LQ}. substantial progress in limiting M_{LQ} will be difficult. Similar arguments can be made for the decays $K^+ \to \pi^+\mu^+e^-$ and $K_L^0 \to \pi^0\mu^+e^-$. However, it is important to note that the searches for these two decay modes are complementary given that $K_L^0 \to \mu e$ must proceed through axial–vector or pseudoscalar interaction and $\pi\mu^+e^-$ modes require a vector or scalar interaction.

Given the wealth of possibilities for new physics, it is important to cast a wide net, keeping in mind that generic constraints implied by limits on one process may be evaded in a particular class of models and lead to observable rates for another process. The impressive limits set for lepton flavor violation with muons and K mesons do not necessarily exclude the possibility of observable rates for lepton flavor violating decays of tau leptons, mesons or baryons containing charm or beauty, or even of the top quark. Several models for new physics include fermion mass–dependent (Higgs–like) couplings that enhance the rate for various τ, D, and B decays relative to the expectations for those of muons and K mesons. Also the space–time nature of the new interaction currents (S, P, V, A, T) plays a crucial role in determining the relative rates for specific decay modes (*e.g.*, whether $B \to \mu e$ suffers a helicity suppression with respect to $B \to \pi\mu e$ or $K\mu e$).

It should be noted that in perhaps the simplest extension of the Standard Model, where neutrinos are given a mass and neutrino oscillations are allowed, lepton flavor violating decays can occur, albeit at rates well below the limits already placed: branching ratios of order 10^{-20} would be expected for $\mu \to e\gamma$ and of order 10^{-15} for $K_L^0 \to \mu e$ assuming neutrino masses and mixing at the current experimental limit (see Working Group report on neutrinos). A branching ratio of 10^{-16} could be expected for $\mu A \to eA$.

4.3 Loop Induced Rare Decays

In the Standard Model, flavor–changing neutral currents (FCNC) are forbidden to first order and strongly suppressed at higher orders by the Glashow–Illiopoulos–Maiani (GIM) mechanism [65]. The measurement of GIM suppressed decays offers an opportunity to study higher–order weak interactions, particularly the contributions of the top quark in loop processes and the CKM couplings V_{td} and V_{ts}. The highly suppressed nature of the FCNC decays also offers a potential window on new physics. In what follows several specific classes of suppressed decay modes are discussed.

4.3.1 $K, D, B \to l^+l^-$

The decay $K_L^0 \to \mu^+\mu^-$ is probably the most famous of the GIM suppressed decays. The suppression of its rate relative to $K^+ \to \mu^+\nu_\mu$ was a serious problem for the early Standard Model, and required the invention of the GIM mechanism and the existence of the charm quark. With GIM suppression the decay rate is expected to be dominated by contributions from a two–photon intermediate state, with the contributions from real photons resulting in a "unitarity bound" of 6.8×10^{-9} for the $K_L^0 \to \mu^+\mu^-$ branching ratio [66]. Recent efforts at BNL and KEK have yielded a measurement of the decay rate to better than 10% and the result is consistent with the "unitarity bound." Further improvements (perhaps approaching a 1% measurement) are expected. Further progress in extracting the contributions of the higher–order weak interactions and contributions from the top quark will require understanding the contributions from the two–photon intermediate state with virtual photons. If the uncertanties from these contributions can be eliminated, the $K_L^0 \to \mu^+\mu^-$ process can provide a measurement of the CKM parameter ρ. Progress on this front has been made recently with measurements at BNL, CERN, KEK and FNAL on the decays $K_L^0 \to \gamma e^+e^-$, $K_L^0 \to e^+e^-e^+e^-$ and $K_L^0 \to \gamma\mu^+\mu^-$ and improved measurements are expected in the future.

The decay $K_L^0 \to e^+e^-$ shares much of the same physics with the decay $K_L^0 \to \mu^+\mu^-$, but contributions from the Standard Model are suppressed by a factor proportional to m_e^2/m_μ^2. The recent limit established at BNL, $\mathrm{BR}(K_L^0 \to e^+e^-) < 4.1 \times 10^{-11}$ [67], is an order of magnitude above the "unitary bound" for this mode of 3×10^{-12}. This still leaves room for contributions from physics beyond the Standard Model, particularly from a pseudoscalar interaction that would avoid the helicity suppression.

The decays of neutral B and D mesons to $\mu^+\mu^-$ final states also offer a means to test the predictions of the Standard Model for one–loop decays. Figure 3 illustrates loop diagrams contributing to the process $B \to \mu^+\mu^-$. The possibility of observing $D \to \mu^+\mu^-$ seems remote due to the presence of helicity suppression, small CKM elements and the absence of the heavy top quark in the one–loop diagrams, particularly remembering that normal D decays are not even Cabibbo suppressed. New physics could of course evade these constraints. With the B meson decays, especially $B_s \to \mu^+\mu^-$ the situation is more favorable. In this case the top quark loop effects are more important for $b \to s$ than $s \to d$ transitions (relevant to K decays), and $B_s \to \mu^+\mu^-$ is expected to be dominated by the higher–order weak interactions rather than long–distance intermediate $\gamma\gamma$ states. The expected branching ratio in the Standard Model is about 10^{-9} [68].

0140695-008

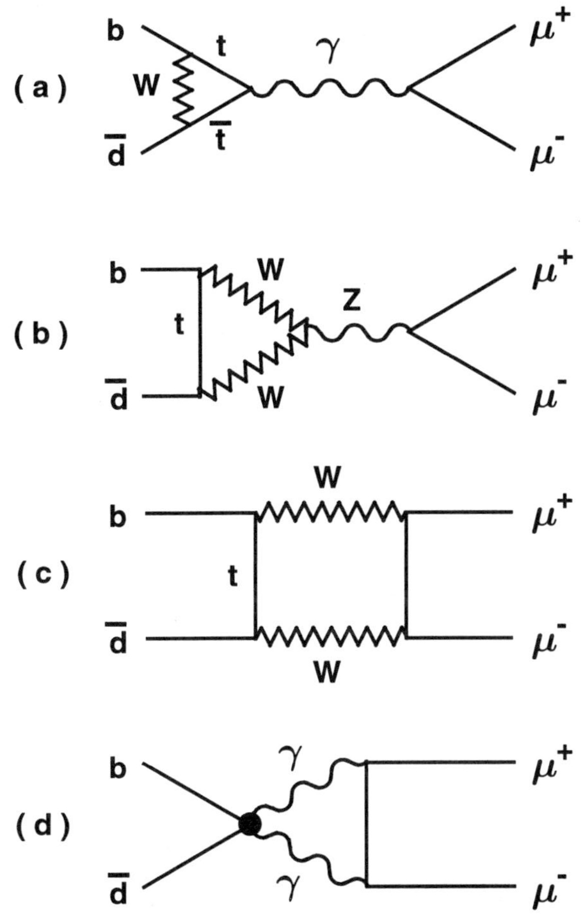

Figure 3: Diagrams contributing to the rare decays $B^0 \to \mu^+\mu^-$ (or $K^0 \to \mu^+\mu^-$ with $b \leftrightarrow s$.)

0140695-010

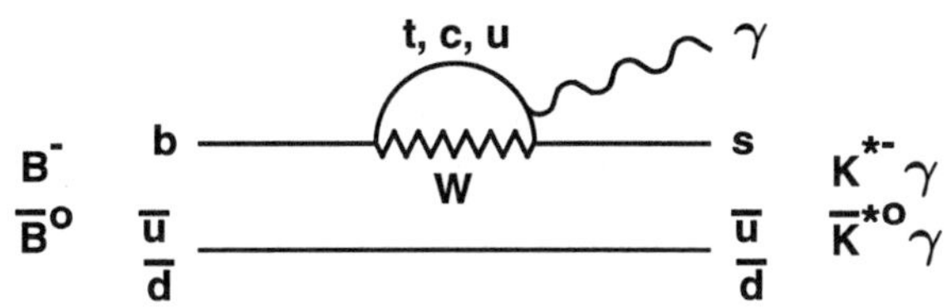

Figure 4: Penquin decay of a B meson to $K^*\gamma$.

4.3.2 Penguin Decays: $K \to \pi l\bar{l}$, $b \to s\gamma$, $b \to sl\bar{l}$.

This class of decays receives contributions from penguin diagrams as well as from the W box diagrams at the one–loop level. Also the helicity suppression of the previously discussed decays is avoided. Correspondingly, rates expected in the Standard Model are enhanced.

The observation [69] of the decay $B \to K^*(892)\gamma$, reported by the CLEO experiment, is the first direct evidence for the penguin diagram. This decay is described by the electromagnetic transition $b \to s\gamma$, which is a $b \to s$ penguin loop accompanied by the radiation of a photon from either the loop, or the initial or final state quarks. (Figure 4 illustrates the penguin diagram for $B \to K^*\gamma$.)

Unfortunately, due to the theoretical uncertainties in translating the branching ratio for the exclusive process into a rate for the parton level process, this result cannot be used to quantitatively test the Standard Model. Recently, CLEO has succeeded in measuring the rate for the inclusive process $b \to s\gamma$ [70]. The measured rate is consistent with the Standard Model and constrains many extensions of the Standard Model. These include charged Higgs bosons below about 250 GeV anomalous $W - W - \gamma$ couplings, SUSY Higgs, and so on [71]. Theoretical uncertainties for the Standard Model rate are currently $\pm 30\%$, but could be reduced considerably with a full next-to-leading-log calculation [72]. This should be given high priority. If these calculations are done, improved experimental precision will provide additional constraints on new physics (or see evidence of it). As noted in other sections, indirect constraints on rare processes can often provide more stringent constraints than direct searches at high-energy colliders.

The process $b \to sl^+l^-$ is mediated at the parton level by the process $b \to s\gamma^*$ or $b \to sZ^*$ where the virtual particle decays to lepton pairs (here γ^*,Z^* indicates a virtual photon or Z). Exclusive final states such as $B \to K^*l^+l^-$ are expected. Unlike $b \to s\gamma$ processes, there are no significant QCD corrections in this case [73]. Heavy neutral particles that occur in extensions of the Standard Model may enhance the rate for this type of process.

Unfortunately, significant long distance contributions are also expected to give rise to the same final state. These may be separated with high statistics observations of the detailed kinematic properties in the decay.

The decay $K^+ \to \pi^+\nu\bar{\nu}$ is expected to occur at a branching fraction of about 10^{-10}; the current experimental limit from BNL is 3×10^{-9}. As discussed in the section on CKM parameters, the decay $K^+ \to \pi^+\nu\bar{\nu}$ offers a relatively clean way of measuring $| V_{td} |$ as the theoretical uncertainties can be controlled to 10% or better. A byproduct of the search for $K^+ \to \pi^+\nu\bar{\nu}$ are limits placed on the process $K^+ \to \pi^+X^0$. Here X^0 is an essentially invisible particle such as an axion, familon, or hyperphoton.

The decays $K_L^0 \to \pi^0e^+e^-$ and $K_L^0 \to \pi^0\nu\bar{\nu}$ have been discussed in the section on *CP* violation, by virtue of their sensitivity to direct *CP* violation in the Standard Model. The current branching ratio sensitivities achieved (10^{-9} for $K_L^0 \to \pi^0e^+e^-$) allow a considerable window of opportunity for non–Standard Model (*CP* violating contributions) above the predicted Standard Model ratio of $O(10^{-12})$.

5 Strong Interaction Effects in Heavy Flavor Physics

5.1 Theoretical Background

Studies of the masses, static properties, and decays of hadrons containing heavy quarks have provided a fertile testing ground for our understanding of QCD dynamics. In addition, to extract the electroweak parameters for quarks from the observations of the mixing and decays of hadrons, the nonperturbative effects of QCD cannot be ignored. In particular, the determinations of the quark masses and CKM parameters from the weak decays of K, D, and B mesons have significant theoretical components to the present errors. Improving these determinations will require both experimental and theoretical efforts.

The principal theoretical tools available to help disentangle these nonperturbative QCD effects are:

- Lattice gauge theory. This is the only existing technique which allows the systematic study of all aspects of QCD and will be a critical ingredient in the theoretical effort to understand heavy flavor physics [74]. A significant present limitation of the lattice approach is that calculations including internal quark loops require more computing resources than now exist. If the rate of improvement in numerical methods and computing cost remains the same, this situation should improve dramatically over the next 5-10 years.
- QCD Sum Rules. This approach exploits the duality in QCD between the description in terms of physical states (at low energy) and elementary quark and gluon excitations (at high energy). Such models for the dynamics can shed considerable light on the properties of heavy systems.
- Heavy Quark Effective Theory(HQET). The dynamics of QCD simplifies as $m_Q \to \infty$. In this limit the dynamics becomes independent of the spin and flavor of the heavy quark [11]. These heavy quark symmetries (HQS) can be exploited to relate masses, decay rates, form factors, and many other physical quantities within various heavy flavor systems.
- Chiral Perturbation Theory. The effective lagrangian for interactions involving pions and kaons is constrained by chiral symmetry [75]. Quark masses explicitly break this symmetry and can be treated as correction terms for π and K interactions.

5.2 Spectroscopy and Strong Decays

5.2.1 Heavy-Heavy systems

Heavy quark systems allow the study of the strong dynamics of QCD in simple limits. The most familiar heavy quark systems are the $Q\bar{Q}$ mesons. As the mass of the quark, m_Q, becomes much heavier than the QCD scale (Λ_{QCD}), the dynamics becomes nonrelativistic. The ratio of the typical momentum transfer (between the heavy quark and antiquark) to the heavy quark mass, m_Q, remains small as $m_Q \to \infty$. Hence a systematic expansion in powers of this ratio (the relative velocity $\vec{v}$) can be employed. The quarks and antiquarks move in an effective potential $V(R)$ generated by the QCD interactions. Nonrelativistic potential models of the ($\bar{c}c$) and ($\bar{b}b$) family of resonances(*i.e.*, the J/ψ and Υ resonances) have been very successful [76]. Relativistic corrections and hadronic transition rates have all been extensively studied. Existing lattice calculations agree well with the phenomenologically determined potential $V(R)$. Lattice techniques are now being used to directly calculate the masses and splittings in the $c\bar{c}$ and $b\bar{b}$ systems and extract the heavy quark mass.

The B_c meson spectrum would be a particularly interesting $Q\bar{Q}$ system to study because of its unequal heavy quark masses. Some of the B_c decay modes are experimentally striking (such as $B_c \to J/\psi\pi^-$ and $J/\psi a_1^-$). Such modes make the study of B_c mesons possible at a high luminosity hadron collider.

5.2.2 Heavy-Light Systems

The top quark is so heavy that it decays weakly before hadronic states have time to fully form; for the remaining heavy quarks, however, the weak interactions can be treated as a small perturbation. For hadrons that contain a single heavy quark, the QCD interactions again takes a simple form as $m_Q \to \infty$. The momentum transfers between the heavy quark and the rest of the system are typically of the scale of the light quark masses and Λ_{QCD} and thus remains bounded as the heavy quark mass goes to infinity. Hence in the $m_Q \to \infty$ limit, the heavy quark is on shell and at rest relative to the hadron and the QCD dynamics becomes independent of m_Q. All heavy quark mass dependence of the bound state can be extracted analytically to produce an effective action (HQET)for a static quark and the remaining light degrees of freedom. This effective action makes transparent symmetry and scaling relations between heavy-light systems which differ by heavy quark spin or flavor [77]. Furthermore, for finite heavy quark masses, the effective action can be improved order by order in $1/m_Q$, and QCD corrections calculated systematically as a expansion in $\alpha(m_Q)$.

These theoretical developments have immediate physical applications for B physics. Since the b quark mass is significantly heavier than the other mass scales (the QCD scale and the light quarks masses) which enter into the dynamics of B hadrons, it is a good approximation to treat the b quark in $m_Q \to \infty$ limit within B hadrons. It is also fruitful to consider the charm mesons and baryons as heavy-light systems, although corrections are larger in this case.

The Heavy Quark Symmetry (HQS) predicts that the leading corrections to the spectrum are inversely proportional to the heavy quark mass. This relates D meson masses to B meson masses, and predicts an interesting spectra of B^{**} and D^{**} mesons. Furthermore the decays of these excited heavy-light mesons satisfy certain symmetry relations in $m_Q \to \infty$ limit.

One application of the existence of such excited B states has been suggested. If significant numbers of B mesons are produced through one or more narrow excited states, then the strong decay $B^{**\pm} \to B^{(*)0}\pi^{\pm}$ tags the neutral B mesons as $(\bar{b}d)$ or $(\bar{d}b)$, respectively [78]. This would improve the situation for *CP* violation studies in hadron machines.

The spectroscopy of heavy baryons, particularly baryons with two heavy quarks, can also provide important additional insight into the dynamics of QCD.

5.3 Weak Leptonic and Semileptonic Decays

A heavy quark decays via the weak charged current into its electroweak partners (with couplings determined by appropriate CKM element) and a virtual W. The virtual W, in turn, decays to either a quark-antiquark pair or into leptons. The quark-antiquark pair decay is enhanced in rate relative to the $l\nu$ decay by the color factor.

To date no evidence for lowest-order flavor- changing neutral decay currents has been observed. This puts stringent constraints on models of physics beyond the standard model. However the limits for heavy flavors are much less stringent than those on the $s \to d$ transition that come from the small $K_L^0 - K_S^0$ mass difference.

5.3.1 Leptonic Decays

The fully leptonic decay probability for a pseudoscalar meson, $P^+ = (q'\bar{q})$, to a charged lepton and neutrino is described by

$$\Gamma(P^+ \to l^+\nu_l) = \frac{G_F^2}{8\pi} f_P^2 \; m_P \; m_l^2 |V_{q'q}|^2 (1 - \frac{m_l^2}{m_P^2})^2. \quad (22)$$

The pseudoscalar decay constants f_P of the heavy-light mesons are of interest for the study of mixing and *CP* violation for **both** the charm and beauty sector. In the large m_Q limit, HQET requires $\sqrt{M_P} f_P$ to be independent of the heavy quark mass up to calculable perturbative corrections [79]. For finite heavy quark mass the corrections are $O(1/m_Q)$.

No data exists for f_D, but during the past year two results on f_{D_s} were reported [80]. One technique uses emulsion to measure the direction of the D_s^+ and then measures the $p_{T\mu}$ of the muon relative to the D_s^+ direction. The second technique uses the decay chain $D_s^{*+} \to \gamma D_s^+$ and examines $\mu\gamma$ correlations. The results of both techniques have large errors and only agree at the one standard deviation level. Over the next several years the value of f_{D_s} will improve. A good measurement of f_D and especially f_B will be difficult, forcing us to rely on theoretical calculations.

Lattice calculations using a variety of methods allow the determination of decay constants in the heavy quark limit as well as directly for the strange, charm, and beauty mesons. The present theoretical range for f_B [81], from 150-200 MeV, implies factor of 2 variation in B mixing. Ratios of decay constants are better determined. Significant improvements in the accuracy of the lattice calculations of f_B and f_D can be expected in the next few years.

5.3.2 Semileptonic Decays

After the fully leptonic decays, the semileptonic decays are the next simplest theoretically. Both decays to inclusive and exclusive hadronic final states are theoretically important. For example, from B meson inclusive measurements, the ratio of CKM matrix elements $|V_{ub}|/|V_{cb}|$ can be determined. There is some dependence on the QCD strong dynamics and theoretical model differences account for a sizable component of the total uncertainty in this measurement. To reduce these theoretical uncertainties we need to understand the dynamics of $b \to u$ transitions near the lepton endpoint for the $b \to c$ transitions.

At present, there is even disagreement between theory and experiment in the B meson system on the total semileptonic branching ratio. The measured value is $(10.35 \pm 0.17 \pm 0.35)\%$ from a model independent dilepton analysis and comparable but more precise rates are obtained from the analysis of the single lepton spectrum [82]. These measurements are significantly smaller than the expected theoretical lower bound $BR_{sl} > 12.5\%$. Whether the resolution of this discrepancy is due to an unexpected enhancement in the non-leptonic B partial rates or other large corrections to the theoretical expectation is presently unresolved.

Semileptonic decays to exclusive single hadron final states are under better theoretical control but require much larger data samples to fully study. They occur only via the spectator diagram and there are no final state interactions. The hadronic physics enters through a single matrix element which is expressed in terms of form factors. The momentum transfer in the decay is given by $q^2 = M^2(l\nu)$. The heavy-light meson (P) decays are usually separated into pseudoscalar (P') and vector ($P^{*\prime}$) (final hadronic state) decays.

The pseudoscalar decay $P \to P'l^+\nu$ decay is governed by two form factors called $f_+(q^2)$ and $f_-(q^2)$. Generally the $f_-(q^2)$ term is neglected as it is multiplied by the lepton mass squared.

For the vector decay $P \to P^{*\prime}l^+\nu$ there are 3 form factors, $A_1(q^2)$, $V(q^2)$, and $A_2(q^2)$ which describe the decay. (Actually there is a fourth, but it appears multiplied by the lepton mass squared, and so is irrelevant experimentally.)

For P' and $P^{*\prime}$ heavy-light mesons, the decay amplitudes take a particularly simple form in the heavy quark limit. All explicit dependence on heavy quark masses vanishes (up to calculable perturbative QCD correction), and only a single form factor, the Isgur-Wise function [11] ξ which is a function of the product of the initial and final hadron four velocities $(v \cdot v')$ remains. The HQET does not determine ξ. Additional form factors appear in order $1/m_Q$.

Thus far the form factors have been measured assuming a pole mass-shape form factor as given in the pseudoscalar decay. The values for $A_1(0)$ and $V(0)$ are reasonably well determined while the error on $A_2(0)$ needs to be reduced. Experiments need to improve statistics in order to gain sensitivity to the shape of the form factors.

The q^2 dependence is usually given as $f_+(q^2) = f_+(0)/(1 - q^2/M^2_{D_s^*})$ although HQET assumes an exponential q^2 dependence.

It is important to measure extremely well the form factors of $D \to Kl^+\nu$ and $D \to K^*l^+\nu$ and then relate them to $B \to \pi l^+\nu$ and $B \to \rho l^+\nu$. This extrapolation might be accomplished with HQET, quark sum rules, or lattice gauge calculations. If the form factors were understood, then it would be possible to make a solid measurement of $|V_{ub}|$.

- $D \to$ Pseudoscalar Decay

In the $D \to Kl^+\nu$ decay one can learn two pieces of information. First one can fit and determine the pole mass and compare with theoretical predictions. Second, one can assume a pole mass shape and measure the CKM element $|V_{cs}|$. For the future it will be important to accurately measure the decays of $D \to Ke\nu$ and $D \to K\mu\nu$. A comparison of the rates will determine $f_-(q^2)$ and should make possible the best direct measurement of $|V_{cs}|$.

Similar measurements are needed for the D_s^+ decays and for the Cabibbo suppressed decay $D \to \pi\mu\nu$. $D \to \pi\mu\nu$ decay can be compared to $D \to K\mu\nu$ decay to determine a ratio of $|V_{cd}|^2/\ |V_{cs}|^2$.

- $B \to$ Pseudoscalar Decay

While the decay $B \to D\ell\nu$ is theoretically much simpler than $B \to D^*\ell\nu$, from an experimental perspective it is much more difficult. $B \to D\ell\nu$ has backgrounds from both $B \to D^*\ell\nu$ (large) and $B \to D^{**}\ell\nu$ (small), while $B \to D^*\ell\nu$ has only a small background from $B \to D^{**}\ell\nu$. Further, D^0 and D^+ have substantial backgrounds under their mass peaks, while D^{*0} and D^{*+}, can be cleanly identified. Consequently, the study of $B \to D\ell\nu$ has not progressed very far. At such time as its form factor can be measured and compared with those from $B \to D^*\ell\nu$, it will provide an interesting test of HQET. In the nearer term, a precise measurement of the $B \to D\ell\nu$ branching ratio is needed, as a step towards determining what fraction of the B semileptonic decay is *not* due to D or D^*. (Unlike the situation on D semileptonic decays, higher resonances and non-resonant states account for a significant piece, approximately $\frac{1}{3}$.)

- $D \to$ Vector Decay

Comparison of measurements of D and B to $\rho l\nu$ (and $\pi l\nu$) could provide one approach to accurate measurement of $|V_{ub}|$. HQET together with approaches like QCD sum rules or lattice calculation can be used to classify and limit the $1/M_Q$ corrections to this quantity.

For D_s^+ decays there are two experimental measurements of the vector and axial-vector form factors. Both measurements have sufficiently low statistics such that one cannot see whether the D_s^+ form factors are the same as those for the D^+ or not. This test needs to be accurately demonstrated if the extrapolation to the B sector is to be made reliably.

- $B \to$ Vector Decay

The decay $B \to D^*l\nu$ has recently been studied by CLEO [83]. They performed a 4-parameter fit, determining the three form factors at the zero-recoil point as well as the slope of the form factors (assumed to be the same for all three). The results are in agreement with the expectations of HQET, but there is need for considerable reduction in experimental error before this can be considered a sensitive test of HQET.

The rate for $B \to D^*l\nu$ extrapolated to the zero-recoil point provides a relatively model-free determination of $|V_{cb}|$, as discussed in Section 2.1.1.

- Baryon Semileptonic Decays

Charmed and beauty baryon semileptonic decays provide another opportunity to test predictions of HQET. The Λ_c^+ decay should proceed via a Λ or a Λ^* as the isospin must be 0. Thus it is expected that the $l = 0$ decay $\Lambda_c^+ \to \Lambda l\nu$ will dominate the semileptonic decay of the Λ_c^+. The information can be used to relate the decay to the D mesons and thereby determine a branching fraction for $\Lambda_c^+ \to pK^-\pi^+$ [84]. This technique may be an excellent method for determining absolute branching ratios for the Ξ_c^+, Ξ_c^0, and Ω_c^0.

- Other Weak Transitions

There are a number of other weak decays involving the flavor transition of only a single quark in an initial hadron. These processes change the quark isospin by 1/2 or 0 and don't change the charge. Examples of these processes are $b \to s\gamma$ inclusive and exclusive decays (discussed in Section 4.3.2) and Kaon $\Delta I = 1/2$ and $K \to \pi e^+e^-$ amplitudes. The strong dynamics issues in these cases are similar to those encountered in semileptonic decays.

5.4 Weak Nonleptonic Decays

The study of nonleptonic weak decays and mixing of charm and beauty mesons is both rich and complicated. Six types of weak diagrams for a heavy-light meson (P) are possible [85]. These are denoted external spectator, color-suppressed (or internal) spectator, annihilation, exchange, penguin, and mixing. Disentangling the underlying quark processes and their relative strengths from the observed decays and mixing requires a detailed understanding of the strong interaction effects.

The character of hadronic charm decays is qualitatively well-understood in the meson sector. More than 90% of all meson decay channels have been observed. In contrast, for the B mesons approximately 90% of the decay channels are yet to be observed.

The charm meson states are dominated by spectator and color-suppressed spectator diagrams, while in the charm baryon sector the decays appear to be dominated by the exchange mechanism. The annihilation and exchange diagrams are expected to be quite small for meson decays because of helicity suppression. Both penguin and mixing diagrams are suppressed in the charm meson system. For B meson states, the dominant weak decay diagrams are spectator diagrams.

Most theoretical work is based on the assumption of factorization; *i.e.*, the amplitude for the weak decay of a heavy meson can be expressed as the product of two current matrix elements: one current between the initial meson and some of the final state hadrons and the other between the vacuum and the remaining final hadrons [86]. Factorization can be justified in special circumstances. For a two-body final state associated with a heavy meson weak decay, this occurs when the quark pair (produced as a color singlet from a virtual W) has high energy (and low Q^2) and hence travels fast enough to leave the interaction region without influencing the second hadron (formed from the remaining quark and the spectator antiquark).

Within the framework of factorization, HQET provides relations between decays. Lattice methods may be able to provide information on two-body final states. Chiral symmetry may also be used to relate decays. However most specific predictions for partial rates are quite model dependent. Here QCD Sum Rule methods provide the best present guide.

There are many topics to investigate even in the charm system. For both the D^0 and the D^+ mesons we know that the decays occur primarily via two body decays. For the D_s^+ meson the situation is not so clear. The weak annihilation process for the D_s^+ has yet to be confirmed [87], with the only evidence for the decay arising from the non-resonant decay of $D_s^+ \to \pi^+\pi^-\pi^+$. The lifetime for the D_s^+ is larger than the D^0 and this again is not expected. The decay of $D^+ \to \phi K^+$ is not only doubly suppressed, but requires an annihilation diagram.

There remains a need to provide a detailed understanding of the strength of the W-exchange process, to provide more information on final-state interactions, to identify SU(3) violations, and to observe penguin decays in charm sector.

Factorization should work better for heavier mesons, but model dependence is still unavoidable. Systematic study of B decays and comparison of B and D decays should help clarify some of the unresolved issues.

5.4.1 Factorization and External Spectator Decays

It is expected that factorization will be more reliable in B meson decays than in the equivalent D meson decays due to the larger characteristic energy transfers and the consequent suppression of final state interactions. For a limited class of decay modes the factorization ansatz has been experimentally confirmed to an accuracy of about 15%.

Factorization is likely a poorer approximation for decays with smaller energy release or larger momentum transfer, q^2. for example in the $B \to D^{(*)}D_s^{(*)}$ decay. Using τ spectral functions, data on $e^+e^- \to n\pi$, and $d\Gamma(B \to D^*l\nu)/dq^2$ factorization can be tested using $B \to D^*(n\pi)^-$ decays (n>2) over the wide q^2 range covered by the $(n\pi)^-$ system.

If factorization holds, measurements of hadronic B decays can be compared to theoretical models and used to extract Standard Model parameters. For instance, the CKM matrix element $|V_{ub}|$ could be obtained from $\bar{B}^0 \to \pi^+\pi^-$ or $\bar{B}^0 \to D_s^-\pi^+$. The wave function overlap of the D_s mesons, parametrized in the decay constant f_{D_s} could be extracted from $\bar{B}^0 \to D_s^- D^{*+}$.

Factorization breaks down in the charm sector due to the presence of large final state interactions. The strength of these long distance effects in the B system can be determined by performing an isospin analysis of related decay channels such as $B^- \to D^0\pi^-$, $\bar{B}^0 \to D^0\pi^0$, and $\bar{B}^0 \to D^+\pi^-$ as was done in the past for $D \to K\pi$ and $D \to K^*\pi$ systems [84].

Although factorization has been used to calculate the rates for color suppressed and penguin decays, it is not known whether this is a correct assumption for these diagrams. Experimental verification is needed before model calculations can be trusted.

5.4.2 Inclusive B Decays and the Color Suppressed Amplitude

For Cabibbo favored B decays, the "color suppressed" or internal spectator diagram has the antiquark from the virtual W^- hadronize with the c quark from the b decay. A simple color counting argument then leads to a "color suppressed" rate for this process. So far, the only compelling experimental evidence for this suppression is the observation of inclusive and exclusive $B \to$ charmonium decays. A first order QCD correction involving a hard gluon exchange can defeat this simple color argument. Further information on the size of the color suppressed contribution could be obtained from $\bar{B}^0 \to D^0$(or D^{*0})X_0 decays, where X_0 is a neutral meson.

QCD radiative corrections also produce corrections to the weak vertices themselves. In particular, they generate a color octet term in the effective weak Hamiltonian. from the QCD radiative corrections to the weak vertices themselves. There is experimental evidence for the color octet contribution from $B \to \chi_{c2}$ decays.

Many of the color suppressed decay modes, in particular $B^0 \to \psi K^0$ and $B^0 \to \psi K^{*0}$, will be important in *CP* violation studies. *CP* asymmetries in some other channels such as $B \to D^0 K^-$ arise from the interference of a color suppressed amplitude and a $(b \to u)$ spectator amplitude.

By considering decays of B^- (where both external and internal spectator diagrams contribute) the magnitude and sign of the color suppressed amplitude can be determined. By comparing rates for corresponding B^- and $\bar{B}^0$ in the the decay modes $B \to D^{(*)}\pi^-$ and $B \to D^{(*)}\rho^-$ this interference is found to be constructive [88]. Since the evidence for constructive interference is based on a set of decay modes which comprises only 3% of the total width, the generality of this result is unknown. This differs from the pattern observed in hadronic charm decays.

A variety of other decays might shed light on "color suppressed" spectator decays. Hadronic B decays to pairs of baryons as well as $b \to c\bar{c}s$ decays may well provide new windows to understanding the dynamics.

5.4.3 B Lifetimes and W Exchange

If the decays of b flavored hadrons are dominated by the external spectator diagram, and our current explanation of the lifetime differences in the charm sector is correct, then the lifetimes of B mesons and baryons should be the same to corrections of order a few percent. Measurements at LEP and the Tevatron find $\tau_{B^+} \approx \tau_{B^0} \approx \tau_{B_s} \approx 1.6$ ps confirming this expectation. However, the lifetime for the Λ_b baryon is found to be significantly shorter, $\tau_{\Lambda_b} \approx 1.0 - 1.2$ ps [89]. This indicates a larger than expected contribution from internal W exchanges to the Λ_b decay width. In contrast to baryons, W exchange is helicity suppressed in mesons and a significant contribution from multibody final states complicates the search for these decays even further. So far, no evidence for the presence of this mechanism has been found in B meson decay. Measurements of the lifetimes for the other b baryons as well as more precise measurements of the b mesons from LEP and hadron colliders may clarify this issue.

5.4.4 Hadronic Penguins

The inclusive rates for the hadronic penguin diagrams $b \to sg$ (where g denotes a gluon) and $b \to sq\bar{q}$ are estimated to be about 1% from the parton model, but predictions for the hadronization into exclusive final states are uncertain because the simple factorization assumption used for the spectator diagram may not be valid. It is important to systematically measure the exclusive final states where the gluonic penguin contribution is dominant, for example $B \to K_s\phi, K^-\pi^+$ [90]. In future studies at B Factories, some of these modes will be used to search for asymmetries from direct *CP* violation arising from the interference of the penguin and $b \to u$ spectator amplitudes. Although it is not experimentally straightforward, the inclusive $b \to sg$ rate should be also measured and compared to the theoretical expectation. This may be accomplished by searching for a mass peak combining kaon candidates and n pions, where $1 < n < 7$, or by studying inclusive $B \to \phi$ production in the momentum range above the kinematic limit for production from $b \to c$.

5.5 Hadronic Tau Decays

The semi-hadronic decays of the tau can be used to probe the hadronic structure of the electroweak charged current; to measure the parameters of low-mass hadronic resonances with great precision; and to study α_s, QCD sum rules, and spin-sum rules in a new regime [26].

- Hadronic final states are produced by the vector or axialvector component of the W. The conserved vector component of the W can be related, via an isospin rotation, to the electromagnetic current, for which much data exists (from low energy e^+e^- collisions). The vector components of the semi-hadronic decays of the tau (primarily the decays to an even number of pions) can be used as a test of vector current conservation (CVC). There is no comparable body of data in which the partially conserved axial vector hadronic current is cleanly resolvable. Studies of this current may yield new insights into low energy hadronic dynamics. Tau decays into odd numbers of pions provide a unique, clean laboratory to study these hadronic currents.

- Tau decays permit the study of resonances (the ρ, a_1, ρ', etc, and the strange resonances K^*, K_{1A}, K_{1B}, etc) with very clean, high statistics samples. The decays via the strange K_1 resonances allow for interesting studies of the interplay between meson mixing (between the K_{1A} and K_{1B} resonances) and the $SU(3)_f$-violating second-class current (which produces the K_{1B}). The decay to $\pi^-\eta\pi^0\nu_\tau$ proceeds in the Standard Model via the chiral anomaly (the Wess-Zumino coupling), which flips the parity of the three meson system. It is related via CVC and isospin to the cross section for $e^+e^- \to \eta\pi^+\pi^-$. This decay has only been seen by CLEO, with very limited statistics; a larger sample will permit tests of the chiral anomaly and CVC.
- There is a wide range of tests of perturbative QCD and QCD sum rules accessible when studying tau decay spectral functions. The total semi-hadronic decay rate has been used to measure the strong coupling constant at the tau mass scale $\alpha_S(m_\tau)$, and therefore the hadronic scale Λ_{QCD}, with precision comparable to that measured at M_Z. The inclusive semi-hadronic differential decay rate $d\Gamma/dq^2$ also provides tests of perturbative QCD, including an independent estimate of the significance of non-perturbative effects. Other QCD predictions can be tested, including the study of polarization in inclusive semi-hadronic tau decays.

6 Facilities

Heavy Flavor Physics is an active part of the program at every existing accelerator laboratory and no doubt will continue to be so no matter what future facilities are supported. This section of our report reviews the programs of existing facilities and approved upgrades, and in addition discusses a number of as-yet-unfunded proposals for extensions, upgrades and new facilities. No attempt has been made to establish priorities for these future efforts. Serious study of the physics reach and costs of each option is needed before that can be attempted, and such a study is beyond the scope of this project.

One area of heavy flavor experiment is not discussed here, and that is possible top-antitop production at hadron colliders or at a future e^+e^- linear collider. See the reports of the Electroweak and Accelerator Working Groups for discussion of these topics.

6.1 K Physics Facilities

6.1.1 Rare K Decay Searches at Brookhaven AGS

The Brookhaven AGS is a 30 GeV proton synchrotron currently producing $> 3 \times 10^{13}$ protons in a 1.3 second spill every 3.8 seconds. Over the past few years the AGS has been undergoing a series of upgrades, which include a new 1.5 GeV Booster, revamped vacuum and RF systems, etc. When these upgrades are completed in FY 1996, the AGS will be able to produce between 6 and 10×10^{13} protons/pulse, making it the most intense source of protons in its energy range. Beam lines will have access to at least five times more protons than available in previous experiments. Rare K decay experiments currently collecting data are designed to reach sensitivities of 10^{-12}/event; these include dedicated searches for the decays $K^+ \to \pi^+\mu^+e^-$, $K_L^0 \to \mu^+e^-$ and $K^+ \to \pi^+\nu\bar{\nu}$. With improvements in beams and detectors, 10^{-13}/event is possible. Such sensitivities are possible at the AGS not only because of the intense fluxes available but because of the powerful particle identification techniques that have been developed to work in the AGS energy range.

The search for $K^+ \to \pi^+\nu\bar{\nu}$ is probably most readily done in stopped K beams. The AGS will soon provide a very pure beam of $> 3 \times 10^6$ stopped K^+/pulse. Experiments involving measurements of muon polarization, such as the currently topical search for T-violation in $K_{\mu 3}$ decay also require low energy beams. The important K mode, $K_L^0 \to \pi^0\nu\bar{\nu}$, also needs further study; at low energy determination of the K_L^0 momentum can perhaps augment the otherwise poor signature of this process. Low energy for this experiment also provides a reduction of hyperon backgrounds and the possibility of combining the decay and detection regions for greater acceptance.

With the advent of RHIC, the base costs of maintaining the AGS will be transferred to nuclear physics, making high-energy physics utilization of the machine possibly very economical for the high-energy physics community. However, no decisions have been made.

6.1.2 K Physics at Fermilab

A variety of high precision, high sensitivity Kaon decay experiments can be performed with high energy beams. At high energies, gamma ray energy resolutions become better so that many modes with γ's or π^0's in the final state are well studied. For example, the best sensitivities for the three $K_L^0 \to \pi^0 l\bar{l}$ decays come from Fermilab and these modes are important for the study of *CP* violation. New searches for flavor violation in both K_L^0 and π^0 have been made [91]. A value for the direct *CP*-violation parameter ϵ'/ϵ has been measured as well as the best value of ϕ_{+-}, providing a very sensitive test of CPT symmetry.

A new facility to study kaon decay physics is now under construction at the Fermilab Tevatron (KTeV) [92]. The facility should be ready for data taking in late 1995. The center piece of the new detector is a large pure CsI crystal array for electromagnetic calorimetry. A run spanning 1996 and 1997 is planned. Further runs of the facility at even greater intensity are possible at the Tevatron. The facility will eventually be targeted with the even higher intensity but lower energy Main Injector beam as well.

This facility has two thrusts: very high precision studies of two pion decays of both the long and short-lived kaons as a means of determining ϵ'; and studies with high sensitivity for a variety of rare kaon decays.

A measurement of ϵ'/ϵ with a precision near 0.0001 is expected; the Standard Model predicts a signal larger than this value. KTeV, CERN NA48, and DAPHNE are all focusing on this physics and all should have results by 1998.

The rare decay part of KTeV should reach 10^{-11} sensitivity for a variety of multi-body modes. In particular, this allows the continued study of the $\pi^0\mu^+\mu^-$ and $\pi^0e^+e^-$ modes for which predictions indicate that the *CP* violating contributions are roughly at this level. Also, hundreds of events of $K_L^0 \to e^+e^-e^+e^-$ and $K_L^0 \to \pi^+\pi^-e^+e^-$ will be collected. These and other modes give sensitivity, through angular distributions, to non-standard *CP* violating mechanisms.

The Fermilab Main Injector is a rapid cycling 150 GeV accelerator which will replace the existing Main Ring in all its functions. For fixed target experiments, a 120 GeV slow spill capability is included in the design which would allow $3-5\times10^{13}$ protons to be spilled over 1.0 second with a 2.9 seconds cycle time, yielding an average duty factor of 34%. These 120 GeV protons could be used to produce intense neutral or charged kaon beams. The kaon intensities will be such that a statistical sensitivity of 1×10^{-13} can be reached in a one year run [93]. A major focus of Main Injector kaon physics could be the three $K_L^0 \to \pi^0 l\bar{l}$ decay modes. Such measurements would provide a new breadth of information on *CP* violating kaon decay physics. In particular, the parameter η could possibly be measured directly to better than 10%. Such experiments require a highly "hermetic" detector to recognize and veto unwanted decays. Studies indicate that the required levels of sensitivity can probably be achieved, but that this exciting physics is highly challenging.

A further subject of study is whether it is possible that a next generation $K^+ \to \pi^+\nu\bar{\nu}$ experiment can be performed at the Main Injector. Lower-energy experiments with stopping kaons require the $\pi-\mu-e$ decay chain to occur for particle identification. The challenge is to be able to do as well with particle identification at high energy, NOT relying on the $\pi-\mu-e$ decay chain. To make a significant impact, such an experiment should have a sensitivity which reaches the 10^{-12} level.

6.1.3 Kaon Physics at Other Facilities

The detailed discussion of the kaon program of Brookhaven and Fermilab should not be taken to imply that there is no competition from overseas laboratories. KEK, Daphne and CERN all have active programs in this area. For the direct *CP* violation (ϵ'/ϵ) experiments the comparison of CERN and Fermilab results has been important, and both laboratories are now building improved experiments (NA48, KTeV). At CERN the CPLEAR project has been studying a variety of K decays, particularly for tests of CPT. At DAPHNE the ϕ Factory under construction may provide a complementary program to the K physics efforts and also offers the possibility of some unique CPT tests. At KEK an experiment on $K_L^0 \to \pi^0e^+e^-$ is in progress, and a study of $K_L^0 \to \pi^0\nu\bar{\nu}$ is proposed.

6.2 Tau and Charm Physics Facilities

6.2.1 Tau and Charm Physics at e^+e^- Colliders

Currently, the only e^+e^- facility running in the tau and charm threshold region is the Beijing Electron Positron Collider (BEPC). There is a single detector at this facility, the Beijing Electron Spectrometer (BES), with a design very similar to the earlier Mark III detector at SPEAR. There is a significant U.S. component to the collaboration working with this detector, and continued U.S. involvement in any future upgrade is likely.

The peak luminosity of the BEPC storage ring is $4-5\times10^{30}$ cm^{-2}s^{-1} at an energy of 4 GeV in the center-of-mass. An upgrade to both the accelerator and the detector has been approved, with completion anticipated in the fall of 1995. The luminosity after the upgrade is expected to be 1.5×10^{31} cm^{-2}s^{-1}. Preliminary studies for a possible future "tau-charm factory" are also actively underway in Beijing, with a luminosity goal of order 1×10^{33} cm^{-2}s^{-1}, and 5-10 times higher if a crab crossing beam separation scheme is used.

In addition to threshold region experiments, the e^+e^- colliders discussed below under b-quark physics will also be excellent producers of tau's and particularly of charmed particles, since continuum hadronic events are 40% $c\bar{c}$ and, in addition, B's decay preferentially to charm quarks. CLEO has already produced significant contributions to our knowledge of charmed hyperon, D and τ decays and will continue to do so. Asymmetric

B Factories will produce large D physics samples, and the vertex separation of the B decays plus good particle identification capability will make these excellent laboratories for the systematic study of charmed mesons. Likewise these facilities will produce of order 3×10^7 tau-pairs per year, and thus compete with a tau-charm factory as a place to study the details of tau decays and lifetime. LEP also produces a significant sample of tau events, and has made important contributions to the study of tau decays.

6.2.2 Fixed Target Charm Experiments

Photo- and hadro-production of charm in fixed target experiments with extracted beams at Fermilab and CERN continue to play an important role in the study of charm production and decay. Two experiments are approved for the next fixed target run at Fermilab: E831 (photoproduction) and E781 (hadroproduction). These experiments are expected to provide as many as 10^5 reconstructed events per decay mode. Future experiments with higher interaction rates and using present day technology could reconstruct at least an order of magnitude more decays. This would make possible more sensitive searches for D mixing and rare, effective flavor-changing neutral current decays.

Another unique aspect of fixed target experiments is the ability to study production mechanisms of charm. The recently established leading charm particle production [94] is an example of the type of interesting effects that can be studied. The production of the different charm species may show other differences between fragmentation in hadronic interactions compared to e^+e^- annihilation. The use of different beam types (pions, kaons, protons, photons) to elucidate the production mechanisms has just "scratched the surface" of what would be possible in future experiments which could cover a large kinematic range (in x_F and p_t) with good statistical power. For example the production of leading charmed baryons provides a test of models of diquark production, at present a poorly understood area.

The potential for charm physics in an internal target experiment has not been explored much, but cross sections are high so this may turn out to be interesting, if such facilities are constructed for B physics.

6.2.3 Hadron Collider Charm Experiments

Hadron colliders offer an excellent opportunity to perform high sensitivity charm studies. Most of the charm production occurs at very low transverse momentum. The proper instrument for carrying out these studies is a forward magnetic spectrometer with planar vertex detectors located transverse to the interaction region and within a few millimeters of it. No such experiment has been constructed at the CERN S$p\bar{p}$S or at the Fermilab Tevatron so there are no good measurements of the charm cross section at collider energies. QCD calculations are uncertain because of the low charm quark mass but indicate that 1-3% of the events will have a charm pair at Fermilab or RHIC energies. At a luminosity of 10^{31} cm^{-2} sec^{-1} (already surpassed at FNAL and well within the capabilities of RHIC), more than 10^4 charm events will be produced per second. A partial study of detection efficiencies indicates that between 10^8 and 10^9 fully reconstructed charm decays could be obtained in about 10^7 seconds of running. Triggering, tracking, particle identification, and data acquisition represent significant challenges. However, the high signal to background make this an attractive alternative to experiments at existing fixed target facilities. Such experiments must increase their sensitivity by an order of magnitude to take advantage of the higher luminosity expected to be available at hadron colliders towards the year 2000.

6.3 b Physics Facilities

6.3.1 Symmetric e^+e^- Colliders

Most of what has been learned to date about the b quark has come from symmetric e^+e^- colliders: CESR and DORIS operating at the $\Upsilon(4S)$ resonance, PEP and PETRA operating at center-of-mass energies near 40 GeV, LEP and SLC operating at the Z. For those studies that do not benefit from measuring time development, the $\Upsilon(4S)$ energy is generally preferable. At this energy, CESR is being upgraded to a luminosity of $1-3 \times 10^{33}$ cm^{-2}s^{-1}, (comparable to a B Factory) status and the CLEO detector is being modified to take advantage of the luminosity increase. The upgraded facility is well-suited for studying rare B decays, for CKM matrix determination, and for searching for direct CP violation in B decays.

Measurement of B lifetimes requires observing time development, as will measurement of $B_s - \bar{B}_s$ mixing. LEP and SLC are the relevant symmetric facilities here. (For the mixing measurement, the polarization of SLC compensates for its lower luminosity relative to LEP.) Further running at these facilities will improve their B event samples and thus their reach for these measurements.

6.3.2 Asymmetric e^+e^- Colliders

Two asymmetric e^+e^- colliders are under construction, (PEP-II)[95] at SLAC and KEKB[96] at KEK. PEP II reuses the PEP tunnel and much of the PEP ring. This facility will collide 9 GeV electrons with 3.1 GeV positrons, to produce $\Upsilon(4S)$ which decays either to B^+B^- or $B^0\overline{B}^0$. The design luminosity is 3×10^{33} cm^{-2}s^{-1}. Such a facility provides a laboratory well-suited to the study of CP violating asymmetries in the decays of neutral B_d's and searches for such effects in charged B decays. In particular the machine is designed to study the time-dependent asymmetries in the decays of B^0's to *CP* eigenstates such as $J/\psi K_s$, D^+D^-, $\pi^+\pi^-$ and $\rho^0\pi^0$, where the time in question is the time difference between the decay of the two B^0's. These asymmetries then measure two of the three angles of the unitarity triangle, and thus can be directly related to the value of ρ and η in the CKM matrix. While the first of these channels can also be studied at a hadron collider, the remaining channels are very difficult to reconstruct in a hadronic environment because of problems with triggering on such modes and with large backgrounds from other hadrons in the events. Thus there are essential parts of the *CP* puzzle which can probably only be studied at such a facility. The asymmetric B Factory can also achieve a significant program of heavy flavor physics in the b, c and τ sectors [97]. The BABAR detector at PEP-II is presently under construction by an international collaboration including physicists from Canada, China, France, Germany, Great Britain, Italy, Russia, Taiwan and the Ukraine (and possibly other countries) as well as a large US participation.

A similar facility is approved for construction at KEK in Japan. The KEK B Factory is designed to eventually achieve a luminosity of 10^{34} (initially 10^33) $\mathrm{cm}^{-2}\mathrm{s}^{-1}$. The circumference of the Tristan tunnel (3018 m), which will be used for this facility is significantly larger than the PEP tunnel (2304 m). The machine is designed with a lower energy asymmetry (8.0 GeV on 3.5 GeV) than PEP-II. A small US contingent is also involved in the KEK B Factory effort.

6.3.3 Future Options at Colliders

The potential of Fermilab collider experiments for b physics could be important for the future U.S. program and needs further study. Existing Fermilab experiments have demonstrated initial capability to detect B events, and are considering upgrades that may give them improved B Physics capability. RHIC has a similar potential if it is used in the *pp* collider mode. The production rate for B mesons is high in such facilities, and in addition they provide baryons containing b quarks. A number of options have been developed by proponents, and are briefly discussed in the following sections.

There are three ways to design a dedicated bottom/charm experiment at a collider: extracted-beam fixed-target, internal fixed-target, or colliding-beam. The comments below summarize some of the ideas that are under consideration, but a serious comparative study is needed to make an adequate judgement of their relative merits.

SSC and LHC design studies suggest that extracted-beam fixed-target b experiments are quite competitive with and in some cases exceed the capability of collider experiments at the same accelerator at high energies. For earlier, lower-energy experiments at Fermilab this has not been born out by experience. However it is expected to be increasingly true at higher energies since the fixed target b cross section rises faster than linearly with the beam energy.

Indeed the HERA-B experiment [98], approved at DESY, aims to discover *CP* violation in B decays. The HERA-B experiment plans to produce B's with a wire target in the 800 GeV beam at the HERA *ep* machine. Interactions rates on the order of 30 MHz must be tolerated to allow this experiment to probe $\sin(2\beta)$ to a precision of 0.1 in $B^0 \to J/\psi K_S^0$ decays. Internal target experiments at LHC and Fermilab have also been under study.

A dedicated program in B physics at a hadron collider offers complementary opportunities and challenges to those at an e^+e^- B Factory. Whereas measurements at the $\Upsilon(4S)$ will be restricted to the B_u and B_d mesons, measurements with B_s, B_c, and Λ_b, and other species of b hadrons are possible using hadronic collisions. Because of the greater b production cross section in hadronic collisions ($\sigma(b\bar{b}) \simeq 50\mu$b at the Tevatron vs. $\simeq 1$ nb at the $\Upsilon(4S)$), rarer processes can in principle be probed. On the other hand, experiments at an e^+e^- B Factory will operate in a much cleaner environment by virtue of the more favorable ratio of b production cross section to the total cross section (about 1/4 at the $\Upsilon(4S)$ and $\simeq 1/1000$ at the Tevatron) and also due the exclusive $B\bar{B}$ decays of the $\Upsilon(4S)$. The challenge for experiments at hadron machines is in obtaining trigger, tagging, and reconstruction efficiencies sufficiently high to take advantage of the large number of b's produced.

Considerable progress has been made recently in the study of B physics at the Tevatron by CDF. The B_s has been confirmed through the fully reconstructed decay $B_s \to J\psi/\phi$, and exclusive lifetime measurements have been made for the B^+, B^0 and B_s. These measurements have been made possible by the installation of a silicon vertex detector (SVX) in CDF. The lifetime results are still statistically limited but are competitive with results from LEP. Limits have been set on the rare $B \to K^{(*)}\mu^+\mu^-$ and $B \to \mu^+\mu^-$ that are competitive with current results from CLEO. CDF has also measured the polarization correlations in the decay $B^0 \to J/\psi K^*$ with a precision comparable to that achieved by CLEO, demonstrating the ability to study decay dynamics in the hadron collider environment.

Further progress is expected at the Tevatron as more luminosity is delivered. Upgrades of the tracking systems and trigger and data acquisition electronics for CDF and DØ can increase the capabilities of these detectors for B physics, especially with the luminosities expected ($\sim 10^{32}$ $\mathrm{cm}^{-2}\mathrm{s}^{-1}$) when the Main Injector becomes operational.

Initial measurements of $\sin(2\beta)$ from *CP* violation in $B_d \to J/\psi K_S^0$ decays may be possible at the Tevatron with an integrated luminosity of 1 fb^{-1} [97]. The precision of these measurements will depend strongly on the tagging strategy employed; the possibility of tagging B's with pions from B^{**} or fragmentation processes is being actively pursued. Measurements of *CP* violation in $B_s \to J/\psi\phi$ and of $B_s - \bar{B}_s$ mixing will also be pursued.

The two existing detectors at the Tevatron Collider, CDF and DØ, are primarily designed to explore high p_t physics and are not optimized for B physics investigations even though they have significant capabilities in this area. Efforts continue to design detectors are that are specifically optimized to have the tracking, vertexing, triggering, particle identification, and data acquisition capabilities to study *CP* violation, B_s mixing, and rare B decays at the Tevatron. The goal of these designs is to achieve at least an order of magnitude improvement in sensitivity over an upgraded CDF or DØ detector to a wide variety of *CP* violating and rare B decays.

Additional opportunities can be anticipated eventually at LHC ($\sigma(b\bar{b}) \simeq 700\mu b$, $L = 10^{33}$ $\text{cm}^{-2}\text{s}^{-1}$) where a program directed at B physics has been endorsed. Another possibility is RHIC. RHIC is scheduled for commissioning in 1999, and will run primarily for nuclear physics (heavy ions), with an anticipated availability of about 20% for high-energy physics, provided HEP funding was available. Calculations show that RHIC, operating in pp mode with 250 GeV/c per beam could produce 10^9 B's per month if operated at a luminosity of 4×10^{31} $\text{cm}^{-2}\text{s}^{-1}$. RHIC has the attractive features of a short luminous region (11 cm rms), 110 ns between bunch crossings with just over 1 interaction/bunch, 60% of the B's are produced with $\eta < 1.5$. The expected signal-to-background ratio is about a factor of four below that of a Fermilab experiment.

7 Conclusions

The mysteries of mass, flavor, and CP violation, which are likely to have deep connections, compel us to continue pushing an experimental and theoretical program studying the decays of heavy flavor. One goal of this program is to perform sufficient measurements to unambiguously overconstrain the CKM matrix. Another goal is to exploit the sensitivity of searches for rare and forbidden decays to physics beyond the Standard Model governed by mass scales exceeding those that can be directly probed in present or planned accelerators. Because we are unable to predict in which sector (bottom, charm, tau, strange) evidence for physics beyond the Standard Model will appear, our studies need to remain as broad as possible.

Continued progress towards the above goals requires continued experimental and theoretical progress in understanding the interplay of the strong, weak and electromagnetic interactions in decay processes. As examples: the completion of full next-to-leading order QCD corrections to the $b \to s\gamma$ branching ratio will increase the power of that measurement as a search for beyond-the-Standard-Model physics; accurate lattice-gauge calculations of the B decay constant f_B will relate measurements of $B^0 - \bar{B}^0$ mixing to the CKM matrix element V_{td}.

The present U.S. program in heavy flavor physics is healthy and extensive, and includes CLEO at Cornell (bottom, charm, tau), the collider program at Fermilab (bottom), the fixed target program at Fermilab (charm, strange), and rare kaon studies at Brookhaven. In addition, there is substantial U.S. involvement in the LEP experiments at CERN (bottom, taus), and in BES in Beijing (taus, charm). The approved upgrades to existing facilities (CESR-CLEO, CDF, DØ) and the approved SLAC asymmetric B Factory ensure a continued healthy program. A small US group is also participating in the KEK asymmetric B Factory effort.

We recommend a study of the opportunities for dedicated B physics experiments at a collider, or hadronic fixed target experiment, at either Brookhaven or Fermilab. The production rate of b quarks in high-energy hadron-hadron collisions greatly exceeds that in e^+e^- collisions, but triggering, background rejection and resolution all provide challenges. A number of ideas for dedicated B-physics experiments at Brookhaven and at Fermilab have been studied. The desirability of a new facility deserves careful consideration, with comparison of the physics reach it would provide to that given by the e^+e^- B Factories, the Hera-B internal target experiment at DESY, and possible LHC B physics initiatives. If an interaction region currently used for high P_t physics is involved, the loss of that physics should be included in the considerations.

The long-range future of kaon physics needs careful consideration. As Brookhaven's role shifts with the advent of RHIC, the role of kaon experiments at that laboratory will need attention. Similarly at Fermilab the interplay of the kaon program with other initiatives requires on-going attention in resource allocation.

Acknowledgements

The authors/conveners gratefully acknowledge the numerous contributions of many colleagues in assembling this report.

References

[1] L. Montanet *et al.* (Particle Data Group), *Phys Rev. D* **50**, 1315 (1994).

[2] L. Wolfenstein, *Phys. Rev. Lett.* **51**, 1945 (1984).

[3] R.D. Peccei in *CP Violation in Particle Physics and Astrophysics*, edited by J. Tran Thanh Van (Editions Frontieres, Singapore, 1989).

[4] L. Montanet *et al.* (Particle Data Group), *Phys Rev. D* **50**, 1304, 1416 (1994).

[5] M. Luke and M.J. Savage, *Phys. Lett. B* **321**, 88 (1994).

[6] M. Shifman, N. Uraltsev and A. Vainshtein, U. Minnesota preprint TPI-MINN-94/13-T (HEP PH 9405207) 1994.

[7] M. Neubert, CERN preprint CERN-TH 7395/94 (1994).

[8] R. Fulton *et al.* (CLEO), *Phys. Rev. Lett.* **64**, 16 (1990); J. Bartelt *et al.* (CLEO), *Phys. Rev. Lett.* **71**, 4111 (1993).

[9] H. Albrecht *et al.* (ARGUS), *Phys. Lett. B* **234**, 409 (1990).

[10] S. Stone in *B Decays*, edited by S. Stone. (World Scientific, Singapore, 1994)

[11] N. Isgur and M.B. Wise in *B Physics*, edited by S. Stone (World Scientific, Singapore, 1994).

[12] M. Battle *et al.* (CLEO), *Phys. Rev. Lett.* **71**, 3922 (1993).

[13] H. Albrecht *et al.* (ARGUS), *Phys. Lett. B* **192**, 245 (1987); H. Albrecht *et al.* (ARGUS), *Z. Phys. C* **54**, 357 (1992).

[14] M. Artuso *et al.* (CLEO), *Phys. Rev. Lett.* **62**, 2233 (1989); J. Bartelt *et al.* (CLEO), *Phys. Rev. Lett.* **71**, 1680 (1993).

[15] H. Schröder in *B Decays*, edited by S. Stone. (World Scientific, Singapore, 1994)

[16] A. Ali in *B Decays*, edited by S. Stone. (World Scientific, Singapore, 1994)

[17] A. Ali and C. Greub, *Phys. Lett. B* **287**, 191 (1992).

[18] E. Golowich and S. Pakvasa, U. Massachusetts preprint UMHEP-411 (1994).

[19] D. Atwood, B. Blok and A. Soni, SLAC-PUB-6635 (1994).

[20] M. Athenas *et al.* (CLEO), Contribution GSL0159 to 27th Int. Conf. on HEP, Glasgow, July 1994.

[21] J. Alexander *et al.* (CLEO), Contribution GSL0160 to 27th Int. Conf. on HEP, Glasgow, July 1994.

[22] A.J. Buras and M.K. Harlander in *Heavy Flavors*, eds. A.J. Buras and M. Lindner (World Scientific, Singapore, 1992).

[23] A.F. Falk, M. Luke and M.J. Savage, *Phys. Rev. D* **49**, 3367 (1994).

[24] I.I. Bigi, M.A. Shifman, N.G. Uraltsev, and A.I. Vainshtein, U. Minnesota preprint TPI-MINN-94-4-T (HEP PH 9402360) 1994.

[25] J.A. Ernst, Ph.D thesis, University of Rochester (1995).

[26] A. Pich in *Heavy Flavours* eds. A. J. Buras and M. Lindner (World Scientific, Singapore, 1992).

[27] L. Montanet *et al.* (Particle Data Group), *Phys Rev. D* **50**, 1248 (1994).

[28] D. S. Akerib *et al.*, *Phys Rev. Lett.* bf 69, 3610 (1992).

[29] L. Montanet *et al.* (Particle Data Group), *Phys Rev. D* **50**, 1415 (1994).

[30] Y. Nir and H. R. Quinn in *B Physics*, edited by S. Stone (World Scientific, Singapore, 1994).

[31] L. Montanet *et al.* (Particle Data Group), *Phys Rev. D* **50**, 1544 (1994).

[32] C. Jarlskog in *C. P. Violation*, ed C. Jarlskog (World Scientific, Singapore, 1989).

[33] L. Montanet *et al.* (Particle Data Group), *Phys Rev. D* **50**, 1288 (1994).

[34] K. F. Smith *et al.*, *Phys. Lett. B* **234**, 191 (1990); I. S. Altarev *et al.*, *Phys. Lett. B* **276**, 242 (1992).

[35] S.M. Barr and W.J. Marciano in *CP Violation*, edited by C. Jarlskog (World Scientific, Singapore, 1989)

[36] S.M. Barr and W.J. Marciano in *CP Violation*, edited by C. Jarlskog (World Scientific, Singapore, 1988).

[37] L. Wolfenstein, *Phys Rev. Lett.* **13**, 562 (1964).

[38] B.D. Winstein and L. Wolfenstein, *Rev. Mod. Phys.* **65**, 1113 (1993).

[39] G.D. Barr *et al.*, *Phys. Lett. B* **317**, 233 (1993).

[40] L.K. Gibbons et al., *Phys. Rev. Lett.* **70**, 1203 (1993).

[41] A.J. Buras and M.K. Harlander in *Heavy Flavours*, eds. A.J. Buras and M. Lindner (World Scientific, Singapore, 1992).

[42] D.A. Harris *et al.*, *Phys. Rev. Lett.*, **71**, 3914, 3918 (1993).

[43] M. Weaver *et al.*, *Phys. Rev. Lett.*, **72**, 3758 (1994).

[44] G. Gidal *et al.*, Fermilab Proposal 871, 1993.

[45] L. Montanet *et al.* (Particle Data Group), *Phys Rev. D* **50**, 1288 (1994).

[46] *KEK-PS Proposal; E-246*, J. Imazato *et al.*, 1991.

[47] I.I. Bigi, Proceedings of CHARM 2000, Fermilab, 1994, CERN-TH7370/94.

[48] I.I. Bigi and A.I. Sanda, *Nucl. Phys. B* **193**, 85 (1981).

[49] For a review of Standard Model Predictions on *CP* Violation in *B* decays see for example Y. Nir and H. R. Quinn *Ann. Rev. Nucl. Part. Sci.* **42** 211, (1992) and references contained therein.

[50] A. Soni, *Lattice Results for Heavy Light Matrix Elements* to appear in *Proceedings of the 27th International Conference on High Energy Physics*, Glasgow, Scotland, July 1994, BNL-60828 (1994).

[51] A.J. Buras *et al.*, *Phys. Rev. D* **50**, 3433 (1994).

[52] The authors thank M. Witherell for providing Figure 2 based on the recent work of P. Burchat and J. Richman. See also M. Witherell, lectures presented at the NATO 1994 Advanced Study Institute on Techniques and Concepts of High Energy Physics, St. Croix, UCSB-HEP-94-03.

[53] F. Abe *et al.* (CDF), *Phys. Rev. Lett.* **73**, 225 (1994). See also F. Abe *et al.* (CDF), *Phys. Rev. D* **50**, 2966 (1994), F. Abe *et al.* (CDF), *Phys. Rev. Lett.* **74**, 2626 (1995) and F. Abachi *et al.* (DØ), *Phys. Rev. Lett.* **74**, 2632 (1995).

[54] L. Montanet *et al.* (Particle Data Group), *Phys Rev. D* **50**, 1632 (1994).

[55] M. Gronau and D. London *Phys. ReV. Lett.***65**, 3381 (1991).

[56] N.G. Deshpande and Xiao-Gang He, preprint OITS-553 (1994)

[57] Y. Nir and H. Quinn in *B Decays*, edited by S. Stone. (World Scientific, Singapore, 1994)

[58] C.A. Nelson, Third Workshop on Tau Lepton Physics, Montreux, 1994, preprint SUNY-BING/10-15-94.

[59] R. Battiston *et al.*, *Phys. Rep.* **214**, 293 (1992).

[60] L. Littenberg and G. Valencia *Annual Review of Nuclear and Particle Science* **43** (1993)

[61] J.L. Ritchie and S.G. Wojcicki, *Rev. Mod. Phys.* **65** 1149 (1993)

[62] P. Langacker, Invited talk presented at the China Center of Advanced Science and Technology Symposium on TeV Physics, 1990, UPR-468T.

[63] P. Mélèse, *Comm. Nucl. Part. Phys.* **19**, 117 (1989).

[64] S.L. Glashow, J. Illiopoulous and L. Maiani, *Phys. Rev. D* **2**, 1282 (1970).

[65] L. Montanet *et al.* (Particle Data Group), *Phys. Rev. D* **50**, 1537 (1994).

[66] W.J Marciano, Invited talk presented at the Rare Decay Symposium, Vancouver (1988).

[67] R. Ammar *et al.* (CLEO), *Phys. Rev. Lett.* **71**, 674 (1993).

[68] M.S. Alam *et al.*, *Phys. Rev. Lett.*,

[69] S. Bertolini, F. Borzumati and A. Masiero in *B Physics*, edited by S. Stone (World Scientific, Singapore, 1994).

[70] A.J. Buras and M.K. Harlander in *Heavy Flavours*, eds. A.J. Buras and M. Lindner (World Scientific, Singapore, 1992).

[71] A.J. Buras and M.K. Harlander in *Heavy Flavours*, eds. A.J. Buras and M. Lindner (World Scientific, Singapore, 1992).

[72] A. Kronfeld and P. Mackenzie, *Ann. Rev. Nucl. Part. Phys.* **43**, 793 (1993).

[73] E. de Rafael in *CP Violation*, edited by C. Jarlskog (World Scientific, Singapore, 1988).

[74] W. Kwong, C. Quigg and J. Rosner, *Ann. Rev. Nucl. Part. Sci.* **37**, 325 (1987).

[75] M. Wise, Proceedings of the 1991 Lake Louise Winter Institute, preprint QCD161-l25-1991.

[76] M. Gronan, A. Nippe and J.L. Rosner, *Phys. Rev. D* **47**, 1988 (1993).

[77] B. Grinstein, Lectures given at TASI, 1994.

[78] L. Montanet *et al.* (Particle Data Group), *Phys Rev. D* **50**, 1564 (1994).

[79] C.W. Bernard, J.N. Labrenz and A. Soni, *Phys. Rev. D* **49**, 2536 (1994).

[80] L. Montanet *et al.* (Particle Data Group), *Phys Rev. D* **50**, 1602 (1994).

[81] S. Sanghera, *et al.*, *Phys Rev.* D **47**, 791 (1993).

[82] S. Stone in *Heavy Flavours*, eds. A.J. Buras and M. Lindner (World Scientific, Singapore, 1992).

[83] M.S. Witherell *UCSB-HEP-93-06* Invited talk given at the International Symposium on Lepton and Photon Interactions at High Energies, Cornell University, Ithaca, NY (August 1993).

[84] M. Neubert, V. Rieckert, B. Stech and Q.P. Xu, in *Heavy Flavours*, eds. A.J. Buras and M. Lindner (World Scientific, Singapore, 1992).

[85] J.C. Anjos *et al.*, *Phys. Rev. Lett.* **62**, 125,1721 (1989).

[86] T.E. Browder, K. Honscheid and S. Playfer in *B Physics*, edited by S. Stone (World Scientific, Singapore, 1994).

[87] V. Sharma and F. V. Weber in *B Physics*, edited by S. Stone (World Scientific, Singapore, 1994).

[88] N. G. Deshpande in *B Physics*, edited by S. Stone (World Scientific, Singapore, 1994).

[89] D.A. Harris *et al.*, *Phys. Lett. B* **320**, 407 (1994).

[90] K. Arisaka *et al.*, KTeV Design Report, FN-580, (1992).

[91] Fermilab publication FN-568 : Conceptual Design Report: Kaons At the Main Injector. 1991.

[92] G. Alves *et al.*, *Phys. Rev. Lett.* **72** 812 (1994)]

[93] An Asymmetric *B* Factory Based on PEP, SLAC Report 372, February 1991.

[94] A Study of *CP* Violation in *B* Meson Decays, KEK Report 95-1, April 1995

[95] BABAR Collaboration, Letter of Intent for the Study of *CP* Violation and Heavy Flavor Physics at PEP-II, SLAC Report 443, (1994).

[96] *HERA-B Proposal*, DESY-PRC 94/02, May 1994.

[97] *Proceedings of the Workshop on B Physics at Hadron Accelerators*, P. McBride and C. Mishra, editors, Snowmass, Colorado, 1993.

Neutrino Mass and Mixing

Paul Langacker

University of Pennsylvania, Philadelphia, PA 19104

Regina Rameika

Fermi National Accelerator Laboratory, Batavia, IL 60510

Hamish Robertson

University of Washington, Seattle, WA 98195

Neutrino Mass Working Group Members

D. Ayres	*Argonne National Laboratory*	B. Bernstein	*Fermi National Accelerator Laboratory*
B. Blumenfeld	*John Hopkins University*	A. Capone	*Universita La Sapienza/INFN*
M. Diwan	*Brookhaven National Laboratory*	G. Domokos	*Johns Hopkins University*
F. Federspiel	*Los Alamos National Laboratory*	T. Gaisser	*Bartol Research Institute*
M. Goodman	*Argonne National Laboratory*	K. Heller	*University of Minnesota*
A. Hime	*Los Alamos National Laboratory*	T. Kycia	*Brookhaven National Laboratory*
K. Lande	*University of Pennsylvania*	M. Lowry	*Princeton University*
A. Mann	*University of Pennsylvania*	M. Murtagh	*Brookhaven National Laboratory*
P. Rosen	*University of Texas at Arlington*	N. Stanton	*Kansas State University*
R. Steinberg	*Drexel University*	R. Stokstad	*Lawrence Berkeley Laboratory*
B. Vogelaar	*Princeton University*	H. White	*Los Alamos National Laboratory*

1 Introduction

There are strong theoretical and experimental motivations to search for neutrino mass. The Standard Model of electroweak interactions provides the framework in which the fundamental particles are arranged into pairs and generations (or flavors) of quarks:

$$\begin{pmatrix} u \\ d \end{pmatrix} \quad \begin{pmatrix} c \\ s \end{pmatrix} \quad \begin{pmatrix} t \\ b \end{pmatrix}$$

and of leptons:

$$\begin{pmatrix} e^- \\ \nu_e \end{pmatrix} \quad \begin{pmatrix} \mu^- \\ \nu_\mu \end{pmatrix} \quad \begin{pmatrix} \tau^- \\ \nu_\tau \end{pmatrix}$$

Experimentally, the neutrinos appear to be massless, and hence the assumption that they are exactly so is built into the basic Standard Model. However, despite the experimental success of the model, we know that it is incomplete. In particular, many of the major outstanding challenges in particle physics today have to do with the question of mass. For example, we ask, "Why is the top quark so heavy?" (or "Why are the other fermions so light?"), "Is the Higgs particle the origin of mass?", and "Are the neutrinos really massless?" There are also a number of persisting questions in cosmology and astrophysics, such as "Where are the solar neutrinos?", "Are there missing atmospheric neutrinos?" and "What makes up the mass of the universe?".

In the minimal Standard Model and in certain extensions such as SU(5), neutrinos are assumed massless. The Standard Model can, however, easily accommodate massive neutrinos without contradicting firmly established symmetries, so the existence of neutrino mass is an open experimental question. Direct measurements of neutrino masses have yielded only upper limits (≤ 5.1 eV, ≤ 0.27 MeV, and ≤ 31 MeV for electron, muon, and tau neutrinos, respectively [1]). Non-zero neutrino masses below these limits contradict neither theory nor experiment. The exciting possibility is that massive, flavor-mixed neutrinos (*e.g.*, $\nu_e \rightarrow \nu_\mu$) may answer many of the above questions. The next generation of oscillation and direct mass measurements will extend sensitivities into ranges of mass and mixing parameters for which a positive result is likely. This could well constitute the first break with the standard model. Alternatively, the experiments should be able to exclude cleanly most of the interesting parameter range. In either case, the implications for particle physics, astrophysics, and cosmology will be significant.

Almost all extensions of the standard model predict non-zero neutrino masses. Both general directions for new physics – grand unification/superstrings and compositeness – are likely to lead to neutrino masses. Many models predict seesaw-type masses, $m_\nu \sim m_D^2/M$, where m_D is a typical quark or charged lepton mass and M is the scale of new physics. Such theories naturally explain why the neutrino masses are so small. Measuring these masses and the leptonic mixings would probe the physics at the underlying mass scale M and would be an excellent complement to searches at high-energy colliders. Large values for M (*e.g.*, 10^{12} GeV, as expected in many grand unified theories), predict masses relevant to the solar neutrinos (*e.g.*, $\nu_e \to \nu_\mu$) and to hot dark matter (*e.g.*, ν_τ).

Observation of non-zero neutrino masses could well shed light not only on new physics, but also on the origin of fermion masses in general, which are among the least understood aspects of the standard model.

During the last decade, several scientific developments have given us reasons to believe that neutrinos may indeed be massive. Recent results by COBE and other groups on angular variations of the temperature of the cosmic background radiation [2] as well as the observed clustering of galaxies on very large scales may be explained by a significant component of hot dark matter [3]. The most economical hypothesis for providing such hot dark matter is to assume that τ neutrinos, which are a component of the primordial radiation field, have a mass somewhere in the range from 5 to 35 eV. The precise mass value depends on the Hubble constant and on the fraction of the cosmic energy density contributed by hot dark matter, but a typical estimate is $m_{\nu_\tau} \sim 7$ eV. If true, neutrinos could well be among the dominant constituents of the universe! As we shall see later in this report, most of this interesting parameter range can be probed in planned short baseline oscillation experiments, supplemented with direct mass searches and neutrinoless double beta decay experiments. The latter is also sensitive to the nature of the neutrino mass. Observation of a nearby supernova would extend the sensitivity to very small or zero mixing.

A second and perhaps more direct experimental indication that neutrinos have mass comes from the detection of solar neutrinos. All solar neutrino experiments observe a deficit of neutrinos compared to the expectations of standard solar models. The gallium results are now significantly below the predictions of almost all solar models. Even more compelling, the relative suppressions observed in the Homestake and Kamiokande experiments imply a distortion of the ^{8}B spectrum, which is inconsistent with any conventional astrophysical explanation, and strongly suggestive of non-standard neutrino properties. Neutrino oscillation effects are expected to be exaggerated by passage through solar matter. This "MSW effect" provides the simplest and most natural explanation, adequately describing all data for Δm^2 around 10^{-5} eV^2. Planned and proposed solar neutrino experiments have the potential for cleanly confirming or refuting MSW and other hypotheses, independent of solar model uncertainties. They should also allow a model-independent determination of the major neutrino flux components, even in the presence of oscillations, and therefore serve as a powerful tool for solar astronomy.

Three experiments indicate a deficit in the ratio of muon and electron neutrinos produced in cosmic ray interactions. This ratio is relatively free of theoretical ambiguities in the incident fluxes, and hints at neutrino oscillations with $\Delta m^2 \sim 1 - 10^{-3}$ eV^2 and large mixing angles. These possibilities could be definitely established or excluded by proposed long baseline experiments.

LEP results on the Z lineshape indicate that there are only three light active (left-handed) neutrinos, and limits from big bang nucleosynthesis severely constrain the parameter space allowed for sterile neutrinos. These results, combined with the experimental and theoretical clues described above, significantly limit the possibilities and the likely parameter space for neutrino mass and oscillations. The program of experiments that are planned or proposed for the next decade could unambiguously establish neutrino masses or exclude almost all of the likely possibilities.

A serious program to probe neutrino masses and mixings in the ranges suggested by hot dark matter, atmospheric neutrinos, and solar neutrinos, as well as many well-motivated theoretical ideas (such as grand unification through supersymmetry), has a real possibility of observing positive signals. These could well be the first break with the standard model of particle physics, could have profound consequences for cosmology, and would serve as a quantitative probe of the solar core. Even negative results would severely constrain the possibilities for new physics beyond the standard model, would exclude one of the major possibilities for the dark matter, and would be a powerful tool for studying the sun.

2 Basic Properties of Neutrinos

There are literally hundreds of models of neutrino mass [4]. The nomenclature, notation, and variety of models are very confusing. In this section, we give an overview of the basic concepts and principal classes of models.

The first concept is that of a Weyl neutrino. A Weyl fermion is the basic two-component fermionic degree of freedom. A left-handed state ν_L is necessarily related by a CP or CPT transformation to a right-handed antiparticle state by

$$\nu_L \underset{\text{CP}}{\leftrightarrow} \nu_R^c \equiv C\bar{\nu}_L^T, \tag{1}$$

where C is the charge conjugation matrix, given by $C = i\gamma^2\gamma^0$ in the Bjorken-Drell conventions.[a] Thus the L and R states are essentially adjoints of each other. It is a matter of convenience which is defined as the particle and which is the antiparticle. Sometimes it is useful to define all right-handed states as antiparticles; when there is a conserved lepton number it is convenient to label the leptons as particles and the antileptons as antiparticles.

An important distinction is between active (SU_2-doublet) and sterile[b](SU_2-singlet) neutrinos. The former have normal weak interactions, while the latter have no weak interactions except those induced by mixing. In the standard model with an additional sterile Weyl neutrino N_R, one has

$$\begin{pmatrix} \nu_e \\ e^- \end{pmatrix}_L \underset{\text{CP}}{\leftrightarrow} \begin{pmatrix} e^+ \\ \nu_e^c \end{pmatrix}_R \qquad N_R \underset{\text{CP}}{\leftrightarrow} N_L^c. \tag{2}$$

The ν_e and its partner are active, associated in weak doublets with the electron, while the extra N_R is sterile with no weak interactions. In the ordinary standard model one introduces three active massless neutrinos ν_{eL}, $\nu_{\mu L}$, $\nu_{\tau L}$ and no sterile states.

Mass terms induce transitions between right and left-handed states. A Dirac mass term, which conserves lepton number, involves transitions between two different Weyl neutrinos, ν_L and N_R (or ν_R^c and N_L^c). That is, the right-handed state N_R is different from ν_R^c. The form is

$$-\mathcal{L}_{\text{Dirac}} = m_D(\bar{\nu}_L N_R + \bar{N}_R \nu_L) = m_D \bar{\nu}\nu, \tag{3}$$

where the Dirac field is defined as $\nu \equiv \nu_L + N_R$. Thus a Dirac neutrino has four components ν_L, ν_R^c, N_R, N_L^c, and the mass term allows a conserved lepton number $L = L_\nu + L_N$. This and other types of mass terms can easily be generalized to three or more families, in which case the masses become matrices. In that case, the charged current transitions involve a leptonic mixing matrix (analogous to the Cabibbo-Kobayashi-Maskawa (CKM) quark mixing matrix), which can lead to neutrino oscillations between the light neutrinos.

For an ordinary Dirac neutrino the ν_L is active and the N_R is sterile. The transition between these states is $\Delta I = \frac{1}{2}$, where I is the weak isospin. The mass requires SU_2 breaking and is generated by a Yukawa coupling

$$-\mathcal{L}_{\text{Yukawa}} = h_\nu (\bar{\nu}_e \bar{e})_L \begin{pmatrix} \varphi^0 \\ \varphi^- \end{pmatrix} N_R + H.C. \tag{4}$$

One has $m_D = h_\nu v/\sqrt{2}$, where the vacuum expectation value (VEV) of the Higgs doublet is $v = \sqrt{2}\langle\varphi^o\rangle = (\sqrt{2}G_F)^{-1/2} = 246$ GeV, and h_ν is the Yukawa coupling. A Dirac neutrino mass is just like a quark or charged lepton mass, but that leads to the question of why neutrino mass is so small. One would require $h_{\nu_e} < 10^{-10}$ in order to have $m_{\nu_e} < 10$ eV. For this reason most people do not take normal Dirac masses seriously. However, some caution is advised since we do not understand the patterns of quark or charged lepton masses either. One interesting idea is to invoke some sort of symmetry that requires $h_\nu = 0$ at tree-level, so that neutrino masses would only be generated at the loop level and might naturally be small. However, there are no compelling or simple models to implement this otherwise attractive idea. Something of this sort may be one possibility for superstring models, in which it is difficult to implement the simplest version of the seesaw mechanism. (Non-renormalizable operators are another possibility.)

A variant form of Dirac neutrino is the Zeldovich, Konopinski, Mahmoud (ZKM) type [5], for which ν_L and N_R are both active states – the N_R is just a right-handed antineutrino associated with a different family than the ν_L. For example, one could have $N_{eR} = \nu_{\tau R}^c$ as the partner of the ν_{eL}. In this case $L_e - L_\tau$ would be conserved and ν_{eL}, ν_{eR}^c, $\nu_{\tau R}^c$, $\nu_{\tau L}$ would be the four components of the ZKM neutrino. Similarly, one could choose $N_{eR} = \nu_{\mu R}^c$, so that $L_e - L_\mu$ would be conserved. The ZKM case is defined as a Dirac neutrino by most authors, because it has a (somewhat unusual) conserved lepton number and has four components. However, a ZKM neutrino requires a more complicated Higgs structure than doublets, and in many ways is more analogous to the Majorana neutrinos to be described below. In fact a ZKM neutrino is a limiting case of Majorana neutrinos, and not all authors agree that it should be referred to as a Dirac neutrino.

A Majorana mass, which violates lepton number by two units ($\Delta L = \pm 2$), makes use of the right-handed antineutrino, $N_R = \nu_R^c$, rather than a separate Weyl neutrino. It is a transition from an antineutrino into a neutrino. Equivalently, it can be viewed as the creation or annihilation of two neutrinos, and if present it can therefore lead to neutrinoless double beta decay. The form of a Majorana mass term is given in Eq. 5.

$$\begin{aligned} -\mathcal{L}_{\text{Majorana}} &= \frac{1}{2} m(\bar{\nu}_L \nu_R^c + \bar{\nu}_R^c \nu_L) \\ &= \frac{1}{2} m(\bar{\nu}_L C \bar{\nu}_L^T + H.C.) \end{aligned}$$

[a] The subscripts L and R really refer to the left and right chiral projections. In the limit of zero mass these correspond to left and right helicity states.

[b] Sterile neutrinos are often referred to as "right-handed" neutrinos, but that terminology is confusing and inappropriate when Majorana masses are present.

$$= \frac{1}{2} m \bar{\nu} \nu \qquad (5)$$

$\nu = \nu_L + \nu_R^c$ is a self-conjugate two-component state satisfying $\nu = \nu^c = C\bar{\nu}^T$. If ν_L is active then the above transition involves $\Delta I = 1$ and m must be generated by either an elementary Higgs triplet or by an effective operator involving two Higgs doublets arranged to transform as a triplet.

For an elementary triplet $m \sim h_T v_T$, where h_T is a Yukawa coupling and v_T is the triplet VEV. The simplest implementation is the Gelmini-Roncadelli (GR) model [6], in which lepton number is spontaneously broken by v_T. The original GR model is now excluded by the LEP data on the Z width. The decay $Z \to M\chi$, in which M is the Majoron (the Goldstone boson associated with the spontaneous breaking of lepton number) and $\chi \simeq Re\ \varphi_T^0$ is a physical scalar field with mass $m_\chi = O(v_T)$, occurs rapidly and contributes $\Delta N'_\nu = 2$ to the effective number of neutrino flavors. Since, $N'_\nu = 2.988 \pm 0.023$ [7] there is no room for this decay. Variant models involving explicit lepton number violation or in which the Majoron is mainly a weak singlet (*invisible* Majoron models) are still possible.

If the mass, m, is generated by an effective operator, one expects $m \sim Cv^2/M$, where C is a dimensionless constant and M is the scale of the new physics which generates this operator. The most familiar example is the seesaw model, to be discussed below.

All of the effects of lepton number violation associated with Majorana masses are proportional to the mass, and there is no distinction between Dirac and Majorana neutrinos in the limit $m \to 0$ unless there are new interactions. For $m \to 0$ both reduce to a Weyl neutrino with the standard model interactions, and the N_R decouples in the Dirac case.

It is also possible to consider mixed models in which both Majorana and Dirac mass terms are present. For two Weyl neutrinos one has a mass term

$$-L = \frac{1}{2} \left(\bar{\nu}_L \bar{N}_L^c\right) \begin{pmatrix} m_T & m_D \\ m_D & m_S \end{pmatrix} \begin{pmatrix} \nu_R^c \\ N_R \end{pmatrix} + H.C., \qquad (6)$$

where $\nu_L \leftrightarrow \nu_R^c$ and $N_L^c \leftrightarrow N_R$ are the two Weyl states. m_T and m_S are Majorana masses which transform as weak triplets and singlets, respectively (assuming that the states are respectively active and sterile), while m_D is a Dirac mass term. Diagonalizing this 2×2 matrix one finds that the physical particle content is given by two Majorana mass eigenstates $n_i = n_{iL} + n_{iR}^c$, $i = 1, 2$; as usual these are two-component objects ($n_i = n_i^c$) constructed from the Weyl neutrinos $n_{iL} \leftrightarrow n_{iR}^c$.

There are a number of important special cases of the general mixed model:

- The simplest is that of pure Majorana masses, $m_D = 0$, in which the two Majorana eigenstates are just the original active and sterile neutrinos: $n_{1L} = \nu_L,\ n_{2L} = N_L^c$.
- Another important case is the Dirac limit, $m_T = m_S = 0$. When one diagonalizes the mass matrix in the Dirac limit the two Majorana mass eigenstates

$$\begin{aligned} n_1^D &= \frac{1}{\sqrt{2}}(\nu_L + N_L^c) + \frac{1}{\sqrt{2}}(\nu_R^c + N_R) \\ n_2^D &= \frac{1}{\sqrt{2}}(\nu_L - N_L^c) - \frac{1}{\sqrt{2}}(\nu_R^c - N_R) \end{aligned} \qquad (7)$$

are degenerate with mass m_D, and can be combined to form a Dirac neutrino with a conserved lepton number. Thus, a Dirac neutrino can be viewed as a pair of degenerate Majorana neutrinos with 45° mixing and opposite CP parity.

- The pseudo-Dirac case $m_T \ll m_D,\ m_S \ll m_D$ is a perturbation on the Dirac limit. The Majorana mass terms lead to a small splitting of the mass eigenstates, *i.e.*, one has two almost-degenerate Majorana neutrinos and a small breaking of L. For example, consider the case $m_S = 0,\ m_T \ll m_D$. One obtains two Majorana neutrinos

$$n_1 = n_1^D + \frac{m_T}{4m_D} n_2^D, \quad n_2 = -\frac{m_T}{4m_D} n_1^D + n_2^D, \qquad (8)$$

with masses

$$m_{1,2} = m_D \pm m_T/2. \qquad (9)$$

Thus, one can have $\nu_L \to N_L^c$ oscillations with near maximal mixing.

- In the seesaw limit [8], $m_T = 0,\ m_D \ll m_S$, one obtains two Majorana neutrinos

$$\begin{aligned} n_{1L} &\simeq \nu_L - \frac{m_D}{m_S} N_L^c \\ n_{2L} &\simeq \frac{m_D}{m_S} \nu_L + N_L^c \end{aligned} \qquad (10)$$

with masses

$$\begin{aligned} m_1 &\sim \frac{{m_D}^2}{m_S} \ll m_D \\ m_2 &\sim m_S. \end{aligned} \qquad (11)$$

Thus, there is one heavy neutrino and one neutrino much lighter than the typical Dirac scale. Such models are a popular and natural way of generating neutrino masses much smaller than the other fermion masses.

There are literally hundreds of versions of the seesaw and related models [4]. The heavy scale m_S can range anywhere from the TeV scale to the Planck scale. The TeV scale models are motivated, for example, by left-right symmetric models [9]. Typically, the Dirac masses m_D are of the order of magnitude of the corresponding charged lepton masses, so that one expects masses of order 10^{-1} eV, 10 keV, and 1 MeV for the ν_e, ν_μ, and ν_τ, respectively. (The latter two violate cosmological bounds unless they decay rapidly and invisibly.) Intermediate scales, such as $10^{12} - 10^{16}$ GeV, are motivated by grand unification and typically yield masses in the range relevant to hot dark matter, and solar and atmospheric neutrino oscillations. The grand unified theories often imply Dirac masses $m_D \sim m_u$, where m_u is the mass of the up-type quark of the corresponding family. Depending on whether there is also a family hierarchy of heavy masses m_S, the light masses

$$m_{\nu_i} \sim C_i \frac{m_{u_i}^2}{m_{S_i}}, \tag{12}$$

of the ith family may vary approximately quadratically with m_{u_i} (the quadratic seesaw) or linearly (the linear seesaw) [10]. $C_i \sim (0.05 - 0.4)$ in (12) is a radiative correction. Typical light neutrino masses in the quadratic seesaw are (10^{-7} eV, 10^{-3} eV, 10 eV) for $M_{S_i} \sim 10^{12}$ GeV (the intermediate seesaw, expected in some superstring models or in grand unified theories with multiple breaking stages). Such masses would cause $\nu_e \to \nu_\mu$ oscillations in the sun, and make ν_τ be a dark matter candidate (or, for a somewhat smaller ν_τ mass, $\nu_\mu \to \nu_\tau$ atmospheric neutrino oscillations). Similarly, for $M_{S_i} \sim 10^{16}$ GeV (the grand unified seesaw, expected in old-fashioned grand unified theories with large Higgs representations) one typically finds smaller masses around (10^{-11} eV, 10^{-7}, 10^{-2} eV), suggesting $\nu_e \to \nu_\tau$ oscillations in the sun. In such models one often (but not always) finds that the lepton and quark mixing matrices are similar.

One can easily generalize the mixed models in Eq. (6) to three or more fermion families. For three families there will be three active and three sterile neutrinos. These will mix to form six Majorana mass eigenstates and a 6×6 mixing matrix. One would in general expect both oscillations of the active neutrinos into each other (known as first class or flavor oscillations) and also (second class) oscillations of the active neutrinos into sterile neutrinos. However, the latter are only significant if the Dirac and Majorana mass terms are of comparable magnitude (*i.e.*, the mass eigenstates that are predominantly active and sterile have comparable masses). This does not occur naturally in most models (it does not occur at all for pure Dirac or Majorana masses, and is strongly suppressed in the seesaw models), so that oscillations between active and sterile neutrinos, while a logical possibility, are perhaps less likely than flavor oscillations.

A very different class of models are those in which the neutrino masses are zero at the tree level (typically because no Weyl singlets or elementary Higgs triplets are introduced), but only generated by loops [11], *i.e.*, radiative generation. Such models are very attractive in principle and explain the smallness of m_ν. However, the actual implementation generally requires the *ad hoc* introduction of new Higgs particles with nonstandard electroweak quantum numbers and lepton number-violating couplings. There are a great variety of these models. When phenomenological constraints are imposed they often have a pseudo-Dirac neutrino, *e.g.*, in which $L_e - L_\mu + L_\tau$ is approximately conserved. Many of the models were motivated by attempts to generate a large value for the neutrino magnetic moment μ_ν [12] (suggested by hints of possible time dependence in the Homestake solar neutrino rate) without generating a large value of the effective neutrino mass $\langle m_{\nu_e} \rangle$ relevant to neutrinoless double beta decay.

3 Neutrino Counting

- There are $N_\nu \geq 2$ Weyl neutrinos in Nature, because the ν_e and ν_μ have been directly observed.
- We also know that $N_\nu \geq 3$, because the interactions of the τ are not suppressed by mixing. There must therefore be a separate ν_τ. (SU_2 gauge invariance requires that each doublet involves a different neutrino state.)
- One has $N'_\nu \equiv \Gamma_{Z\to \text{inv}}/\Gamma_{Z\to\nu\bar{\nu}} = 2.988 \pm 0.023$ from the LEP measurements of the invisible Z width $\Gamma_{Z\to\text{inv}} = \Gamma_Z - \Gamma_{Z\to\text{vis}}$ [7]. N'_ν counts the active neutrinos with masses $m \ll M_Z/2$, and also includes contributions from heavy active neutrinos up to the kinematic limit with an additional phase space suppression. It does not count sterile neutrinos. Other light invisible states also contribute. For example, a triplet Majoron [6] yields $\Delta N'_\nu = 2$, which is clearly excluded, while a doublet Majoron [13] or a light scalar neutrino in supersymmetry each contribute $\Delta N'_\nu = \frac{1}{2}$.
- There is a limit $N''_\nu < 3.3$ at 95% CL from the observed abundance of helium and other light elements [14]. This bound is complementary to the LEP limit. It includes neutrinos only for masses up to $m < O(30\ MeV)$. However, in addition to active neutrinos it includes sterile neutrinos ν_S for a wide range of masses and mixing with active states [15], for which the sterile neutrinos would have been produced in the early universe prior to nucleosynthesis. In particular, for oscillations into sterile neutrinos the parameter range relevant to laboratory experiments and atmospheric neutrinos is excluded, as can be seen in Figure 1. Nucleosynthesis allows $\nu_e \to \nu_S$ for solar neutrinos via the MSW effect on the small

angle (nonadiabatic) but not the large angle branch[c]

4 Cosmology and Astrophysics

There are many constraints from cosmology and astrophysics on neutrino masses, life-times, and other properties [17]. One of the most important is the limit

$$\sum m_i < 35\,\text{eV}, \tag{13}$$

where the sum extends over the light, stable, active neutrinos. (Also included are sterile neutrinos with a significant mixing with the ordinary neutrinos [15]). If (13) does not hold, then there would be too much energy density in the present universe. If neutrinos are unstable there are a large variety of other constraints [17]. If the neutrinos decay via normal weak interactions then the decays would be either $\nu_2 \to \nu_1\gamma$ or $\nu_1 e^+e^-$, depending on the masses. There are many limits on such decays from the present energy density, the properties of the microwave background radiation, supernovae and nucleosynthesis. The result is that for such unstable neutrinos one must still have $m_{\nu_\mu,\nu_\tau} < 35$ eV.

The limit in (13) can be evaded if the neutrinos decay or annihilate into invisible decay products. In this case, one has the weaker limit [17],

$$m_\nu < 35\left[\frac{T_0}{\tau_\nu}\right]^{\frac{1}{2}}\ \text{eV}, \tag{14}$$

where T_0 is the age of the present universe and τ_ν is the neutrino lifetime, to ensure that their decay products will have redshifted to a low enough energy. Other constraints follow from the fact that structure can only grow in a matter-dominated universe [18]. In principle, one could have decays of a heavier neutrino into three lighter neutrinos, but it is difficult to find models in which the decay is sufficiently rapid. More promising are decays $\nu_2 \to \nu_1 G$ or $\bar{\nu}_1 G$, where G is a Goldstone boson, such as a Majoron[d](for a spontaneously broken lepton number) or a familon (for a broken family symmetry). There are many laboratory and astrophysical restrictions on such models [4,17].

5 Supernova Constraints on Neutrino Masses

The detection of neutrinos[e] from SN 1987A in the Kamiokande II (KII) [19] and Irvine-Michigan-Brookhaven (IMB) [20] detectors on February 23, 1987

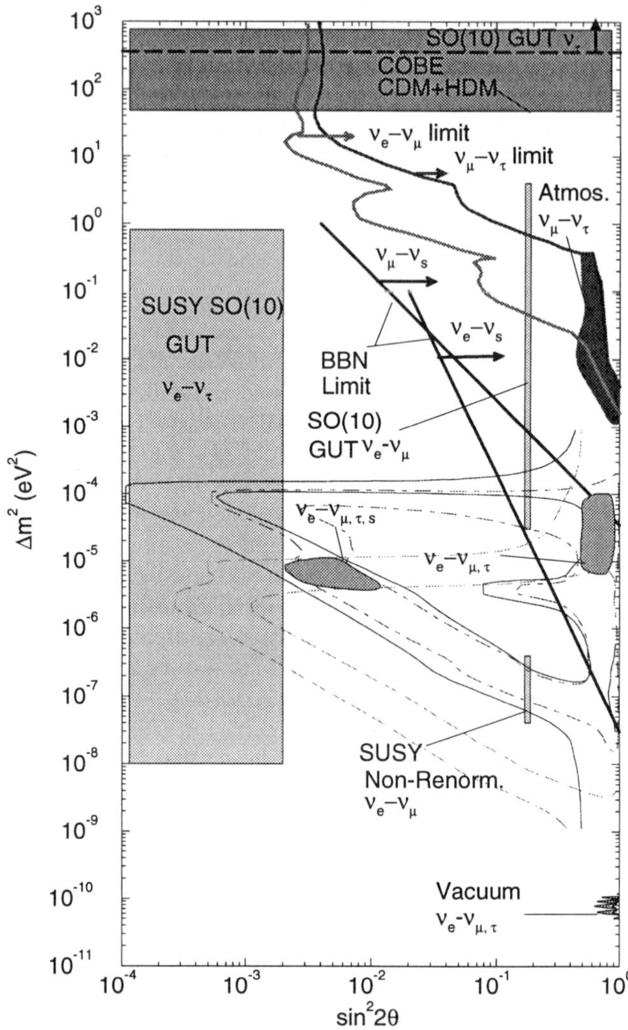

Figure 1: Allowed and excluded regions in the neutrino oscillation parameters Δm^2 - $\sin^2\theta$. Shown are the regions excluded by laboratory experiments (upper right), and by big bang nucleosynthesis for oscillations into sterile neutrinos (to right of solid lines). Also indicated are the regions suggested by solar neutrinos (MSW or vacuum oscillations), atmospheric neutrinos, and a hot dark matter component of the universe, as well as predictions of some specific seesaw models. From [16].

[c]In fact, the present solar neutrino data are most likely inconsistent with a large angle solution for sterile neutrinos.

[d]One can also have the annihilation of two neutrinos into two Majorons.

[e]This section is based on materials prepared by Andrew Hime, LANL.

confirmed the characteristic duration, energetics, and overall gross structure including the antielectron neutrino content of the burst [21]. A total of some 19 events attributable to the supernova were recorded, and hundreds of interpretations of the data in terms of upper and lower limits on neutrino mass were offered. The variety of conclusions speaks to the considerable model-dependence of the supernova burst structure as well as uncertainties related to experimental sensitivity, timing, and background. In a thoughtful retrospective, Lamb and Loredo [22] concluded that an upper limit of about 25 eV at the 95% confidence level could be set on the mass of ν_e from the supernova data. Neutral current signatures from muon and tau neutrinos were absent since the IMB and KII detectors were not sufficiently sensitive.

A neutrino with a mass of a few eV would contribute significantly to the closure density of the universe. Precision measurements of the tritium endpoint rule out the electron neutrino as the dominant dark matter in the universe, unless significant amounts of both hot and cold dark matter exist [3], as does the limit on neutrinoless double beta decay for Majorana ν_e. Muon and tau neutrino masses in the cosmologically interesting regime are unlikely to be determined directly in any foreseeable terrestrial experiment. Consequently, the detection of neutrino bursts from supernovae may present our best hope for directly inferring a neutrino mass of cosmological significance. (Short baseline laboratory oscillation experiments will probe this range if there is non-negligible $\nu_\mu - \nu_\tau$ mixing.)

5.1 Neutrinos from Supernovae

In *The Future of Supernova Detection*, Burrows, Klein, and Ghandi [23] describe generic models of supernovae in our own galaxy, and discuss the response for a number of representative Ĉerenkov detectors and scintillation neutrino telescopes to a nearby neutrino burst. The supernova model employed by Burrows *et al.* assembles a number of modern calculations. While not representative of a specific event it is meant to be realistic in terms of gross structure, temperature, energy, and time scale. Figure 2 displays the initial electron neutrino spike due to the release of trapped neutronization neutrinos as the initial shock breaks out of the core. The ringing that follows is not necessarily a feature that will occur in most supernovae. There then follows a longer period of thermal decay during which the neutrino flux drops according to a power law until the neutron star becomes transparent to neutrinos, at which time ($\sim$ 50 seconds) neutrino emission ceases. Burrows *et al.* point out that limits on the electron neutrino mass as low as 1 eV could be set if the initial breakout were observed; otherwise, uncertainties in supernova modeling make it difficult to derive limits competitive with direct terrestrial experiments.

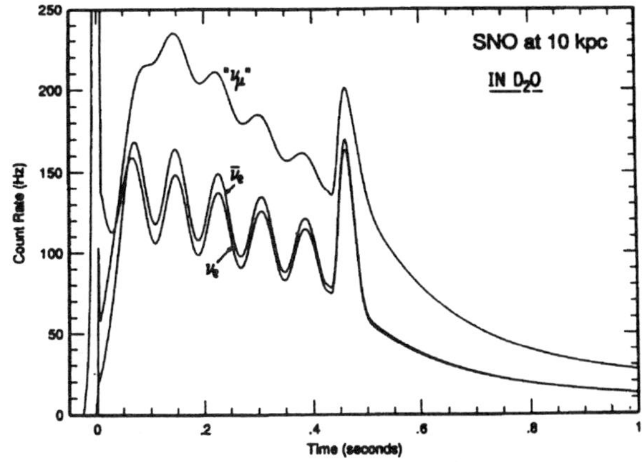

Figure 2: Interaction rates for all neutrino flavors in the SNO detector using a generic supernova model at 10 kpc. The designation "ν_μ" denotes the total contributions from both muon and tau neutrinos and their antiparticles. From Burrows *et al.* [23].

5.2 Supernova Neutrinos in SNO

The unique capability of the Sudbury Neutrino Observatory (SNO) [24] (see also Section 9.3.1) to measure the neutral-current interaction of all active neutrino species makes it possible to determine a muon or tau neutrino mass in the cosmologically significant range by studying the lengthening in flight time of a massive neutrino species from a distant supernova. The effect of a 20 eV and 50 eV tau neutrino mass is shown in Figure 3a for a generic supernova at 10 kpc (the center of our galaxy). Even if only one neutrino species has a significant mass, all species and their antiparticles will contribute to the total neutral-current rate in SNO. The effect of a single massive species is shown in Figure 3b, and can be recovered with sufficient statistics by comparison to the charged-current data from electron and antielectron neutrinos. The comparison is complicated by the very different temperatures of the neutrino sources of different flavors [25].

Given a galactic supernova, about 270 neutral-current and 150 charged-current events are expected to be registered in the heavy water component of SNO as well as 350 charged-current events in the outer light water shield. Approximately one third of these events would arrive during the first half-second, during which the delayed arrival of 20 eV neutrinos would be most evident. Assuming about 30% neutron detection efficiency, a neutrino mass in the 20 - 50 eV range would constitute a $\sim$ 3 standard-deviation effect. If neutral current events are converted to Čerenkov light by radiative neutron capture, then the charged and neutral-current spectra need

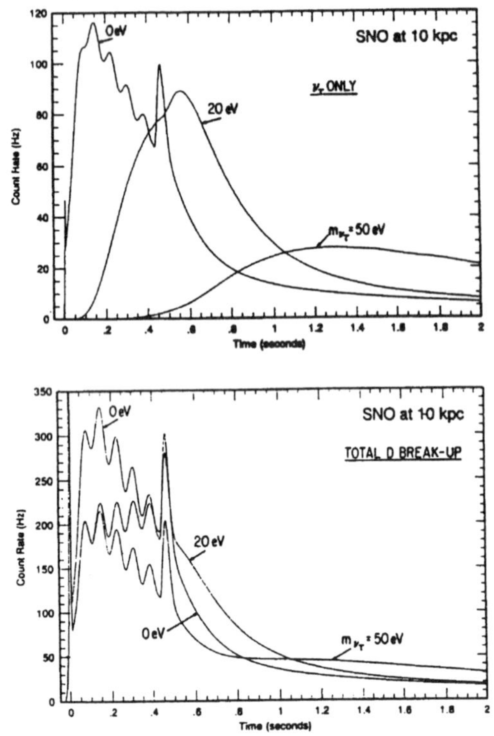

Figure 3: Neutral-current rates in the SNO detector for the generic supernova model of Burrows *et al.* [23]. In (a) the spectrum is shown for the case of a single massive neutrino species. In (b) the total deuteron disintegration rate is shown for the case of a single massive neutrino and two massless species.

to be decomposed via the light water charged-current signal. It is therefore planned to deploy ^{3}He proportional counters in the SNO detector [25] to obtain a continuous and distinct measure of the neutral-current signal.

5.3 An International Network of Detectors

The upgrade of KII to SuperKamiokande (SK) and construction of the Large Volume Detector (LVD) [26] will allow very high statistics measurements of the ν_e and $\bar{\nu}_e$ signal during the first second of a supernova burst. The characteristic rise-time of 50 msec in SK and LVD for a massless neutrino is lengthened to 300 msec for a 10 eV electron neutrino [23]. This method of neutrino mass determination cannot be applied to the muon and tau neutrinos without implementing a neutral-current based detector such as SNO or the proposed Supernova Burst Observatory (SNBO) [27]. It is desirable to combine the unique features of a variety of future neutrino detectors to optimize our ability to extract physics from a supernova burst. For example, muon and/or tau neutrino masses could be determined by employing the combined signals from different detectors. The large mass and high sensitivity to electron antineutrinos in SK [28] would provide the time of stellar collapse, while the subsequent time shift associated with the onset of the ν_μ and/or ν_τ signal in a neutral-current based detector could be employed to constrain or determine the ν_μ and/or ν_τ mass [29].

A number of unique detectors, either under construction or proposed, have been discussed in references [23] and [29], both with respect to their ability in elucidating our understanding of stellar collapse and to the prospects for detection of a cosmologically significant neutrino mass. The next generation of underground neutrino detectors can serve to determine a neutrino mass in the 15 to 100 eV range. Depending upon one's confidence in the predictive power of supernova burst models the lower end of this range might be pushed down to 5 eV [29]. Pennypacker *et al.* estimate a galactic core-collapse supernova at least every 31 ± 7 years [30]. Hence, over the next decade the probability of such an occurrence is ~20-30%, a sufficiently high probability that every effort must be made to capitalize upon the physics associated with such an event.

6 Direct Mass Measurements

The most obvious way to explore the question of neutrino mass is to consider the kinematics of the weak decays of the e, μ, and τ leptons. In this section we will review the experimental efforts to measure the masses of ν_e, ν_μ, and ν_τ using kinematics.

6.1 Direct Measurement of m_{ν_e}

A neutrino with mass around 7 eV is a leading candidate [3] for the hot dark matter component of the universe suggested by the COBE and galaxy distribution results [2]. Barring a new supernova, direct (kinematic) determinations of the masses of the neutrinos are within striking distance of this mass only in the case of ν_e.

For ν_e, the current generation of tritium beta decay experiments set upper limits on the $\bar{\nu}_e$ mass at about this level. Thus improved experiments, down to the 1 or 2 eV level, as well as admixture searches in this mass range, must be a significant priority in future efforts. At the same time note must be taken of a curious feature of the current experiments: the central value of neutrino mass-squared is negative in all cases. The average of 5 experiments yields $m^2_{\nu_e} = -96 \pm 21$ eV^2, almost a 5-σ effect. (The parametrization in terms of $m^2_{\nu_e}$ is only a convenient way to describe a "bump" that appears near the endpoint of the tritium spectrum.) Such a statistically unlikely result suggests the presence of some as yet unidentified systematic error common to the experiments, such as an unexpectedly large problem in the calculation of the excitation spectrum of the residual molecular ion THe^+, or the presence of new particle physics. Speculations on new physics include tachyonic neutrinos, capture from a (very dense) relic cosmic sea of neutrinos which would produce a monoenergetic peak at or above the endpoint, and admixture of a massive neutrino with strong final-state interactions that would produce a bump at or below the endpoint. Such possibilities are exciting, even if unlikely, and demand that improved experiments be performed to verify or reject them.

Another approach [31] is a study of the beta spectrum of ^{187}Re (endpoint of 2.64 keV *vs* ^{3}H of 18.6 keV) using the developing cryogenic calorimetric technique. It offers the exciting capability of eliminating two of the major sources of systematic uncertainty, energy loss in the source, and atomic final state distributions. There are, however, significant technical challenges that must be overcome before a successful measurement can be carried out.

New experimental work on the tritium spectrum is nearing completion in Russia [32] with an intense molecular tritium source and a very high-resolution retarding-field analyzer.

A summary of the existing experimental limits on the direct m_{ν_e} is given in Table 1.

There is an anomaly seen at the endpoint of the tritium spectrum in three experiments on pure T_2 that far exceeds statistical fluctuations. The current published experimental results (*post* 1988) are consistent with each other, but not with $m_{\nu_e} = 0$ and standard Fermi β decay. In this circumstance, it is difficult to assign a confidence level to an upper limit on the mass. Qualitatively, a mass greater than 10 eV can confidently be ruled out, but the situation is clearly unsatisfactory.

A plan of action might include the following ingredients:

- Continuing current laboratory work on the tritium decay. Experiments are still in progress at Troitsk [32], Mainz [37], Austin [40], and Yorktown Heights [39].
- Evaluation of the possibility of reactivating the Los Alamos and Livermore experiments.
- Experiments to measure atomic-physics parameters of the decay of T_2 to THe^+, such as the branch to (at least) the ground state, radiative decays from excited states, 1- and 2-electron shakeoff probabilities, infrared spectra, and resonances in H-He collisions as a check on the crucially important final-state spectrum calculations.
- New measurements of the T-^{3}He mass difference.
- An atomic T experiment, where the final-state spectrum becomes trivial. Spin-polarized tritium at low temperatures might form the basis for a compact, intense atomic source.
- Bolometric (calorimetric) measurements on tritium or ^{187}Re.
- Continued theoretical examination of exotic possibilities, such as a gross error in the theory of the two-electron molecule, high densities of relic neutrinos, and non-standard electroweak models.

6.2 Direct Measurement of m_{ν_μ}

The most precise data on the mass of ν_μ come from measurements of the μ momentum following the decay of stopped pions. The most recent determination of the mass of the π^+ by Jeckelmann *et al.* [41], coupled with the muon data of Abela *et al.* [42], give

$$m^2_{\nu_\mu} \leq -0.097(72)\ \text{MeV}^2, \tag{15}$$

which, with the Bayesian procedure described by the Particle Data Group [1], yields an upper limit of 270 keV at 90% confidence on the mass of ν_μ. A result slightly more than 1-σ negative is not unexpected if the true value is zero.

However, a new round of experiments at the Paul Scherrer Institute has now reached such a high level of precision that a serious problem in this approach has been discovered. Table 2 gives the recent history of these measurements.

As indicated, with the mass of the charged pion published by Jeckelmann *et al.*, the central value for m^2_ν is over 5σ negative. The reason for this is now, it appears, understood. The mass of the π^+ has been deduced from pionic X-ray spectra. The X-ray lines have a fine structure corresponding to the presence of zero, one, or two K-shell vacancies. The data on the $4f - 3d$ line in ^{24}Mg show only two components of significant intensity, allowing for two possible assignments of structure. Auger widths, which depend sensitively on the K-vacancy probability, were used to resolve the ambiguity in favor of the "B" solution, but both new results on pionic X-ray intensity ratios and new calculations of the strong absorption widths now favor the opposite choice [46], *i.e.*, the "A" solution. With the "A" solution, the central value for m_{ν_μ} is (by chance) exactly 0, and an upper limit of 270 keV on m_{ν_μ} at 90% CL can be derived.

Stopped π decay is approaching a limit where theoretical uncertainty (in deducing the masses of the π^+ and the μ^+) is limiting the accuracy with which m_ν can be determined. One novel approach was that of Anderhub *et al.* [47], who used a magnetic 'racetrack' and π decay in flight to obtain,

$$m^2_{\nu_\mu} = -0.14(20)\ \text{MeV}^2 \tag{16}$$

$$m_{\nu_\mu} \leq 500\ \text{keV (90\%CL)} \tag{17}$$

The method is (by design) relatively insensitive to m_π and m_μ, and could potentially offer an avenue for further

Table 1: Existing limits on the square of the electron neutrino mass.

Institute/Location	Year	Source State	Mass^2, eV^2 $\pm$ stat. $\pm$ syst.	Limit, eV (95%CL)	Ref.
ITEP, Moscow	88	Valine-T	$+676\pm 235$	$= 26^{6}_{5}$	[33]
Los Alamos	91	T_2 gas	$-147\pm 68 \pm 41$	≤ 9.3	[34]
Univ. of Zürich	91	OTS-T	$-24\pm 48 \pm 61$	≤ 11	[35]
INS, Tokyo	91	Cd Arach.-T	$-65\pm 85 \pm 65$	≤ 13	[36]
Mainz Univ.	93	T_2 solid	$-39\pm 34 \pm 15$	≤ 7.2	[37]
LLNL (Chem.)	94	T_2 gas	$-130\pm 20 \pm 15$	-	[38]
INR, Moscow	Run'g	T_2 gas			[32]
IBM, Yktn Hts	Stdby	Metal-T			[39]
UTx/Brandeis	R&D	T_2 gas			[40]

Table 2: Limits on mu neutrino mass from $\pi^+ \to \mu^+ + \nu_\mu$ at rest.

Collaboration	Ref.	m_μ, MeV	p_μ, MeV/c	m_π, MeV	m_ν^2, MeV^2
Abela *et al.* 84	[42]	105.65932(29)	29.79139(83)	139.56761(77)	-0.163(80)
Jeckelmann *et al.* 86	[41]			139.56871(53)	-0.097(72)
PDG 92[a]	[43]	105.658387(34)		139.56737(33)	
Daum *et al.* 91	[44]		29.79206(68)	139.56996(67)	
Frosch *et al.* 92	[45]		29.79144(20)		-0.127(25)
Jeckelmann *et al.* 94	[46]	"A" solution:		139.56995(35)	0.000(45)
		"B" solution:		139.56782(37)	-0.127(46)

[a] Electron mass down 8 ppm

improvement. No direct measurement scheme is known that can access the cosmologically interesting mass range below 100 eV, other than time-of-flight from a supernova.

6.3 Direct Measurement of m_{ν_τ}

Of the three neutrinos now known to exist, the tau neutrino is arguably the least well understood. Not only has it never been directly observed, but the limits on its mass are orders of magnitude less stringent than those on its electron and muon neutrino cousins [1].

Studies placing the most stringent limits on the tau neutrino mass, m_{ν_τ}, have used the spectrum of the invariant mass of the measured tau decay products, M_X. The maximum attainable M_X in a given tau decay mode is a function of m_{ν_τ}. This measurement is most sensitive to events with $M_X \sim M_\tau$ since this is where the spectrum varies most dramatically as a function of m_{ν_τ}. The tau decay final states with the least energy available for the tau neutrino are generally more sensitive to m_{ν_τ} and recent studies have concentrated on high-multiplicity tau decays largely for this reason. Previous limits on m_{ν_τ} are given in Table 3.

In the next few years, any improvement in the limit on m_{ν_τ} at e^+e^- colliders will come from improved statistics using techniques that rely on obtaining an event near the kinematic limit $M_X = M_\tau$. While experiments at LEP and CESR can all do these analyses, the poorer mass resolution on M_X and smaller (though purer) sample sizes at the LEP experiments places them at a competitive disadvantage relative to CLEO-II.

CLEO-II will continue to accumulate large amounts of data, and will probably update the analyses of the $\tau \to 5\pi\nu_\tau$ and $\tau \to 3\pi\, 2\pi^0\, \nu_\tau$ decays in an attempt to set an improved limit on m_{ν_τ}. As with earlier analyses of these decay modes, luck plays a regrettably large role in the stringency of the limit one can set, since the probability of getting an event very close to the kinematic limit is exceedingly small. The tau decay to the final state $3\pi\pi^0\, \nu_\tau$, while peaked lower in invariant mass than the five pion decay modes, has a much larger branching fraction and therefore may have a similar sensitivity to m_{ν_τ}.

Decay modes with kaons are potentially more sensitive to m_{ν_τ} by virtue of the reduced q^2 available to the ν_τ. Although the CLEO-II detector has relatively poor K–π separation, sufficient statistics make using the fi-

Table 3: Existing limits on the tau neutrino mass.

Expt.	Year	Final State	Events	Limit (95% CL, MeV)	Ref.
DELCO	85	$KK\pi\nu_\tau$,$K\pi\pi\nu_\tau$	4	157	[48]
MARK-II	85	$5\pi\pi^0\,\nu_\tau$	3	125	[49]
MARK-II	85	$3\pi\,(\pi^0\,)\nu_\tau$	22	143	[50]
ARGUS	85	$3\pi\,\nu_\tau$	1536	70	[51]
HRS	86	$5\pi(\pi^0\,)\nu_\tau$	13	76	[52]
CLEO	87	$3\pi\,(\pi^0\,)\nu_\tau$	9135	85	[53]
ARGUS	88	$5\pi\nu_\tau$	12	31	[54]
CLEO-II	93	M_τ	–	75	[55]
CLEO-II	93	$5\pi\nu_\tau$, $3\pi\,2\pi^0\,\nu_\tau$	113	33	[56]

nal states $K\pi\pi\nu_\tau$ and $KK\pi\nu_\tau$ feasible. The final state $K_sK_s\pi\nu_\tau$ may also be usable. Decay modes with kaons may be *less* sensitive, however, if their spectral functions cause the 3-prong invariant mass spectrum to be peaked at lower masses than that of the 5-pion decays.

With the advent of very high statistics (10^7-10^8 τ pairs) event samples at either B or τ–c (Tau–Charm) factories, the technique described above will be superseded by one which relies on statistical subtraction of backgrounds. Improved detector resolutions and particle identification techniques will also play important roles.

Monte Carlo studies assuming 10^8 τ pairs predict a sensitivity to a τ neutrino with a mass as low as 2 MeV at a τ-c Factory, and as low as 2.5 MeV at a B Factory [57]. These estimates are the combined result of $5\pi\nu_\tau$, $3\pi\,2\pi^0\,\nu_\tau$, and $KK\pi\nu_\tau$ decay mode analyses, and assume resolutions of 2-5 MeV on the invariant masses of the final state particles.

7 Double Beta Decay

Double beta decay[f] is a very rare process, with a half-life of order 10^{20} years, in which a metastable nucleus (A, Z) transforms into its next nearest neighbor $(A, Z+2)$ and emits two electrons [4]. In the standard electroweak model, two antineutrinos always accompany the electrons and share the $1-3$ MeV energy released on a roughly equal basis. In theories going beyond the standard model, it becomes possible for the two electrons to be emitted either by themselves alone, or accompanied by a light scalar particle, the Majoron. These two double beta decay modes violate lepton number conservation, and their existence would imply that at least one neutrino is a Majorana particle with non-zero mass.

The half-life for the two-neutrino decay mode can be calculated directly from the standard model, and can be written

$$\frac{1}{T_{2\nu}} = G_{2\nu}|M_{2\nu}|^2, \tag{18}$$

where $G_{2\nu}$ is a phase space integral and $M_{2\nu}$ is the nuclear matrix element. Much theoretical progress has been made in recent years in our understanding of the two-neutrino matrix elements, which tend to be small. Since the no-neutrino decay mode vanishes in the absence of neutrino mass, its half-life depends directly upon an "effective" mass $\langle m_{\nu_e}\rangle$,

$$\frac{1}{T_{0\nu}} = G_{0\nu}|M_{0\nu}|^2 \left[\frac{\langle m_{\nu_e}\rangle}{m_e}\right]^2, \tag{19}$$

where

$$\langle m_{\nu_e}\rangle = \sum_i m_i \lambda_i U_{ei}^2 \tag{20}$$

involves a sum over all light ($\ll$ 30 MeV) Majorana neutrino mass eigenstates, m_i, weighted by the CP parities ($\lambda_i = \pm 1$) and the squares U_{ei}^2 of the mixing matrix elements[g] For heavier m_i there is an additional propagator suppression in the contribution to $\langle m_{\nu_e}\rangle$.

An important diagnostic tool in distinguishing between different decay modes is the spectrum of energy carried by the two electrons. For two-neutrino decay, the spectrum is continuous with a broad peak just below the mid-point and a very flat approach to zero at the end-point; in fact, it vanishes for all practical purposes in the last 20% of the range. For no-neutrino decay, the spectrum is a spike at the end-point because the two electrons carry all the available energy. The spectrum for the Majoron mode is continuous, but peaks towards the end-point, in the region where the two-neutrino spectrum

[f] This section is based on material prepared by Peter Rosen, Univ. Texas at Arlington.

[g] $\langle m_{\nu_e}\rangle$ vanishes for Dirac neutrinos, which can be interpreted as two degenerate Majorana neutrinos ($m_1 = m_2 = m_D$) with $U_{e1} = U_{e2} = \frac{1}{\sqrt{2}}$ and opposite CP eigenphases.

vanishes. Thus any events occurring in the last 20% of the spectrum must either be background, or signals for non-standard double beta decay modes.

Double beta decay has been the object of experimental searches for more than forty years. The earliest experiments invoked the "geochemical method" in which one probes ancient ores rich in the parent nucleus (for example, ^{82}Se or ^{130}Te) searching for traces of the noble gas daughter nuclei (^{82}Kr or ^{130}Xe). This method allows one to establish the existence of the double beta decay transition and measure its lifetime, but does not determine which mode, or modes are actually occurring. In the most recent experiment of this kind [58], the half-life of ^{128}Te was measured to be $(7.7 \pm 0.4) \times 10^{24}$ years.

Searches specifically aimed at detecting the no-neutrino mode have been performed with germanium crystals, which serve as both source (^{76}Ge, 8% natural abundance) and detector, and which measure the total energy of the emitted electrons. The most recent experiments use 90% enriched crystals [59] and give a lower limit on the half-life of 3×10^{24} years, which corresponds to an upper bound on $\langle m_{\nu_e} \rangle$ of about 1 eV. This is the best limit available to date. Actual observation of the decay would imply that at least one of the Majorana mass eigenstates must be heavier than

$$m_i^{\max} > 1 \text{ eV} \left[\frac{10^{24}\text{yrs}}{T_{0\nu}(^{76}\text{Ge})} \right]^{1/2} . \qquad (21)$$

The first direct observation of two-neutrino double beta decay in the laboratory was made in 1986 with a xenon-filled time-projection chamber [60]. Electron tracks were observed directly and timing was used to exclude background events. Spectra consistent with two-neutrino decay were observed in ^{82}Se, ^{100}Mo, and ^{150}Nd, and the corresponding half-lives measured. In the case of ^{82}Se, the result was consistent with the geochemical measurement. More elaborate versions of this technique are being developed.

Future experiments will concentrate on the search for the no-neutrino mode because of its implications for physics beyond the standard model. As the bounds on the effective mass $\langle m_{\nu_e} \rangle$ go down, so the experiments will have to grow bigger and more ambitious, as in the case of the NEMO collaboration in France, or they will have to develop new cryogenic techniques, as is happening in the area of solar neutrinos.

8 Neutrino Oscillations

A direct consequence of neutrino mass is that the flavor eigenstates[h] ν_e, ν_μ, ν_τ, are actually a linear combinations of massive neutrinos, ν_1, ν_2 and ν_3. For simplicity, we use a two component formulation such that

$$\nu_e = \nu_1 \cos\theta + \nu_2 \sin\theta$$
$$\nu_\mu = -\nu_1 \sin\theta + \nu_2 \cos\theta$$

This is analogous to the well understood phenomena of mass mixing in the quark sector. Because neutrinos are only weakly interacting and ultra relativistic, the mass eigenstates, ν_1 and ν_2 will evolve in time. Therefore, one can derive a probability that a neutrino that starts out as flavor a can be found at a later time as flavor b. This probability is typically expressed in terms of the mixing angle, θ, the mass-squared difference $\Delta m^2 = m_2^2 - m_1^2$ (in eV^2) between the two mass states, and an experimental ratio L/E :

$$P(\nu_a \to \nu_b) = \sin^2 2\theta \sin^2 \left(1.27 \Delta m^2 L/E\right), \qquad (22)$$

where L is the distance (in km) from the original source of the neutrinos to where the probability is being measured, and E is the energy of the neutrinos in GeV. This concept of oscillation is familiar to the particle physicist from the oscillation phenomena observed in the neutral kaon system.

For a given θ and Δm^2 the oscillation probability varies periodically with time or distance. One does not know *a priori* the values of θ and Δm^2, and, because the probability depends on the ratio L/E, there are many regimes of particle physics in which oscillation searches can take place.

There are two types of experiment to search for neutrino oscillations. Most, including the solar and reactor measurements, are called "disappearance" experiments. This means that a source of neutrinos of a single flavor is created and the flux is measured at two different places separated by a distance. Differences in the flux at the two locations will indicate that neutrinos have oscillated into a different flavor. Alternatively, in an "appearance" experiment a source of neutrinos of a single flavor will travel a distance to a target where they can interact. Observation of a lepton of flavor different from the original beam will indicate that neutrinos have oscillated. Results of experimental searches for neutrino oscillations are displayed on plots in $\sin^2 2\theta$ vs Δm^2 parameter space, as in Figure 1. The oscillation channels most easily explored are $\nu_\mu \leftrightarrow \nu_e$, $\nu_\mu \to \nu_\tau$, $\bar{\nu}_e \to X$, and $\nu_\mu \to X$, where X is any neutrino type, including one which does not interact (sterile). In the following four sections we will review the status of experimental searches for neutrino oscillations. We begin with the long-standing solar-neutrino puzzle and progress being made toward confirming or denying neutrino oscillations as an explanation of the puzzle. We will then address the more recently discovered anomaly in the atmospheric neutrino experiments. Finally we will

[h] As described in Section 2, it is also possible in principle to have oscillations between active and sterile neutrinos. However, these are less likely theoretically, and most of the interesting parameter range (except for solar neutrinos) is excluded by primordial nucleosynthesis.

discuss the status of oscillation searches that can be performed with neutrino beams produced at nuclear reactors and particle accelerators.

9 Masses Below 10^{-2} eV (Solar Neutrinos)

Solar neutrinos provide, at present, the only available way to reach masses in the range 10^{-2} to 10^{-5} eV. Neutrino oscillations which occur in vacuum [61] or which may be amplified by passage through solar material (the MSW effect [62,63]), can deplete the flux of ν_e by converting some of them into ν_μ, ν_τ, or a sterile species. The observations made with the four experiments performed so far indicate a flux at the earth that is consistently below the results of astrophysical calculations. Moreover, the relative rates in the different types of experiment indicate that the largest depletion is in the middle of the spectrum (the ^{7}Be line and the low energy part of the ^{8}B spectrum), which is inconsistent with any known astrophysical or nuclear explanation [64,65,66,67,68,69,70]. On the other hand, the data from all four experiments can be described well in terms of calculations which include the MSW-enhanced neutrino-oscillation effect. The deficit of electron-type neutrinos thus provides suggestive evidence for a nonzero neutrino mass and nonconservation of lepton family number.

The solar neutrino spectrum is dominated by the first step in nuclear burning, the *pp* reaction,

$$p + p \to d + e^+ + \nu \tag{23}$$

with a continuous neutrino spectrum ending at 420 keV. From there, several competing pathways lead to ^{4}He. In addition, a catalytic process involving carbon, nitrogen and oxygen isotopes (CNO) accounts for a few percent of hydrogen burning. The fluxes and energies of neutrinos from the sun in the standard Bahcall-Pinsonneault [71] model are given in Table 4.

Experimentally, the higher energy neutrinos from ^{8}B decay are the easiest to detect, but they originate in a reaction branch that plays a negligible role in energy production in the sun.

9.1 Experiments Now in Operation

The low energies of solar neutrinos have severely taxed the ingenuity of experimentalists. Cross sections are small and the events compete with ubiquitous backgrounds. Two general types of experiment have been devised to overcome these difficulties, the radiochemical method and 'active' (*i.e.*, electronic) detection. Detectors must be large, and they must be deep underground.

We review briefly the four currently-operating experiments and then describe the experiments that are being developed and which will begin to operate in about two years. Finally, we summarize some of the possibilities among proposed future experiments.

9.1.1 The Chlorine Experiment

In this, the first solar neutrino experiment [72], neutrinos are detected in a target of 615 tons of perchloroethylene (C_2Cl_4) 1500 meters underground in the Homestake Gold Mine in Lead, South Dakota. The target isotope, ^{37}Cl, can capture a neutrino, producing a radioactive isotope ^{37}Ar:

$$\nu_e + {}^{37}\mathrm{Cl} \to {}^{37}\mathrm{Ar} + e^- . \tag{24}$$

The reaction can occur for electron neutrinos with energy greater than 0.814 MeV. The neutrino reactions are registered by extracting and counting the decays of ^{37}Ar (τ = 35 days) in a proportional counter. The present result from runs 18 through 120 over the years 1970-93 is 2.55(23) SNU, whereas the expected rate [71] is 8.0(10). (1 SNU = 10^{-36} interactions per target atom per second.) Over the 25 years this experiment has been in operation, many refinements and tests have been made. The most significant was the introduction in 1970 of the rise-time of the ionization pulse as a means of distinguishing the point-like decays of ^{37}Ar from the longer tracks produced by background events – it was this strategy that disclosed a true solar neutrino signal above background. More recently, events have been analyzed according to probability distributions rather than cuts in energy-rise-time space, the muon-induced background has been remeasured [the current best estimate [72] is 0.047(16) atoms per day, somewhat lower than the value in general use, 0.08(3)], counter efficiencies have been redetermined with attention to losses in handling the argon (leading to a 3% increase in the flux) [73], and the cross section for detection of ^{8}B neutrinos has been re-evaluated [74].

Because the Cl-Ar experiment (and its relationship to Kamiokande) plays an important role in conclusions about the likelihood of neutrino oscillations, a popular pastime has been to search for internal inconsistencies in the experiment, particularly in the form of apparent time dependence or nonstatistical behavior. By selecting certain groups of data points, it is possible to find disagreements with the average, but little evidence for nonstatistical behavior has been demonstrated. The experiment is also frequently characterized as 'uncalibrated', and it is true that a test with an artificial neutrino source has not been done. However, extensive calibrations in which, for example, ^{37}Ar has been produced *in situ* by neutron reactions and measured, show that the experiment behaves as expected.

Table 4: Solar neutrino fluxes at the earth calculated by Bahcall and Pinsonneault.

Source	Flux, 10^{10} cm^{-2}s^{-1}	1σ, %	E$_{max}$, MeV
pp	6.0	0.7	0.420
pep	0.014	1.7	1.442 (line)
^{7}Be	0.49	5	0.862 (line, 90%)
			0.384 (line, 10%)
^{8}B	0.00057	12	15
^{13}N	0.049	17	1.199
^{15}O	0.043	19	1.732
^{17}F	0.00054	15	1.740

9.1.2 The Kamiokande Experiment

Kamiokande (Kamioka Nucleon Decay Experiment) is a 3,000-ton water Čerenkov detector with a fiducial volume of 680 tons for solar neutrinos [75,76]. The reaction observed is the elastic scattering of neutrinos by electrons,

$$\nu + e^- \rightarrow \nu + e^- \ . \tag{25}$$

At the energies of interest for solar neutrino detection, reaction (25) is about six times more likely for ν_e than for ν_μ or ν_τ. Despite a significant background from radioactivity of the rocks in the mine, from radioactivity in the water, and from cosmic-ray induced events, neutrino interactions can be detected at the rate of about one per week above a threshold of 7.5 MeV because the electrons struck by neutrinos recoil preferentially in the direction from the sun to the earth. Kamiokande, the first "active" (electronic rather than radiochemical) solar-neutrino experiment, has shown three very important things: first, the detected neutrinos do indeed come from the sun, second, the observed flux is about half that predicted by the standard solar model; and, third, that the spectrum is consistent with the expected ^{8}B β spectrum. The current result [77] is 0.51±0.04±0.06 of the Bahcall-Pinsonneault [71] flux, 5.69 x 10^6 cm^{-2}s^{-1}.

9.1.3 The Gallium Experiments

The *pp* neutrinos are too low in energy to be detected by either the chlorine-argon or the Kamiokande experiment. Since *pp* neutrinos are the most abundant, their flux is more reliably predicted than any other [78], being determined mainly by the solar luminosity and only weakly dependent on details of the sun's interior. Consequently, the measurement of the flux of *pp* neutrinos has long been seen [79,80] as potentially a decisive method for searching for neutrino disappearance.

Two radiochemical experiments that can detect *pp* neutrinos are currently in operation; they use ^{71}Ga as their target via the process

$$\nu_e \ + \ ^{71}\text{Ga} \ \rightarrow \ ^{71}\text{Ge} \ + \ e^-, \tag{26}$$

which has an energy threshold of 0.233 MeV. As in the chlorine experiment, radioactive atoms of ^{71}Ge are extracted and counted. The standard-model prediction is that after a month of operation about 16 atoms of ^{71}Ge will be present in 30 tons of gallium.

One of the gallium experiments, SAGE [81,82] (the Soviet-American Gallium Experiment) is operating in the Baksan Laboratory in the Russian Caucasus Mountains. SAGE currently uses 60 tons of gallium metal. The other experiment, Gallex [83,84], is a European-American-Israeli collaboration operating in the Gran Sasso tunnel near Rome, Italy. Gallex uses 30 tons of gallium in a GaCl$_3$-HCl solution.

The current SAGE result [85] is $69 \pm 11^{+5}_{-7}$ SNU, and the current Gallex result [86] is $79 \pm 10 \pm 6$ SNU. The experiments are in good agreement (gratifying in view of the differences in method), with an average value of 74.3 ± 8.5 SNU, far below the standard-model prediction [71] of 132 ± 7 SNU. An intense artificial neutrino source (1.7 MCi of ^{51}Cr) has been used to verify the proper operation of Gallex; the ratio of observed to expected events was 1.04 ± 0.12 [87]. A similar test for SAGE is in progress.

9.2 Discussion of the Solar Neutrino Problem

The long-standing discrepancy between the measured rate in the Cl-Ar experiment and the predictions of astrophysical models engendered intensive experimental and theoretical work over the years to see if an explanation could be found within the framework of conventional physics. As a result, the "standard solar models" have been quite carefully checked with respect to both their conceptual framework and their experimental inputs. Small effects continue to be found, and the remote possibility of a gross, but still undetected, error, remains.

The MSW effect [62,63] can describe consistently the results of all four of the experiments [88,89,90,91,92,93]. Two allowed regions (see Figure 4) in oscillation space fit the data, within the framework of the standard solar model. An alternative particle physics solution is vacuum oscillations with large mixing ($\sin^2 2\theta \geq 0.6$) and small $\Delta m^2 \sim 10^{-10} eV^2$, for which the earth is at or near an oscillation node. This requires fine-tuning the earth-sun distance, but is otherwise marginally allowed [94,95]. Other solutions, such as neutrino decay [96] or large neutrino magnetic moments [97,98] are disfavored by the current data.

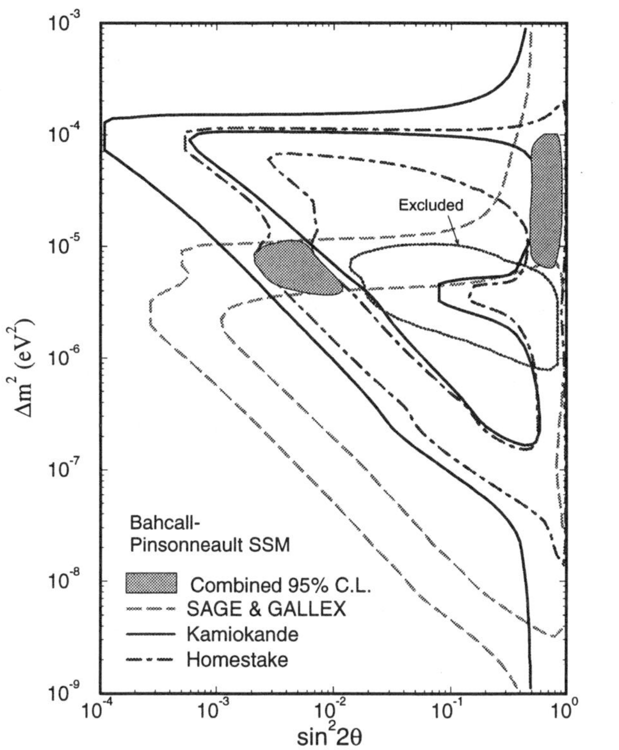

Figure 4: Allowed regions in Δm^2 - $\sin^2\theta$ space for the four solar neutrino experiments and the Bahcall-Pinsonneault solar model. From [16].

The specific oscillation parameters derived depend on standard solar models (and, of course, contradict the Standard Model of particle physics). However, the existence of data from 3 distinctive spectral regions in 4 separate experiments invite us to distance ourselves from the detailed challenges of astrophysical theory. With a very limited set of assumptions, namely that the only type of neutrino emitted by the sun is the electron neutrino, and the only significant sources of such neutrinos are the *pp* and *pep* reactions, and the beta (or inverse) decays of ^{7}Be, CNO isotopes, and ^{8}B, one can ask if the total measured flux of neutrinos is consistent with the known power output of the sun. It is possible to address this question without reference to detailed models because each termination in which 4 protons fuse to ^{4}He generates 26.732 MeV and precisely 2 electron neutrinos, independent of the reaction chain leading to termination. Some energy is carried away by the neutrinos, and is not available to heat the sun, but that ranges from 2.0% (PPI) to 6.4% (CNO). (The exception is terminations through ^{8}B – 26% energy loss by neutrinos – but we know experimentally that that branch is orders of magnitude from playing a significant role in energy production.)

With current values for the experimental rates seen in the Cl-Ar, Kamiokande, and gallium experiments, one concludes that the total neutrino flux is (marginally) consistent (at about the 1.5σ level) with the elementary assumption set given, but that the ^{7}Be + CNO flux must be far smaller than calculated [64,65,66,67,68,69,70].

A particle-physics solution to the solar neutrino problem is therefore not absolutely required (yet), although there is manifest and unexplained inconsistency with "standard solar models." This situation has been driven home by the gallium experiments, which were hoped to provide a clear decision for or against oscillations, but which have fallen in the "grey" region between forced particle-physics solutions and probable astrophysical solutions. To some degree, the fact that there is fair consistency with the total expected electron-neutrino flux tends to contraindicate a particle-physics solution. Elegant though the MSW mechanism is, it nevertheless is describing three data points with two parameters.

The solar neutrino problem does, however, suggest neutrino mass and oscillations by virtue of fundamental disagreement with astrophysical calculations, calculations that give a good account of many other solar parameters, such as the spectrum of mechanical oscillations of the sun and the Hertzsprung-Russell diagram main sequence. The disagreement is not just in the difficult-to-calculate flux of ^{8}B neutrinos, but at the basic level that ^{7}Be seems to be almost absent in spite of the demonstrated presence of ^{8}B, which can only be made from ^{7}Be. No plausible astrophysical mechanism has been proposed to account for this. The possibility of failure of an experiment cannot be overlooked, but at least two of the three classes of experiment (chlorine, Kamiokande, gallium) would have to be in error [66,67,69] in addition to there being a failure in astrophysical theory. Final resolution of the issue must await new experiments that measure specific indicators of neutrino oscillations in a way independent of astrophysical models. New experiments are also of importance for astrophysics. With or without neutrino oscillations, they hold out the possibility of determining the initial neutrino flux components in a model independent way [65,66,68,69]. This and helioseismology are the only known direct probes of the solar core.

9.3 Forthcoming Experiments

9.3.1 The Sudbury Neutrino Observatory (SNO)

SNO is a neutrino detection facility sited at the 6800-foot level of the INCO nickel mine in Sudbury, Canada [99]. Built with support from four Canadian agencies, the US Department of Energy, and the UK Science and Engineering Research Council, SNO will consist of 1000 tons of 99.92%-enriched ultra-pure heavy water (D_2O) in an acrylic sphere surrounded by 7000 tons of light water as a shield.

Deuterium has unique nuclear properties that make it ideal for the study of neutrino interactions. There are four principal modes by which neutrinos can interact with heavy water:

$$\mathrm{d} + \nu_e \rightarrow \mathrm{p} + \mathrm{p} + e^- - 1.44\ \mathrm{MeV} \tag{27}$$

$$\mathrm{d} + \nu_x \rightarrow \mathrm{p} + \mathrm{n} + \nu_x - 2.22\ \mathrm{MeV} \tag{28}$$

$$\mathrm{d} + \bar{\nu}_e \rightarrow \mathrm{n} + \mathrm{n} + e^+ - 4.03\ \mathrm{MeV} \tag{29}$$

$$e^- + \nu_x \rightarrow e^- + \nu_x \tag{30}$$

The first of these reactions proceeds by the charged-current (CC) interaction of electron neutrinos specifically. The second is the neutral-current (NC) disintegration of deuterium and can be initiated with equal probability by any of the left-handed neutrinos (ν_e, ν_μ or ν_τ) and their antiparticles. The third is the charged-current interaction of electron antineutrinos (CA), which are not expected to be emitted from the sun except in certain very exotic scenarios. The fourth is the elastic scattering of neutrinos (ES). For each of these reactions, respectively, 9750, 5000, 0 and 1100 events per year are expected.

If neutrino oscillations occur, then the neutrinos may reach earth as other flavors. SNO therefore has the capability to reveal the presence of neutrino oscillations largely independent of solar properties. If the NC rate exceeds the CC rate (suitably normalized for cross sections), then neutrinos must be oscillating. If the CC energy spectrum differs from the known ^{8}B spectrum, then neutrinos must be oscillating via the MSW mechanism [62,63], vacuum oscillations [94,95], the Voloshin-Vysotsky-Okun mechanism [97][i] or neutrino decays [96], all of which predict different distortion patterns.

Unlike the kinematically convolved ES spectrum, the CC electron spectrum measured in SNO directly reflects the incoming ν_e spectrum (with quite good energy resolution) simply shifted down by 1.44 MeV. Being a real-time experiment, SNO will also be able to search for day-night differences due to regeneration in the earth (expected for some regions of the MSW parameters) [100,101,102], and seasonal variations expected, for example, for vacuum oscillations [94,95]. SNO is a high-rate experiment. The interaction rates listed with the reactions above are for a ^{8}B ν_e flux of 6 x 10^6 cm^{-2} s^{-1}, but no detection threshold. The NC threshold is the binding energy of the deuteron.

The Čerenkov light from charged-current and elastic-scattering events is detected in SNO by an array of 9,500 photomultipliers surrounding the acrylic sphere that holds the D_2O.

Neutral-current interactions on deuterium release a free neutron, and a number of strategies for detecting the neutron can be devised. One is to dissolve in the heavy water a fraction of a percent of chloride ion. When a neutron captures on 75% abundant ^{35}Cl, it emits an 8.6-MeV γ, which showers. The resulting Čerenkov radiation can be detected by the PMT array in the same way CC events are detected.

Another method is to detect the neutron in ^{3}He proportional counters distributed in the heavy water so that the NC and CC events become completely distinct and no longer represent "backgrounds" to each other. One may also compare the ES rate with the CC rate to deduce the neutral-current contribution to the former. There are, then, 3 completely distinct methods for determining the NC rate, which will afford a check on systematic effects.

SNO will begin operation in 1996, and may provide a resolution of the solar neutrino problem.

9.3.2 SuperKamiokande

SuperKamiokande is a proton-decay detector with a total mass of 50,000 tons [77]. Like its predecessor, Kamiokande, 'SK' will also have solar-neutrino detection capabilities via the elastic scattering from electrons. However, the enormous fiducial mass (22,000 tons) means that, if the threshold can be brought down to 5 MeV, approximately 8,000 events per year will be recorded, 160 times more than with Kamiokande. A detailed measurement of the shape of the y-distribution can be made to see if the primary spectrum has been distorted by the MSW effect, time-dependent (day-night and seasonal) rates can be determined with precision, and an independent determination of the neutral-current rate can be made by comparison with the pure charged-current data from SNO (that can be done within SNO itself, but with much lower statistical precision). Any of those measurements can potentially provide model-independent evidence of neutrino oscillations.

SuperKamiokande is being constructed in Japan at the same depth as Kamiokande (3,000 mwe), and uses 11,300 50-cm photomultipliers plus 1,000 smaller tubes recovered from the IMB experiment in the veto. New electronics and careful control of radon will keep the

[i] The VVO mechanism and its variants [98] postulate a large magnetic moment (possibly excluded by other astrophysical effects [1]) to explain a possible anticorrelation of the Homestake rate with the sunspot activity. This has not been supported by the most recent data.

thresholds low. SK is scheduled for completion in 1996.

9.4 Experiments Under Development

9.4.1 Iodine

There is now under construction in the Homestake Mine an iodine solar neutrino detector. When fully completed, the detector will contain 100 tons of ^{127}I (225 tons total detector mass). The total solar neutrino detection rate of this detector should be fairly similar to that of the present chlorine detector at Homestake (135 tons of ^{37}Cl, 615 tons total). It is estimated [103] that the iodine detection rate per target nucleus is 4.5 times that of the chlorine detector. More importantly, the detection of neutrinos from ^{7}Be (see Table 4) by ^{127}I is almost half of the total rate, 14 SNU, while the detection rate for these neutrinos is only 1.1 SNU in the chlorine detector.

The fits to the neutrino flux measurements of the presently operating detectors indicate that the observed ^{8}B rate is about 40% of that predicted by the conventional solar model while that for ^{7}Be is less than 10% of the prediction. While it may be possible to reduce the predicted ^{8}B flux by modifications of nuclear input parameters and thus reduce or even eliminate the difference between predicted and observed neutrino fluxes, there have been no suggestions of any procedure that could significantly lower the expected ^{7}Be neutrino rate. Thus, the most critical present issue in the solar neutrino observations is the apparent absence of ^{7}Be electron neutrinos. It is generally assumed that this absence is an indication of neutrino flavor transitions, either in matter or in vacuum, into other neutrino flavors. Solar electron neutrino flux measurements with an iodine detector combined with those from the other detectors should be able to provide a clear measure of the electron neutrino flux from ^{7}Be in the sun. That, when combined with the multi- flavor neutrino measurements from Borexino, when that detector becomes operational, will then provide a measure of the non-electron neutrino flux from ^{7}Be, a crucial issue in searching for neutrino-flavor transitions.

The iodine detector will be modular with 10 tons of iodine per module. This modular structure permits fast extraction, 2 hours per cycle. Computer controlled extractions will be carried out daily at 6 a.m. and 6 p.m. The resulting ^{127}Xe will be stored in two charcoal traps, one for the daytime produced ^{127}Xe and the other for the nighttime produced ^{127}Xe. This procedure will permit a search for any day/night differences in the ^{7}Be neutrino flux caused by earth regeneration. In addition, since the detector will always be free of accumulated solar neutrino ^{127}Xe, it will always be fully sensitive to supernova signals.

All of the detector operating characteristics have been tested with one module of the detector. Construction of the full 100 ton, 10 module detector is now underway. The detector should be operational before the end of 1995.

A calibration program to determine the sensitivity of ^{127}I to ^{7}Be and ^{8}B neutrinos is also underway. The most readily available source of high-energy electron neutrinos is from the decay of stopped positive muons. The integral of the cross section for $^{127}\text{I}\rightarrow^{127}$Xe has been measured at LAMPF using the same radiochemical technique as that developed for the Homestake detector.

The ^{7}Be neutrino sensitivity can best be determined by calibrating the detector with a ^{37}Ar source. This source generates 814-keV neutrinos, very close to the 862-keV neutrinos emitted by ^{7}Be. Such a source can be produced in a reactor by fast neutron interactions on ^{40}Ca. It is also possible to determine the required matrix elements from the ^{127}I(p,n)^{127}Xe reaction.

9.4.2 Borexino

Present data indicate that a measurement of the ^{7}Be flux is of special interest. Astrophysical solutions would have to find a method to severely suppress the ^{7}Be flux, while MSW may convert most of the ν_e into ν_μ or ν_τ , which have an ES cross section 1/5 of the ν_e cross section. Some particle physics solutions predict a large time dependence. An Italian-US-German collaboration is carrying out the R&D necessary to construct a 100-tonne liquid-scintillator detector in Gran Sasso [104]. Extremely pure materials are required if radioactive backgrounds are not to mask the electron-scattering signal at 250 keV. No Čerenkov threshold protects the data from low-energy backgrounds in this experiment. A 4-tonne test detector is nearing completion. Given success, the INFN, NSF, and DFG will fund construction of the full-scale detector.

9.4.3 Icarus

A 5,000-ton liquid argon time-projection chamber is planned for Gran Sasso with a number of physics objectives including solar neutrinos. 'Icarus' can record both the elastic scattering from electrons and the charged-current inverse beta decay of ^{40}Ar to the isobaric analog state in ^{40}K, whose subsequent gamma decay provides a clear signature of the CC reaction. A 3-ton prototype has demonstrated the feasibility of drifting tracks 40 cm and reconstructing them [105].

9.4.4 Heron

A need to go to real time detection for the lower energy reactions (*pp* and ^{7}Be) requires the handling of events with energy depositions in the few-hundred-kilovolt range. The physical mechanisms for energy extraction (*e.g.*, phonons) in many materials at low temperature would facilitate this. Because the fluxes are higher for these reactions detectors are expected to be of lower mass but significant complexity. They will require low temperature but low-power-capacity refrigeration. As in all solar neutrino detectors, extreme attention must be paid to backgrounds by shielding, implementation of distinctive event signatures, and radioactive purity of the target material and detector components. Helium is the purest material known with respect to radioactivity.

"HERON" (for Helium:Roton detection of Neutrinos) is a project under development by a group at Brown University which has invented a technique based on liquid helium in the superfluid state. The goal is to provide capability for a real-time detector with a threshold of 10 keV suitable for measuring the recoil-electron spectrum and rate from neutrino-electron scattering of both the *pp* and ^{7}Be neutrinos. A detector with a 10-ton fiducial volume would yield 18 events per day with the full standard-model flux of electron-neutrinos from those two sources.

In superfluid helium, energy transport out of the liquid can occur by ballistic propagation of rotons (phonons), evaporation of helium from the surface, and the subsequent detection of heat pulses on an array of bolometers above the liquid. The high multiplicity of rotons relative to ions is expected to be an important advantage in achieving a low threshold. Extensive tests of the detection method have been carried out by the Brown group using alpha, beta and gamma radioactive sources in a 3-liter prototype at 30 mK with a variety of bolometers. With 5-MeV alpha particles the details of the mechanism have been proven and sensitivity to event position and recoil direction established through a measured asymmetry in the radiated rotons. Energy, position and direction resolutions have not yet been measured for electrons; further work on electron detection and on bolometry is needed. The discovery that determination of recoil direction might be possible, a capability currently possessed only by the water Čerenkov detectors, is a significant development for studying the low energy solar neutrinos [106].

9.4.5 Hellaz

The contrasting approach utilizing gaseous helium under high pressure at low temperature ("Hellaz") has been suggested to access the low-energy fluxes by a group from CERN/Collège de France/Gran Sasso/World Laboratory [107]. Two versions of a possible detector share a common basic technical implementation. A 20-ton mass of helium gas would be pressurized to 25 (or 50) atmospheres and held at 77 K. Operated as a time projection chamber, the detector cells would do pattern recognition as well as measure energy by track length and recoil-energy direction. The choice of TPC technology precludes the use of liquid helium owing to the its poor ion-drift properties. A successful application of the TPC method could be very powerful as a signature for solar neutrinos if the calculated energy and angle resolution can be achieved, and the radioactivity of the internal TPC hardware can be kept down. The basic physics processes are known but research is needed to provide laboratory evidence as to its feasibility, in view of the complexity of the device.

10 Atmospheric Neutrinos

Cosmic-ray interactions in the upper atmosphere produce pions and kaons whose subsequent decays produce a significant flux of neutrinos with energies below one GeV and moderate fluxes above this. Having traveled over path lengths of ≥ 20 km, these neutrinos are well suited for oscillation searches via the channel $\nu_\mu \rightarrow X$ in the range Δm^2 between 10^{-2} and 10^{-4} eV^2. There is much recent interest in the results of experiments measuring the flux of atmospheric neutrinos.[j] The atmospheric neutrino flux is a mixture of approximately $\frac{1}{3}$ $(\nu_e + \bar{\nu}_e)$ and $\frac{2}{3}$ $(\nu_\mu + \bar{\nu}_\mu)$. It is therefore possible, in principle, to do both ν_μ and ν_e disappearance and appearance experiments. It is useful to define various classes of atmospheric neutrino interactions as follows:

- Fully contained events, with all tracks inside the fiducial volume of the detector.
- Partially contained events, which have the vertex inside the fiducial volume but at least one exiting track.
- Upward, stopping muons are produced from neutrinos interacting outside the detector, and contain a muon which stops inside a fiducial volume.
- Upward, throughgoing muons are also from external neutrino interactions but the muons pass through the detector, producing than some minimum track-length inside a fiducial volume.

Low statistical precision limits the reach in $\sin^2\theta$, while the neutrino energy (E_ν) and the earth's radius (R_{earth}) limit the reach in Δm^2 of neutrino oscillation experiments with the atmospheric neutrino flux. Uncertainties in normalization of the primary cosmic rays, together with uncertainties in the production of pions and kaons in proton collisions with light nuclei, translate into significant uncertainties in the normalization and shape of the atmospheric neutrino spectrum at production. For this reason, conclusions based on ratios are safest.

10.1 Current Status

10.1.1 Contained Events

The neutrino flavor ratio from experiments [108,109] which measure contained events is shown in Table 5,

[j] This section is based on materials prepared by T. Gaisser, Bartol Research Institute and M. Goodman, Argonne National Lab.

Table 5: The atmospheric neutrino flavor ratio from various experiments. R' is defined in the text.

Experiment	Exposure kT-year	R'
IMB1	3.8	0.68 ± 0.08
Kamiokande ring	7.70	$0.60 \pm 0.06 \pm 0.05$
Kamiokande decay		$0.69 \pm 0.06 \pm 0.05$
IMB-3 ring	7.70	$0.54 \pm 0.05 \pm 0.05$
IMB-3 decay		$0.69 \pm 0.06 \pm 0.07$
Fréjus contained	2.0	$0.87 \pm 0.13 \pm 0.07$
Soudan 2	1.01	$0.64 \pm 0.17 \pm 0.09$
NUSEX	0.5	0.99 ± 0.29
Uncontained events included:		
Kamiokande Multi-GeV	7.70	$0.57 \pm 0.08 \pm 0.07$
Frejus total	2.00	$1.00 \pm 0.15 \pm 0.08$

while the data and calculations themselves are shown in Table 6. The flavor ratio is

$$R \equiv \frac{(\nu_\mu/\nu_e)^{data}}{(\nu_\mu/\nu_e)^{MC}}, \tag{31}$$

while the experiments actually measure

$$R' \equiv \frac{(track/shower)^{data}}{(track/shower)^{MC}}. \tag{32}$$

These only differ if there are neutrino oscillations in the presence of neutral current backgrounds to the track and shower samples. Systematic errors in Table 6, where shown are added quadratically. IMB has shown their systematic errors added linearly [110].

The high-statistics water Čerenkov detectors in Table 5 have two ways to distinguish quasi-elastic ν_μ and ν_e events. In one method, the shape of the Čerenkov ring on the wall of the phototubes is used. In the other method, excess phototube hits in a window several microseconds after the event indicate μ or π decay. The two methods have different systematic errors, but are statistically correlated. Experimental tests of e/μ detection efficiency are underway at KEK. Results from the Kamioka group presented at Eilat [111] show no sign of misidentification. Results from the IMB exposure at KEK can be expected in 1995. The Fréjus experiment prefers to analyze their contained and partially contained events together. This is discussed in the next section.

The results on contained atmospheric neutrino events are consistent with each other, and with a 30-40% deficit of ν_μ events. Such a deficit could be the result of either $\nu_\mu \leftrightarrow \nu_\tau$ or $\nu_\mu \leftrightarrow \nu_e$ oscillations. Since the interpretation of the atmospheric neutrino anomaly in terms of neutrino oscillations depends on the normalization of the neutrino flux, it is desirable to check the normalization as well as possible. In a conventional two-detector neutrino oscillation experiment, this is done by measuring the interaction rate in the neutrino beam close to the source. The nearest analog for the atmospheric neutrino beam of a close-in detector is a measurement of the intensity of muons at the same altitudes where both the neutrinos and muons are produced (10 to 20 km). New, higher precision measurements of this flux are being made [112,113], and preliminary results [112] are consistent with a relatively high normalization of the neutrino flux [114]. This favors an interpretation of the atmospheric neutrino puzzle in which ν_μ disappear, rather than transform into ν_e. (Note that the neutrino energies are below τ production threshold.)

Table 6: The data and Monte Carlo predictions on the atmospheric neutrino deficit from several experiments. TR refers to single tracks, SH to single showers.

Experiment	TR^{see} (ν_μ^{see})	TR^{MC} (ν_μ^{MC})	SH^{see} (ν_e^{see})	SH^{MC} (ν_e^{MC})
IMB1	104	136	297	265
Kam. ring	234	356.8	248	227.6
Kam. decay	182	277.5	300	313.9
IMB-3 ring	182	268	325	257.3
IMB-3 decay	208	261.5	402	348.5
Fréjus contained	94	100	89	82
Soudan 2	33.5	42.1	35.3	28.7
NUSEX	32	36.8	18	20.5

10.1.2 Partially Contained Events

The most recent new result on atmospheric neutrinos is the Kamiokande analysis of their "multi-GeV" events, which include fully contained events with $E_{\text{vis}} >$

1.33 GeV as well as partially contained events [115]. The mean neutrino energy responsible for this latter class of events is ~ 6 GeV. This is nearly an order of magnitude higher than for the independent sample [116] of contained events with $E_{\rm vis} < 1.33$ GeV. In this new data sample, there is again evidence for a ν_μ deficit, with a measurement of $R' = 0.59 \pm 0.08$.

The most remarkable feature of the data is the appearance of an angular dependence of the muons of the kind that would be expected if downward muons have path-lengths too short to oscillate; that is, the upward muons are suppressed relative to the downward muons. Given the range of E_ν and R involved for the multi-GeV sample, Eq. (22) implies an upper limit on Δm^2. In contrast to the muons, the angular dependence of the electrons is closer to what is expected if the ν_e are not affected by oscillations. The angular-dependence alone therefore suggests an interpretation in terms of ν_μ disappearance.

On the other hand, a preliminary re-analysis of the Fréjus data presented at Snowmass [117] seems to be even more divergent from the Kamiokande and IMB results of Refs. [116,118] than the original Fréjus publication [119]. A simultaneous analysis of their contained, partially contained, and throughgoing muons leads to a neutrino oscillation exclusion plot which contradicts a neutrino oscillation interpretation of the data in Tables 5 and 6. The Fréjus fully contained data are not, by themselves, in strong disagreement with the other experiments. However, their partially contained events and neutrino induced throughgoing muons seem to be in conflict with the new Kamiokande result. An analysis of all Fréjus neutrino data leads to the result $R' = 0.96 \pm 0.18$.

One puzzling feature of the new Kamiokande result is that there appears to be a large excess of electron neutrinos and a relatively small deficit of muon neutrinos as compared to the calculation. Using the neutrino flux of Ref. [120], they find

$$e-\text{like}\ \left(\frac{\text{measured}}{\text{calculated}}\right) = 1.47 \pm 0.15 \qquad (33)$$

and

$$\mu-\text{like}\ \left(\frac{\text{measured}}{\text{calculated}}\right) = 0.83 \pm 0.07. \qquad (34)$$

The corresponding numbers for the low-energy, fully contained event sample are 1.09 (electrons) and 0.66 (muons). Thus, while the low energy sample suggests $\nu_\mu \leftrightarrow \nu_\tau$, the high-energy sample looks more like $\nu_\mu \leftrightarrow \nu_e$.

10.1.3 Upward Stopping Muons

The measured ratio of stopping to throughgoing upward neutrino-induced muons is consistent with expectation [121]. The data are shown in Table 7. This means *either* that there are no oscillations of ν_μ with mixing angle and Δm^2 large enough to be seen with upward neutrinos *or* that there may be large mixing of ν_μ, but with Δm^2 large enough so that the lower energy stopping muons as well as the higher energy throughgoing muons are affected. The measured stopping/throughgoing ratio is used to rule out a range of $10^{-3} < \Delta m^2 < 10^{-2}$ eV^2 at large mixing angle for ν_μ disappearance [121].

The ratio of upward stopping muons to throughgoing muons is sensitive to the absence of background from events that are not initiated by neutrinos. In particular, downgoing muons which initiate hadronic cascades in the rock could mimic upward going stopping muons in some cases. There is evidence for such events from Soudan, Kamioka, and MACRO [122,123,124], but it is not straightforward to calculate the possible contamination of the IMB data sample. A background of 12 events out of the 85 would lead to the absence of any neutrino oscillation limit from the IMB analysis. A comparison of upward showers to expectation in IMB may be evidence that such a background in fact exists.

Table 7: Neutrino induced upward event rates in IMB.

	Observed	Calculated
Stopping tracks	85	84
Exiting tracks	532	516
Showers	49	17

10.1.4 Upward Muons

The implications of upward muon measurements have been somewhat controversial. The IMB group [121] calculated the rate of upward, throughgoing muons assuming a particular neutrino flux [125] and a particular representation of the neutrino cross section [126]. Their calculated rate agreed with their measured rate, apparently ruling out most of the available parameter space needed for a ν_μ-disappearance interpretation of the low energy, contained event anomaly. Frati *et al.* [127] showed that the Kamiokande data on upward, throughgoing muons [128] is similarly consistent with a calculation assuming the same input. On the other hand, they found that, starting from different, more recent calculations of the neutrino flux [129,130] and a different representation of the cross section [131], the predicted rate is some 15% higher than the observed rate. They further showed that a discrepancy of this size is consistent with a range of parameters needed to explain the contained event anomaly as ν_μ-disappearance. Recently reported results from MACRO [132] similarly show a significant deficit of the measured upward rate as compared to a calculation with the "high" input assumptions. This matrix of measurements and calculations is indirect evidence that these experiments are all consistent. The same cannot be said for Baksan [133] where there is agreement between measurement and expectation for "high" input assumptions, but an *excess* of measurements over calculation when the low input is used. In all cases the differences are only of marginal statistical significance, as indicated by the summary of measurements/calculations in Table 8.

Table 8: Upward muons. For Kamioka the units are upward muons in units of $10^{-13}\text{cm}^{-2}\text{s}^{-1}\text{sr}^{-1}$. For IMB, upward events per day and for Baksan and MACRO total number of observed upward events.

Experiment	Observed	Calculated	
		"high"	"low"
IMB [121]	0.47 ± 0.02	—	0.455
KAM [128]	2.04 ± 0.13	2.36	2.18
Baksan [133]	161	162	142
MACRO [132]	$74 \pm 9 \pm 8$	101 ± 15	—

10.1.5 Summary of Atmospheric Neutrino Data

Experiments with atmospheric neutrinos can be ranked in the order of their potential for discovery of new physics as follows:

1. Measurements of the ratio ν_e/ν_μ. Here, the principal uncertainties in the input to the calculation of this ratio all cancel. Three completely independent calculations [134,120,135] for the expected value of this ratio agree among themselves within 5%.
2. Angular distributions of neutrino fluxes in a particular energy region. Again, the principal uncertainties cancel in the calculation, but the statistical significance is poorer because the data must be subdivided into angular bins.
3. The ratio of stopping to throughgoing neutrino-induced upward muons. The cancellation of uncertainties in the calculations here is less complete because the two classes of events in this ratio correspond to different ranges of energy.
4. The absolute intensity of ν_μ or ν_e.

It is noteworthy that the experiments with the greatest statistical significance [116,118] show a significant deviation from the most secure feature of the atmospheric neutrino beam—the ν_e/ν_μ ratio. The measured value is apparently greater than the ratio at production. The possibility that the discrepancy might result from an oversimplified use of the Fermi Gas Model (FGM) in calculating the neutrino interaction rate is not borne out by careful investigations [136], which conclude that the energy is high enough that any deviation from the FGM affects both flavors of neutrinos similarly. On the other hand, recent reports from the Los Alamos LSND experiment indicate that the ν_μ cross section on ^{12}C is a factor 3 below FGM expectations while the ν_e cross section is not. The energies in this experiment, however, are considerably lower than those relevant to the atmospheric neutrino signal.

If the observed effect is a signal of neutrino oscillations, the lack of observed angular dependence for low energy ($\lesssim$ 1 GeV) events [116,118] implies $\Delta m^2 > 10^{-3}$ eV^2. The large size of the discrepancy requires $\sin^2 2\theta > 0.5$. Much, but not all, of the available phase space for a $\nu_e \leftrightarrow \nu_\mu$ explanation is ruled out by reactor limits. New long-baseline reactor experiments at San Onofre and Chooz are under construction to close the gap completely (see Section 11).

A ν_μ disappearance interpretation (*e.g.*, $\nu_\mu \leftrightarrow \nu_\tau$) of the ratio ν_e/ν_μ in low-energy contained events with $\Delta m^2 > 10^{-2}$ eV^2 and large mixing angle would imply disappearance of $\sim$ 100 GeV ν_μ with path-lengths $L > R_{earth}$. This in turn predicts a deficit of ν_μ-induced, upward, throughgoing muons. Measurements of this rate are inconclusive because of uncertainty in the normalization of the calculated neutrino flux. There is also some uncertainty in the neutrino cross section. Combining all their data (sub-GeV and multi-GeV), the Kamiokande group find a rather limited allowed region, either for $\nu_\mu \leftrightarrow \nu_\tau$ or for $\nu_\mu \leftrightarrow \nu_e$ with nearly full mixing and with $\Delta m^2 \sim 1$ to 2×10^{-2} eV^2. Simultaneous analysis of all atmospheric and solar neutrino data sets (as of 1993) in the space of 3-flavor neutrino oscillations [137] favors oscillation predominantly in the $\nu_\mu \leftrightarrow \nu_\tau$ sector to explain the atmospheric[k] and in the $\nu_\mu \leftrightarrow \nu_e$ sector for the solar neutrinos. This in turn implies a "natural" order for the neutrino masses, with $m_{\nu_\tau} \sim 0.1$ eV and $m_{\nu_\mu} \sim 0.003$ eV.

10.2 The Next Five Years

Between now and 1999, Soudan 2's exposure will grow from 1.8 to 5.0 kt-years. The present result, 0.69 ± 0.19 (based on 1.0 kt-year) could change to 0.69 ± 0.08 or 1.0 ± 0.12. The low value would confirm the deficit in a detector with much different systematic errors. However, compared to 14 kt-years of present Čerenkov detector data, neither Soudan answer would resolve the situation. Systematic studies of a water Čerenkov detector using IMB and Kamioka photomultiplier tubes are taking place

[k]Oscillations into sterile neutrinos are disfavored by nucleosynthesis. See Section 3.

in a KEK test beam. This will check their thresholds, efficiencies, and pattern recognition capabilities. In 1996, Superkamiokande (see Section 9.3.2) will take data and, in another year or two, these data will compete statistically with the present world sample. Since the muon neutrino deficit is unlikely to be due to a statistical fluctuation, this may not resolve the situation either. The angular distribution of contained events and Multi-GeV events in Superkamiokande might provide additional evidence concerning an oscillation interpretation for values of Δm^2 below $\sim 10^{-2}$ eV^2.

If the atmospheric neutrino deficit is an experimental artifact, it is hoped that the cause can be identified in the next few years. However, if the cause is neutrino oscillations, the evidence may be limited to the anomalous flavor ratio and low statistics angular distributions unless new atmospheric or accelerator experiments are built.

10.3 Beyond 1999

Three new detector projects have the potential to bring atmospheric neutrinos to the realm of a high statistics and well measured field of study during the first decade of the 21st century. These are the ICARUS high resolution Argon drift chamber for the study of contained events, a large new underground long baseline neutrino detector for the study of semi-contained events and upward stopping muons, and a km^3 array in water or ice for the study of throughgoing neutrino induced muons. In addition, the phenomena can be studied directly in long baseline accelerator experiments, as discussed in Section 12.

10.3.1 ICARUS

The ICARUS collaboration plans to build a 5-kiloton liquid-argon projection chamber in the Gran Sasso Laboratory. As they have demonstrated with a 3-ton prototype, such a detector would have unprecedented pattern recognition capabilities for atmospheric neutrinos. The technique can provide a high electron to muon discrimination for both quasi-elastic and deep inelastic atmospheric neutrino events, a precise energy measurement, a good neutrino direction reconstruction, and a partial neutrino to antineutrino separation (by means of the different capture to decay ratio for μ^+ and μ^- in argon.) With about 4 kT-yrs per year of fiducial exposure, this should provide dramatic confirmation if the atmospheric neutrino anomaly is due to oscillations.

10.3.2 Long Baseline Detector Capabilities for Atmospheric Neutrinos

The motivation and capabilities of long baseline neutrino detectors will be discussed in the next section. However, a 15 kTon underground magnetized detector, such as envisaged for a Fermilab or CERN experiment, would be a new powerful tool in understanding the semi-contained atmospheric neutrinos. The magnetic field would allow the charge ratio to be measured unambiguously, and good timing would allow separation of semicontained neutrino interactions from other cosmic-ray backgrounds. The combination of measuring the hadron shower in the calorimeter, and the muon momentum with the magnetic field, would allow the full reconstruction of atmospheric charged-current neutrino events. If the neutrino-oscillation interpretation of the atmospheric anomaly is correct, there would be many new signals that could be confirmed in such detectors.

10.3.3 Cubic Kilometer Neutrino Detectors

IMB measured the upward going muon flux with an effective area around 300 m^2. Kamioka, Baksan and Soudan 2 have similar effective areas, while MACRO is larger. However a square kilometer effective area (as part of a cubic kilometer detector) would measure the flux and angular dependence of the upward going muons with 3000 times the present statistics. The DUMAND, NESTOR, AMANDA and Lake Baikal collaborations are providing the first generation of large underwater or deep ice experiments. To date, none of them have been able to separate atmospheric neutrino induced muons from the cosmic ray backgrounds. But each has overcome considerable technical obstacles, and hope to demonstrate feasibility of deep water/ice muon detection and directionality soon. If further economies of scale can be shown to exist in the course of those studies, a km^3 detector could be planned for ten years from now. As discussed above, very high statistics measurement of the angular distribution of neutrino induced muons would be sensitive to large neutrino mixing over a wide range of Δm^2.

11 Reactor Neutrinos

Since the discovery of the electron antineutrino by Reines and Cowan in 1956, reactor neutrino experiments[l] have continued to provide an essential tool in the elucidation of the properties of the neutrino.

The nuclear β-decay of the fission products of nuclear reactors produce a very large flux of low energy (up to ~ 8 MeV) $\bar{\nu}_e$'s. This allows for a study of the disappearance channel $\bar{\nu}_e \to X$. One can search for oscillations by measuring the energy spectrum of antineutrinos at a distance, L, from their source and comparing with expectations. The occurrence of oscillations will cause this spectrum to be modified from the predicted "no oscillation" spectrum. Such a measurement is non-trivial

[l]This section is based on material prepared by Richard Steinberg, Drexel University.

since the comparison depends on understanding the initial spectrum, which depends on the yields of each of the fission products. Uncertainties in the calculations which lead to these predictions limit the ultimate accuracy. One method to reduce the uncertainties involves comparing neutrino yields at two or more distances from the reactor source. In most reactor experiments, neutrinos are detected using the charged current reaction $\bar{\nu}_e + \mathrm{p} \rightarrow \mathrm{e}^+ + \mathrm{n}$. The signature for this reaction is a time correlated pair of positron and neutron signatures.

11.1 Current Status

The current best limits for neutrino-oscillation parameters from reactor neutrino experiments are derived from a generation of experiments using detectors of several hundred kilogram mass at distances up to about 100 meters from the source. Although these distances are relatively short compared to those possible with accelerator neutrino experiments, the L/E ratios are more than competitive.

The lower energies, together with excellent understanding (at the level of $\sim 2\%$) of the initial neutrino flux spectrum and flavor composition, permit the reactor neutrino oscillation experiments to provide convincing information about neutrino oscillations in an interesting region in the space of Δm^2 vs $\sin^2 2\theta$. The current limits from reactor experiments at both the Gosgen and Bugey reactors rule out neutrino oscillations involving electron neutrinos to a Δm^2 level of $10^{-2}\,\mathrm{eV}^2$ at maximal mixing, and to $\sin^2 2\theta$ values greater than about 0.05 for large Δm^2.

11.2 Experiments Under Construction

The continuing puzzle with respect to solar neutrinos together with an apparent anomaly in the atmospheric neutrino fluxes has stimulated interest in a new generation of reactor neutrino experiments. Two of the proposed experiments have been approved and are now under construction. The sensitivity of each will completely cover the region of parameter space indicated by the atmospheric neutrino anomaly and extend down to the level of $\Delta m^2 = 10^{-3}\,\mathrm{eV}^2$.

Both experiments will utilize a gadolinium-loaded liquid scintillator neutrino target of about 5-10 tons mass located about one kilometer from a large nuclear power station. The reaction to be measured is the classic Reines-Cowan inverse β-decay reaction, $\bar{\nu}_e + p \rightarrow e^+ + n$. Neutrino oscillations coupling $\bar{\nu}_e$ to any other neutrino would be inferred from the $\bar{\nu}_e$ disappearance.

The first and most sensitive method for extracting the signal would be an absolute comparison of the measured integrated count rate with the rate calculated from knowledge of the neutrino source and of the absolute detector efficiency.

An additional signature for neutrino oscillations would result from the study of distortions in the measured positron spectrum. In certain regions of the parameter space, this study could lead to a very clear, perhaps unmistakable signature for neutrino oscillations.

The detector design philosophies in the two experiments, however, are quite different. This situation is fortunate, since the different systematic effects in the two detectors should provide an extra measure of confidence in any jointly observed result. History teaches that this approach is indeed wise.

The San Onofre experiment [138] is being performed by a collaboration of groups from Caltech, the Institute for Nuclear Research (Moscow) and Stanford University. The experiment will be located at the depth of 25 meters water equivalent (mwe). To reduce the substantial cosmic ray-induced backgrounds at this depth, a segmented detector design was developed, utilizing a three-fold prompt coincidence signature for the positron, together with a delayed coincidence requirement for the positron-neutron capture sequence. The price paid for the complex geometry and coincidence requirements is a relatively low overall efficiency of about 20% as well as substantial difficulty in precise determination of the absolute detector efficiency.

The Chooz Experiment [139,140] is being performed by a collaboration of groups from Annecy, the Collège de France, Drexel University, the Kurchatov Institute, Pisa, Trieste, and the Universities of California (Irvine) and New Mexico. The experiment utilizes a site one kilometer from an 8.2 GW nuclearpower station nearing completion in northern France. The site was selected because of its substantial underground depth of 300 mwe. With a 200-fold cosmic-ray rate reduction provided by the overburden, an unsegmented nearly-spherical detector design was adopted. The efficiency of the detector for positrons from neutrino inverse beta decay events is above 80% and relatively easy to model and to calibrate.

Each of the experiments expects to begin data taking before the end of 1995.

11.3 The Future

Future work on neutrino oscillation studies with reactor neutrinos could be aimed in at least three complementary directions.

Improvement of our knowledge of the reactor neutrino spectrum would allow improved sensitivity to small values of $\sin^2 2\theta$. This direction may be especially appropriate as a means to verify results from the LSND experiment discussed in Section 12. Sensitivity to mixing angles as small as 1% would be a reasonable goal.

A second opportunity for further work would be to

improve the sensitivity of the long baseline reactor experiments to smaller values of Δm^2. It has been suggested that it is possible to extend the basic technique of the Chooz experiment to neutrino baselines of at least ten kilometers [141,142]. With such an experiment, which would require a deep underground site and a kiloton-scale detector, sensitivity to Δm^2 lower than 10^{-4} eV^2 can be reached. The decision whether to pursue such an experiment probably should await the results of the present generation of one-kilometer neutrino oscillation experiments.

Finally, one should explore and encourage the possibility of using other neutrino reactions, such as neutral- and charged-current reactions in deuterium to provide sensitive oscillation tests.

12 Accelerator Neutrinos

The regions of parameter space which appear "interesting" because of the outstanding questions of dark matter and atmospheric neutrinos extend just beyond the reach of completed first generation accelerator experiments. This tantalizing situation has recently spurred a renewed interest in accelerator based searches and led the way to a second generation, some of which are currently operational and will take data in the near future. In this section we will review the status of on-going experiments and discuss options for new initiatives to significantly extend the reach in parameter space.

12.1 Large Δm^2

12.1.1 The $\nu_\mu \to \nu_e$ Channel

LSND The LSND experiment at Los Alamos is searching for $\nu_\mu \to \nu_e$ and $\bar{\nu}_\mu \to \bar{\nu}_e$ oscillations using muon neutrinos produced in the decay of pions produced by the LAMPF proton beam. A signal in their detector, composed of 200 tons of dilute liquid scintillator which is sensitive to the appearance of low energy electrons, would indicate the occurrence of a $\nu_e\,(\bar{\nu}_e)$ charged current interaction resulting from the oscillation of $\nu_\mu \to \nu_e$ ($\bar{\nu}_\mu \to \bar{\nu}_e$). Preliminary reports suggest that such signals are indeed seen, with strong statistical significance in the $\bar{\nu}_\mu$ channel[143]. The regions of parameter space in which an oscillation may be indicated is for Δm^2 of a few eV^2, and $\sin^2 2\theta \sim 10^{-2}$.

KARMEN The KARMEN experiment is also sensitive to the $\bar{\nu}_\mu \to \bar{\nu}_e$ channel. KARMEN is a 56 ton liquid scintillator calorimeter which like LSND, measures neutrinos from a beam dump for 800 MeV protons from the ISIS storage ring at Rutherford Laboratory. The experiment has been taking data for four years and anticipates a further two years of data taking. At the present time the experiment does not report any significant number of events above the expected background[144].

12.1.2 The $\nu_\mu \to \nu_\tau$ Channel

The recently renewed realization that tau neutrinos may play a critical role in cosmological dynamics has revived interest in searching for $\nu_\mu \to \nu_\tau$ oscillations for mass differences in the 5-100 eV range. This search, when done with a definitive "appearance" experiment, requires intense primary proton beams with energies above 100 GeV but only "short" baselines, *i.e.*, of less than a kilometer. These parameters are ideally suited for carrying out (relatively inexpensive) oscillation searches at Fermilab and CERN.

Tau neutrino appearance experiments are designed to detect the τ lepton from ν_τ interactions, either by direct observation of a short-lived τ track and kinematic information, or by use of kinematic information alone. These highly-instrumented but relatively low-mass detectors have high sensitivity (ultimately, a few times 10^{-5}) in $\sin^2 2\theta$ for $\nu_\mu \to \nu_\tau$ oscillations because of their excellent background rejection. Sensitivity to $\nu_e \to \nu_\tau$ is typically a factor 50-100 worse because ν_e is a minority component in accelerator-produced beams. (Fortunately, the ν_τ content of such beams is negligible.) These experiments have maximum sensitivity in the mass range which is cosmologically interesting. However, with short baselines they cannot access Δm^2 below a few tenths of an eV^2, even at maximum mixing.

The CERN Laboratory is now beginning a new generation of $\nu_\mu \to \nu_\tau$ oscillation experiments, CHORUS [145] and NOMAD [146], that promise to extend the relevant mixing-angle sensitivity an order of magnitude by identifying one or more channels of tau lepton decay.

CHORUS The first experiment, CHORUS, will use a hybrid emulsion spectrometer to directly search for the production of τ leptons from ν_τ charged current interactions. The distance of the detector from the proton target ($\sim$ 800 m) and the average beam neutrino energy makes the experiment sensitive to tau neutrino mass as low as 1 eV. Limits in the mixing angle between tau and muon neutrinos have been measured previously (in this range of masses) by the Fermilab E-531 experiment[147] at the level of $\sin^2 2\theta = 4\text{x}10^{-3}$. The CHORUS experiment expects to improve this limit by a factor 10. For an integrated intensity of 2.4 x 10^{19} protons on target about 500,000 muon neutrino interactions are expected. This should be achievable over two calendar years.

NOMAD The second experiment, NOMAD, will attempt to identify the occurrence of τ production by the kinematic distribution of particles produced in the interaction. Unlike CHORUS, which has the ability to definitively identify the tau track, NOMAD, relies totally on the signature of missing transverse momentum. Hence, to make a case that a signal has been seen, it will be crucial to demonstrate that the missing transverse momentum is not due to other sources. The NOMAD experiment will measure the momentum of charged particles by tracking them in a magnetic field, and will measure the momentum of photons with a preshower and lead glass electromagnetic calorimeter. A hadronic calorimeter is also being added to the experiment.

Data taking began with the new CERN wide band neutrino beam in April 1994, and ended in October 1994. After a several-month hiatus data taking will resume and continue until the end of 1995. Further data taking in 1996-97 has been approved.

FNAL E-803 In early discussions of the Fermilab Main Injector project it was realized that this new machine would provide the opportunity to produce a very high-intensity, moderate energy neutrino beam. A beam of 120 GeV protons could be extracted at least every two seconds at an intensity of 3-6 x 10^{13} protons per pulse. In 1990 the first proposal submitted to the Laboratory to use such a beam, E-803 [148], proposed to use a high-resolution hybrid emulsion spectrometer to do a definitive tau appearance experiment. Delays in the approval of the Main Injector project resulted in a several-year deferral of serious consideration of E-803. In the meantime the opportunity to do a similar experiment at CERN (discussed above) was proposed and approved. The major difference between CHORUS and E-803 is in E-803's significantly greater resolution for actually identifying τ decays. In contrast to the CERN experiments which emphasize the $\mu\nu\bar{\nu}$ and the $e\nu\bar{\nu}$ modes, E-803 will concentrate on the ability to identify τ decays in the two-body and quasi-two-body modes $\pi^-\nu$ and $\rho^-\nu$ (with $\rho^- \rightarrow \pi^-\pi^0$), which have distinctive kinematic signatures and measurable proper decay times. E-803 was approved in November 1993. As will be discussed below, the present Fermilab plan is to perform E-803 in conjunction with a long-baseline experiment, hence maximizing the reach into parameter space with a single neutrino beam. In the absence of a signal, E-803 would extend the limits in mixing angle by a factor of five to ten over the CHORUS experiment at very large Δm^2, and by more than a factor of thirty at values of Δm^2in the few electron volt range. These limits are summarized in Figure 5.

FNAL E-872 Though there is a great deal of evidence that the ν_τ exists as a unique lepton and that its couplings are not anomalous at the present level of sensitivity, interactions of ν_τ have yet to be observed, as have those of the ν_e and ν_μ. To accomplish this, an experiment requires high proton intensities, high energy, and

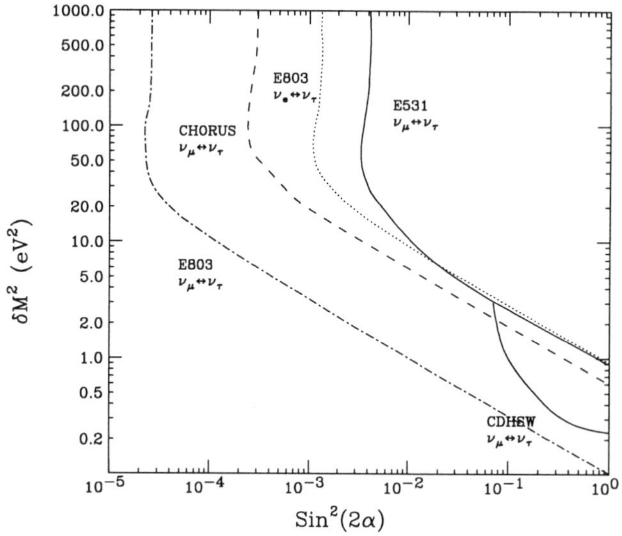

Figure 5: Present limits in the channel $\nu_\mu \rightarrow \nu_\tau$, as well as those expected from the on-going CHORUS experiment, and the E-803 experiment.

extremely good detector resolution. The Fermilab Tevatron in conjunction with a high resolution hybrid emulsion spectrometer meets these requirements. A newly approved experiment, E-872, will produce τ neutrinos in a beam dump and directly measure ν_τ charged current interactions by seeing τ decays in an emulsion target using the same hybrid emulsion technique as CHORUS and E-803. This experiment will be performed during the 1996-97 Fermilab fixed-target run prior to the completion of the Main Injector. E-872 should see the interaction that the oscillation experiments hope to see, if the ν_τ is "standard". E-872 will be also be sensitive to and measure the processes that give rise to backgrounds for the experiments seeking ν_τ oscillation.

12.2 Extending the Search to Small Δm^2

Conceivably, neutrinos could have the small, non-zero masses and relatively large mixing angles suggested by the apparent deficit of atmospheric muon neutrinos. Definitive "disappearance" experiments can be carried out at a number of accelerator laboratories to test this hypothesis directly. Such experiments require intense neutrino beams and at least two detectors separated by tens to hundreds of kilometers, depending on the neutrino energy and the difference in neutrino masses that is being explored. Long base-line "disappearance" experiments require careful detector alignments and neutrino flux measurements. They tend to be relatively expensive because of the required infrastructure (downward-sloping beams, two detectors with the far one containing a large target mass). However, they represent a unique opportunity to discover neutrino oscillations in the range from about 10^{-3} to several eV in a controlled laboratory setting.

In the following subsections we will briefly discuss several proposals for accelerator based long-baseline oscillation searches. Reactor searches were described in Section 11.

12.2.1 $\nu_\mu \to \nu_e$ and $\nu_\mu \to X$

BNL The Brookhaven AGS has a long history of providing high intensity proton beams which are then used to produce high flux beams of neutrinos. Present plans to increase the intensity of the AGS by a factor of four, to 6 x 10^{13} protons per AGS cycle of 1.6 sec, have made it possible to consider extending the oscillation search into the small Δm^2 region as well [149].

In Spring 1993, the Brookhaven AGS PAC recommended physics approval for P899, a *Proposal for a Long Baseline Neutrino Oscillation Experiment at the AGS.* The principal aim of the experiment is to explore the region of parameter space suggested by the atmospheric neutrino deficit. The experiment would use a neutrino beam produced by the 26 GeV protons from the AGS. The average neutrino energy would be about 1 GeV, which is similar to the energies of cosmic neutrinos which show evidence for oscillations. By placing detectors at several locations (1, 3 and 24 km), the disappearance of neutrinos (and possibly the appearance of ν_e) in the mass region of a few 10^{-3}eV^2 will be detectable if the mixing is relatively large. For an integrated delivered proton intensity of $\sim$ 2 x 10^{20} protons, more than 4000 contained muon events should be detectable in the far detector. As proposed, the far detector would be located in a tank farm on the north shore of Long Island. To produce a neutrino beam in this direction will require construction of a new proton transport line from the AGS. The detectors for E899 are identical, massive water Cerenkov detectors, 15 meters in diameter by 15 meters in height. Each tank will contain 1900 20-cm diameter photomultiplier tubes.

As of late 1994, the E899 collaboration was preparing an upgraded proposal. In addition to the detectors at 1, 3, and 24 km, as originally proposed, a fourth detector site consisting of two additional tanks is proposed at a distance of 67 km, to be located at a government facility on Plum Island in Long Island Sound. The detectors would be displaced slightly from the center of the neutrino beam, which has the advantage of suppressing unwanted higher energy neutrinos. The four-detector scheme would allow the experiment to be constructed in stages. The first stage, with the furthest detector at 24 km, would be sensitive to ν_μ disappearance at a statistically compelling (3-5 σ) level down to $\Delta m^2 \sim 5 \times 10^{-3}$ eV^2 for $\sin^2 2\theta = 1$, and $\Delta m^2 \sim 10^{-2}$ eV^2 for $\sin^2 2\theta \sim 0.5$, covering the entire range suggested by the atmospheric neutrino anomaly. The second stage, including the detectors at 67 km, would extend the range of Δm^2 to values smaller by a factor of ~ 2. The statistical significance for $\nu_\mu \to \nu_e$ appearance would be much larger. The sensitivity of this experiment in both the ν_μ disappearance mode and the $\nu_\mu \to \nu_e$ appearance channel are shown in Figure 6 and Figure 7, respectively.

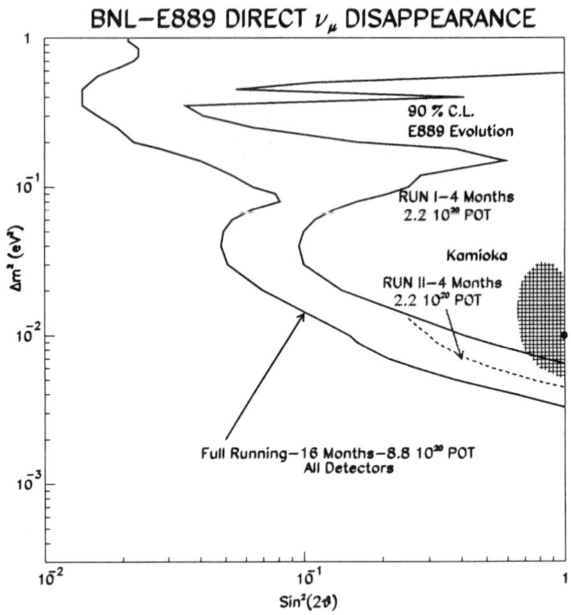

Figure 6: BNL E889 limit curves for ν_μ disappearance.

KEK The KEK proton synchrotron can produce extracted proton beams of 12 GeV at an intensity of 4 x 10^{12} every 2.5 seconds. A modest upgrade under consideration can provide an intensity of 7.5 x 10^{12} every 1 second. A proposal is under discussion of using an upgraded PS beam to send neutrinos to the SuperKamiokande detector, located 250 km from the KEK site.

12.2.2 $\nu_\mu \to \nu_\tau$

FNAL To do both a short and long baseline experiment using the same beamline is an ideal strategy for exploring the question of neutrino mass. The unique feature of such a program at Fermilab is that both the short and long-baseline experiments can be designed to exploit the high-energy feature of the Main Injector - τ appearance. Over the past few years several options for a long-baseline site from the Fermilab beam have been considered, including that of the IMB detector in Cleveland, Ohio, the DUMAND detector in Hawaii, and the Soudan 2 detector in Tower-Soudan, Minnesota. In 1990 members of the Soudan 2 collaboration submitted a pro-

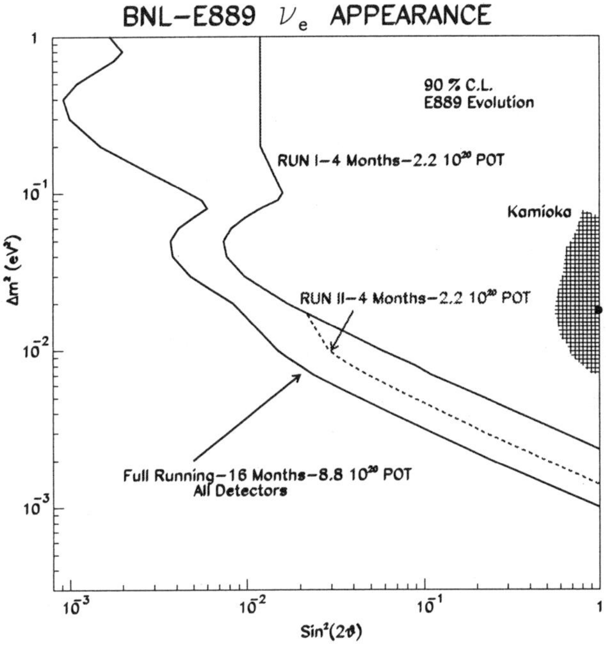

Figure 7: BNL E889 limit curves for $\nu_\mu \to \nu_e$ oscillations.

posal, P-822, to use the existing Soudan 2 detector to probe the region of parameter space suggested by the atmospheric-neutrino deficit.

A far detector distance between 500 and 1000 km is appropriately matched to the average neutrino energy of 15-20 GeV which can be produced by a horn focussed neutrino beam from the Main Injector 120 GeV protons. The major complication with such a beam is that it must be extracted and brought to a 58 mr downward bend angle, requiring the beamline enclosures and the experimental hall to be significantly below grade, and hence at increased cost due to the civil construction aspect of the project. Because of the financial investment required for such a beam, considerable debate has gone into the question of what the potential reach of an experiment using such a beam should be.

In spring 1994, Fermilab called for Expressions of Interest from the user community for long baseline experiments. Three were received, and though there were technical differences, all three emphasized the advantage that the high-energy Main Injector provides for doing appearance experiments in the channel $\nu_\mu \to \nu_\tau$. At the June 1994 Fermilab PAC meeting, the PAC strongly endorsed the idea that Fermilab proceed with further development of the Conceptual Design Report for a beamline which will provide for both short and long baseline experiments. Members of the long baseline EOI's were encouraged to join forces and form a strong collaboration to build a detector with the capability of "seeing" τ leptons.

Such a collaboration has been formed and the group has submitted a full proposal, P-875 (a.k.a MINOS) for a 10 kton detector to be located in a new cavern adjacent to the existing Soudan 2 detector [150]. The sensitivity of this experiment to both $\nu_\mu \to \nu_\tau$ and $\nu_\mu \to \nu_e$ oscillations are shown in Figure 8 and Figure 9, respectively.

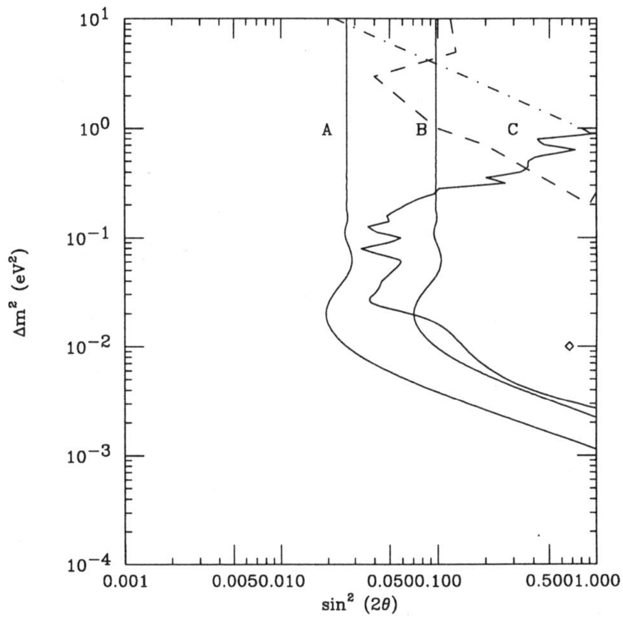

Figure 8: MINOS two year 90% CL limit curves for $\nu_\mu \to \nu_\tau$ oscillations. Curves A, B and C represent different rate comparisons and energy measurements which can be made in the experiment. The dashed curve shows measured limits from the CDHS experiment, and the dot-dashed curve show FNAL E-531 limits. The diamond is the Kamiokande best-fit point.

CERN At CERN studies have been done to show that it is technically feasible to send a neutrino beam to the Gran Sasso Laboratory in Italy or even to the Superkamiokande detector in Japan, which both happen to be in favorable directions for using existing or planned beamlines. The possibility of pointing a beam toward the NESTOR detector in Greece is also being discussed. The most cost-effective of these options is the beam to Gran Sasso, 732 km from CERN. Such a beamline would share facilities with the planned transfer line from the SPS to the LHC. The detector would be the proposed ICARUS detector [151].

The optimal conditions for $\nu_\mu \to \nu_e$ oscillations with ICARUS and a CERN SPS beam are obtained with a low proton energy (120 GeV, 3×10^{13} protons on target every 3.6 seconds). For the $\nu_\mu \to \nu_\tau$ oscillations, a higher beam energy is better suited. If the CERN SPS can deliver 3×10^{13} 450 GeV protons on target every 7.2

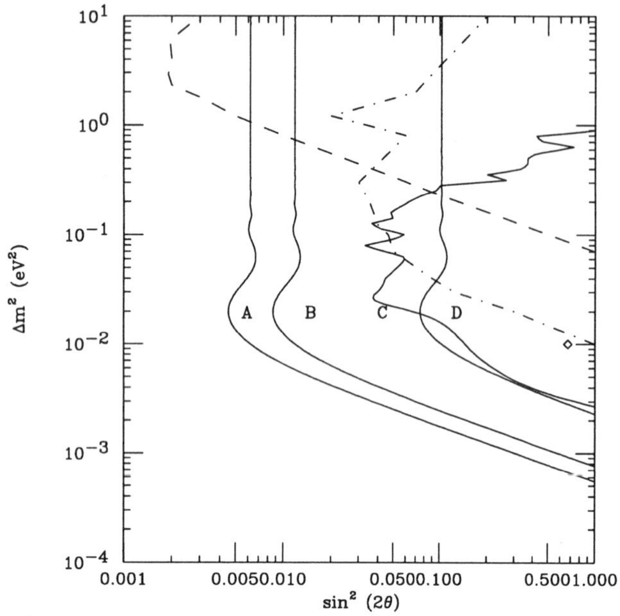

Figure 9: MINOS two year 90% CL limit curves for $\nu_\mu \to \nu_e$ oscillations. Curves A, B, C and D represent different rate comparisons and energy measurements which can be made in the experiment. The dashed curve shows measured limits from the BNL E-776 experiment, and the dot-dashed curve shows the Bugey 4 reactor experiment limits. The diamond is the Kamiokande best-fit point.

sec, the sensitivity of ICARUS for one year of data taking extends to $\sin^2 2\theta = 5 \times 10^{-2}$ and $\Delta m^2 = 2 \times 10^{-3}$ eV^2.

Proponents of a long-baseline oscillation program at CERN are proposing that the program could take place naturally in the period between the end of the LEP program and the beginning of the LHC program, *i.e.*, early in the next decade.

12.3 Comparison of Long Baseline Options

A strict comparison of differing options for a neutrino oscillation experiment should be done in the Δm^2-$\sin^2 2\theta$ plane. Even then, one experiment may have a sensitivity to a lower Δm^2 while the other may reach lower $\sin^2 2\theta$. The preference of one over the other is then strictly a matter of physics value judgement. Nevertheless, some qualitative comparisons of the proposed long baseline experiments are possible. An experiment's sensitivity to mixing angle depends on the test and statistics. The mass sensitivity then also depends on L and the neutrino energy distribution.

This can be made a bit more quantitative. An experiment is generally sensitive to oscillations above some minimum probability of oscillation, or can set a limit at that probability, which we call P_{min}. Then the limit on $\sin^2 2\theta$ at high Δm^2 is at $2 \times P_{min}$. Likewise, the mass limit at maximal mixing is approximately

$$\Delta m^2 = \frac{\langle E_\nu \rangle \sqrt{P_{min}}}{1.27L}, \tag{35}$$

where Δm^2, L, and E_ν are respectively in eV^2, km, and GeV. Since the slope of the limit curve at $\sin^2 2\theta = 1$ is $-1/2$, this almost entirely defines the limit curve, up to the energy distribution and some oscillatory behavior near the first few oscillations at the average E_ν. Then the value of P_{min} depends only on the test, the statistics, and possible systematic limitations. For the most favorable tests, $P_{min} \sim 1.5/\sqrt{N}$ at 90% CL, where N is the number of events measured at the far detector. For tests which involve kinematic cuts or particular decay modes, P_{min} is worse (larger) by the detection efficiencies and branching ratios.

The statistics which an experiment can achieve depends on the neutrino flux, the detector size, the distance, and the neutrino energy distribution. The event rate is proportional to the integrated neutrino flux, which in turn depends on both the duration of the experiment and the fraction of accelerator cycles that can be devoted to a neutrino program.

A higher energy beam gives more neutrino events due to the rising ν cross section. There are also more secondaries produced, and on average they are boosted more in the forward direction. On the other hand, a higher energy beam has more time between spills, less secondary decays in a fixed length beam pipe, and the Δm^2 sensitivity decreasing as $1/\langle E_\nu \rangle$. For tests that require the presence of a τ lepton in the final state, the energy must be significantly above the ν_τ charged current threshold. A 120 GeV proton beam, for example, provides neutrinos with an energy distribution such that $\langle \sigma_\tau/\sigma_\mu \rangle \sim 0.25$, while a 400 GeV beam is required for that ratio to reach 0.50. The lower energy beams at BNL or KEK will not make $\tau's$ by charged current processes.

When realistic detector sizes, running conditions, and accelerator parameters are used, the Fermilab proposal, CERN ICARUS proposal, BNL E899, and KEK to SuperKamiokande proposal all reach similar sensitivities of 5-8% in $\sin^2 2\theta$ and $2-3 \times 10^{-3}$eV^2 in Δm^2. The BNL proposal as written has a small disadvantage in Δm^2 sensitivity (limits will only improve as $N^{1/4}$ as opposed to $N^{1/2}$ in $\sin^2 2\theta$), but recent thoughts about placing a far detector at 67 km would overcome that difference. All of these long baseline proposals would cover the region of neutrino oscillation parameter space suggested by the atmospheric neutrino deficit.

A much larger 10-20 kT new detector for the Fermilab to Soudan long baseline experiment, as endorsed by the Fermilab PAC, would increase the sensitivity to lower mixing angle. However, the angles motivated by the atmospheric neutrinos are much larger and in fact are near maximal. Of course, the addition of such a massive new detector will need to be balanced against cost and available resources.

There are two other important qualitative difference between the low energy experiments (BNL, KEK), for which $\langle E_\nu \rangle \sim 1$ GeV, and higher energy $\langle E_\nu \rangle \sim 15$ GeV (FNAL, CERN) long baseline proposals. The first is that the lower energy experiments are more similar to the energy of the atmospheric neutrino beam. It is possible, for example, that a non-neutrino-oscillation explanation of the atmospheric anomaly could better be tested at a similar energy. The other involves the number of independent tests of $\nu_\mu \leftrightarrow \nu_\tau$ oscillations which can be performed above and below ν_τ charged current threshold. Each test will have its own unique backgrounds and systematic challenges.

13 Summary

We have attempted to review the broad based attack presently being made by both the theoretical and experimental communities on the intriguing and outstanding questions of neutrino mass and mixing. From the outset of this project it was recognized that this would be a topic which unabashedly crosses the traditional interdisciplinary boundaries of high-energy and nuclear physics, cosmology and astrophysics. The conveners of this group as well as the many contributors knew that while boundaries are usually easy to cross, barriers most often are not. We found that, from the standpoint of the physics questions, there were no fundamental boundaries. Barriers imposed by competition for funding and resources are, however, difficult to ignore and extremely difficult to cross.

Neutrino physics, particularly the study of the intrinsic properties of neutrinos, is a subject that has important implications for the disciplines of elementary particle physics, nuclear physics, astrophysics, and cosmology. Forty years after the first direct detection of neutrino interactions with matter, our knowledge of these elusive elementary particles may be summarized as follows. The three neutrino flavors ν_e, ν_μ and ν_τ (and their antiparticles) carry linear and angular momentum (left-handed ν and right-handed $\bar{\nu}$) and an assigned quantum number to specify the lepton family to which they belong. All other fundamental properties, *e.g.*, mass, charge, magnetic dipole moment, and whether they mix with one another, are known only by upper limits. Indeed, it is not known whether the neutrino is a Dirac particle with a distinct antiparticle or a Majorana particle, *i.e.*, its own antiparticle. In the search for physics beyond the Standard Model, new information in the neutrino sector is likely to make an important, perhaps vital, contribution.

For astrophysics, neutrinos probe the interior of stars early and late in their lifetimes. They are the coolant of neutron stars left from supernovae, and provide a potential thermometer for a precision temperature measurement of our sun's core. Neutrino astrophysics has become a recognized discipline in the last decade.

In the cosmological Big Bang Model, neutrinos decoupled relatively early in the evolution of the universe. This guarantees that they have a substantial relic abundance today if they are stable and do not annihilate. Their intrinsic properties are constrained by cosmological consequences, and, in turn, improved knowledge of neutrino properties would constrain cosmological models of the early universe. If any neutrinos have rest masses in the electron-volt region, they would contribute to the non-baryonic dark matter, and perhaps significantly to the total mass of the universe.

The suggestion was offered in the 1960's that oscillations among neutrino flavors might occur with rates dependent on neutrino mass and mixing amplitude, For the first time, there are reliable data indicating the occurrence of resonant and vacuum neutrino oscillations from studies of solar and atmospheric neutrinos. In an effort to confirm or repudiate these empirical clues, new experiments are being mounted and proposals for new experiments are being made. Because the potential importance of the envisioned results is well recognized, the proposals must be treated seriously even at a time of restricted funding. The relevant ranges of Δm^2 and $\sin^2 2\theta$ are now well defined, and the planned and proposed experiments should be able to establish or exclude them definitively.

For the first time, direct measurements placing limits on the mass of the electron-neutrino in the region of electron-volts have been made in the past few years.. A common feature of all of these experimental results is an anomalous excess of counts near the endpoint. This anomaly is indicative of a misconception—perhaps a serious one—on our part, and experiments to probe its nature and pursue the hint it offers are crying to be done.

Also for the first time, compelling direct measurements of the rates for double beta decay involving two neutrinos have been made recently. These measurements set the stage for incisive attempts to observe double beta decay in the absence of neutrinos and possibly discover the Majorana identity of the neutrino. This is an area of neutrino physics that is not costly but is of fundamental importance.

The field of neutrino physics is diverse and carried out non-centrally in self-contained experiments in university laboratories, deep underground sites, and accelerators. In part for this reason, it appears to involve a small community with few needs. But in truth, it is a world-wide effort pursued by many physicists, which often requires well-instrumented, massive apparatus. Experiments, whether large or small, require adequate long term support. Given the great scientific potential of physics involving neutrino mass and lepton flavor mixing, the enterprise deserves support beyond the level at which it now functions.

References

[1] *Review of Particle Properties*, L. Montanet *et al.*, *Phys. Rev. D* **50**, 1173 (1994).

[2] G. F. Smoot *et al.*, *Astrophys. J.* **396**, L1 (1992); A. de Oliviera Costa and G. F. Smoot, astroph-9412003.

[3] A. Klypin *et al.*, *Astrophys. J.* **416**, 1 (1993); M. Davis, F. Summers, and D. Schlegel, Nature **359**, 393 (1992); D. Schlegel *et al.*, *Astrophys. J.* **427**, 512 (1994); Y. P. Jing *ct al.*, *Astron. Astroph.* **284**, 703 (1994); R. Nolthenius, A. Klypin, and J. Primack, *Astrophys. J. Lett.* **422**, L45 (1994); J. Primack *et al.*, SCIPP 94/28.

[4] For detailed reviews, see G. Gelmini and E. Roulet, UCLA/94/TEP/36; P. Langacker in *Testing The Standard Model*, ed. M. Cvetic and P. Langacker (World, Singapore, 1991) p. 863, and in *Beyond the Standard Model III*, ed S. Godfrey and P. Kalyniak (World, Singapore, 1993) p. 91; B. Kayser, F. Gibrat-Debu, and F. Perrier, *The Physics of Massive Neutrinos*, (World Scientific, Singapore, 1989).

[5] Y. Zeldovich, DAN SSSR **86**, 505 (1952); E. S. Konopinski and M. Mahmoud, *Phys. Rev.* **92**, 1045 (1953).

[6] G. B. Gelmini and M. Roncadelli, *Phys. Lett. B* **99**, 411 (1981); H. Georgi *et al.*, *Nucl. Phys. B* **193**, 297 (1983).

[7] D. Schaile, plenary talk presented at the *27th International Conference on High Energy Physics*, Glasgow, July 1994.

[8] M. Gell-Mann, P. Ramond, and R. Slansky, in *Supergravity*, ed. F. van Nieuwenhuizen and D. Freedman, (North Holland, Amsterdam, 1979) p. 315; T. Yanagida, *Proc. of the Workshop on Unified Theory and the Baryon Number of the universe*, KEK, Japan, 1979; S. Weinberg, *Phys. Rev. Lett.* **43**, 1566 (1979).

[9] R. N. Mohapatra and G. Senjanovic, *Phys. Rev. D* **23**, 165 (1981).

[10] See S. A. Bludman, D. C. Kennedy, and P. Langacker, *Nucl. Phys. B* **374**, 373 (1992) and *Phys. Rev. D* **45**, 1810 (1992), and references therein.

[11] A. Zee, *Phys. Lett. B* **93**, 389 (1980); **161**, 141 (1985); *Nucl. Phys. B* **264**, 99 (1986). For later references, see [4].

[12] For a review of limits on neutrino magnetic moments, see [1].

[13] S. Bertolini and A. Santamaria, *Nucl. Phys. B* **310**, 714 (1988); **315**, 558 (1988).

[14] T. P. Walker *et al.*, *Astrophys. J.* **376**, 51 (1991).

[15] P. Langacker, UPR-0401T; R. Barbieri and A. Dolgov, *Phys. Lett. B* **237**, 440 (1990); *Nucl. Phys. B* **349**, 743 (1991); K. Kainulainen, *Phys. Lett. B* **244**, 191 (1990); K. Enqvist, K. Kainulainen, and J. Maalampi, *Phys. Lett. B* **244**, 186 (1990); **249**, 531 (1990); M. J. Thomson and B. H. J. McKellar, *Phys. Lett. B* **259**, 113 (1991); X. Shi, D. Schramm, and B. Fields, *Phys. Rev. D* **48**, 2563 (1993).

[16] N. Hata, private communication. For updated figures, see "http://dept.physics.upenn.edu/~www /neutrino/solar.html".

[17] For recent reviews and discussions, see S. Bludman, *Phys. Rev. D* **45**, 4720 (1992); G. Gelmini, *Neutrino Physics*, H. V. Klapdor and P. Povh, eds. (Springer-Verlag, Berlin, 1988) p. 44; E. W. Kolb and M. Turner, *The Early universe* (Addison-Wesley, Redwood City, 1989).

[18] G. Steigman and M. S. Turner, *Nucl. Phys. B* **253**, 375 (1985).

[19] K. Hirata *et al.*, *Phys. Rev. Lett.* **58**, 1490 (1987).

[20] R. M. Bonita *et al.*, *Phys. Rev. Lett.* **58**, 1494 (1987).

[21] W. D. Arnett *et al.*,*Ann. Rev. Astron. Astrophys.* **7**, 629 (1989); A. Burrows, *Ann. Rev. Nucl. Part. Sci.* **40**, 181 (1990).

[22] T. J. Loredo and D. Q. Lamb, Proceedings of *Relativistic Astrophysics*, Dallas, 1988, p. 601.

[23] A. Burrows, D. Klein, and R. Gandhi, *Phys. Rev. D* **45**, 3361 (1992).

[24] G. T. Ewan *et al.*, in *Proceedings of the Supernova Watch Workshop*, Santa Monica, California, (1990).

[25] R. G. H. Robertson *et al.*, in *Perspectives in Neutrinos, Atomic Physics, and Gravitation*, XIII'th Moriond Workshop, Villars-sur-Ollon, Switzerland, (1993).

[26] E. S. Hafen *et al.*, in *Proceedings of the Supernova Watch Workshop.*

[27] D. B. Cline *et al.*, in the International Conference on Neutrino Astrophysics,Takayama, Kamioka, Japan (1992).

[28] L. M. Krauss, P. Romanelli, D. Schramm, R. Lehrer, *Nucl. Phys. B* **380**, 507 (1992).

[29] D. B. Cline *et al.*, *Phys. Rev. D* **50**, 720 (1994).

[30] C. Pennypacker *et al.*, in *Preliminary Estimates of Core-Collapse Supernova Rates from the Berkeley Automated Supernova Search*, Lawrence Berkeley Laboratory Preprint, LBL-3059 (1991); *Ap. Astro. J* **384**, L9 (1992).

[31] M. M. Lowry, private communication.

[32] V. M. Lobashev *et al.*, Proc. 16th Int. Conf. Neutrino Physics and Astrophysics, *Neutrino '94*, Eilat, Israel, May 29 - June 3, 1994, to be published.

[33] V. A. Lyubimov, Proc. 13th Int. Conf. Neutrino Physics and Astrophysics, *Neutrino '88*, ed. J. Schneps *et al.*(World Scientific, Singapore, 1989) p. 2.

[34] R. G. H. Robertson *et al.*, *Phys. Rev. Lett.* **67**, 957 (1991).

[35] E. Holzschuh, M. Fritschi, and W. Kündig, *Phys. Lett. B* **287**, 381 (1992).

[36] H. Kawakami *et al.*, *Phys. Lett. B* **256**, 105 (1991).

[37] Ch. Weinheimer *et al.*, *Phys. Lett. B* **300**, 210 (1993).

[38] D. Decman and W. Stoeffl, submitted to *Phys. Rev. Lett.* (1994).

[39] G. J. Clark and M.A. Frisch, Proc. Santa Fe Meeting of Div. of Particles and Fields of the APS, ed. T. Goldman and M. M. Nieto (World Scientific, Singapore 1984), p. 245.

[40] M. Fink and H. Wellenstein, unpublished.

[41] B. Jeckelmann *et al.*, *Phys. Rev. Lett.* **56**, 1444 (1986).

[42] R. Abela *et al.*, *Phys. Lett. B* **146**, 431 (1984).

[43] K. Hikasa *et al.*, *Phys. Rev. D* **45**, S1 (1992).

[44] M. Daum *et al.*, *Phys. Lett. B* **265**, 425 (1991).

[45] R. Frosch, private communication (1992).

[46] B. Jeckelmann, P. F. A. Goudsmit, and H. J. Leisi, *Phys. Lett. B* **335**, 326 (1994).

[47] H. B. Anderhub *et al.*, *Phys. Lett. B* **114**, 76 (1982).
mohapatra

[48] G. B. Mills *et al.*, *Phys. Rev. Lett.* **54**, 624 (1985).

[49] P. Burchat *et al.*, *Phys. Rev. Lett.* **54**, 2489 (1985).

[50] C. Matteuzzi *et al.*, *Phys. Rev. D* **32**, 800 (1985).

[51] H. Albrecht *et al.*, *Phys. Lett. B* **163**, 404 (1985).

[52] S. Abachi *et al.*, *Phys. Rev. Lett.* **56**, 1039 (1986).

[53] S. Csorna *et al.*, *Phys. Rev. D* **35**, 2747 (1987).

[54] H. Albrecht *et al.*, *Phys. Lett. B* **202**, 149 (1988).

[55] R. Balest *et al.*, *Phys. Rev. D* **47**, 3671 (1993).

[56] D. Cinabro *et al.*, *Phys. Rev. Lett.* **70**, 3700 (1993).

[57] *Sensitivity of Future High-Luity* e^+e^- *Colliders to a Massive* τ *Neutrino*, J. J. Gomez-Cadenas, CERN-PPE/94-12, Jan. 1994.

[58] T. Bernatowicz *et al.*, *Phys. Rev. Lett.* **69**, 2341 (1992).

[59] A. Balysh *et al.*, *Phys. Lett. B* **283**, 32 (1992).

[60] S. R. Elliott *et al.*, *Phys. Rev. Lett.* **56**, 2582 (1986).

[61] V. Gribov and B. Pontecorvo, *Phys. Lett. B* **28**, 493 (1969); S. M. Bilenky and B. Pontecorvo, *Phys. Rep.* **41**, 225 (1978).

[62] L. Wolfenstein, *Phys. Rev. D* **17**, 2369 (1978); *Phys. Rev. D* **20**, 2634 (1979).

[63] S. P. Mikheyev and A. Yu. Smirnov, *Sov. J. Nucl. Phys.* **42**, 913 (1986); *Nuovo Cimento* C **9**, 17 (1986).

[64] W.-K. Kwong and S. P. Rosen, *Phys. Rev. Lett.* **73**, 369 (1994).

[65] N. Hata, S. Bludman, and P. Langacker, *Phys. Rev. D* **49**, 3622 (1994).

[66] N. Hata and P. Langacker, UPR-0625T.

[67] J. N. Bahcall, IASSNS-AST 94/37.

[68] V. Castellani *et al.*, ASTROPH-9405018, *Phys. Rev. D* **50**, 4749 (1994).

[69] S. Degl'Innocenti, G. Fiorentini, and M. Lissia, INFNFE-10-94.

[70] X. Shi, D. Schramm, and S. P. Dearborn, *Phys. Rev. D* **50**, 2414 (1994).

[71] J. N. Bahcall and M. H. Pinsonneault, *Rev. Mod. Phys.* **64**, 885 (1992) and references therein; S. Turck-Chièze and I. Lopes, *Astrophys. J.* **408**, 347 (1993).

[72] R. Davis Jr., *Frontiers of Neutrino Astrophysics*, ed. by Y. Suzuki and K. Nakamura (Universal Academy Press, Inc., Tokyo, Japan, 1993), p. 47.

[73] K. Lande, invited talk, *Solar Neutrinos and Neutrino Astrophysics*, Institute for Nuclear Theory, University of Washington, Feb. 22 - May 31, 1994 (unpublished).

[74] A. Garcia, E. G. Adelberger, P. V. Magnus, H. E. Swanson, O.Tengblad, and the Isolde Collaboration, *Phys. Rev. Lett.* **67**, 3654 (1991).

[75] K. S. Hirata, *et al.*, *Phys. Rev. Lett.* **63**, 16 (1989).

[76] K. S. Hirata, *et al.*, *Phys. Rev. D* **44**, 2241 (1991).

[77] A. Suzuki, KEK Preprint 93-96, 1993.

[78] J. N. Bahcall, *Neutrino Astrophysics* (Cambridge University Press, Cambridge, England, 1989).

[79] V. A. Kuz'min, *Sov. Phys. - JETP* **22**, 1059 (1966).

[80] J. N. Bahcall, B. T. Cleveland, R. Davis, Jr., *et al.*, *Phys. Rev. Lett.* **40**, 1351 (1978).

[81] A. I. Abazov *et al.*, *Phys. Rev. Lett.* **67**, 3332 (1991).

[82] J. N. Abdurashitov *et al.*, *Phys. Lett. B* **328**, 234 (1994).

[83] P. Anselmann *et al.*, *Phys. Lett. B* **285**, 376 (1992).

[84] P. Anselmann, *Phys. Lett. B* **314**, 445 (1993).

[85] J. S. Nico, Proc. Int. Conf. on High-Energy Physics, *ICHEP 1994*, Glasgow, July 18-29, 1994 (to be published).

[86] P. Anselmann *et al.*, *Phys. Lett. B* **327**, 377 (1994).

[87] P. Anselmann *et al.*, Gallex preprint GX65 -1994.

[88] N. Hata and P. Langacker, *Phys. Rev. D* **50**, 632 (1994); S. Bludman *et al.*, *Phys. Rev. D* **47**, 2220 (1993).

[89] X. Shi, D. Schramm, and J. Bahcall, *Phys. Rev. Lett.* **69**, 717 (1992); X. Shi and D. Schramm, *Phys. Lett. B* **283**, 205 (1992).

[90] J. M. Gelb, W.-K. Kwong and S. P. Rosen, *Phys. Rev. Lett.* **69**, 1864 (1992).

[91] P. I. Krastev and S. T. Petcov, *Phys. Lett. B* **299**, 99 (1993).

[92] G. L. Fogli, E. Lisi, and D. Montanino, *Phys. Rev. D* **49**, 3626 (1994).

[93] G. Fiorentini *et al.*, *Phys. Rev. D* **49**, 6298 (1994).

[94] P. I. Krastev and S. T. Petcov, *Phys. Rev. Lett.* **72**, 1960 (1994) and SISSA-41-94-EP, and references therein.

[95] N. Hata, UPR-0605T.

[96] A. Acker and S. Pakvasa, *Phys. Lett. B* **320**, 320 (1994).

[97] M. B. Voloshin, M. I. Vysotskii, and L. B. Okun, *Yad. Fiz.* **44**, 677 (1986) [*Sov. J. Nucl. Phys.* **44**, 440 (1986)].

[98] E. K. Akhmedov, FTUV-94-26; E. K. Akhmedov, A. Lanza, and S. T. Petcov, *Phys. Lett. B* **303**, 85 (1994), and references therein.

[99] H. H. Chen *et al.*, *Phys. Rev. Lett.* **55**, 1534 (1985); G. Aardsma *et al.*, *Phys. Lett. B* **194**, 321 (1987); G. T. Ewan *et al.*, Sudbury Neutrino Observatory Proposal SNO-87-12 (1987).

[100] A. J. Baltz and J. Weneser, BNL-60387, *Phys. Rev. D* **35**, 528 (1987).

[101] E. D. Carlson, *Phys. Rev. D* **34**, 1454 (1986).

[102] N. Hata and P. Langacker, *Phys. Rev. D* **48**, 2937 (1993), and references therein.

[103] J. Engel, S. Pittel, and P. Vogel, *Phys. Rev. C* **50**, 1702 (1994).

[104] G. Ranucci for the Borexino Collaboration, *Nucl. Phys.* (Proc. Suppl). **32**, 149 (1993); C. Arpasella *et al.*, in *Borexino at Gran Sasso: Proposal for a real-time detector for low energy solar neutrinos*, Volumes. I and II, University of Milan, INFN report (unpublished).

[105] P. Cennini *et al.*, *Nucl. Instrum. Meth. A* **345**, 230 (1994).

[106] Heron Collaboration, *Phys. Rev. Lett.* **58**, 2498 (1987); *Phys. Rev. Lett.* **68**, 2429 (1992); *Rev. Sci. Instrum.* **63**, 230 (1992); *J. Low Temp. Phys.* **93**, 709 (1993); *ibid* **93**, 785 (1993); *ibid* **93**, 715 (1993).

[107] J. Seguinot, T. Ypsilantis, and A. Zichini, *A High Rate Solar Neutrino Detector with Energy Determination*, LPC92-31, College de France, 12/8/92.

[108] M. Goodman (Soudan 2), *Nucl. Phys.* (Proc. Suppl). **38**, 337 (1995).

[109] M. Aglietta *et al.*, (NUSEX), Europhysics Letters **8**, 611 (1989).

[110] R. Becker-Szendy *et al.*, *Phys. Rev. D* **49**, 3720 (1992); and D. Casper, Ph. D. thesis, University of Michigan, 1990.

[111] Kamiokande presentation at Neutrino 1994 (Eilat).

[112] M. Circella *et al.*, *Proc. 23rd Int. Cosmic Ray Conf.* (Calgary) **4**, 503 (1993).

[113] D. Müller *et al.*, (HEAT Collaboration), *Proc. 22nd Int. Cosmic Ray Conf.* (Dublin), **2**, 177 (1991).

[114] D. H. Perkins, Astroparticle Physics **2**, 249 (1994).

[115] Y. Fukuda *et al.*, (Kamiokande collaboration), *Phys. Lett. B* **335**, 237 (1994).

[116] K. S. Hirata *et al.*, (Kamiokande Collaboration), *Phys. Lett. B* **280**, 146 (1992). This paper reported on a 4.92 kT yr exposure. The exposure for sub-GeV, fully contained events is now 7.7 kT yr [115].

[117] H. Meyer, for the Fréjus collaboration, talk presented at Snowmass, July 7, 1994.

[118] R. Becker-Szendy *et al.*, (IMB Collaboration), *Phys. Rev. D* **46**, 3720 (1992).

[119] Ch. Berger *et al.*, (Frejus), *Phys. Lett. B* **245**, 305 (1990), **227**, 489 (1989).

[120] M. Honda, K. Kasahara, K. Hidaka, and S. Midorikawa, *Phys. Lett. B* **248**, 193 (1990).

[121] R. Becker-Szendy *et al.*, *Phys. Rev. Lett.* **69**, 1010 (1992).

[122] M. Goodman, *Proceedings of the Atmospheric Neutrino Workshop*, 6-8 May 1993, LSU-HEPA-93-6.

[123] T. Kajita, *Proceedings of the Atmospheric Neutrino Workshop*, 6-8 May 1993, LSU-HEPA-93-6.

[124] D. Michael, private communication.

[125] L. V. Volkova, *Sov. J. Nucl. Phys.* **31**, 1510 (1980).

[126] E. Eichten, I. Hinchcliffe, K. Lane, and C. Quigg, Rev. Mod. Phys. **56**, 579 (1984) [**56**, 1065E' (1986)].

[127] W. Frati, T. K. Gaisser, A. K. Mann, and T. Stanev, *Phys. Rev. D* **48**, 1140 (1993).

[128] M. Mori *et al.*, *Phys. Lett. B* **270**, 89 (1991).

[129] A. V. Butkevich, L. G. Dedenko, and I. M. Zheleznykh, *Sov. J. Nucl. Phys.* **50**, 90 (1989).

[130] V. Agrawal, T. K. Gaisser, P. Lipari, and T. Stanev (unpublished).

[131] J. F. Owens, *Phys. Lett. B* **266**, 126 (1991).

[132] D. Michael for the MACRO Collaboration, talk presented at Snowmass, July 7, 1994.

[133] M. M. Boliev *et al.*, (Baksan Collaboration), in *Proc. 3rd Int. Workshop on Neutrino Telescopes*, Venice, 1991 (ed. M. Baldo-Ceolin) p. 235.

[134] G. Barr, T. K. Gaisser, and T. Stanev, *Phys. Rev. D* **39**, 3532 (1989).

[135] E. V. Bugaev and V. A. Naumov, *Phys. Lett. B* **232**, 391 (1989).

[136] J. Engel, E. Kolbe, K. Langanke, and P. Vogel, *Phys. Rev. D* **48**, 3048 (1993).

[137] G. L. Fogli, E. Lisi, and D. Montanino, *Phys. Rev. D* (to be published).

[138] The San Onofre Collaboration, *Proposal for the San Onofre Neutrino Oscillation Experiment*, Caltech, January 1994, unpublished.

[139] The Chooz Collaboration, *Proposal to Search for Neutrino Vacuum Oscillations to* $\Delta m^2 = 10^{-3}\,\mathrm{eV}^2$ *Using a 1 Km Baseline Reactor Neutrino Experiment*, Drexel University, June 1993, unpublished.

[140] The Chooz Collaboration, *Chooz Neutrino Experiment Home Page*, http://duphy4.physics.drexel.edu/chooz_pub/.

[141] F. Boehm *et al.*, *A Large, Low-Energy Neutrino Detector for Neutrino Oscillations and Supernovae Watch*, *Nucl. Instrum. Meth. A* **300** 395 (1991).

[142] R. Steinberg *et al.*, *PERRY: A Reactor Neutrino Oscillation Experiment Sensitive to* $\Delta m^2 = 10^{-4}\,\mathrm{eV}^2$, in *The FERMILAB Meeting DPF92*, ed. C.H. Albright, (World Scientific, Singapore, 1993), p. 1296.

[143] D.H. White, Invited Paper, Long-Range Planning Town Meeting on Electroweak Interactions, Astrophysics, and Non-Accelerator Experiments, Berkeley, CA, Feb. 4-5, 1995 (unpublished).

[144] KARMEN Collaboration, B. Armbruster *et al.*, *Nucl. Phys.* (Proc. Suppl). **38**, 235 (1995).

[145] CHORUS Collaboration, N. Armenise *et al.*, CERN-SPSC/90-42 (1990).

[146] NOMAD Collaboration, P. Astier *et al.*, CERN-SPLC/91-21 (1991).

[147] N. Ushida *et al.*, *Phys. Rev. Lett.* **57**, 2897 (1986).

[148] E-803, Muon Neutrino to Tau Neutrino Oscillations (October 1990); Update (october 1993), N. W. Reay, spokesperson.

[149] *Proposal for a Long Baseline Neutrino Oscillation Experiment at the AGS*, 1993, A. Mann, spokesperson.

[150] P-875: A Long-baseline Neutrino Oscillation Experiment at Fermilab, February 1995, S. Wojcicki, spokesperson.

[151] Icarus II Proposal,*A Second-generation Proton Decay Experiment and Neutrino Observatory at the Gran Sasso Laboratory*, LNGS-94/99-II, May 1994.

Electroweak Symmetry Breaking and New Physics at the TeV Scale

Tim Barklow

Stanford Linear Accelerator Center, Stanford, CA 94309

Sally Dawson

Brookhaven National Laboratory, Upton, NY 11973

Howard E. Haber

University of California, Santa Cruz, CA 95064

James Siegrist

Lawrence Berkeley Laboratory, Berkeley, CA 94720

Working Group Members

H. Aihara,	*Lawrence Berkeley Laboratory*
H. Baer,	*Florida State University*
U. Baur,	*SUNY, Buffalo*
R.S. Chivukula,	*Boston University*
M. Cvetič,	*University of Pennsylvania*
A. Djouadi,	*University of Montreal*
M. Drees,	*University of Wisconsin*
S. Errede,	*University of Illinois*
S. Godfrey,	*Carleton University*
M. Golden,	*Harvard University*
J.F. Gunion,	*University of California, Davis*
T. Han,	*University of California, Davis*
J.L. Hewett,	*Stanford Linear Accelerator Center*
C.T. Hill,	*Fermilab*
D. London,	*University of Montreal*
T.W. Markiewicz,	*Stanford Linear Accelerator Center*
S.P. Martin,	*University of Michigan*
H. Murayama,	*Lawrence Berkeley Laboratory*
J. Ng,	*TRIUMF, Vancouver BC*
S.J. Parke,	*Fermilab*
M.E. Peskin,	*Stanford Linear Accelerator Center*
H. Pois,	*University of California, Davis*
T.G. Rizzo,	*Stanford Linear Accelerator Center*
R. Rosenfeld,	*Boston University*
E.H. Simmons,	*Boston University*
A. Stange,	*Brookhaven National Laboratory*
T. Takeuchi,	*Fermilab*
X. Tata,	*University of Hawaii*
J. Terning,	*Boston University*
S. Thomas,	*Stanford Linear Accelerator Center*
S. Willenbrock,	*University of Illinois*
D. Zeppenfeld,	*University of Wisconsin*

1 Introduction

The development of the Standard Model of particle physics is a remarkable success story. Its many facets have been tested at present day accelerators; no significant unambiguous deviations have yet been found. In some cases, the model has been verified at an accuracy of better than one part in a thousand. This state of affairs presents our field with a challenge. Where do we go from here? What is our vision for future developments in particle physics? Are particle physicists' recent successes a signal of the field's impending demise, or do real long-term prospects exist for further progress? We assert that the long-term health and intellectual vitality of particle physics depends crucially on the development of a new generation of particle colliders that push the energy frontier by an order of magnitude beyond present capabilities. In this report, we address the scientific issues underlying this assertion.

Despite the success of the Standard Model as a description of the properties of elementary particles, it is clear that the Standard Model is not a fundamental theory. Apart from any new physics phenomena that might lie beyond the Standard Model, one knows that the Standard Model neglects gravity. A more complete theory of elementary particle interactions (including gravity) contains the Planck scale ($M_P \simeq 10^{19}$ GeV) as its basic energy scale. The Standard Model at best is a good low-energy approximation to this more fundamental theory of particle interactions, valid in a limited domain of "low-energies" of order 100 GeV and below. In order to extend the Standard Model to higher energies, one must address three basic questions.

1. *What is the complete description of the effective theory of fundamental particles at and below the electroweak scale?*

In the Standard Model with one physical Higgs scalar, the scale of electroweak symmetry breaking is characterized by $v = 246$ GeV, the magnitude of the Higgs vacuum expectation value. More complicated models of particle physics can be easily constructed which closely approximate the Standard Model at present day energies, but have a much richer spectrum in the mass range above 100 GeV and below a few TeV (henceforth referred to as the TeV scale). If one wishes to extrapolate physics from the electroweak scale and eventually deduce the nature of physics near the Planck scale, one must know the complete structure of the particle theory at the electroweak scale. The Standard Model possesses three generations of quarks and leptons (and no right-handed neutrinos) and an SU(3)×SU(2)×U(1) gauge group. Is that all? Might there be additional particles that populate the TeV scale? Some possibilities include:

(a) an extended electroweak gauge group (*e.g.*, extra U(1) factors, left-right (LR) symmetry as in $SU(2)_L \times SU(2)_R \times U(1)$ models, *etc.*);
(b) new fermions (*e.g.*, a fourth generation, mirror fermions, vector-like fermions, massive neutrinos, *etc.*);
(c) new bosons (CP-odd neutral and charged scalars in models with extended Higgs sectors, Goldstone or pseudo-Goldstone bosons, other exotic scalars, leptoquarks, vector resonances, *etc.*).

Clearly, in order to extrapolate our particle theories to high energy with any confidence, one must know the complete TeV-scale spectrum.

2. *What is the mechanism for electroweak symmetry breaking?*

The Standard Model posits that electroweak symmetry breaking (ESB) is the result of the dynamics of an elementary complex doublet of scalar fields. The neutral component of the scalar doublet acquires a vacuum expectation value and triggers ESB; the resulting particle spectrum contains three massive gauge bosons and the photon, massive quarks and charged leptons (and massless neutrinos) and a physical Higgs scalar. However, if one attempts to embed the Standard Model in a fundamental theory with a high energy scale (M_P), then the masses of elementary scalars naturally assume values of order the high energy scale. This is due to the quadratic sensitivity of squared scalar masses to the high-energy scale of the underlying fundamental theory. In contrast, theoretical considerations (*e.g.* unitarity) require the mass of the Standard Model Higgs boson to be of order the electroweak scale. To accomplish this, one has three choices:

(a) "Unnaturally" adjust the parameters of the high-energy theory (a fine tuning of 34 orders of magnitude in the scalar squared mass parameter is required) to produce the required light elementary scalar field.
(b) Invoke supersymmetry to cancel quadratic sensitivity of the scalar squared masses to M_P. Electroweak symmetry breaking is triggered by the dynamics of a weakly-coupled Higgs sector.
(c) Elementary Higgs scalars do not exist. A Higgs-like scalar state (if it exists) would reveal its composite nature at the TeV scale, where new physics beyond the Standard Model enters. Electroweak symmetry breaking is triggered by non-perturbative strong forces.

In order to implement (b), one must first note that supersymmetry is not an exact symmetry of nature (otherwise, all known particles would have equal-mass supersymmetric partners). If supersymmetry is to explain why the Higgs boson mass is of order the electroweak scale, then supersymmetry-breaking effects which split the masses of particles and their super-partners must be roughly of the same order as the electroweak scale. Thus, if supersymmetry is connected with the origin of ESB, one expects to discover a new spectrum of particles and interactions at the TeV scale or below. In addition, such models of "low-energy" supersymmetry are compatible with the existence of weakly-coupled elementary scalars with masses of order the ESB scale. No direct experimental evidence for the supersymmetric particle spectrum presently exists. But there is tantalizing indirect evidence. Starting from the known values of the SU(3), SU(2), and U(1) gauge couplings at m_Z and extrapolating to high energies, one finds that the three gauge couplings meet at a single point if one includes the effects of supersymmetric particles (with masses at or below 1 TeV) in the running of the couplings. Unification of couplings then takes place at around 10^{16} GeV, only a few orders of magnitude below M_P. The three gauge couplings do not meet at a single point if only Standard Model particles contribute to the running of the couplings. This could be a hint for low-energy supersymmetry, and suggests that the theory of fundamental particles remains weakly coupled and perturbative all the way up to energies near M_P.

New physics is also invoked to explain the origin of electroweak symmetry breaking in choice (c). For example, in technicolor models (which make use of the mechanism analogous to the one that is responsible for chiral symmetry breaking in QCD), electroweak symmetry breaking occurs when pairs of technifermions condense in the vacuum. One then identifies a new scale, $\Lambda_{ESB} \simeq 4\pi v \sim \mathcal{O}(1 \text{ TeV})$, where new physics beyond the Standard Model must enter. Other approaches, such as effective Lagrangian descriptions of the strongly interacting Higgs sector, preon models, top-mode condensate models, composite Higgs models, *etc.*, also fall into this category. Unfortunately, due to the presence of non-perturbative strong forces, it is often difficult to make reliable detailed computations in such models. Moreover, phenomenological difficulties inherent in the simplest examples often require additional structure (*e.g.*, an extended technicolor sector is needed in technicolor models to generate fermion masses). Unfortunately, a completely phenomenologically viable fundamental model of strongly-coupled ESB has not yet been constructed.

There is one important theoretical lesson that one should draw from the above remarks. Namely, *the nature of ESB dynamics must be revealed at or below the 1 TeV energy scale.* This is not a matter of conjecture. The relevant scale is implicitly built into the Standard Model through $v \equiv 2m_W/g$. If one wishes to take the most conservative approach, one would say that the study of the TeV scale must reveal the Higgs boson, but with no guarantees of anything beyond. However, if one takes the naturalness argument seriously, it follows that the study of the TeV scale must reveal a new sector of physics beyond the Standard Model responsible for ESB dynamics—either a super-particle spectrum or a new strongly-interacting sector of fermions, scalars, and/or vector resonances.

3. *What is the origin of the Standard Model parameters?*

I.I. Rabi asked the famous question concerning the unexpected discovery of the muon: "Who ordered that?!" In the context of the Standard Model, we ponder the generalizations of Rabi's question:

(a) Why are there three generations? [Or – what is the origin of flavor?]
(b) Where do the quark and lepton masses come from? [Or – what is the origin of the fermion mass matrices?]
(c) What is the origin of CP-violation? [Or – is the phase in the Cabibbo-Kobayashi-Maskawa (CKM) mass matrix the only source of CP-violation? and why is θ_{QCD}, which would give rise to CP-violation from the strong interactions, so small?]

Note that although the minimal supersymmetric extension of the Standard Model addresses the origin of electroweak symmetry breaking, it provides little insight into any of the above questions. In fact, one would simply add yet another question:

(d) What is the origin of the (soft)-supersymmetry-breaking parameters that characterize the model of low-energy supersymmetry?

In contrast to the problem of ESB discussed above, the Standard Model does *not* provide any clues as to the energy scale responsible for addressing the origin of the Standard Model parameters (or a supersymmetric extension thereof). The relevant energy scale could lie anywhere from 1 TeV to M_P. For example, in conventional theories of low-energy supersymmetry, the origin of the Standard Model and supersymmetric parameters lies at or near the Planck scale. In contrast, theories of extended technicolor typically attempt to solve the flavor problem at energy scales below 100 TeV. The latter approach is more ambitious; perhaps this is one reason why no compelling model of this type has been constructed. Nevertheless, any model of flavor and fermion masses requires a definite framework for electroweak symmetry breaking. Thus, the elucidation of TeV-scale physics will have significant ramifications for our understanding of physics at even higher energies.

Other future particle physics experiments may also address some of these issues. B Factories will thoroughly test the CKM-parametrization of CP-violation. Any evidence for new sources of CP-violation would constitute physics beyond the Standard Model. The next generation of neutrino mixing and mass experiments will try to verify recent experimental hints for massive neutrinos. If massive neutrinos are confirmed, then the Standard Model will require an extension. The simplest possibility is to include right-handed neutrino states. Majorana mass terms for the right-handed neutrinos are $SU(2)\times U(1)$ conserving and therefore constitute a completely new energy scale (M_R) characterizing the extended Standard Model. At present, M_R is undetermined and can lie anywhere between the electroweak scale and M_P. If $M_R \gg m_Z$, then the effect of this new scale will be simply to add the neutrino mass and mixing parameters to the list of Standard Model parameters whose origin will be difficult to discern directly from particle physics experiments in the foreseeable future.

In this sense, the search for electroweak symmetry breaking presents the unique guarantee—the dynamics that generates ESB must be revealed by the next generation of colliders that can directly explore the TeV scale. *Thus, it is crucial to explore thoroughly the TeV energy scale in order to elucidate the dynamics of electroweak symmetry breaking and the associated fundamental particle spectrum. This program must be a primary focus of particle physics research in the long-term future plans of the field.*

In this report, we explore the capabilities of the approved future facilities and a variety of hypothetical future machines, summarized in Table 1. We have also begun to study the physics opportunities at more speculative future colliders, such as a 60 TeV hadron supercollider (which would supersede the LHC) and a multi-TeV $\mu^+\mu^-$ collider, although the overall capabilities of these machines have yet to be addressed in a comprehensive fashion.

Table 1:

Name	Type	$\sqrt{s}$	Yearly $\int \mathcal{L}dt$
Approved Projects			
LEP 2	e^+e^-	~ 190 GeV	150 pb^{-1}
Tevatron (Main Injector)	$p\bar{p}$	2 TeV	1 fb^{-1}
LHC	pp	14 TeV	10–100 fb^{-1}
Tevatron Upgrades (under consideration)			
TeV*	$p\bar{p}$	2 TeV	10 fb^{-1}
DiTevatron	$p\bar{p}$	4 TeV	20 fb^{-1}
e^+e^- Linear Collider			
JLC, NLC, TESLA	e^+e^- †	0.5–1.5 TeV	50–200 fb^{-1}
	(† with $e\gamma, \gamma\gamma, e^-e^-$ options)		

In this report, we do not directly address the origin of the Standard Model parameters. Nor do we address the possibility that new and interesting physics beyond the Standard Model might populate the energy interval *above* the TeV scale. We focus directly on TeV-scale physics accessible to the next generation of colliders. Namely, we examine the requirements for completing the TeV-scale particle spectrum, and for unravelling the dynamics of electroweak symmetry breaking. Particularly in the latter case, we expect that future colliders will discover new physics beyond the Standard Model (BSM) associated with this dynamics. Ten ESB&BSM subgroups were established to address these issues. Two additional subgroups were added to examine experimental issues associated with the search for ESB&BSM physics at future hadron and e^+e^- colliders, respectively. Conveners of the twelve subgroups prepared executive summaries which form the basis of this document. More detailed expositions will be published separately in Ref. [1], which will also contain a more comprehensive list of references.

The subgroups and their conveners are listed below. Three subgroups addressed the case of weakly-coupled ESB dynamics:

- Weakly-Coupled Higgs Bosons (J.F. Gunion, A. Stange, and S. Willenbrock)
- Low-Energy Supersymmetry—Implications of Models (M. Drees, S.P. Martin, and H. Pois)
- Low-Energy Supersymmetry Phenomenology (H. Baer, H. Murayama, and X. Tata)

Two subgroups addressed the case of strongly-coupled ESB dynamics:

- Strongly-Coupled ESB Sector—Model Independent Study (M. Golden and T. Han)
- Strongly-Coupled ESB Sector—Implications of Models (R.S. Chivukula, R. Rosenfeld, E.H. Simmons, and J. Terning)

Two subgroups addressed the question of identifying the interactions and full spectrum of the TeV-scale particle theory:

- New Gauge Bosons Beyond the $W^\pm$ and Z (M. Cvetič and S. Godfrey)
- New Particles and Interactions (A. Djouadi, J. Ng and T.G. Rizzo)

Three subgroups addressed the search for ESB&BSM physics via precision measurements of Standard Model processes:

- Anomalous Gauge Boson Couplings (H. Aihara, U. Baur, D. London and D. Zeppenfeld)
- Top-Quarks as a Window to Electroweak Symmetry Breaking (S.J. Parke, C.T. Hill and M.E. Peskin)
- Virtual Effects of New Physics (J.L. Hewett, T. Takeuchi, and S. Thomas)

Two subgroups focussed on the interplay of ESB&BSM physics and detector and machine issues for future colliders:

- Experimental Issues in the Search for ESB&BSM at Future Hadron Colliders (S. Errede and J. Siegrist)
- Experimental Issues in the Search for ESB&BSM at Future e^+e^- Colliders (T. Barklow and T.W. Markiewicz)

2 Weakly-Coupled Higgs Bosons

2.1 The Standard Model and MSSM Higgs Sectors [2]

In the Standard Model (SM), the electroweak symmetry is broken by an elementary weak doublet of scalar fields, whose neutral component acquires a non-zero vacuum expectation value. This model predicts the existence of a single CP-even neutral scalar called the Higgs boson, H_{SM}, of unknown mass but with fixed couplings to other particles proportional to their masses. This is the minimal model of electroweak symmetry breaking. Detailed studies of the phenomenological profile of H_{SM} are important, since they provide significant benchmarks for any experimental search for electroweak symmetry breaking.

The underlying physics responsible for electroweak symmetry breaking is only weakly constrained by present-day experimental data. Thus, in the search for electroweak symmetry breaking, one must be open to more complicated possibilities. For example, it is easy to generalize the minimal Higgs model by simply adding additional scalar doublets. Models of this type possess additional CP-even neutral scalars, new CP-odd neutral scalars, and charged scalars. One of the simplest extensions of the SM Higgs sector is the two-Higgs-doublet model (with a CP-invariant scalar potential). The scalar spectrum of this model contains two CP-even neutral scalars, h^0 and H^0 (with $m_{h^0} \leq m_{H^0}$), one CP-odd neutral scalar, A^0, and a charged Higgs pair, $H^\pm$. The importance of this model will be clarified below.

In Section 1, a theoretical overview was presented of the possible mechanisms that could be responsible for electroweak symmetry breaking. The Standard Model with its minimal Higgs sector or with an extended Higgs sector (as described above) is theoretically unsatisfactory, since there is no "natural" way of understanding the origin of the electroweak symmetry breaking scale. Two generic solutions to this theoretical problem have been put forward. In the first solution, the Standard Model is extended to include supersymmetric partners. In the second solution, new strongly-interacting forces are invoked, whose dynamics are responsible for triggering electroweak symmetry breaking. In the former case, the Higgs bosons are elementary weakly-coupled scalars, while in the latter case, any physical scalar state is probably strongly-coupled and composite. The supersymmetric approach is treated in Sections 3 and 4, while the strongly-coupled scalar sector is considered in Sections 5 and 6.

The focus of this section is the search for weakly-coupled elementary Higgs bosons. In the Standard Model, the Higgs self-coupling is proportional to the square of the Higgs mass. For Higgs masses below (roughly) 700 GeV, H_{SM} is a weakly-coupled scalar. In supersymmetric models, the Higgs bosons are also elementary and weakly-coupled. Moreover, as reviewed in Section 3, the Higgs sector of the minimal supersymmetric extension of the Standard Model (MSSM) consists of two scalar doublets (interacting via a CP-conserving potential). Thus, the MSSM Higgs spectrum consists of h^0, H^0, A^0, and $H^\pm$, introduced above. As a result, the phenomenology of this Higgs sector is particularly important, since it introduces new kinds of Higgs phenomena that can be observed in future experiments, while being a theoretically well-motivated model for the scalar sector associated with electroweak symmetry breaking.

This section considers the prospects for the discovery of weakly-coupled Higgs bosons at present and future colliders. Both the SM Higgs boson, H_{SM}, and the scalars of the MSSM Higgs sector are considered. Some aspects of the phenomenology of h^0, H^0, A^0, and $H^\pm$ do not strongly depend on the fact that they arise in a supersymmetric model. However, the general parameter space of an arbitrary two-Higgs doublet model is quite large (with four independent mass parameters and at least two other additional dimensionless parameters). In contrast, the two-doublet Higgs sector of the MSSM possesses numerous relations among Higgs masses and couplings due to the underlying supersymmetry. In particular, the MSSM tree-level Higgs masses and couplings are determined in terms of two free parameters: m_{A^0} and $\tan\beta \equiv v_2/v_1$ [where v_2 (v_1) is the vacuum expectation value of the Higgs field that couples to up-type (down-type) fermions]. One remarkable consequence of supersymmetry is that the mass of the lightest CP-even neutral scalar, h^0, is bounded from above by a mass value of order m_Z. However, in obtaining the precise upper bound, there are important radiative corrections that cannot be neglected. Through the radiative corrections, the Higgs masses also acquire dependence on the details of the MSSM spectrum. The most important piece of these corrections depends primarily on the masses of the top-quark (taken here to be $m_t = 175$ GeV) and its supersymmetric partners [3]. To be conservative, in this section the effects of the radiative corrections have been maximized by assuming that all supersymmetric particle masses (and mass mixing parameters) are of order 1 TeV. In this case, the light CP-even Higgs boson mass is bounded by $m_{h^0} \leq m_{h^0}^{\rm max} \simeq 120$ GeV. The following properties of the MSSM Higgs sector are noteworthy. If $m_{A^0} \gg m_Z$, then $m_{A^0} \simeq m_{H^0} \simeq m_{H^\pm}$ (in addition, if $\tan\beta \gg 1$ then $m_{h^0} \simeq m_{h^0}^{\rm max}$). Moreover, the couplings of h^0 to fermions and gauge bosons are indistinguishable from those of the Standard Model Higgs boson, while the couplings $H^0W^+W^-$, H^0ZZ and ZA^0h^0 are severely suppressed. Finally, assuming $\tan\beta > 1$, the couplings of A^0 and H^0 to $b\bar{b}$ and $\tau^+\tau^-$ ($t\bar{t}$) are enhanced (suppressed). In contrast, for the case $m_{A^0} \lesssim \mathcal{O}(m_Z)$ and $\tan\beta > 1$, A^0 and h^0 exhibit the latter coupling pattern to fermions. The couplings $h^0W^+W^-$, h^0ZZ and ZA^0H^0 are severely suppressed, while the H^0 couplings to fermions and gauge bosons are SM-like.

We now outline the capabilities of present and proposed future colliders to search for the SM Higgs boson and the MSSM Higgs bosons [2]. The SM Higgs discovery limits for the most useful discovery modes are summarized in Table 2. The corresponding MSSM Higgs discovery limits are summarized in Table 3. For b-tagged modes, a tagging efficiency of 30% and purity of 1% are assumed. Since the MSSM Higgs discovery limits are rather complex, additional information is provided here to assist the reader in interpreting the results presented in Table 3. The MSSM Higgs sector is most conveniently parametrized in terms of m_{A^0} and $\tan\beta$. All values of $m_{A^0} \leq 1000$ GeV and $1 \leq \tan\beta \leq 60$ have been surveyed (as preferred in the MSSM). There is also some dependence of the discovery limits on the radiative corrections, the main effect of which is to increase the mass of h^0 above its tree-level value. The discovery regions of Table 3 have been obtained assuming that the effects of the radiative corrections are maximal, as noted above.

A few examples are now given to clarify the meaning of the discovery regions presented in Table 3. At LEP 2 at $\sqrt{s} = 176$ GeV and 500 pb^{-1} integrated luminosity, either h^0 or A^0 (or both) can be discovered via $e^+e^- \to$

Table 2: Standard Model Higgs boson discovery modes. All masses are specified in GeV.

Machine ($\sqrt{s}$, $\int \mathcal{L}\,dt$)	**Mode**	**Discovery Region**
LEP 1	$e^+e^- \to Z^* H_{SM}$	$m_{H_{SM}} \lesssim 65$
LEP 2, (176 GeV, 500 pb^{-1})	$e^+e^- \to Z H_{SM}$	$m_{H_{SM}} \lesssim 80$
LEP 2, (190 GeV, 500 pb^{-1})	$e^+e^- \to Z H_{SM}$	$m_{H_{SM}} \lesssim 95$
Tevatron, (2 TeV, 5 fb^{-1})	$p\bar{p} \to W H_{SM};\ H_{SM} \to b\bar{b}$	$m_{H_{SM}} \lesssim 60 - 80$
TeV*, (2 TeV, 30 fb^{-1})	$p\bar{p} \to W H_{SM};\ H_{SM} \to b\bar{b}$	$m_{H_{SM}} \lesssim 95$
	$p\bar{p} \to W H_{SM};\ H_{SM} \to \tau^+\tau^-$	$110 \lesssim m_{H_{SM}} \lesssim 120$
DiTeV, (4 TeV, 30 fb^{-1})	$p\bar{p} \to W H_{SM};\ H_{SM} \to b\bar{b}$	$m_{H_{SM}} \lesssim 95$
	$p\bar{p} \to ZZ \to 4\ell$	$m_{H_{SM}} \sim 200$
LHC, (14 TeV, 100 fb^{-1})	$pp \to H_{SM} \to ZZ^{(*)} \to 4\ell$	$120 \lesssim m_{H_{SM}} \lesssim 700$
	$pp \to H_{SM} \to \gamma\gamma$	$80 \lesssim m_{H_{SM}} \lesssim 150$
	$pp \to t\bar{t}H_{SM}, W H_{SM};\ H_{SM} \to \gamma\gamma$	$80 \lesssim m_{H_{SM}} \lesssim 120 - 130$
	$pp \to t\bar{t}H_{SM};\ H_{SM} \to b\bar{b}$	$m_{H_{SM}} \lesssim 100 - 110$
NLC, (500 GeV, 50 fb^{-1})	$e^+e^- \to Z H_{SM}$	$m_{H_{SM}} \lesssim 350$
	$WW \to H_{SM}$	$m_{H_{SM}} \lesssim 300$
	$e^+e^- \to t\bar{t}H_{SM}$	$m_{H_{SM}} \lesssim 120$
NLC, (1 TeV, 200 fb^{-1})	$e^+e^- \to Z H_{SM}$	$m_{H_{SM}} \lesssim 800$
	$WW \to H_{SM}$	$m_{H_{SM}} \lesssim 700$

Table 3: MSSM Higgs boson discovery modes. All masses are specified in GeV. The ordered pair $(m_{A^0}, \tan\beta)$ fixes the MSSM Higgs sector masses and couplings. The parameter regime $m_{A^0} \leq 1000$ GeV and $1 \leq \tan\beta \leq 60$ is surveyed. If a range of $\tan\beta$ values is specified below, then the first (second) number in the range corresponds to the appropriate minimal (maximal) value of m_{A^0}. See text for further clarifications.

Machine($\sqrt{s}$, $\int \mathcal{L}\,dt$)	**Mode**	**$(m_{A^0}, \tan\beta)$ Discovery Region**
LEP 1	$e^+e^- \to Z^* h^0, h^0 A^0$	$(\lesssim 25, \gtrsim 1)$ or $(\lesssim 45, \gtrsim 2)$
LEP 2 (176 GeV, 500 pb^{-1})	$e^+e^- \to Z h^0, h^0 A^0$	$(\lesssim 80, \gtrsim 1)$ or $(\lesssim 150, \lesssim 4\text{–}1)$
LEP 2 (190 GeV, 500 pb^{-1})	$e^+e^- \to Z h^0, h^0 A^0$	$(\lesssim 85, \gtrsim 1)$ or $(\gtrsim 85, \lesssim 5\text{–}1.5)$
TeV* (2 TeV, 30 fb^{-1})	$p\bar{p} \to W h^0$; $h^0 \to b\bar{b}$; 2b-tag	$(\gtrsim 130\text{–}150, \geq 1)$
DiTeV (4 TeV, 30 fb^{-1})	$p\bar{p} \to W h^0$; $h^0 \to b\bar{b}$; 2b-tag	$(\gtrsim 130\text{–}150, \geq 1)$
LHC (14 TeV, 100 fb^{-1})	$pp \to W h^0, t\bar{t}h^0$; $h^0 \to b\bar{b}$; 2, 3b-tag	$(\gtrsim 130\text{–}150, \geq 1)$
	$pp \to t\bar{t}$; $t \to H^+ b$; 1b-tag	$(\lesssim 130, \geq 1)$
	$pp \to H^0$; $H^0 \to ZZ^{(*)} \to 4\ell$	$(\lesssim 350, \lesssim 10\text{–}2)$
	$pp \to h^0, W h^0, t\bar{t}h^0$; $h^0 \to \gamma\gamma$	$(\gtrsim 180, \geq 1)$
	$pp \to b\bar{b}A^0, H^0$; $A^0, H^0 \to \tau^+\tau^-$	$(\gtrsim 100, \gtrsim 5\text{–}50)$
	$pp \to b\bar{b}A^0, H^0$; $A^0, H^0 \to \mu^+\mu^-$	$(\gtrsim 100, \gtrsim 5\text{–}50)$
	$pp \to b\bar{b}h^0$; $h^0 \to b\bar{b}$; 3b-tag	$(\lesssim 125, \gtrsim 5\text{–}10)$
	$pp \to b\bar{b}H^0$; $H^0 \to b\bar{b}$; 3b-tag	$(\gtrsim 125, \gtrsim 10\text{–}60)$
	$pp \to b\bar{b}A^0$; $A^0 \to b\bar{b}$; 3b-tag	$(\lesssim 125, \gtrsim 5\text{–}10)$ or $(\gtrsim 125, \gtrsim 10\text{–}60)$
NLC (500 GeV, 50 fb^{-1})	$e^+e^- \to Z h^0$	visible unless $(\lesssim 90, \gtrsim 7)$
	$WW \to h^0$	visible unless $(\lesssim 80, \gtrsim 12)$
	$e^+e^- \to h^0 A^0$	$(\lesssim 120, \geq 1)$
	$e^+e^- \to Z H^0, WW \to H^0$	$(\lesssim 140, \geq 1)$
	$e^+e^- \to H^0 A^0$	$(\lesssim 220, \geq 1)$, unless $(\lesssim 90, \gtrsim 7)$
	$e^+e^- \to H^+ H^-$	$(\lesssim 230, \geq 1)$
NLC (1 TeV, 200 fb^{-1})	$e^+e^- \to H^0 A^0$	$(\lesssim 450, \geq 1)$, unless $(\lesssim 90, \gtrsim 7)$
	$e^+e^- \to H^+ H^-$	$(\lesssim 450, \geq 1)$

Zh^0, h^0A^0 if $m_{A^0} \lesssim 80$ GeV and $\tan\beta \gtrsim 1$. A second discovery region also exists, beginning at $\tan\beta \lesssim 4$ (at $m_{A^0} = 80$ GeV) and ending at $m_{A^0} \lesssim 150$ GeV (at $\tan\beta = 1$). If $\sqrt{s}$ is increased to 190 GeV, both discovery regions become larger; in particular, the latter region now begins at $\tan\beta \lesssim 5$ (at $m_{A^0} = 85$ GeV) and ends at $\tan\beta \lesssim 1.5$ for the maximal value of m_{A^0} considered. At the NLC-500, h^0 can be detected via $e^+e^- \to Zh^0$ unless $\tan\beta \gtrsim 7$ and $m_{A^0} \lesssim 90$ GeV. A similar discovery region exception appears for $WW \to h^0$. The reason such restrictions arise is that in the indicated region of parameter space, H^0 is SM-like in its couplings, while the h^0ZZ and $h^0W^+W^-$ couplings are very suppressed. The ZH^0A^0 coupling is likewise suppressed in the same parameter regime, which explains the other two discovery region exceptions listed in Table 3.

2.2 A Tour of Higgs Search Techniques at Future Colliders

The goal of the Higgs search at present and future colliders is to examine the full mass range of the SM Higgs boson, and the full parameter space of the MSSM Higgs sector. The LHC can cover the entire range of SM Higgs boson masses from the upper limit of LEP 2 ($m_{H_{SM}} = 80$–95 GeV, depending on machine energy) to a Higgs mass of 700 GeV (and beyond). The most difficult region for high luminosity hadron colliders is the intermediate-mass range $m_{H_{SM}} = 80$–130 GeV. The LHC detectors are being designed with the capability of fully covering the intermediate Higgs mass region. The TeV* and DiTevatron may also have some discovery potential in this mass region. In contrast, the intermediate mass Higgs search at the NLC (which makes use of the same search techniques employed at LEP 2) is straightforward. At design luminosity, the NLC discovery reach is limited only by the center-of-mass energy of the machine.

In the search for the Higgs bosons of the MSSM, two objectives are:

- the discovery of h^0 and
- the discovery of the non-minimal Higgs states (H^0, A^0, and $H^\pm$).

Two theoretical results play a key role in the MSSM Higgs search. First, the mass of h^0 is bounded by $m_{h^0}^{\max} \simeq 120$ GeV, as noted above. Second, if the properties of h^0 deviate significantly from the SM Higgs boson, then $m_{A^0} \lesssim \mathcal{O}(m_Z)$ and the H^0 and $H^\pm$ masses must lie in the intermediate Higgs mass region. As a result, experiments that are sensitive to the intermediate Higgs mass region have the potential for detecting at least one of the MSSM Higgs bosons over the entire MSSM Higgs parameter space (parametrized by $\tan\beta$ and m_{A^0}, as noted above). LEP 2 does not have sufficient energy to cover the entire MSSM Higgs parameter space, since $m_{h^0}^{\max}$ lies above the LEP 2 Higgs mass reach. The TeV*, DiTevatron and LHC all possess the capability of detecting Higgs bosons in the intermediate mass range, and consequently can probe regions of the MSSM Higgs parameter space not accessible to LEP 2. As of this writing, no set of experiments at any future hadron collider considered in this report can guarantee the discovery (or exclusion) of at least one MSSM Higgs boson for *all* values of m_{A^0} and $\tan\beta$ not excluded by the LEP 2 search. However, it may be possible to close this gap in the MSSM parameter space if sufficiently reliable b quark identification in hadron collider events is possible. On the other hand, because of the relative simplicity of the NLC Higgs search in the intermediate mass region, the NLC is certain to discover at least one MSSM Higgs boson (either h^0 or H^0) if the supersymmetric approach is correct. If h^0 is discovered and proves to be SM-like in its properties, then one must be in the region of MSSM Higgs parameter space where the non-minimal Higgs states are rather heavy and approximately degenerate in mass. In this case, the LHC may not be capable of discovering any Higgs bosons beyond h^0, while H^0, A^0 and $H^\pm$ can be detected at the NLC only if its center-of-mass energy $\sqrt{s} \gtrsim 2m_{A^0}$.

Below is a detailed description of the Higgs potential of present and future colliders. A Higgs boson is deemed observable if a 5σ-excess of events can be detected in a given search channel.

- LEP 1

The current lower bound on the SM Higgs mass is 64.5 GeV, and will increase by at most a few GeV. The lower bounds on the MSSM Higgs masses are $m_{h^0} > 44.5$ GeV, $m_{A^0} > 24.3$ GeV (for $\tan\beta > 1$), and $m_{H^\pm} > 45$ GeV.

- LEP 2

For $\sqrt{s} = 176$ GeV, the SM Higgs is accessible via ZH_{SM} production up to 80 GeV for an integrated luminosity of 500 pb^{-1} [4]. For $\sqrt{s} = 190$ GeV, $m_{H_{SM}}$ values as high as m_Z can be probed with 200 pb^{-1} of data, using b-tagging in the region $m_{H_{SM}} \sim m_Z$. The reach for $\sqrt{s} = 205$ GeV is about 105 GeV. In the MSSM, the same mass reaches apply to h^0 if $m_{A^0} \gtrsim m_Z$, since in this case Zh^0 is produced at about the same rate as ZH_{SM}. If $m_{A^0} \lesssim m_Z$, then the h^0ZZ coupling (which controls the Zh^0 cross section) becomes suppressed while the ZA^0h^0 coupling becomes maximal. In the latter parameter region, $e^+e^- \to A^0h^0$ can be detected for all values of $m_{A^0} \lesssim \sqrt{s}/2 - 10$ GeV, assuming that $\tan\beta > 1$. Since $m_{h^0} \lesssim 120$ GeV, increasing the LEP 2 energy to roughly $\sqrt{s} = 220$ GeV while maintaining the luminosity would be be sufficient to guarantee the detection of at least one Higgs boson (via Zh^0 and/or A^0h^0 final states) over the entire MSSM Higgs parameter space [4,5].

- Tevatron ($\sqrt{s} = 2$ TeV, $\mathcal{L} = 10^{32}$ cm^{-2} s^{-1} with the Main Injector)

The most promising mode for the SM Higgs is WH_{SM} production, followed by $H_{SM} \to b\bar{b}$ [6]. With 5 fb^{-1}

of integrated luminosity, this mode could potentially explore a Higgs mass region of 60–80 GeV, a region which will already have been covered by LEP II via the ZH_{SM} process.

- TeV* ($\sqrt{s} = 2$ TeV, $\mathcal{L} \geq 10^{33}$ cm^{-2} s^{-1})

The Higgs mass reach in the WH_{SM} mode, with $H_{SM} \to b\bar{b}$, is extended over that of the Tevatron [6,7]. With 30 fb^{-1} of integrated luminosity, a Higgs of mass 95 GeV is potentially accessible, but the peak will not be separable from the WZ, $Z \to b\bar{b}$ background. The mode $(W,Z)H_{SM}$, followed by $H_{SM} \to \tau^+\tau^-$ and $W, Z \to jj$ (where j stands for a hadronic jet), is potentially viable for Higgs masses sufficiently far above the Z mass [7]. With 30 fb^{-1} of integrated luminosity, a Higgs in the mass range 110 – 120 GeV may be detectable. However, the Higgs peak will be difficult to recognize on the slope of the much larger Zjj, $Z \to \tau^+\tau^-$ background. Both of these modes are of particular significance for h^0. For $\tan\beta > 1$ and $m_{A^0} \lesssim 1.5 m_Z$, the enhanced coupling of the h^0 to b's and τ's makes it unobservable at the LHC via $h^0 \to \gamma\gamma$ or $h^0 \to ZZ^* \to 4\ell$.

- DiTevatron ($\sqrt{s} = 4$ TeV, $\mathcal{L} \geq 10^{33}$ cm^{-2} s^{-1})

For the SM Higgs, the "gold-plated" mode, $H_{SM} \to ZZ \to 4\ell$, requires about 30 fb^{-1} for the optimal Higgs mass in this mode, $m_{H_{SM}} = 200$ GeV [6]. The 4ℓ mode is not useful for any of the MSSM Higgs bosons. The Higgs mass reach in the WH_{SM}, $H_{SM} \to b\bar{b}$ mode is only marginally better than at the TeV*, due to the increase in the top-quark backgrounds relative to the signal [6,7]. The mode $(W,Z)H_{SM}$, with $H_{SM} \to \tau^+\tau^-$ and $W, Z \to jj$, has less promise than at the TeV* due to the relative increase in the background [7].

- LHC ($\sqrt{s} = 14$ TeV, $\mathcal{L} = 10^{33}$–10^{34} cm^{-2} s^{-1})

For the SM Higgs boson, the "gold-plated" mode, $H_{SM} \to ZZ^{(*)} \to 4\ell$, including the case where one Z boson is virtual, covers the range of Higgs masses 130–700 GeV and beyond with 100 fb^{-1} [8,9]. For $m_{H_{SM}} > 700$ GeV, the Higgs is no longer "weakly coupled", and search strategies become more complex — see Section 5.

The CMS detector is planning an exceptional electromagnetic calorimeter ($PbWO_4$ crystal) which will enable the decay mode $H_{SM} \to \gamma\gamma$ to cover the Higgs mass range 85–150 GeV with 100 fb^{-1} of integrated luminosity [9]. This range overlaps the reach of LEP II (with $\sqrt{s} = 190$ GeV) and the lower end of the range covered by the gold-plated mode. The ATLAS detector covers the range 110-140 GeV with this mode [8]. Both CMS and ATLAS find that the modes $t\bar{t}H_{SM}$ and WH_{SM}, with $H_{SM} \to \gamma\gamma$, are viable in the mass range 80–120 GeV with 100 fb^{-1} of integrated luminosity [8,9]. Since backgrounds are smaller, these modes do not require such excellent photon resolutions and jet-photon discrimination as does the inclusive $\gamma\gamma$ mode. CMS has also studied the production of the Higgs in association with two jets, followed by $H_{SM} \to \gamma\gamma$, and concludes that this mode covers the Higgs mass range 70–130 GeV [9].

The modes $t\bar{t}H_{SM}$ and WH, with $H_{SM} \to b\bar{b}$, are potentially useful in the intermediate-mass region [6,7]. The reach with 100 fb^{-1} of integrated luminosity is 100 GeV, reduced to 80 GeV with 30 fb^{-1} [8]. Overall, ATLAS will cover the Higgs mass region 80–140 GeV with 100 fb^{-1} using a combination of $H_{SM} \to \gamma\gamma$; $t\bar{t}H_{SM}$, WH_{SM} with $H_{SM} \to \gamma\gamma$; and $t\bar{t}H_{SM}$ with $H_{SM} \to b\bar{b}$.

The main search mode for the lightest MSSM Higgs boson is $h^0 \to \gamma\gamma$, which is viable when m_{h^0} is near $m_{h^0}^{\max}$ [8,9,5,10,11]. The processes $t\bar{t}h^0$ and Wh^0, with $h^0 \to b\bar{b}$, can potentially extend h^0 detection to the somewhat lower m_{h^0} values (corresponding to somewhat lower m_{A^0}) that would assure that at least one MSSM Higgs boson can be detected over all of the MSSM parameter space [6]. Adapting the ATLAS study for the SM Higgs to h^0 suggests that the required sensitivity could be achieved in the $t\bar{t}h^0$ mode. Due to the large top-quark background, the Wh^0 with $h^0 \to b\bar{b}$ mode at the LHC has little advantage over this mode at the TeV* or the DiTevatron for the same integrated luminosity.

The other MSSM Higgs bosons are generally more elusive. There are small regions of parameter space in which H^0 can be observed decaying to $\gamma\gamma$ or $ZZ^{(*)}$ [8,9]. The possibility exists that the charged Higgs can be discovered in top decays [8,9]. For $m_{A^0} > 2m_Z$ and large $\tan\beta$, H^0 and A^0 can have sufficiently enhanced b quark couplings that they would be observed when produced in association with $b\bar{b}$ and decaying via $H^0, A^0 \to \tau^+\tau^-$ [8,9,11,6]. For very enhanced couplings of the Higgs bosons to b quarks (which occurs for very large $\tan\beta$), the modes $b\bar{b}(h^0, H^0, A^0)$, with $h^0, H^0, A^0 \to b\bar{b}$, and $\bar{t}bH^+$, with $H^+ \to t\bar{b}$, are potentially viable [6].

Let us suppose that the modes involving b-tagging (especially $t\bar{t}h^0$ production with $h^0 \to b\bar{b}$) eventually prove to be viable. Then, the LHC Higgs search using the b-tagging modes in combination with the 4ℓ and $\gamma\gamma$ final state modes should be capable of discovering at least one of the MSSM Higgs bosons (or in absence of a discovery rule out the MSSM) [6]. For much of parameter space only the h^0 will be detectable. Detection of all the MSSM Higgs bosons at the LHC is only possible for very small regions of parameter space at moderate values of m_{A^0}.

- e^+e^- linear collider ($\sqrt{s} = 500$ GeV to 1 TeV, $\mathcal{L} \geq 5 \times 10^{33}$ cm^{-2} s^{-1})

At $\sqrt{s} = 500$ GeV with an integrated luminosity of 50 fb^{-1}, the SM Higgs boson is observable via the ZH_{SM} process up to $m_{H_{SM}} = 350$ GeV [12]. Employing the same process, a Higgs boson with $m_{H_{SM}} = 130$ GeV (200 GeV) would be discovered with 1 fb^{-1} (10 fb^{-1}) of data. Other Higgs production mechanisms are also useful. With 50 fb^{-1} of data, the WW-fusion process is observable up to $m_{H_{SM}} = 300$ GeV [12,13,14]. The $t\bar{t}H_{SM}$

process is accessible for $m_{H_{SM}} \lesssim 120$ GeV, thereby allowing a direct determination of the $t\bar{t}H_{SM}$ Yukawa coupling [13]. The $\gamma\gamma$ collider mode of operation does not extend the reach of the machine for the SM Higgs, but does allow a measurement of the $H_{SM} \to \gamma\gamma$ partial width up to $m_{H_{SM}} = 300$ GeV [13].

In the MSSM, for $m_{A^0} \lesssim m_Z$, the h^0 and A^0 will be detected in the h^0A^0 mode, the H^0 will be found via ZH^0 and WW-fusion production, and H^+H^- pair production will be kinematically allowed and easily observable [15,13,14]. At higher values of the parameter m_{A^0}, the lightest Higgs boson is accessible in both the Zh^0 and WW-fusion modes. However, the search for the heavier Higgs boson states is limited by machine energy. The mode H^0A^0 is observable up to $m_{H^0} \sim m_{A^0} \sim 220$ GeV, and H^+H^- can be detected up to $m_{H^\pm} = 230$ GeV [15,13,14]. The $\gamma\gamma$ collider mode could potentially extend the reach for H^0 and A^0 up to 400 GeV if $\tan\beta$ is not large [5].

At $\sqrt{s} = 1$ TeV with an integrated luminosity of 200 fb^{-1}, the SM Higgs boson can be detected via the WW-fusion process up to $m_{H_{SM}} = 700$ GeV [13,14]. In the MSSM, H^0A^0 and H^+H^- detection would be extended to $m_{H^0} \sim m_{A^0} \sim m_{H^\pm} \sim 450$ GeV [5,15,13,14]. Measurements of cross sections, branching ratios, and angular distributions of the Higgs events will determine model parameters and test the underlying theory of the scalar sector [13].

- Influence of MSSM Higgs decays into supersymmetric particle final states

For some MSSM parameter choices, the h^0 can decay primarily to invisible modes, including a pair of the lightest supersymmetric neutralinos or a pair of invisibly decaying sneutrinos [6]. The h^0 would still be easily discovered at e^+e^- colliders in the Zh^0 mode using missing-mass techniques [13]. At hadron colliders, the Wh^0 and $t\bar{t}h^0$ modes may be detectable via large missing energy and lepton plus invisible energy signals, but determination of m_{h^0} would be difficult [6]. The H^0, A^0, and $H^\pm$ decays can be dominated by chargino and neutralino pair final states and/or slepton pair final states [2,5,10,6]. Such modes can either decrease or increase the chances of detecting these heavier MSSM Higgs bosons at a hadron collider, depending upon detailed MSSM parameter choices [5,10,6]. At e^+e^- colliders H^0, A^0 and $H^\pm$ detection up to the earlier quoted (largely kinematical) limits should in general remain possible.

- Hadron collider beyond the LHC

It could provide increased event rates and more overlap of discovery modes for the SM Higgs boson. However, its primary impact would be to extend sensitivity to high mass signals associated with strongly-coupled electroweak symmetry breaking scenarios (see Section 5). In the case of the MSSM Higgs bosons, the increased event rates would significantly extend the overlap of the various discovery modes, expanding the parameter regions where several and perhaps all of the MSSM Higgs bosons could be observed.

- e^+e^- collider beyond $\sqrt{s} = 1$ TeV

Such an extension in energy is not required for detecting and studying a weakly-coupled SM Higgs bosons, but could be very important for a strongly-coupled electroweak-symmetry-breaking scenario. Detection of H^0A^0 and H^+H^- production in the MSSM model would be extended to higher masses.

2.3 Precision Measurements of Higgs Properties

Detailed studies of the properties of Higgs bosons can be carried out at e^+e^- colliders. The measurements of cross sections, branching ratios, and angular distributions of Higgs events will determine model parameters and test the underlying theory of the scalar sector. In this discussion, h denotes the lightest CP-even neutral Higgs. It may be H_{SM} or h^0 (of the MSSM), or perhaps a scalar arising from a completely different model. Once discovered, determining the identity of h is precisely what one hopes to be able to address based on the considerations presented here. Below we list some examples of the measurements that can be made at $\sqrt{s} = 500$ GeV with an integrated luminosity of 50 fb^{-1}, assuming that the Higgs mass is less than 350 GeV and that the process $e^+e^- \to Zh$ occurs at approximately the Standard Model rate.

The single event Higgs mass resolution using $e^+e^- \to Zh$ is approximately 4 GeV [14]. The number of Higgs events provided by an integrated luminosity of 50 fb^{-1} will then allow the determination of the central value of the Higgs mass with a statistical accuracy of $\pm$ 180 MeV. Since most Standard Model parameters only have logarithmic dependence on the Higgs mass, this accuracy is far greater than what is needed to check the self-consistency of the Standard Model. One could use this mass determination to make accurate predictions for the Higgs production cross sections and branching fractions.

For a SM Higgs boson, signals can be selected with small backgrounds and one can easily measure the angular distributions of the Higgs production direction in the process $e^+e^- \to Zh$ and the directions of the outgoing fermions in the Z rest frame [14], and verify the expectations for a CP-even Higgs boson [16]. For a more general Higgs eigenstate, since only the CP-even part couples at tree-level to ZZ, these measurements would still agree with the CP-even predictions unless the Higgs is primarily (or purely) CP-odd, in which case the cross section will be much smaller and the event rate inadequate to easily verify the mixed-CP or CP-odd distribution forms. Analysis of photon polarization asymmetries in $\gamma\gamma \to h$ production rates, and of angular correlations among secondary decay products arising from primary $h \to \tau\tau$ or

$h \to t\bar{t}$ final state decays in Zh production, can both provide much better sensitivity to a CP-odd Higgs component in many cases [17].

By choosing a decay channel whose reconstruction is independent of the Higgs mass and decay mode, such as $e^+e^- \to Zh$, where $Z \to \ell^+\ell^-$, one can measure the total cross section for Higgs production. For an integrated luminosity of 50 fb^{-1}, this can be determined with a statistical accuracy of better than 10% [14]. This is one possible technique for measuring the hZZ coupling.

A specific study [18] has been conducted to consider the experimental problems of determining the production cross section σ times the branching fractions of a discovered Higgs boson into all of its possible (Standard Model) final states. The resulting expected statistical precisions are strong functions of the mass of the Higgs, as the branching fractions change dramatically in the mass range 100 GeV$< m_h <$ 180 GeV. For a Standard Model Higgs of mass 120 GeV, it is found that statistical precisions of 7%, 14%, 39%, and 48% can be achieved for $\sigma \times BR(h \to b\bar{b})$, $\sigma \times BR(h \to \tau^+\tau^-)$, $\sigma \times BR(h \to c\bar{c} + gg)$, and $\sigma \times BR(h \to WW^*)$ respectively.

Reasonably precise measurements of certain $\sigma \times BR$ combinations will also be possible at hadron colliders. For example, at the LHC a good determination of $\sigma(gg \to h) \times BR(h \to \gamma\gamma)$ should be possible for a SM-like Higgs in the 80 to 150 GeV mass range, and of $\sigma(gg \to h) \times BR(h \to ZZ)$ for $m_h > 130$ GeV.

3 Low Energy Supersymmetry: Implications of Models

3.1 *Elements of Minimal Low-Energy Supersymmetry*

A truly fundamental theory of particles and their interactions must account for the large hierarchy of energy scales inherent in the theory—from the scale of electroweak symmetry breaking [characterized by $(\sqrt{2}G_F)^{-1/2}$ = 246 GeV] up to energies as large as the Planck scale (of order 10^{19} GeV). Supersymmetry (SUSY) provides a mechanism for stabilizing the large hierarchy of scales, and thus is a natural framework for addressing such problems. The simplest example of such an approach is the minimal supersymmetric extension of the Standard Model (MSSM). The MSSM possesses a new class of particles—supersymmetric partners of the quarks, leptons, gauge and Higgs bosons of the Standard Model. A particle and its supersymmetric partner would be degenerate in mass if supersymmetry were an exact symmetry of nature. Since this degeneracy is not realized in nature, supersymmetry must be broken. But, if the breaking is "soft" (with supersymmetry-breaking masses of order 1 TeV or less), then the stability of the hierarchy can be successfully maintained.

Unification of gauge couplings provides an intriguing hint for an underlying supersymmetric theory of particle interactions. If one extrapolates the SU(3)×SU(2)×U(1) gauge couplings from their low-energy values measured at LEP to very large energy scales, one finds that the three couplings meet at approximately the same energy scale if the renormalization group equations (RGE's) of the MSSM are used [19]. In contrast, if Standard Model RGE's are used, unification of gauge couplings fails by more then seven standard deviations! As a result, it is possible to construct a model of particle interactions such that:

- the strong and electroweak forces are unified at very large energies (at a scale $M_X \sim 2 \times 10^{16}$ GeV, which is not too far from the Planck scale, $M_P = 10^{19}$ GeV);
- the couplings of all particles with non-trivial SU(3)×SU(2)×U(1) quantum numbers are perturbative in strength at all energies below the Planck scale; and
- the stability of the large hierarchy between the electroweak and grand unified energy scales is guaranteed by the underlying supersymmetry.

In this approach, the origin of the electroweak scale is tied to the scale of supersymmetry breaking. In particular, in the models considered later in this section, electroweak symmetry breaking is generated through radiative corrections. Briefly stated, the squared Higgs mass parameter is positive at the high energy scale, M_X. However, due to the one-loop radiative effects generated by the large top quark Higgs Yukawa coupling (*i.e.*, as a consequence of the large top quark mass), the squared Higgs mass parameter is driven negative in its evolution from M_X down to the low energy electroweak scale [20]. These effects trigger electroweak symmetry breaking at the proper mass scale (of order $v = 246$ GeV) and establish the large hierarchy of scales between M_X and m_Z.

Although the electroweak symmetry breaking scale has now been related to the supersymmetry breaking scale, one has not yet explained where the supersymmetry-breaking masses come from. As noted above, these masses must be of order 1 TeV or less in order to generate the electroweak symmetry breaking scale without an excessive ("unnatural") fine-tuning of model parameters. The fundamental origin of supersymmetry breaking is one of the most important unsolved problems of supersymmetry theory. Nevertheless, one can take a purely phenomenological approach and parametrize the MSSM in terms of a general set of a priori unknown supersymmetry breaking parameters. Ultimately, these parameters would be measured in experiment once supersymmetric particles are discovered. A detailed knowledge of these parameters could provide important clues as to their origins.

A brief synopsis of the parameters of the MSSM is now presented. The SUSY-conserving part of the MSSM is fixed by the three gauge coupling constants of the Standard Model, the Higgs-fermion Yukawa couplings (denoted by h_U, h_D and h_E below), and one new supersymmetric Higgs mass parameter μ. The particle content of the MSSM consists of three families of quarks and leptons and their scalar partners; the vector gauge bosons and their fermionic gaugino partners corresponding to the gauge group SU(3)×SU(2)×U(1); and *two* SU(2)-doublet Higgs scalars and their fermionic higgsino partners. The introduction of a second Higgs doublet is necessary in order for the gauge anomalies associated with the higgsinos to cancel, and to give masses to both "up-type" quarks and "down-type" quarks and charged leptons (the latter cannot be done with one Higgs scalar without violating SUSY). As a result, the Higgs sector of the MSSM is a two-Higgs doublet model, whose phenomenology was discussed in great detail in Section 2. The colorless gauginos and higgsinos with the same electric charge can mix. The resulting mass eigenstates are called charginos and neutralinos, with corresponding electric charges ± 1 and 0, respectively. With the particle content and the Higgs-fermion Yukawa couplings fixed as described above, all SUSY-preserving interactions (which correspond to all the dimensionless couplings of the model) are determined by the SU(3)×SU(2)×U(1) gauge invariance and/or by the supersymmetry.

However, in contrast to the Standard Model, the Higgs-fermion Yukawa couplings do not exhaust all possible scalar–fermion interactions. The MSSM possesses new scalars—squarks and sleptons—which carry baryon number (B) and lepton number (L), respectively. Consequently, it is possible to write down additional scalar–fermion interaction terms which violate either B or L (but preserve supersymmetry). In the MSSM, one simply assumes that these terms are not present. This can be implemented with a discrete symmetry, called R-parity which assigns $(-1)^{2S+3(B-L)}$ to particles of spin S [*i.e.*, +1 to all Standard Model particles and −1 to their supersymmetric partners], and is multiplicatively conserved by all interactions of the model. R-parity conservation guarantees the existence of a stable lightest supersymmetric particle (LSP), which is an excellent candidate for dark matter. Although there are some strong phenomenological constraints on R-parity violating interactions, it is possible to construct R-violating models which are not inconsistent with present data. Such models lie outside the scope of the MSSM, and will be briefly mentioned at the end of this section.

The SUSY-breaking part of the MSSM is where most of the new parameters of the MSSM lie. In addition to breaking the mass degeneracy between particles and their supersymmetric partners, the SUSY-breaking terms are required in order to break the SU(2)×U(1) electroweak gauge symmetry. The Higgs potential, with soft-SUSY breaking mass terms included, can have a non-trivial minimum. Both neutral Higgs fields acquire vacuum expectation values; this introduces an important new parameter, $\tan\beta \equiv v_2/v_1$, where $v_i \equiv \langle H_i^0\rangle$. Note that after electroweak symmetry breaking, the Yukawa coupling matrices are proportional to the corresponding quark and lepton mass matrices. (In the limit where only third generation Yukawa couplings are considered, $h_\tau = \sqrt{2}m_\tau/(v\cos\beta)$, $h_b = \sqrt{2}m_b/(v\cos\beta)$, and $h_t = \sqrt{2}m_t/(v\sin\beta)$, where $v = (v_1^2+v_2^2)^{1/2} = 246$ GeV.)

Supersymmetric breaking parameters are not completely arbitrary, since they must preserve the stability of the hierarchy discussed above. Such terms, called "soft", were classified in Ref. [21], and consist of masses for scalars (but not their fermionic partners), masses for the gauginos, and trilinear scalar couplings. If there were no reason for the scalar masses and trilinear couplings to be diagonal in the same basis as the Yukawa couplings, then there would be a huge number (> 100) of possible new SUSY-breaking parameters. In models based on supergravity [22], a dramatic simplification occurs. SUSY is presumed to be broken in a "hidden sector" of particles which have only gravitational interactions with the "visible sector" of quarks, leptons, gauge and Higgs bosons, and their supersymmetric partners. The SUSY breaking is then transmitted to the visible sector by gravitational interactions, which results in the soft-SUSY breaking terms of the MSSM, characterized by a mass scale $M_{\rm SUSY} \lesssim \mathcal{O}(1~{\rm TeV})$. Moreover, in the minimal approach, the soft-SUSY breaking terms are taken to be flavor independent and can be summarized in terms of a few universal parameters, regardless of the unknown physics in the hidden sector. The SUSY-breaking scalar interactions which arise take the form

$$\begin{aligned}\mathcal{L}_{\rm soft} &= m_0^2\Big[|H_1|^2+|H_2|^2 \\ &+ \sum_{\rm families}(|Q_L|^2+|U_R|^2+|D_R|^2+|L_L|^2+|E_R|^2)\Big] \\ &+ A_0(h_UQ_LH_2U_R+h_DQ_LH_1D_R+h_EL_LH_1E_R) \\ &+ B\mu H_1H_2 + \text{h.c.} \qquad (1)\end{aligned}$$

This simple form of the SUSY-breaking Lagrangian, which depends on three new SUSY-breaking parameters: m_0^2, A_0, and B, holds at a mass scale close to the Planck mass. In addition, three soft-SUSY-breaking Majorana mass terms can be introduced for the fermionic gaugino partners of the SU(3)×SU(2)×U(1) gauge bosons, respectively. If one adds the additional assumption of minimal grand unification at a high energy scale M_X, then one finds a further reduction of parameters. Both the gauge couplings and the gaugino masses unify at M_X. There may be additional relations among the Yukawa couplings, although these depend on the details of the superheavy particle spectrum and their interactions associated with the grand unified gauge group. The supersymmetry model parameters can then be taken to be:

- the common mass parameter $m_{1/2}$ for the gauginos,
- a common scalar mass parameter m_0,
- a common trilinear scalar coupling parameter A_0,
- the ratio of the Higgs vacuum expectation values, $\tan\beta$, and
- the sign of μ.

(The parameters $|\mu|$ and B can be eliminated in favor of $\tan\beta$ and m_Z by minimizing the scalar potential.) Here $m_{1/2}$, m_0, and A_0 are all typically of order $M_{\rm SUSY}$.

To find the Lagrangian relevant for physics at the electroweak scale, one must evolve the parameters of the model according to their renormalization group equations from M_X down to m_Z. This splits the gaugino masses and the various scalar masses at low energies. The most striking result of this running is that radiative corrections due to the large top Yukawa coupling push the squared mass of H_2 negative at scales far below M_X, resulting in electroweak symmetry breaking [20]. This is a remarkable effect, connected with the large top quark mass, and exhibits the intimate relation between low-energy supersymmetry and electroweak symmetry breaking. The large top quark mass also has an impact on a number of other MSSM parameters via the effects of the large top-quark Yukawa coupling in the RGE's. For example, one finds that the squared masses of the top squarks and left-handed bottom squark are reduced relative to other squark masses, while the mass of the lightest CP-even neutral Higgs boson may be substantially increased above m_Z.

3.2 Features of the Superpartner Spectrum

With the masses of all the superpartners determined only by the parameters ($m_{1/2}$, m_0, A_0, $\tan\beta$, sign μ), the spectrum is highly constrained. With a universal m_0, squarks (or sleptons) with the same gauge quantum numbers will remain degenerate even at low energies, unless they have large Yukawa couplings. This degeneracy is a success of the model, since it automatically maintains the suppression of FCNCs in low-energy physics. (This remains true even if one relaxes the usual minimal assumptions by introducing a larger gauge group at some intermediate scale or allowing the gaugino mass parameters to be independent.) So the superpartner spectrum will contain seven distinct groups of degenerate scalar states ($\tilde{u}_L, \tilde{c}_L$); ($\tilde{d}_L, \tilde{s}_L$); ($\tilde{u}_R, \tilde{c}_R$); ($\tilde{d}_R, \tilde{s}_R, \tilde{b}_R$); ($\tilde{e}_L, \tilde{\mu}_L, \tilde{\tau}_L$); ($\tilde{\nu}_e, \tilde{\nu}_\mu, \tilde{\nu}_\tau$); ($\tilde{e}_R, \tilde{\mu}_R, \tilde{\tau}_R$). If $\tan\beta \gg 1$, then the bottom and tau Yukawa couplings are non-negligible, so that $\tilde{b}_R$, $\tilde{\tau}_L$, $\tilde{\tau}_R$, and $\tilde{\nu}_\tau$ need not be degenerate with their colleagues. With the minimal assumptions, one finds for the physical squark and slepton masses of the first two families approximately

$$m^2_{\tilde{e}_R} = m_0^2 + 0.15 m_{1/2}^2 - 0.23 m_Z^2 \cos 2\beta; \quad (2)$$

$$m^2_{\tilde{e}_L} = m_0^2 + 0.52 m_{1/2}^2 - 0.27 m_Z^2 \cos 2\beta; \quad (3)$$

$$m^2_{\tilde{\nu}} = m_0^2 + 0.52 m_{1/2}^2 + 0.50 m_Z^2 \cos 2\beta; \quad (4)$$

$$m^2_{\tilde{q}} = m_0^2 + (4.5 \text{ to } 6.5) m_{1/2}^2. \quad (5)$$

Eq. (5) holds for first and second generation squark mass. The range in the coefficient of $m_{1/2}^2$ is partly due to the uncertainty in α_3; moreover, this coefficient is smaller for larger squark masses. Eqs. (2)–(5) imply $m_{\tilde{q}} > m_{\tilde{e}_L} > m_{\tilde{e}_R}$. At least one of the top squark mass eigenstates and usually a bottom squark are below the main squark band implied by (5), because of radiative effects from the top Yukawa coupling. This ordering of squark masses should lead to enhanced production of third generation squarks at hadron colliders, and to enhanced branching ratios of gluinos, charginos and neutralinos into third generation quarks.

In the minimal approach, the physical gluino mass is related to $m_{1/2}$ by

$$M_{\tilde{g}} = (2.2 \text{ to } 3.5) m_{1/2} \quad (6)$$

Therefore, the squarks whose corresponding Higgs–quark Yukawa couplings are small cannot be much lighter than the gluino. When correlations among parameters are taken into account, one finds $m_{\tilde{q}} > 0.8 M_{\tilde{g}}$. With $m_t \simeq$ 175 GeV, radiative electroweak breaking requires that $|\mu|$ be large compared to the electroweak gaugino masses. The lighter two neutralinos and the lighter chargino will then be gaugino–like, and the heavier neutralinos and chargino will only rarely be produced in the decays of squarks and gluinos. This also implies that the heavier Higgs mass eigenstates A, H^0 and $H^\pm$ will be quite heavy and almost degenerate; they will usually also be able to decay into charginos and neutralinos. The lightest Higgs scalar, h^0, will have couplings close to those of the Standard Model Higgs boson. There is an upper bound on the h^0 mass of about 120 GeV in the MSSM. A similar h^0 mass bound of about 150 GeV is found in a completely general class of low-energy supersymmetric models [23] (assuming all model couplings are perturbative up to M_X).

3.3 Beyond the Minimal Planck Scale Assumptions

Many model builders have begun to explore the consequences of relaxing the minimal assumptions outlined above. For example, the boundary condition of universal scalar masses at M_X can be modified by non-trivial string effects, threshold effects associated with the grand unified model, evolution effects between M_X and M_P, or additional contributions arising from the effects of a larger gauge symmetry group which is broken at a scale above M_X [24]. Generally, small or moderate changes of the boundary conditions at M_X will only be important near the border of the allowed region of parameter space. For example, if the mass squared of H_1 is larger than that of H_2 at the scale M_X, $\tan\beta$ of order m_t/m_b or larger becomes possible without strong fine-tuning of soft breaking parameters. Non-universal scalar masses may also reduce $|\mu|$, possibly allowing for light higgsinos.

One can even entertain the possibility of large non-universalities among sfermions with the same electroweak quantum numbers. One often finds the statement in the literature that near–degeneracy of sfermion masses at the scale M_X is required by bounds on FCNC. This statement is not entirely correct; the potentially most dangerous gluino loop contributions will always vanish if quark and squark mass matrices are aligned [25], *i.e.*, can be diagonalized simultaneously. This is true *independent* of the (differences between) squark masses. The same is also true for neutralino loops. In such models only chargino loops contribute to FCNC. Notice that in general these do *not* constrain masses of $SU(2)$ singlet squarks. The most sensitive bound comes from K^0–$\bar{K}^0$ mixing. Demanding that the chargino loop contribution to Δm_K be smaller than the experimental value of 3.5×10^{-15} GeV gives

$$|\delta_{\tilde{u}\tilde{c}}| \leq 0.8\mathrm{TeV}^{-1}(1 + 6m_{1/2}^2/m_0^2)\sqrt{4.3m_{1/2}^2 + 0.4m_0^2}, \tag{7}$$

where m_0^2 is the average and $\delta_{\tilde{u}\tilde{c}} \cdot m_0^2$ the difference between the squared $\tilde{u}_L$ and $\tilde{c}_L$ masses at scale M_X. We see that $\delta_{\tilde{u}\tilde{c}} \simeq 1$ is allowed even for $m_0 = m_{1/2} = 100$ GeV. Highly non–degenerate squarks are therefore allowed *if* quark and squark mass matrices are (approximately) aligned. This might make squark hunting at hadron colliders difficult, since the signal from any one squark state might be too weak to be visible above backgrounds.

One should also consider extensions of the MSSM that go beyond simple modifications of the MSSM boundary conditions at the scale M_X. The "Next–to–Minimal" Supersymmetric extension of the Standard Model introduces a gauge–singlet Higgs chiral supermultiplet in addition to the usual MSSM particles. This variation does not seem to have a major effect on the spectrum of the MSSM particles [26], but it does impact on discovery channels for the Higgs and neutralinos. In particular, one may have a light Higgs scalar with reduced couplings to the Z boson. One can also contemplate low-energy supersymmetric models in which R-parity conservation is absent. For example, R-parity could be broken spontaneously by a sneutrino vacuum expectation value $\langle\tilde{\nu}\rangle \neq 0$. One can also replace R-parity by an alternative discrete symmetry [27] which allows either baryon number or lepton number violation (but not both!) among the renormalizable interactions. The phenomenology of R-parity violating models is quite different from that of the MSSM. This is mainly due to the fact that lightest supersymmetric particle is no longer stable, but can decay via an R-parity violating interaction. No systematic study of R-parity violating models in the context of supergravity and superunified theories has yet been undertaken.

4 Low Energy Supersymmetry: Phenomenology

If supersymmetry is associated with the origin of the electroweak symmetry breaking scale, then low-energy supersymmetry must exist at a scale of order 1 TeV or below. That is, supersymmetry must be discovered at either upgrades of existing colliders or at the LHC. Due to its mass reach, the LHC is the definitive machine for discovering or excluding low-energy supersymmetry. Section 4.1 summarizes the supersymmetry mass discovery potential for hadron colliders. A variety of signatures are explored. The conventional models of low-energy supersymmetry (which are R-parity conserving) possess a stable lightest supersymmetric particle (LSP) which is electrically (and color) neutral and therefore interacts very weakly in matter. It follows that the LSP behaves like a heavy neutrino which does not interact in collider detectors. Its presence would be signaled by excess missing transverse energy (which could not be attributed to neutrinos). Many of the supersymmetry searches rely on the missing energy signature as an indication of new physics beyond the Standard Model. Multi-leptonic signatures also play an important role in supersymmetry searches at hadron colliders. (Such signals can also be exploited in the search for R-parity-violating low-energy supersymmetry.)

Suppose that a signal is observed in one of the expected channels. This would not be a confirmation of low-energy supersymmetry, unless there is confirming evidence from other expected signatures. This presents a formidable challenge to experimenters at the LHC. Can they prove that a set of signatures of new physics is low-energy supersymmetry? Can they extract parameters of the supersymmetric models with any precision and test the details of the theory? These are questions that have only recently attracted seriously study. It is in this context that a future e^+e^- collider can be invaluable. As shown in Section 4.2. if the lightest supersymmetric particles were produced at LEP 2 or the NLC, precision measurement could begin to map out in detail the parameter space of the supersymmetric model. In particular, beam polarization at the NLC provides an invaluable tool for studying the relation between chirality and the properties of supersymmetric particles. For example, one can begin to prove that there is a correlation between the left and right-handed electrons and their slepton partners as expected in supersymmetry. The elucidation of the details of the light chargino and neutralino spectrum at the NLC could also play a pivotal role in untangling complex decay chains of heavier supersymmetric particles produced at the LHC.

4.1 Supersymmetry at Hadron Colliders

The detection of supersymmetry at a hadron collider is a complex task. It requires the ability to extract signals of new physics out of data samples with potentially large Standard Model backgrounds. Realistic simulations are required that can accurately model the physics of QCD jets and the theoretical predictions of low-energy supersymmetry. The physics of low-energy supersymmetry can be treated either in the context of specific theoretical models such as those discussed in Section 3, or in a more phenomenological approach which makes as few theoretical assumptions as is practical. Both approaches have been followed in this section. The minimal supersymmetric model (MSSM) [29] has recently been embedded in the event generator program ISAJET [30], so that all lowest order sparticle production processes and decay chains can be adequately simulated. In addition, ISAJET now allows the input of a mixed GUT and weak scale parameter set $(m_0, m_{1/2}, A_0, \tan\beta, sgn(\mu), m_t)$, for which a complete superparticle spectrum is generated using renormalization group equations with radiative electroweak symmetry breaking, under the assumption of gauge coupling unification at $M_{GUT} \sim 2 \times 10^{16}$ GeV, and universal GUT scale soft supersymmetry breaking terms. In addition, versions of ISAJET past 7.11 include all lowest order $e^+e^- \to$ SUSY reactions, so the effect of cascade decays can be explored at future e^+e^- colliders [31].

Much effort has gone into examining the superparticle mass reach and discovery potential of various proposed future colliders, to aid in decision making and prioritization for future collider facilities. A summary of these results can be found in Table 4. First, we consider the case of gluino and squark production, followed by their cascade decays via three-body $\widetilde{\chi}^{\pm}$ and $\widetilde{\chi}^0$ decay modes. Kamon *et al.* [32] have examined the mass reach in $m_{\tilde{g}}$ at the Tevatron and DiTevatron colliders using the missing transverse energy ($\not{E}_T$) signature, and found gluino mass reaches of $m_{\tilde{g}} \sim 270(450)$ GeV for the Tevatron Main injector (DiTevatron), assuming that $m_{\tilde{q}} \gg m_{\tilde{g}}$. In addition, Ref. [33] has examined the $\tilde{g}$ mass reach via *multi-lepton* topologies. The LHC reach for gluinos and squarks has recently been examined in considerable detail by ATLAS studies [8]. It is generally expected that the lighter of the two top squarks ($\tilde{t}_1$) is significantly lighter than the other squarks, due to the large top quark Yukawa coupling which drives diagonal stop masses to small values and enhances the off-diagonal mixing of $\tilde{t}_L$ and $\tilde{t}_R$. In this case, the present squark mass collider limits do not apply to $\tilde{t}_1$. In Ref. [34], it is shown that Tevatron collider experiments with 100 pb^{-1} of data can explore $m_{\tilde{t}_1} \sim$ 80–100 GeV, considerably further than the best current bounds of $\sim$ 45 GeV from LEP.

Next, we examine neutralino and chargino production. A significant portion of the parameter space of minimal supergravity models can be explored at the Tevatron and DiTevatron colliders by searching for the clean trilepton plus $\not{E}_T$ signal from $\widetilde{\chi}_1^{\pm}\widetilde{\chi}_2^0 \to 3\ell$ production [28]. In Table 4, we present the mass reach of the neutralino and chargino searches in terms of an equivalent gluino mass. The latter is obtained under the assumption of unified soft gaugino masses at the GUT scale (although there is some additional dependence on other supersymmetric model parameters). Note that the DiTevatron does not significantly increase the mass reach in the 3ℓ channel beyond that of the Tevatron Main Injector, although further optimization of cuts may be possible. This is in contrast to the mass reach for gluinos and squarks via $\not{E}_T$ searches, which is significantly higher for the DiTevatron. The clean trilepton signal has also been examined in Ref. [28] for the LHC. Note that it is not possible to probe $m_{\tilde{g}} >\sim$600–700 GeV via $3\ell + \not{E}_T$ mode at the LHC due to the turn on of the $\widetilde{\chi}_2^0 \to \widetilde{\chi}_1^0 h$ or $\widetilde{\chi}_2^0 \to \widetilde{\chi}_1^0 Z$ decay modes which spoil the signal. However, the dilepton spectrum from $\widetilde{\chi}_2^0 \to \widetilde{\chi}_1^0 \ell\bar{\ell}$ decay can be used at the LHC to measure $m_{\widetilde{\chi}_2^0} - m_{\widetilde{\chi}_1^0}$ with significant precision. Finally, if R-violating $\widetilde{\chi}_1^0$ decays are present (via baryon or lepton number violating interactions), then sparticle mass reaches can change significantly to either higher or lower values [35].

The search for sleptons at hadron colliders was addressed in Ref. [36]. It was shown that it would be difficult to search at the Tevatron for $m_{\tilde{\ell}}$ much beyond 50 GeV (and beyond 100 GeV at TeV*), due to low signal rates and irreducible backgrounds. In contrast, at the LHC stiffer cuts may be made; in this case slepton masses up to 250–300 GeV can be detected.

Tevatron signals within the constrained supergravity framework have been discussed in Ref. [37]. Because all signals are determined by just four parameters, the various SUSY processes are strongly correlated, and may provide an experimental probe of GUT scale assumptions. For example, Ref. [38] argues that in the parameter regime where sleptons are much lighter than squarks, the Main Injector discovery limit may reach $m_{\tilde{g}} \sim 500$ GeV via the enhanced multi-lepton signals from supersymmetric particle production and their subsequent decay chains.

Table 4: Discovery reach of various options of future hadron colliders. The numbers are subject to ±15% ambiguity. Also, the clean trilepton signals are sensitive to other model parameters; we show representative ranges from Ref. [28] where $|\mu|$ is typically much larger than the soft-breaking electroweak gaugino masses. For $\mu > 0$, the leptonic decay of $\widetilde{\chi}_2^0$ may be strongly suppressed so that 3ℓ signals may not be observable even if charginos are just above the LEP bound ($m_{\tilde{g}} \sim 150$ GeV).

Signal	Tevatron I 0.01 fb^{-1} 1.8 TeV	Tevatron II 0.1 fb^{-1} 1.8 TeV	Main Injector 1 fb^{-1} 2 TeV	Tevatron* 10 fb^{-1} 2 TeV	DiTevatron 1 fb^{-1} 4 TeV	LHC 10 fb^{-1} 14 TeV
$\not{E}_T(\tilde{q} \gg \tilde{g})$	$\tilde{g}(150)$	$\tilde{g}(210)$	$\tilde{g}(270)$	$\tilde{g}(340)$	$\tilde{g}(450)$	$\tilde{g}(1300)$
$\ell^\pm\ell^\pm(\tilde{q} \gg \tilde{g})$	—	$\tilde{g}(160)$	$\tilde{g}(210)$	$\tilde{g}(270)$	$\tilde{g}(320)$	$\tilde{g}(1000)$
$all \to 3\ell\ (\tilde{q} \gg \tilde{g})$	—	$\tilde{g}(150–180)$	$\tilde{g}(150–260)$	$\tilde{g}(150–430)$	$\tilde{g}(150–320)$	
$\not{E}_T(\tilde{q} \sim \tilde{g})$	$\tilde{g}(220)$	$\tilde{g}(300)$	$\tilde{g}(350)$	$\tilde{g}(400)$	$\tilde{g}(580)$	$\tilde{g}(2000)$
$\ell^\pm\ell^\pm(\tilde{q} \sim \tilde{g})$	—	$\tilde{g}(180–230)$	$\tilde{g}(325)$	$\tilde{g}(385–405)$	$\tilde{g}(460)$	$\tilde{g}(1000)$
$all \to 3\ell\ (\tilde{q} \sim \tilde{g})$	—	$\tilde{g}(240–290)$	$\tilde{g}(425–440)$	$\tilde{g}(550)$	$\tilde{g}(550)$	$\gtrsim\tilde{g}(1000)$
$\tilde{t}_1 \to c\widetilde{\chi}_1^0$	—	$\tilde{t}_1(80–100)$	$\tilde{t}_1(120)$			
$\tilde{t}_1 \to b\widetilde{\chi}_1^\pm$	—	$\tilde{t}_1(80–100)$	$\tilde{t}_1(120)$			
$\Theta(\tilde{t}_1\tilde{t}_1^*) \to \gamma\gamma$	—	—	—	—	—	$\tilde{t}_1(250)$
$\tilde{\ell}\tilde{\ell}^*$	—	$\tilde{\ell}(50)$	$\tilde{\ell}(50)$	$\tilde{\ell}(100)$		$\tilde{\ell}(250–300)$

4.2 Supersymmetry at e^+ e^- Colliders

At an e^+e^- collider, discovery of superparticles is relatively easy up to the kinematic limit. There is a vast literature concerning the discovery reach at LEP 2 [39]. Recent works discuss measurement of SUSY parameters rather than discoveries alone. For instance with a chargino pair production alone, one can put stringent constraints on the parameter space $(M_1, M_2, \mu, \tan\beta, m_{\tilde{l}})$ using the cross section, forward-backward asymmetry and branching fractions, where the sensitivity on particular parameters depends on the actual parameter values [40].

Experimentation at linear e^+e^- colliders would be richer due to the higher energies and beam polarization with the following characteristics:

- small beam energy spread (better than 1 %) even after including initial state radiation and beamstrahlung effects,
- clean environment basically without underlying events even with photon induced hadronic processes, and
- high beam polarization of 90% and beyond.

The virtues of polarization are two-fold. First, use of polarization can suppress the background substantially by as much as two orders of magnitudes, resulting in a very pure signal sample appropriate for precision studies. Second, one basically doubles the number of experimental observables by using both polarization states, which enables the efficient measurements of various parameters.

A detector simulation of $e^+e^- \to \tilde{\mu}^+\tilde{\mu}^-$ based on an ALEPH-type detector at $\sqrt{s} = 500$ GeV and $\int \mathcal{L}dt = 20$ fb^{-1} gives a 5 σ signal for $\tilde{\mu}$ for $m_{\tilde{\mu}}$ below 230 GeV [41]. A mass difference of 25 GeV between slepton and the LSP is enough for detection even without employing beam polarization. The case for the chargino search is similar. It is noteworthy that both purely hadronic and mixed hadronic-leptonic modes can be used for discovery and study of charginos [42].

We show here the sample results of Monte Carlo analysis with detector simulation [43] based on JLC-I detector [44]. Fig. 1 shows the power of the polarization to obtain a pure sample (>99%) of $\tilde{\mu}$ pair signals with high efficiency ($\epsilon > 50\%$). Since the signature of a slepton pair is an acoplanar lepton pair, the main background is W^+W^-, which can be substantially suppressed by employing right-handed electron beam. A kinematic fit to the end points of lepton energy distribution allows us the determination of both slepton and LSP masses better than 1%. The typical sensitivity of the chargino and neutralino mass measurements are shown in Fig. 2. Here the mixed hadronic-lepton mode is used, and the measurement is based on total dijet energy distribution. Using a technique to combine information from the tracking detector and calorimeters, one achieves mass measurements of the chargino and LSP at the 3% level. Similar sensitivities are expected for top and bottom squark signals [45].

The determination of superparticle masses allows us to test theoretical assumptions at various levels. For example, generation independence of slepton masses (universality) can be tested at the 1% level. More experimentally challenging is the test of the GUT-relation among gaugino masses, since the gauginos mix with higgsinos to form chargino and neutralino mass eigenstates. However, one can still test the GUT-relation at a few percent level combining slepton and chargino signals, based on

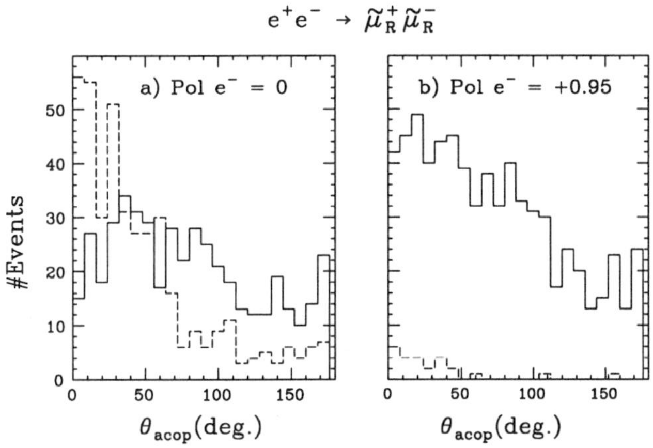

Figure 1: Acoplanarity distribution of muon pair for $E_{CM} = 350$ GeV, $m_{\tilde{l}} = 142$ GeV, $\int \mathcal{L}dt = 20$ fb^{-1}, for the signal (solid) and standard model backgrounds (dashed). Fig. b) shows the dramatic reduction of the backgrounds using the right-handed electron beam. Taken from Ref. [43].

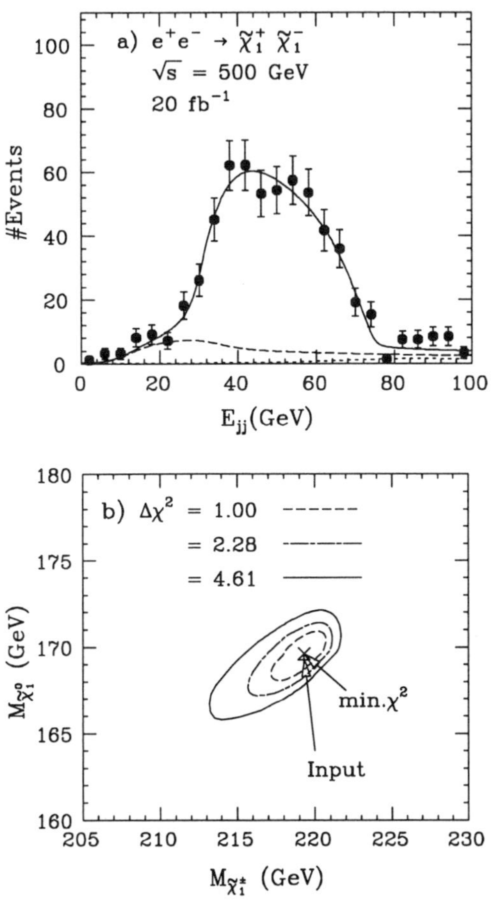

Figure 2: a) Dijet energy distribution from chargino pair production, for the Monte Carlo generated events (points), polynomial fit to the Monte Carlo (solid), contributions from the W-pair background (dashed) and the sum of other standard model backgrounds (dotted). b) Obtained contour on the masses of chargino and LSP. Taken from Ref. [43].

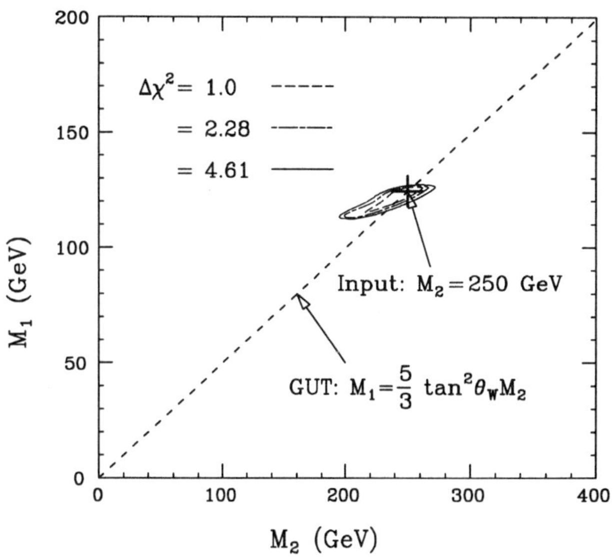

Figure 3: Test of the GUT-relation of the gaugino masses. The contours are obtained by a four-dimensional fit of $(M_1, M_2, \mu, \tan\beta)$ plane, using the LSP and chargino masses, and the production cross sections for selectrons and charginos with a right-handed electron beam. Taken from Ref. [43].

the measured masses and polarization dependence of the cross sections (see Fig. 3).

4.3 Complementarity

Typically, e^+e^- colliders have lower center-of-mass energies compared to hadron machines of the same era, but their complementary character in superparticle searches is easily understood. The colored superparticles which are the main targets of the hadron colliders are supposed to be 3–4 times heavier than the colorless ones under GUT assumptions. For example, the constraints on the supersymmetric parameter space from experiments at LEP ($\sqrt{s} = 90$ GeV) and the Tevatron ($\sqrt{s} = 1.8$ TeV) are comparable. This continues to be true for LEP 2 and the Tevatron with Main Injector, especially in the context of models with minimal supergravity boundary condition constraints [37,38]. The discovery reach of an e^+e^- linear collider with $\sqrt{s} = 1$ TeV will be again comparable to that at the LHC. But their complementary character will be more important *after* the discoveries of superparticles. Suppose that signals from squarks and gluinos are seen at the LHC. Then it would be a formidable task to demonstrate that they are superparticles due to the simultaneous production of many new particles and their complicated decay chains. The e^+e^- collider might not reach the energies where squarks could be produced; however, it would study the low-lying spectrum of superparticles (the sleptons, neutralinos and charginos) as described in the previous section. Analysis of data from the hadron collider with inputs from the e^+e^- collider could sort out the complex decay chains, revealing the mass spectrum and parameters of the heavier superparticles. These results would augment the measurements of the properties of light superparticles made at the e^+e^- collider, and would provide additional tests of GUT and/or supergravity predictions and test other model assumptions.

5 Strongly Coupled ESB Sector: Model-Independent Approaches

If a Higgs boson or technirho meson or some other direct indication of electroweak symmetry breaking is not found at some future accelerator, one would like to know whether it is still possible to learn something about electroweak symmetry breaking by studying WW scattering. In this scenario, whatever is responsible for giving the W and Z bosons their masses occurs at a high energy scale and most probably gives rise to a strongly interacting sector where perturbation theory is not applicable. Fortunately, one can still analyze the low energy behavior of the theory with a generalization of the $\pi\pi$ scattering theorems derived by Weinberg cast in the form of chiral perturbation theory.

The low energy theorems can easily be generalized from pions to the longitudinal components of the W and Z gauge bosons. These scattering theorems are only valid at low energy, that is $\sqrt{s} << (M_{\rm Res}, 4\pi v)$, where $M_{\rm Res}$ is the mass of the lowest lying resonance interacting with the longitudinal W's and Z's. The attraction in this approach is that it is valid for *any* mechanism of electroweak symmetry breaking. It is also possible to couple the chiral Lagrangian describing the low energy interactions of the gauge bosons with a (color-singlet) vector meson (the BESS model-equivalence) or with a scalar particle (a Higgs scalar type of models). This description of vector mesons and scalars coupled to longitudinal gauge bosons relies only on the symmetry of the theory and so is independent of the details of specific models.

Lastly, we discuss a model with a "hidden" symmetry breaking sector, one in which the electroweak symmetry breaking sector has a large number of particles. The theory can be strongly interacting and the *total* W and Z cross sections in this model are large: most of the cross section is for the production of particles other than the W or Z. In such a model, discovering the electroweak symmetry breaking sector depends on the observation the other particles and the ability to associate them with symmetry breaking.

5.1 Strongly Coupled ESB Sector at the LHC

In what follows, we assume the LHC runs at 14 TeV with a luminosity of 10^{34} cm^{-2} sec^{-1} (an annual integrated luminosity 100 fb^{-1}).

Vector Resonance Signal

A vector meson with a mass up to $m_\rho = 2.5$ TeV can be found at the LHC through the $W^\pm Z$ channel [46]. One would need to run the LHC for about 2 years to collect about 10 leptonic events, after kinematical cuts to effectively suppress the backgrounds.

Complementarity of Vector Resonant and Non-resonant WW Scattering:

Ref. [47] considered a vector-dominance model for three sets of mass-width parameters: $(m_\rho, \Gamma_\rho) =$ (1.78,0.33), (2.52,0.92), and (4.0,0.98). Their criterion for a significant signal (S) with a background (B) is $\sigma^\uparrow \equiv S/\sqrt{B} \geq 5$; $\sigma^\downarrow \equiv S/\sqrt{S+B} \geq 3$; $S \geq B$. The complementarity here implies that one observes either a low mass (I,J)=(1,1) resonance via the $W^\pm Z$ channel, or significant enhancement via the $W^\pm W^\pm$ channel with $I = 2$. From Table 5, one sees that the worst case among those parameter choices is at $m_\rho = 2.52$ TeV, where an integrated luminosity of at least 105 fb^{-1}—slightly more than the LHC annual luminosity—is required for the $W^\pm W^\pm$ channel to meet their discovery criterion. ATLAS collaboration [8] carried out a Monte Carlo study for W^+W^+ channel including detector simulations, obtaining qualitatively similar results.

Table 5: Minimum luminosity to satisfy observability criterion for $W^\pm Z$ and $W^\pm W^\pm$ channels at the LHC. Each entry contains $\mathcal{L}_{MIN}$ in fb^{-1}, the number of signal/background events per 100 fb^{-1}.

m_ρ (TeV)	1.78	2.52	4.0
$W^\pm Z$	44 fb^{-1} 38/20	325 fb^{-1} 5.8/3.4	No Signal
$W^\pm W^\pm$	142 fb^{-1} 12.7/6.0	105 fb^{-1} 15.9/5.8	77 fb^{-1} 22.4/8.9

All-Channel Comparison for WW Scattering

Ref. [48] compared all-channel WW scattering at $\sqrt{s} = 40$ TeV and 16 TeV, and new analysis at 14 TeV is in progress. In particular, the signal for a 1 TeV Higgs boson may be observable through the leptonic channels $H \to WW \to l\nu l\nu$ and $ZZ \to l\bar{l}\nu\bar{\nu}$, to add to the clean but low rate $ZZ \to 4l$ mode.

ATLAS collaboration [8] also studied a 1 TeV Higgs boson signal via the decay mode $H \to WW \to l\nu jj$ and $ZZ \to l\bar{l}jj$. It was found that after identifying one W in the leptonic mode and one W in the di-jet mode with $M_{jj} \sim m_W$, with a tagged forward jet and a veto on all other central jets, the $WW \to \ell\nu jj$ channel gives a viable signal with statistical significance above 6.8σ.

5.2 Strongly Coupled ESB Sector at the NLC

Our standard NLC has an energy $\sqrt{s}$ =1.5 TeV and an integrated luminosity of 190 fb^{-1} per year.

Vector Resonance Signal

At the NLC a vector resonance would be probed most effectively via $e^+e^- \to W_L^+W_L^-$. This can be generically described [49] by an Omnés function with a complex form factor F_T. An infinitely massive vector resonance is equivalent to the case where the $W_L^+W_L^-$ scattering amplitude is described by the Low Energy Theorem (LET).

Fig. 4 shows the results in Ref. [50] for a technirho-like vector resonance in the $e^+e^- \to W^+W^-$ process. In one year at the NLC, technirhos of arbitrarily large mass can be distinguished from the weakly-interacting SM. An integrated luminosity of 225 fb^{-1} (1.2 years at design luminosity) would yield 7.1σ, 5.3σ, and 5.0σ signals for a 4 TeV technirho, a 6 TeV technirho, and LET, respectively. As was the case for the LHC of Table 5, these results assume statistical errors only. The systematic errors for the $e^+e^- \to W^+W^-$ results will be small compared to the statistical error, since the analysis consists of a background-free maximum likelihood fit of angular distributions.

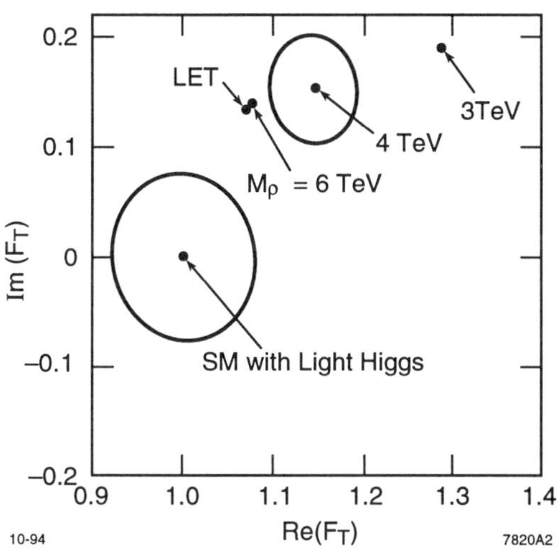

Figure 4: Confidence level contours for $Re(F_T)$ and $Im(F_T)$ at c.m. energy 1.5 TeV and $\mathcal{L}$ = 190 fb^{-1}. The contour about the light Higgs value of F_T = (1,0) is 95% C.L. and the contour about m_ρ = 4 TeV point is 68% C.L..

The processes of $e^+e^- \to W^+W^-$ and $e^+e^- \to f\bar{f}'$ via the vector resonance have also been studied in the context of the BESS model and variables such as the total cross sections, forward-backward and left-right asymmetries are used to constrain the virtual V effects. For $\sqrt{s} \sim m_\rho$, the NLC will be very sensitive to the BESS model parameters [51].

WW Fusion Processes

WW fusion processes open more channels to study the strongly coupled ESB sector beyond a vector resonance. Ref. [52] demonstrates that the signal for a 1 TeV Higgs boson should be observable at a 12σ level, for $\sqrt{s}$ =1.5 TeV and 200 fb^{-1}. Furthermore, the ratio $R \equiv \sigma(W^+W^- \to W^+W^-)/\sigma(W^+W^- \to ZZ)$ may provide hints that help to distinguish different strongly coupled ESB sector models, such as a scalar resonance ($R > 1$), a vector resonance ($R \gg 1$), or a Low Energy Theorem amplitude ($R < 1$).

Due to the relatively clean environment in e^+e^- collisions, cross sections for background processes may be measured rather well for both the normalization as well as the shape of the distribution, so that statistically significant deviation from the SM background may be easily identified as new physics signal. One example is the "Hidden Sector" [53], which may be impossible to observe in hadronic collision environment but possibly feasible in e^+e^- collisions.

Searches at $\gamma\gamma$ Colliders

The heavy Higgs boson and other strong scattering signals in longitudinal gauge boson channels can be investigated at the NLC at $\sqrt{s}$ = 1.5 TeV operating in the $\gamma\gamma$ collider mode [54]. The processes considered here involve two gauge boson final states $\gamma\gamma \to ZZ$, $\gamma\gamma \to W^+W$, and four gauge boson final states $\gamma\gamma \to W^+W^-W^+W^-$, W^+W^-ZZ [55]. The four gauge boson final state processes seem to be more promising in searching for strongly coupled ESB dynamics as compared with the two gauge boson final states, although all the above processes are plagued by large backgrounds consisting of transversely produced gauge bosons [54]. This issue has been addressed recently in Refs. [56,57]. For example, Ref. [56] finds that with luminosity of 200 fb^{-1}, a 1.5 TeV e^+e^- collider operating as a $\gamma\gamma$ collider can successfully observe a heavy Higgs boson with mass up to 700 GeV. To obtain a reach in the Higgs boson mass up to 1 TeV, a 2 TeV e^+e^- collider is required with luminosity of 200 fb^{-1}. The results were obtained taking into account the back-scattered photon luminosity spectrum. It is also found that a 2 TeV linear collider in the $\gamma\gamma$ mode is roughly equivalent to a 1.5 TeV e^+e^- collider in searching for heavy Higgs boson physics (roughly scales as $\sqrt{s_{\gamma\gamma}} \sim 0.8\sqrt{s_{ee}}$). The viability of this signal still requires that the decays of the W and Z bosons be incorporated, and detector simulations be performed to determine a realistic acceptance.

6 Strongly Coupled ESB Sector: Implications of Models

6.1 *Ingredients of Techni-models*

In theories of dynamical electroweak symmetry breaking (ESB), such as technicolor [58], ESB is due to chiral symmetry breaking in an asymptotically-free, strongly-interacting, gauge theory with massless fermions. These theories attempt to understand electroweak symmetry breaking without introducing any scalar particles such as a Higgs boson. Instead, the symmetry breaking is modeled after that which occurs in QCD when the $SU(3)_c$ color gauge group becomes strong. In QCD, the breaking of the chiral symmetry gives rise to pions. In the simplest technicolor theory one introduces a left-handed weak-doublet of "technifermions", and the corresponding right-handed weak-singlets, which transform as the fundamental $[N]$ representation of a strong $SU(N)_{TC}$ technicolor gauge group. The global chiral symmetry of the technicolor interactions is then $SU(2)_L \times SU(2)_R$. When the technicolor interactions become strong, the chiral symmetry is broken to the diagonal subgroup, $SU(2)_V$, producing three Nambu-Goldstone bosons, which then yield the correct W and Z masses.

However, the symmetry breaking sector must also couple to the ordinary fermions, allowing them to acquire mass. In models of a strong ESB sector there must either be additional flavor-dependent gauge interactions, the so-called "extended" technicolor (ETC) interactions, or Yukawa couplings to scalars which communicate the breaking of the chiral symmetry of the technifermions to the ordinary fermions. The most popular type of strong ESB model that attempts to explain the masses of all observed fermions contains an entire family of technifermions with Standard Model gauge couplings.

Since these models contain more than one doublet of technifermions, they have a global symmetry group larger than $SU(2)_L \otimes SU(2)_R$. Therefore chiral symmetry breaking produces additional (pseudo-)Nambu-Goldstone bosons (PNGBs). Furthermore, the models typically possess a larger variety of resonances than the one-doublet model. There will be heavy ($\sim$ 1–2 TeV) vector meson resonance with the quantum numbers of the ρ, ω etc. The phenomenology of color-neutral resonances with the quantum numbers of the ρ is discussed in Section 5. In the case of color-neutral PNGBs the phenomenology is much like that of the extra scalars in "two-Higgs" doublet models. Therefore we will focus on the colored resonances and PNGBs occuring in most technicolor models. These models have a possibly unique signature: resonances associated with the *electroweak* symmetry breaking sector which are *strongly produced*. Hadron colliders will be especially important for searching for signatures of such colored particles associated with ESB.

6.2 *Theoretical Challenges for Technicolor Model Building*

Models incorporating ETC interactions ran into trouble with flavor changing neutral currents (FCNC's) early on. Assuming technicolor behaved like a scaled-up version of QCD it was found that it is impossible to generate a large enough mass for the s quark (much less the c quark) while avoiding FCNC's. One possible solution is to make the TC gauge coupling run slower than in QCD (*i.e.* make it 'walk'). This makes the technifermion condensate much larger than scaling from QCD suggests, so that for a fixed quark or lepton mass, the necessary mass scale for ETC gauge bosons is increased, suppressing FCNC's. A second solution is to build a Glashow-Illiopoulos-Maiani (GIM) symmetry into the ETC theory (sometimes called Techni-GIM).

A further problem for ETC models is producing a large isospin splitting for the t and b quark masses without causing large isospin splitting in the W and Z masses. Isospin splitting in the gauge sector, as described by the radiative correction parameter T is tightly constrained by experiment.[a]

A current challenge for TC models is that models with QCD-like dynamics tend to produce large positive contributions to the electroweak radiative correction parameter S, while experiments tend to prefer small S. Since the discrepancy grows with the number of technicolors and the number of technidoublets, one way to evade the problem is to limit the number of technifermions contributing to ESB. Alternatively one can invoke mechanisms yielding a negative contribution to S: the technifermions may have exotic electroweak quantum numbers or the TC dynamics can be sufficiently unlike QCD so as to invalidate the naive scaling-up of QCD phenomenology. Such alternatives usually also rely on some form of isospin breaking. In multiscale models isospin breaking at the lowest ESB scale can produce a negative contribution to S without infecting the T parameter, which is sensitive to isospin breaking at the highest ESB scale.

The most recent problem for ETC models arises from the measurement of the $Z \to b\bar{b}$ partial width at LEP. The experiments find a partial width which is slightly larger than the Standard Model prediction, while almost all the ETC models that have been constructed so far produce a correction which further reduces this width. The sign of the correction is a consequence of the almost universal assumption that $SU(2)_L$ commutes with the ETC gauge group. The alternative, where $SU(2)_L$ is embedded in the ETC gauge group, may be interesting, since this reverses the sign of the correction.

[a] The radiative correction parameters, S, T and U are defined in the chapter on Precision Tests of Electroweak Physics.

Table 6: Discovery reach (in GeV) of different accelerators for particles associated with realistic models of a strong ESB sector.

Particle	$(SU(3)_C,$ $SU(2)_V)$	Tevatron	LHC	LEP I	LEP II	NLC-1000
$P^{0\prime}$	(1,1)	—	110–150^a	8^b;28^b	—^{c}	—^{c,d}
P^0	(1,3)	—	—	—	—	—^{d}
P^+P^-	(1,3)	—	400^e	45^f	100^g	500^g
$P_8^{0\prime}(\eta_T)$	(8,1)	400–500^h	325^i	—	—	—
P_8^0d	(8,3)	10–20^i	$325^{i,j}$	—	—	—
$P_8^+P_8^-$	(8,3)	10–20^{i,j}	$325^{i,j}$	45^f	100^g	500^g
$P_3^+P_3^-$	(3,3)	—^{j}	—^{j}		100 GeV^g	500 GeV^g

^{a} Decay mode $P^{0\prime} \to \gamma\gamma$, similar to a light neutral Higgs [59].
^{b} Decay mode $Z \to \gamma P^{0\prime}$, assuming an one-family model, with $N_{TC} = 7$ and $N_{TC} = 8$ respectively; no reach for $N_{TC} < 7$; for larger $Z\gamma P^{0\prime}$ couplings, the discovery reach extends to 65 GeV [60].
^{c} No reach for one-family model; possibility of reach for the Lane-Ramana [61] multiscale model in the processes $e^+e^- \to P\gamma$, Pe^+e^- [62].
^{d} The analogous discovery reach for a CP-odd Higgs scalar can be greatly improved using the NLC operating in the $\gamma\gamma$ collider mode (with polarized photon beams) [63].
^{e} Estimated from work on charged Higgs ($gb \to tH^- \to t\bar{t}b$) for $\tan\beta \simeq 1$ [64].
^{f} LEP I kinematical limit.
^{g} Kinematical limits for LEP200 and the NLC-1000.
^{h} Contribution to the $\bar{t}t$ cross section in multiscale models [65].
^{i} QCD pair production of colored PNGBs with decay into 4 jets [66].
^{j} QCD pair production of colored PNGBs with decay into $t\bar{t}$, $t\bar{b}$, $\tau^+\tau^-$ or $\tau\nu_\tau$ may allow higher reach in mass. This has yet to be studied.

6.3 Phenomenology of Pseudo-Goldstone Bosons

Many ETC models are expected to include a spectrum of pseudo-Nambu-Goldstone bosons (P's) and vector resonances (ρ_T's and ω_T) that may carry color quantum numbers. These particles may be classified by their $SU(3)_C$ and $SU(2)_V$ quantum numbers as indicated (for the P's) in the first column of Table 6, using the nomenclature of Ref. [67]. Since these particles are strongly produced, they have large cross sections at the Tevatron and LHC. The color triplet (P_3) and color octet (P_8) pseudo-Nambu-Goldstone bosons will be pair produced from gluon-gluon initial states in hadron colliders and will give rise to 4 jet and $t\bar{t}$ final states, since in many models the PNGBs couple preferentially to heavy fermions. Since P_3 and P_8 couple to the Z boson they can also be pair produced in e^+e^- colliders, which are expected to be sensitive to PNGB masses up to the kinematic limit. The LHC will be sensitive to P_8 masses up to ~ 325 GeV, while an NLC-1000 would reach ~ 500 GeV. Since colored PNGBs receive their masses from QCD interactions and typically have masses in the 100–300 GeV range, both the LHC and NLC-1000 can probe the interesting region. The expected discovery potential of various colliders for the pseudo-Nambu-Goldstone bosons is shown in Table 6.

While as yet no completely realistic technicolor model exists, the models all share the general feature of having a rich spectrum of scalar and vector particles which can be probed at either the LHC or a high-energy e^+e^- collider.

7 New Gauge Bosons

7.1 Introduction

Extended gauge symmetries and the associated heavy neutral (Z') and/or charged (W') gauge bosons are a feature of many extensions of the Standard Model (SM). We survey and compare the discovery potential of the experiments that will be performed over the next decade (Tevatron and LEP 2) and future facilities that are being planned and considered for the period beyond.

We present results for heavy gauge boson physics for the following representative models: Z_χ in $SO_{10} \to SU_5 \times U_{1\chi}$, Z_ψ in $E_6 \to SO_{10} \times U_{1\psi}$, $Z_\eta = \sqrt{3/8}Z_\chi - \sqrt{5/8}Z_\psi$ in superstring inspired models in which E_6 breaks directly to a rank 5 group [68], as well as Z_{LR} and W'_{LR} in the LR symmetric models [69]. For simplicity, we assume that the neutrino produced in the W' decay does not produce visible energy in the detector.

7.2 Discovery Limits of Extra Gauge Bosons

Current limits on the mass of new heavy gauge bosons are relatively weak. In Table 7 the direct bounds on Z' production from the main production channel $p\bar{p} \to Z' \to e^+e^-$ from current Tevatron data as well as the indirect unconstrained (no assumption on the Higgs sector) and constrained (specific assumption on the Higgs sector) bounds from a global analysis of electro-weak data are presented [70]. From the non-observation of $pp \to W' \to e\nu_e$ at the Tevatron, the CDF Collaboration concludes that $M_{W'_{LR}} > 652$ GeV [71]. In contrast, the indirect constrained [unconstrained] Tevatron bounds on $M_{W'_{LR}}$ are in the 1.4 TeV [300 GeV] region.

Discovery limits for new gauge bosons at future colliders are shown in Table 8 [72]. The LHC and a high luminosity 1 TeV e^+e^- collider have discovery limits for a Z' which are in many ways comparable. Both measurements have strengths and weaknesses. The limits obtained from the LHC are robust, in the sense that they are obtained from a direct measurement with little background, but they are, however, dependent on the total width of the Z' which is sensitive to assumptions on the particle content of the model. In contrast, limits obtained for the NLC are indirect, based on statistical deviations from the Standard Model and are therefore both more model dependent and dependent on having the systematic errors under control, but do not depend on the unknown particle content of the model.

If a Z' were discovered, its mass could be measured without any ambiguity at a hadron collider from the invariant mass of the lepton pairs which signalled the Z' discovery. In contrast, at the NLC it is the ratio $(g/M'_Z)^2$ which enters, so that extracting the Z' mass is very model dependent. If evidence were found for a Z' at the NLC, a model independent determination of $M_{Z'}$ would be imprecise at best, except for perhaps a Z' not much greater than the collider energy.

7.3 Z' Diagnostics at the LHC and NLC

An immediate need after a Z' discovery would be to determine its origins by measuring its couplings to quarks and leptons [73]. Assuming family universality and the Z' charge commuting with the $SU(2)_L$ generators, the relevant quantities to distinguish between different models are the fermion charges, $\hat{g}^f_{L2(R2)}$ and the gauge coupling strength, g_2. Because the signs of the charges will be hard to determine at hadron colliders the following four "normalized" observables are used:

$$\begin{aligned}
\gamma^\ell_L &= \frac{(\hat{g}^\ell_{L2})^2}{(\hat{g}^\ell_{L2})^2 + (\hat{g}^\ell_{R2})^2}, \\
\gamma^q_L &= \frac{(\hat{g}^q_{L2})^2}{(\hat{g}^\ell_{L2})^2 + (\hat{g}^\ell_{R2})^2}, \\
\tilde{U} &= \frac{(\hat{g}^u_{R2})^2}{(\hat{g}^q_{L2})^2}, \\
\tilde{D} &= \frac{(\hat{g}^d_{R2})^2}{(\hat{g}^q_{L2})^2}\,. \qquad (8)
\end{aligned}$$

The values of these couplings at the LHC for the typical models and the corresponding statistical uncertainties are given in Table 9 for $M_{Z'}$=1 TeV.

For $M_{Z'} \simeq 2$ TeV a reasonable discrimination between models and determination of the couplings may still be possible, primarily from the forward-backward asymmetry and the rapidity ratio. However, for $M_{Z'} \simeq 3$ TeV there is little ability to discriminate between models. The LHC can also address W' diagnostics for $M_{W'} \lesssim$ 1–2 TeV.

If a Z' were produced on shell at the NLC it would be relatively straightforward to determine its properties. On the other hand, if it is far off-shell (a more likely possibility) its properties could be deduced through interference effects of the Z' propagator with the γ and Z propagators. At the NLC σ^ℓ, $R^{had} = \sigma^{had}/\sigma^\ell$ and A^ℓ_{FB} will be measured and assuming longitudinal polarization of the electron beam is available $A^{\ell,had}_{LR}$, $A^\ell_{FB}(pol)$ will be possible. Although the above quantities can distinguish between different models they do not yield information on all the Z' couplings.

The following four "normalized" charges and an overall gauge coupling strength divided by the "reduced" Z' propagator are probed at the NLC:

$$\begin{aligned}
P^\ell_V &= \frac{\hat{g}^\ell_{L2} + \hat{g}^\ell_{R2}}{\hat{g}^\ell_{L2} - \hat{g}^\ell_{R2}}, \\
P^q_L &= \frac{\hat{g}^q_{L2}}{\hat{g}^\ell_{L2} - \hat{g}^\ell_{R2}}, \\
P^{u,d}_R &= \frac{\hat{g}^{u,d}_{R2}}{\hat{g}^q_{L2}}, \\
\epsilon_A &= (\hat{g}^\ell_{L2} - \hat{g}^\ell_{R2})^2 \frac{g_2^2}{4\pi\alpha} \frac{s}{M^2_{Z'} - s}. \qquad (9)
\end{aligned}$$

Recall that couplings (8) probed by the LHC, do not determine couplings (9) unambiguously. The couplings with statistical uncertainties are given in Table 10. The Z' charges can be determined to around 10–20 %. Relative error bars are about a factor of 2 larger than the corresponding ones at the LHC.

7.4 Complementarity

Among existing facilities currently in operation, the Tevatron offers the highest discovery reach for new gauge bosons with masses up to the 700–900 GeV range although LEP 2 can achieve comparable limits for some models. In the longer term, hadron colliders, *i.e.*, different upgraded versions of the Tevatron and the LHC, as

Table 7: Current constraints on $M_{Z'}$ (in GeV) for typical models (see Ref. [70] for details) from direct production at the Tevatron ($\mathcal{L}_{int} = 19.6$ pb^{-1}), as well as indirect limits (95 % C.L.) from a global electro-weak analysis.

Models	Direct	Indirect (unconstrained)	Indirect (constrained)
χ	425	330	920
ψ	415	170	170
η	440	210	600
LR	445	390	1380

Table 8: Discovery limits for $M_{Z'}$ (in GeV) for typical models achievable at proposed hadron and e^+e^- colliders. At hadron colliders the discovery limits for Z' $[W'^+ + W'^-]$ are for typical models with 10 events in e^+e^- + $\mu^+\mu^-$ $[(e^+\nu_e + e^-\bar{\nu}_e\ ) + (\mu^+\nu_\mu + \mu^-\bar{\nu}_\mu)]$ from Drell-Yan production of the gauge boson. For e^+e^- colliders the discovery limits are the 99% C.L. from a χ^2 fit of the observables: $\sigma(e^+e^- \to \mu^+\mu^-)$, $R^{had} = \sigma^{had}/\sigma^\ell$, A^ℓ_{FB}, and A^ℓ_{LR}.

Collider	$\sqrt{s}$ [TeV]	$\mathcal{L}_{int}$ [fb^{-1}]	χ	ψ	η	LR	W'_{LR}
Tevatron ($p\bar{p}$)	1.8	1	775	775	795	825	920
Tevatron* ($p\bar{p}$)	2	10	1040	1050	1070	1100	1180
DiTevatron ($p\bar{p}$)	4	20	1930	1940	1990	2040	2225
LHC (pp)	14	100	4380	4190	4290	4530	5310
LEP 2 (e^+e^-)	0.2	0.5	695	269	431	493	
NLC (e^+e^-)	0.5	50	3340	978	1990	2560	
NLC (e^+e^-)	1	200	6670	1940	3980	5090	
NLC (e^+e^-)	2	200	9560	3150	5830	7210	

Table 9: Values of the "normalized" couplings (1) for the typical models. The statistical error bars indicate how well the coupling could be measured at the LHC ($\sqrt{s} = 14$ TeV, $\mathcal{L}_{int} = 100$ fb^{-1}) for $M_{Z'} = 1$ TeV.

	χ	ψ	η	LR
γ_L^ℓ	0.9 ± 0.016	0.5 ± 0.02	0.2 ± 0.012	0.36 ± 0.007
γ_L^q	0.1	0.5	0.8	0.04
$\tilde{U}$	1 ± 0.16	1 ± 0.14	1 ± 0.08	37 ± 6.6
$\tilde{D}$	9 ± 0.57	1 ± 0.22	0.25 ± 0.16	65 ± 11

Table 10: The value of the couplings (2) for typical models and statistical error bars as determined from probes at the NLC ($\sqrt{s} = 500$ GeV, $\mathcal{L}_{int} = 20$ fb^{-1}). $M_{Z'} = 1$ TeV. 100% heavy flavor tagging efficiency and 100% longitudinal polarization of the electron beam is assumed for the first set of error bars, while the error bars in parentheses are for the probes without polarization.

	χ	ψ	η	LR
P_V^ℓ	$2.0 \pm 0.08\,(0.26)$	$0.0 \pm 0.04\,(1.5)$	$-3.0 \pm 0.5\,(1.1)$	$-0.15 \pm 0.018\,(0.072)$
P_L^q	$-0.5 \pm 0.04\,(0.10)$	$0.5 \pm 0.10\,(0.2)$	$2.0 \pm 0.3\,(1.1)$	$-0.14 \pm 0.037\,(0.07)$
P_R^u	$-1.0 \pm 0.15\,(0.19)$	$-1.0 \pm 0.11\,(1.2)$	$-1.0 \pm 0.15\,(0.24)$	$-6.0 \pm 1.4\,(3.3)$
P_R^d	$3.0 \pm 0.24\,(0.51)$	$-1.0 \pm 0.21\,(2.8)$	$0.5 \pm 0.09\,(0.48)$	$8.0 \pm 1.9\,(4.1)$
ϵ_A	$0.071 \pm 0.005\,(0.018)$	$0.121 \pm 0.017\,(0.02)$	$0.012 \pm 0.003\,(0.009)$	$0.255 \pm 0.016\,(0.018)$

well as a high luminosity e^+e^- collider, *i.e.*, the NLC, would significantly improve limits on the heavy gauge boson masses. For the typical models such limits are in the 1–2 TeV region for the Tevatron upgrades, in the 4–5 TeV region for the LHC, and roughly 2–10 times the center-of-mass energy for the NLC with 50 fb^{-1}.

The LHC and a high luminosity 1 TeV e^+e^- collider have discovery limits for Z' which are in many cases comparable. However, limits obtained for the LHC and the subsequent measurement of its mass are robust, in the sense that they are obtained from a direct measurement with little background. In contrast, discovery limits and mass measurements at the NLC are indirect and therefore model specific.

Once discovered, the LHC and the NLC possess complementary diagnostic power for the model independent determination of the Z' couplings to quarks and leptons. In conjunction, the two machines may allow for determination of the $M_{Z'}$, an overall Z' gauge coupling strength as well as a unique determination of *all* the quark and lepton charges with error bars in the 10–20% range, provided $M_{Z'} \lesssim$ 1–2 TeV.

8 New Particles and Interactions

Many theories beyond the Standard Model (SM) of the electroweak and strong interactions, such as Grand Unified Theories (GUT) or composite models, require the existence of new matter particles with the possibility of new interactions not contained in the SM. These particles can be cast into three categories: new elementary fermions [68,74,75,76,77], difermions [68,78,79] and excited fermions [80,81,82]. If any signals are seen, it will be crucial to distinguish among the variety of possible new states described in this section. In general, total cross sections, angular distributions and the polarization of final state particles (resulting from new particle decays) can be used to discriminate among the different possibilities [83].

8.1 New Elementary Fermions

The classic examples of new fermions include: sequential fourth generation fermions (with a right-handed neutrino state to allow for neutrino masses larger than $m_Z/2$); vector–like fermions with both left- and right-handed components in weak isodoublets [68], mirror fermions which have the opposite chiral properties as the SM fermions [75] and isosinglet fermions such as the SO(10) Majorana neutrino [76].

Fermions that have the usual lepton/baryon quantum numbers but possess non-canonical $SU(2)_L \times U(1)_Y$ quantum numbers are called exotic fermions. They occur naturally in GUT models that contain a single representation into which a complete generation of SM quarks and leptons can be embedded. For instance, in the group E_6 each fermion generation lies in the **27** representation, which contains 12 new fermions in addition to the 15 chiral fermions of the SM [68].

It is conceivable that these new fermions acquire masses not much larger than the Fermi scale, if these masses are protected by some symmetry. In fact, this is necessary if the associated new gauge bosons (which are present in the same GUT model that contains the new fermions) are relatively light [84]. In the case of sequential and mirror fermions (at least in the simplest models where the ESB pattern is the same as in the SM), unitarity arguments suggest that their masses should not exceed a few hundred GeV [85]. These particles, if they exist, should therefore be accessible at the next generation of colliders.

Except for singlet neutrinos, the new fermions couple to the photon and/or to the weak gauge bosons W/Z (and for heavy quarks, to gluons) with full strength; these couplings allow for pair production with practically unambiguous cross sections. In general, new fermions can mix with SM fermions which have the same conserved quantum numbers (such as color and electric charge). This mixing gives rise to new currents, which determine the decay properties of the heavy fermions and allow for their single production. However, flavor changing neutral currents (FCNC's) will be generated at levels inconsistent with present experiments unless intergenerational mixing is highly suppressed or absent. In the latter case, the mixing pattern simplifies considerably [86]. The remaining angles are restricted by LEP and low energy experiment data to be smaller than $\mathcal{O}(0.04 - 0.1)$ [86]. Note that LEP1 sets bounds of $\sim m_Z/2$ on the masses of these particles [87] (stronger mass bounds from Tevatron can be set for new quarks); masses up to m_Z might be probed at LEP2.

The new fermions decay through mixing into massive gauge bosons plus their ordinary light partners, $F \to fZ/f'W$. For masses larger than $m_w(m_Z)$ the vector bosons will be on–shell. For small mixing angles, $\theta <$ 0.1, the decay widths are less than 10 MeV (GeV) for $m_F = 0.1(1)$ TeV. The charged current decay mode is always dominant and for $m_F \gg m_Z$, it has a branching fraction of 2/3.

New fermions can be pair-produced in e^+e^- collisions, $e^+e^- \to F\bar{F}$, through s–channel gauge boson exchange. The cross sections are of the order of the point–like QED cross section and therefore, are rather large [83]. Because of their clear signatures, the detection of these particles is straightforward in the clean environment of e^+e^- colliders, and masses very close to the kinematical limit can be probed. Charged fermions can also be pair–produced at $\gamma\gamma$ colliders via $\gamma\gamma \to F\bar{F}$. The cross sections are comparable (and may be larger for smaller F masses) to the corresponding rates in e^+e^- collisions.

Heavy quarks can be best searched for at hadron colliders where the production processes, $gg/q\bar{q} \to Q\bar{Q}$, give very large cross sections: at the LHC with $\sqrt{s} = 14$ TeV and a luminosity of 10 fb^{-1} quark masses up to 1 TeV can be reached [88].

If the mixing angles between the SM and heavy fermions are not too small, then singly-produced new fermions in association with their light partners may produce an observable signal. The rate for single production is very model dependent (in contrast to pair-production rates mentioned above); however, it has the potential for significantly increasing the new fermion mass reach of a given collider. In e^+e^- collisions, single production proceeds only via s–channel Z exchange in the case of quarks and second/third generation leptons, leading to small rates. For the first generation leptons, however, one has additional t–channel exchanges which increase the cross sections by several orders of magnitude [83]. A full simulation of the signals and the backgrounds has been recently carried out; see Ref. [77].

Heavy leptons of the first generation can also be singly produced in ep collisions through t–channel exchange. At LEP×LHC with $\sqrt{s} = 1.2$ TeV and $\int \mathcal{L} = 2$ fb^{-1}, the N and E mass reach has been considered in Ref. [89]. In Left–Right models, an additional production mechanism for single N production via t-channel W_R-exchange must be included. A detailed Monte Carlo analysis of this process has been carried out in Ref. [90]. In e^+e^- collisions these neutrinos, can be pair produced with observable cross sections if the W_R bosons are not too heavy ($M_{W_R} < 2$ TeV at $\sqrt{s} = 0.5$ TeV). One can also search for heavy Majorana neutrinos, through their virtual effects in the process $e^-e^- \to W_R^- W_R^-$ [91].

8.2 *Difermions*

Difermions are new bosons (either spin 0 or spin 1) that have unusual baryon and/or lepton quantum numbers. Examples of these particles are leptoquarks (LQ) with B= ±1/3 and L= ±1, diquarks with B= ±2/3 and L= 0 and dileptons with B= 0 and L= ±2. They occur in some GUT models (*e.g.*, in E_6, the color triplet weak isosinglet new particle can be either a LQ or diquark) and in composite models.

In addition to the usual couplings to gauge bosons, difermions have couplings to fermion pairs which determine their decays. These couplings are *a priori* unknown, although one typically assumes that the couplings between different generations are negligible in order to avoid generating tree-level FCNC processes. As an example, consider the case of LQ's. A systematic description of their quantum numbers and interactions can be made by starting from an effective lagrangian with general SU(3)×SU(2)×U(1) invariant couplings and conserved B and L numbers. This leads to the existence of 5 scalar and 5 vector LQ's with distinct SM transformation properties [76]. In general, present data constrain difermions to have masses larger than 50–150 GeV.

Leptoquarks can be produced in pairs at e^+e^- colliders through gauge boson exchange; significant t-channel quark exchange can be present in some channels if the quark-lepton-LQ couplings are not too small. Depending on the charge, the spin and isospin of the LQ, the cross sections can vary widely [92]. LQ's can also be pair produced in $\gamma\gamma$ collisions via t-channel LQ-exchange. Single production of scalar and vector LQs has been also studied in the e^+e^-, $e\gamma$ and $\gamma\gamma$ modes of the NLC [93]. The kinematical reach is thus extended to $\sim \sqrt{s}$, but the production rates are suppressed by the unknown LQ Yukawa coupling to quark–lepton pairs. At $e\gamma$ colliders, first generation LQ's are observable as long as the corresponding squared LQ Yukawa coupling is larger than about $10^{-2}\, e^2$ (where $e^2 \equiv 4\pi\alpha$ is the electromagnetic coupling strength) [93].

Leptoquarks are color-nonsinglets and hence strongly interacting. As a result, LQ's can be produced at hadron colliders with very large rates. At the LHC, pair production in the $gg/q\bar{q} \to$ LQ $\bar{\text{LQ}}$ process leads to cross sections ranging from a few nb for masses $\mathcal{O}(0.1$ TeV) to a few fb for masses $\mathcal{O}(1$ TeV). The rate for vector particles (which depends on their anomalous magnetic moment κ) is substantially higher than for scalars. At the LHC with 100 fb^{-1} the search reach for scalar/vector LQ's is 1.4/2.2(1.8) TeV for $\kappa = 1$ (0), if one assumes a branching fraction of unity for the e^+e^-+two jet final state [94]. Single scalar LQ production, through $gu \to e^+$LQ and $gd \to \bar{\nu}$LQ, can also lead to large cross sections for Yukawa couplings of order unity. In this case masses up to $\sim$ 1.5 TeV can be reached at the LHC [94]. Vector leptoquarks have larger production rates and the discovery reach can be extended to $\sim$ 2 TeV. Although LQ's can also be singly produced in association with a W_R in alternative Left-Right models via $gu \to W_R$+LQ, a detailed study of the signal and backgrounds has shown that the LQ discovery reach in this case is quite limited [95]. For first generation leptoquark searches, ep colliders are particularly well suited. Such LQ's can be produced as s channel resonances in $eq \to$LQ. For example, at LEP×LHC masses up to 1 TeV can be reached, unless the corresponding LQ Yukawa couplings are very small.

Dilepton production has been considered at the NLC in the three collider modes: e^+e^-, $e\gamma$ and $\gamma\gamma$. These particles are accessible up to masses close to $\sqrt{s}/2$ in pair production, $e^+e^-/\gamma\gamma \to X^{++}X^{--}$: the rates (especially in $\gamma\gamma$ collisions because of the charge) are very large and the signatures (four leptons) are spectacular. Dileptons can also be singly produced in the three modes of the collider [96]. Diquarks can be pair produced in e^+e^- and $\gamma\gamma$ collisions for masses smaller than $\sqrt{s}/2$ with appreciable rates, with a signal consisting of an excess of 4 jet events.

They can be also produced at hadron colliders in pairs (or singly for the first generation diquark if its coupling to quark pairs is not too small). However, large QCD backgrounds present a formidable challenge to diquark searches at the LHC.

8.3 Excited Fermions

The existence of excited particles is a characteristic signal of substructure in the fermionic sector [80]: if the known fermions are composite (composed of more elementary constituents called *preons*), then they should be the ground state of a rich spectrum of excited states which decay down to the former states via a magnetic dipole type de-excitation. Since there is not yet a satisfactory and predictive dynamical model for composite fermions, one has to use phenomenological inputs to study this scenario. For simplicity, one assumes that the excited fermions have spin and isospin 1/2; their couplings to gauge bosons are vector–like (form factors and new contact interactions may also be present). Furthermore, the coupling which describes the transition between excited and ordinary fermions is taken to be chiral and inversely proportional to the compositeness scale Λ which is of $\mathcal{O}(1$ TeV).

Excited fermions can be pair-produced in $f\bar{f}$ annihilation via s-channel gauge boson exchange ($V = W$, Z, γ or gluon), although some suppression due to form factors is expected [81]. The excited fermions decay into gauge bosons and their ordinary partners, $f^* \to fV$. The charged leptons have the electromagnetic decay which has at least a branching ratio of 30%, and the excited quarks decay most of the time into quarks and gluons. These two decays constitute a very characteristic signature and discriminate them from the exotic fermions previously discussed.

In e^+e^- and $\gamma\gamma$ collisions, the processes and the cross sections are the same as the ones previously described for vector–like exotic fermions (up to possible form-factor suppressions); hence the detection of f^* masses up to the kinematic limit may be possible. Excited fermions can be singly produced with their light partners, but the rates are suppressed by a factor of $1/\Lambda^2$. At e^+e^- colliders, for quarks and second/third generation leptons, for which the process is mediated by s–channel boson exchange, the cross sections are very small. But for the first generation excited fermions, one has substantial contributions due to additional t–channel diagrams: W exchange for ν_e^* and Z/γ exchanges for e^*, which increase the cross sections by several orders of magnitude. The excited leptons can be also singly produced in $e\gamma$ collisions (e^* as a resonance and ν^* in association with a W) with much larger rates. All excited fermions can be singly produced in $\gamma\gamma$ collisions with appreciable cross sections. For $\Lambda \sim$ few TeV, excited quarks and leptons can easily be found in such machines, if kinematically allowed [83].

Due to the special couplings of the electron to the excited leptons of the first generation, single production of e^* through t-channel γ and Z exchange, and ν_e^* through W exchange are possible processes in ep collisions. The cross sections are large; requiring a few tens of events to establish a signal, one can probe ν^* and e^* masses up to 800 GeV at LEP$\times$LHC for $\int \mathcal{L} = 2$ fb^{-1} [97]. Excited quarks of the first generation can also be produced in ep collisions, however background problems make this possibility less interesting.

Excited quarks can be produced in pp collisions through a variety of mechanisms [81]. The dominant production channels are the gluonic excitation of quarks $g + q \to q^*$ which occurs through the q^*qg magnetic interaction, and the excitation through the preon interactions $qq \to qq^*$ and q^*q^*. The latter is also responsible for excited lepton production, $q\bar{q} \to ee^*$ and e^*e^*. The signatures of excited quarks are dijet mass bumps, jet + gauge boson and jet + lepton pair combinations. Excited leptons would reveal themselves in leptons + gauge particles or leptons + quark jets. QCD backgrounds have been studied and have been shown to be under control. At the LHC with a luminosity of 10 fb^{-1}, a mass range of 5–6 TeV and ~ 4 TeV can be reached for excited quarks and excited leptons, respectively [89].

8.4 Synopsis

The prospects for the discovery of new fermions (of the sequential, exotic, or excited variety) or di–fermions at future high–energy colliders can be summarized as follows:

e^+e^-, $e\gamma$, $\gamma\gamma$, e^-e^- colliders:

- All new particles considered in this section (except for weak isosinglet neutrinos) can be pair produced in e^+e^- collisions up to the kinematical limit of $\sqrt{s}/2$. Similarly for new charged particles in $\gamma\gamma$ collisions.
- Due to their special couplings to the electron, the first generation of exotic leptons, excited leptons, leptoquarks and dileptons can be singly produced if the mixing angles, the compositeness scale or the couplings to the light fermion pairs are not prohibitively small; one can then reach masses close to the center-of-mass energy of the collider. Here, all four potential modes of the NLC can be useful.
- Due to the clean environment of these machines, one can perform precision measurements which allow one to probe the indirect effects of some of these particles for mass scales much higher than the center-of-mass energy.

ep colliders:

- *ep* colliders are well suited for the production of the first generation exotic leptons, excited leptons and especially leptoquarks, if the mixing angles, the compositeness scale or the couplings to quark–lepton pairs, respectively, are not too small. At LEP×LHC, masses around 1 TeV can be reached.

pp colliders:

- Proton colliders are ideal machines for the production of new color non-singlets. At the LHC, particle masses in the TeV range can be probed: ~ 1 TeV for sequential or exotic quarks, $\sim$ 1.5—2 TeV for leptoquarks and 5—6 TeV for excited quarks for reasonable values of the compositeness scale.
- New color-singlet particles can also be produced, but with smaller cross sections than for quarks. However the signatures are cleaner and might compensate for the small rates. For instance, some exotic leptons can be discovered for masses smaller than $\sim 1/2$ TeV; through contact interactions, excited electrons can also be produced for masses in the TeV range.

Finally, it should be noted that the effects of new interactions may be observable even if all the associated new particles beyond the SM are too heavy to be observed directly. If the mass scale of the new physics, Λ, is assumed to be much larger than m_Z, then one can formally integrate out the heavy particles and remove them from the full theory. The resulting low-energy effective theory will differ from the SM by new (anomalous) interactions among SM particles which are suppressed by inverse powers of Λ. For example, new physics beyond the SM could generate four-fermion contact interactions [98], which could be detected as deviations from SM predictions in 2-to-2 scattering processes involving quarks and/or leptons [91]. In this way, one may have experimental sensitivity to much larger values of Λ as compared to direct searches for new particles. The search for anomalous couplings in the SM may be particularly fruitful in the precision study of W and Z self-interactions and top-quark interactions. Since these are the heaviest particles of the SM, it is plausible that their interactions could be modified by the dynamics underlying the generation of mass. The phenomenology of anomalous gauge boson and top-quark couplings is studied in more detail in the next two chapters.

9 Anomalous Gauge Boson Couplings

9.1 Parametrization of Anomalous Gauge Bosons Interactions

One of the most direct consequences of the $SU(2) \times U(1)$ gauge symmetry of the SM, the non-abelian self-couplings of the W, Z, and photon, remains poorly measured to date. Although information on these couplings can be extracted from low energy data and high precision measurements at the Z pole, there are ambiguities and model dependencies in the results. However, a direct measurement of these vector boson couplings is possible in present and future collider experiments, in particular via pair production processes like $e^+e^- \to W^+W^-$ and $\overset{(-)}{p}p \to W^+W^-$, $W\gamma$, WZ. The first and major goal of such experiments will be the verification of the SM predictions at a quantitative level.

For our discussion of experimental sensitivities we employ a parametrization of the WWV couplings ($V = \gamma$, Z) in terms of a phenomenological effective Lagrangian [99]:

$$\begin{aligned} i\mathcal{L}_{WWV} &= g_{WWV}\Big[g_1^V\left(W^\dagger_{\mu\nu}W^\mu V^\nu - W^\dagger_\mu V_\nu W^{\mu\nu}\right) \\ &+ \kappa_V W^\dagger_\mu W_\nu V^{\mu\nu} + \frac{\lambda_V}{m_W^2} W^\dagger_{\lambda\mu} W^\mu_\nu V^{\nu\lambda}\Big] \quad (10) \end{aligned}$$

Here the overall coupling constants are defined as $g_{WW\gamma} = e$ and $g_{WWZ} = e\cot\theta_W$. The couplings g_1^V, κ_V, and λ_V need to be determined experimentally. Within the SM, at tree level, they are given by $g_1^Z = g_1^\gamma = \kappa_Z = \kappa_\gamma = 1$, $\lambda_Z = \lambda_\gamma = 0$. $g_1^\gamma = 1$ is fixed by electromagnetic gauge invariance; g_1^Z, however, may well be different from its SM value 1 and appears at the same level as κ_γ or κ_Z.

The effective Lagrangian of Eq. (10) parametrizes the most general Lorentz invariant and C and P conserving WWV vertex which can be observed in processes where the vector bosons couple to effectively massless fermions. Terms with higher derivatives are equivalent to a dependence of the couplings on the vector boson momenta and thus merely lead to a form-factor behavior of these couplings [99]. Analogous to the general WWV vertex it is possible to parametrize anomalous $Z\gamma V$, $V = \gamma, Z$ couplings in terms of two coupling constants, h_3^V and h_4^V, if CP invariance is imposed. All $Z\gamma V$ couplings are C odd; h_3^V corresponds to a dimension 6, h_4^V to a dimension 8 operator [99].

Tree level unitarity uniquely restricts the WWV and $Z\gamma V$ couplings to their SM gauge theory values at asymptotically high energies. This implies that deviations of g_1^V, κ_V, λ_V and h_i^V, $i = 3, 4$ must be parametrized by form factors which indicates that the effective Lagrangian describing the anomalous vector boson self-interactions breaks down at some high energy scale. SM one-loop effects lead to corrections of $\mathcal{O}(\alpha)$ with a form factor scale, Λ_{FF}, of order the electroweak symmetry breaking scale.

Models which have been considered so far, such as technicolor or supersymmetric models, predict rather small deviations from the SM values, of $\mathcal{O}(10^{-2})$ or less. If we regard the Standard Model as an effective field theory, then new physics can be parametrized in a model independent way by additional gauge-invariant operators.

Regardless of whether Higgs physics is strongly or weakly interacting, there are restrictions on the strength of such contributions which, however, depend on naturalness assumptions. Detailed analyses [100,101] show that these restrictions leave a slim chance only that new physics will be discovered through anomalous couplings at machines which are less capable than the LHC or NLC.

9.2 Search for Anomalous Gauge Boson Couplings at Present and Future Colliders

The direct measurement of WWV and $Z\gamma V$ couplings in present and future collider experiments in general makes use of the high energy behavior of the non-standard contributions to the helicity amplitudes which grow like a power of the parton center of mass energy. In di-boson production, this leads to an excess of events at large W, Z and photon transverse momenta as well as at large invariant masses of the di-boson system. Furthermore, the various non-standard couplings contribute to different helicity amplitudes in the high energy limit. Angular distributions of the W and Z-boson decay products, therefore, are helpful as polarization analyzers, in particular in e^+e^- collisions. Quantitative bounds on $\Delta g_1^Z = g_1^Z - 1$, $\Delta\kappa_V = \kappa_V - 1$ and λ_V are obtained by comparing the shape of the measured and the predicted distributions.

Present and future 95% confidence level (CL) limits on a variety of WWV and $Z\gamma V$ couplings are listed in Table 11. In addition, the limits on the couplings $\Delta\kappa_\gamma$ and λ_γ are displayed in bargraph form in Fig. 5. The present limits are from di-boson production at the Tevatron [102] and from single photon production at LEP [103]. The future limits are estimated for experiments at the Tevatron, LEP II, the LHC and a future linear e^+e^- collider (NLC). In contrast to the bounds on anomalous WWV couplings, the limits on the $Z\gamma V$ couplings h_{i0}^V, $i = 3,4$ depend significantly on the form factor scale Λ_{FF}. Here they are assumed to appear in the form $h_i^V = h_{i0}^V/(1 + \hat{s}/\Lambda_{FF}^2)^{n_i}$, $i = 3,4$ with $n_3 = 3$ and $n_4 = 4$.

A direct measurement of the $WW\gamma$ vertex is achieved via $W\gamma$ production. The best current bounds on WWV couplings, however, depend significantly on information obtained from W^+W^- and WZ production at the Tevatron in the channel $p\bar{p} \to W^+W^-$, $W^\pm Z \to \ell^\pm\nu jj$. In deriving these bounds, $\Delta\kappa_\gamma = \Delta\kappa_Z$ and $\lambda_\gamma = \lambda_Z$ is assumed.

The present limits on anomalous WWV and $Z\gamma V$ couplings can be improved by one order of magnitude or more in future Tevatron experiments if an integrated luminosity of 1 fb^{-1} can be achieved. Raising the integrated luminosity to 10 fb^{-1} typically leads to a factor ~ 2 improvement over the limits which can be achieved with 1 fb^{-1}. Fitting several different distributions simultaneously might lead to further improvements. Bounds on Δg_1^Z and $\Delta\kappa_V$ are expected to be comparable to those which can be obtained at LEP II. For λ_V, the limits from future Tevatron experiments should be better.

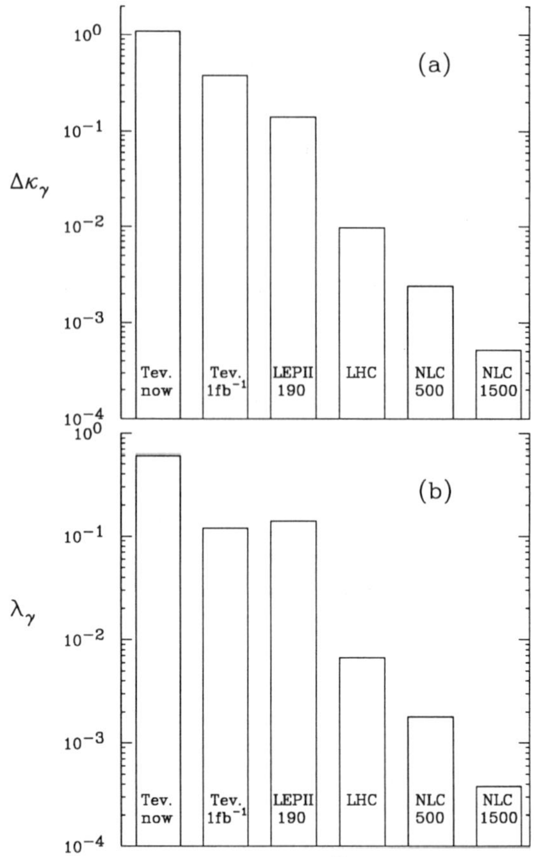

Figure 5: Present and future 95% CL bounds on $\Delta\kappa_\gamma$ (a) and λ_γ (b).

The future bounds listed in Table 11 for W^+W^- and $W^\pm Z$ final states were obtained assuming $\lambda_Z = \lambda_\gamma$, $\Delta g_1^Z = \Delta\kappa_\gamma/2\cos^2\theta_w$ and $\Delta g_1^Z + \Delta\kappa_Z = \Delta\kappa_\gamma$. These relations are motivated by an effective Lagrangian based on a complete set of $SU(2) \times U(1)$ invariant operators of dimension 6 [101]. If one assumes $\kappa_\gamma = \kappa_Z$, $\lambda_\gamma = \lambda_Z$ and $\Delta g_1^Z = 0$ instead, the limit on $\Delta\kappa_\gamma$ can be improved by about 40%.

At the LHC, with an integrated luminosity of 100 fb^{-1}, values of $\Delta\kappa_Z$, Δg_1^Z and λ_Z of $\mathcal{O}(10^{-2})$ can be probed. Similar sensitivities can be achieved for the $WW\gamma$ couplings. A linear e^+e^- collider with a center of mass energy of 500 GeV or more in general can probe anomalous couplings of order 10^{-3} [104,50].

The present bounds on $Z\gamma V$ couplings can be improved only by about a factor 2 at best in $Z\gamma$ production at LEP II. Future experiments at the Tevatron will be

Table 11: Present and future 95% CL bounds on WWV and $Z\gamma V$ couplings.

Experiment	Channel	Limit
DØ	$p\bar{p} \to W^{\pm}\gamma \to \ell^{\pm}\nu\gamma$	$-1.6 < \Delta\kappa_\gamma < 1.8$
(present limit)	$\ell = e,\, \mu$	$-0.6 < \lambda_\gamma < 0.6$
CDF	$p\bar{p} \to W^+W^-,\, W^{\pm}Z \to \ell^{\pm}\nu jj$	$-1.0 < \Delta\kappa_V < 1.1$
(present limit)	$\kappa_\gamma = \kappa_Z,\, \lambda_\gamma = \lambda_Z,\, g_1^Z = 1$	$-0.8 < \lambda_V < 0.8$
DØ	$p\bar{p} \to Z\gamma \to \ell^+\ell^-\gamma$	$-1.9 < h_{30}^V < 1.8$
(present limit)	$\ell = e,\, \mu,\, \Lambda_{FF} = 0.5$ TeV	$-0.5 < h_{40}^V < 0.5$
L3	$e^+e^- \to \bar{\nu}\nu\gamma$	$-0.9 < h_{30}^Z < 0.9$
(present limit)		$-2.3 < h_{40}^Z < 2.3$
Tevatron	$p\bar{p} \to W^{\pm}\gamma \to e^{\pm}\nu\gamma$,	$-0.38 < \Delta\kappa_\gamma < 0.38$
(CDF sim.)	$\sqrt{s} = 2$ TeV, $\int \mathcal{L}dt = 1$ fb^{-1}	$-0.12 < \lambda_\gamma < 0.12$
Tevatron	$p\bar{p} \to W^+W^-,\, W^{\pm}Z \to \ell^{\pm}\nu jj,\, \ell^+\ell^- jj$	$-0.31 < \Delta\kappa_\gamma < 0.41$
(CDF sim.)	$\int \mathcal{L}dt = 1$ fb^{-1}, $\Lambda_{FF} = 2$ TeV	$-0.19 < \lambda_\gamma < 0.19$
Tevatron	$p\bar{p} \to W^{\pm}Z \to \ell_1^{\pm}\nu_1\ell_2^+\ell_2^-$	$-0.26 < \Delta\kappa_Z < 0.70$
(NLO theor.)	$\int \mathcal{L}dt = 1$ fb^{-1}, $\Lambda_{FF} = 1$ TeV	$-0.24 < \lambda_Z < 0.32$
Tevatron	$p\bar{p} \to Z\gamma \to e^+e^-\gamma$	$-0.105 < h_{30}^V < 0.105$
(CDF sim.)	$\int \mathcal{L}dt = 1$ fb^{-1}, $\Lambda_{FF} = 1.5$ TeV	$-0.0064 < h_{40}^V < 0.0064$
Tevatron	$p\bar{p} \to Z\gamma \to e^+e^-\gamma$	$-0.044 < h_{30}^V < 0.044$
(CDF sim.)	$\int \mathcal{L}dt = 10$ fb^{-1}, $\Lambda_{FF} = 1.5$ TeV	$-0.0025 < h_{40}^V < 0.0025$
LEP II	$e^+e^- \to W^+W^- \to \ell^{\pm}\nu jj$,	$-0.13 < \Delta\kappa_\gamma < 0.14$
(L3 sim.)	$\sqrt{s} = 190$ GeV, $\int \mathcal{L}dt = 500$ pb^{-1}	$-0.13 < \lambda_\gamma < 0.14$
LEP II	$e^+e^- \to Z\gamma \to \nu\bar{\nu}\gamma$, $\sqrt{s} = 180$ GeV,	$-0.50 < h_{30}^Z < 0.50$
(L3 sim.)	$\int \mathcal{L}dt = 500$ pb^{-1}, $\Lambda_{FF} = 1$ TeV	$-0.45 < h_{40}^Z < 0.45$
LHC	$pp \to W^{\pm}Z \to \ell_1^{\pm}\nu_1\ell_2^+\ell_2^-$,	$-0.006 < \Delta\kappa_Z < 0.0097$
(NLO theor.)	$\int \mathcal{L}dt = 100$ fb^{-1}, $\Lambda_{FF} = 3$ TeV	$-5.3 \cdot 10^{-3} < \lambda_Z < 6.7 \cdot 10^{-3}$
LHC	$pp \to Z\gamma \to e^+e^-\gamma$	$-5.1 \cdot 10^{-3} < h_{30}^V < 5.1 \cdot 10^{-3}$
(LO theor.)	$\int \mathcal{L}dt = 10$ fb^{-1}, $\Lambda_{FF} = 1.5$ TeV	$-9.2 \cdot 10^{-5} < h_{40}^V < 9.2 \cdot 10^{-5}$
NLC	$e^+e^- \to W^+W^- \to \ell^{\pm}\nu jj$,	$-0.0024 < \Delta\kappa_\gamma < 0.0024$
	$\sqrt{s} = 500$ GeV, $\int \mathcal{L}dt = 80$ fb^{-1}	$-0.0018 < \lambda_\gamma < 0.0018$
NLC	$e^+e^- \to W^+W^- \to \ell^{\pm}\nu jj$,	$-5.2 \cdot 10^{-4} < \Delta\kappa_\gamma < 5.2 \cdot 10^{-4}$
	$\sqrt{s} = 1500$ GeV, $\int \mathcal{L}dt = 190$ fb^{-1}	$-3.8 \cdot 10^{-4} < \lambda_\gamma < 3.8 \cdot 10^{-4}$

sensitive to values of $h_{30}^V \sim 0.11$ and $h_{40}^V \sim 0.006$ (with $\Lambda_{FF} = 1.5$ TeV).

Within the next few years experiments at the Tevatron and at LEP are expected to confirm the non-abelian self-couplings of the SM at the 10% level. The $\mathcal{O}(\alpha)$ deviations predicted by the SM and most models of new physics, however, can only be probed at the LHC or a high energy linear e^+e^- collider. Of course, if the scale of new physics is directly accessible at these machines one would rather focus on the spectrum of new particles and direct measurements of their interactions with the gauge boson sector.

10 Top Quarks as a Window to Electroweak Symmetry Breaking

Among the various matter fermions, the top quark must be considered the most mysterious. It is the heaviest known elementary particle, with mass 30–40 times that of the next heaviest quark. It is more strongly coupled to the mechanism of electroweak symmetry breaking than the weak gauge bosons themselves. It is plausible, in fact, that the top quark is an essential component of this mechanism. On a more mundane level, the top quark is the only quark or lepton whose couplings to the Standard Model are not constrained to percent accuracy by the LEP experiments. Thus, the couplings of the top can easily yield surprises.

Most of this section is concerned with the search for anomalies in the couplings of the top quark to the Standard Model gauge bosons. We will consider first the possibility of large deviations from the standard properties of the top quark and then the precision experiments that could show signs of smaller deviations. Finally, we will discuss the possibility that the top quark has decay modes outside the Standard Model.

10.1 Anomalous Top Couplings – s-Channel Resonance

The original CDF announcement of evidence for the top quark [105,106] came with an anomaly, the measurement of a production cross section 2–3 times larger than the theoretical prediction. The more recent announcements of the observation of the top quark by CDF [107] and DØ [108] give a production cross section of 6.5 ±1.8 pb, which is consistent with the Standard Model unless the top quark mass is above 185 GeV. Nevertheless, the possibility of order-1 deviations of the top quark couplings from the predictions of the Standard Model is extremely tantalizing and might well open a new window into physics of flavor.

At the Tevatron collider, the dominant mechanism of $t\bar{t}$ production is through $q\bar{q}$ annihilation; at a top mass of 175 GeV, the gg fusion cross section is a factor 10 smaller. To dramatically enhance this cross section, it is natural to consider models with an s-channel resonance in $t\bar{t}$ production, with a mass of 500–600 GeV, just above the $t\bar{t}$ threshold. Two groups of authors have proposed explanations of this type based on pre-existing theoretical models [109,65].

Higher luminosity running at the Tevatron collider will clarify this situation. The clearest prediction of the model is an excess of events at large $t\bar{t}$ invariant mass and high t transverse momentum. Taking the model [109] with a 600 GeV resonance for definiteness, event samples of 0.1 fb^{-1}, 3 fb^{-1}, and 100 fb^{-1} lead to roughly 10, 300, and 10,000 observed $t\bar{t}$ events with $m(t\bar{t}) > 600$ GeV, assuming efficiency (1/10) for detection, above a background 5 times lower. These samples correspond to the goals for the current Tevatron collider run, the new Main Injector upgrade, and the TeV* luminosity upgrade. With the same event samples, one could put lower limits on the mass of an s-channel resonance at roughly 700, 1000, 1500 GeV.

At the LHC, the production cross section for $t\bar{t}$ is 90% gg fusion, but the observed rates are large, of order 10^6 $t\bar{t}$ pairs per year. The cross section measurement at the LHC clearly distinguishes the $q\bar{q}$ from the gg model for the s-channel resonance; these models predict cross sections differing by a factor of 10. The mass reach of the LHC with 10 fb^{-1} is about 1000 GeV in the $q\bar{q}$ model and 2000 GeV in the gg model, with stronger constraints at 100 fb^{-1}.

Since both models involve resonances that do not couple to leptons, their direct effect on $t\bar{t}$ production is not easily tested at an e^+e^- collider. However, for both models, an e^+e^- collider with an energy about 1 TeV could provide important complementary information on the associated production of the new resonances with $t\bar{t}$ and on the pair-production of related new particles with electric charge and exotic color.

10.2 Anomalous Top Couplings – Effective Lagrangian

If the top quark production cross section proves not to be anomalous at the zeroth order, how well can the top quark couplings be searched for anomalies at higher order? An appropriate framework for discussing this question is an effective Lagrangian governing the top quark coupling to gauge bosons. For example, the coupling of the W boson to $t\bar{b}$ may be written

$$\Delta\mathcal{L} = \frac{g}{\sqrt{2}} F_1^W(q^2)\bar{t}\gamma^\mu b_L W_\mu + \frac{1}{4m_t} F_2^W(q^2)\bar{t}\sigma^{\mu\nu} b_L F_{\mu\nu}^W. \tag{11}$$

Analogous formulae define the F_1 and F_2 form factors of g, γ, and Z. Phenomenological couplings of this form have been studied by many authors [110]. Electroweak interactions typically give contributions to these form factors at the 1% level or below. However, it is likely that there are larger contributions from models in which the top quark is strongly coupled to the symmetry breaking sector. For example, extended technicolor models give contributions to the various F_1 form factors of t_L, t_R, and b_L which are typically of order 5% but whose pattern depends on the details of the model [111].

We consider first the measurement of these form factors in hadron-hadron collisions. It is possible that these form factor corrections are large; in fact, it has been proposed [112] that a nonzero color anomalous magnetic moment $F_2^g \sim 0.25$ could provide an alternative explanation of the large cross section reported by CDF. On the other hand, if the form factor effect is smaller, one's ability to measure it quickly becomes limited by the 15% QCD uncertainty in the $t\bar{t}$ production cross section. In a one-parameter fit based on the total cross section, the Tevatron could bound the F_2 of the ttg coupling to ± 0.15 (95% conf.) with 0.1 fb^{-1} and reach the systematics limit of ± 0.10 with 0.5 fb^{-1} [113]. The same systematics limit appears at the LHC. Yuan has suggested that the F_1 coefficients of the $Wt\bar{b}$ couplings can be constrained by measuring the cross section for single top production: $gW \to t\bar{b}$; this cross section, however, has a theoretical uncertainty of order 30% [114]. One parameter of $t\bar{t}$ production at hadron colliders which seems very promising for a precision constraint is the measurement of the longitudinal/transverse polarization of W bosons from top decay; a 3 fb^{-1} event sample at the Tevatron should bring the statistical uncertainty below 4% [115].

The environment offered by e^+e^- colliders offers many advantages for a precision study of the top couplings. In addition to low backgrounds and the ability to reconstruct $t\bar{t}$ in the 6-jet mode, top production is characterized by order-1 forward-backward and polarization asymmetries which reflect sizeable interference of the γ and Z exchange diagrams. Top quarks produced in the forward direction carry the electron polarization and thus provide a polarized sample for studies of the decay form factors.

For comparison with the estimate given above, the total cross section measurement at an e^+e^- collider with 20 fb^{-1} should bound a common F_2 for the $tt\gamma$ and $t\bar{t}Z$ couplings, in a one-parameter fit, at ± 0.03. However, much more detailed studies are possible, at least with large event samples. For example, consider a variation in the $Zt\bar{t}$ coupling leaving the $t\bar{t}\gamma$ coupling fixed. The left- and right-handed form factors of $Zt\bar{t}$ coupling can be constrained in a two-parameter fit for an event sample of 100 fb^{-1} at 500 GeV, to deviations of ± 0.08 (95% conf.). (The systematic contribution to this error should be small, and will be analyzed carefully for the final report.) For comparison with the parameter of top decays discussed above, Fujii has found that the fraction of longitudinal W bosons in t decay can be determine with a statistical precision of 2% with 10 fb^{-1} of data at 350 GeV. By studying top pair production with an extra jet, one can obtain [116] a direct limit on the color anomalous magnetic moment F_2^g, to be compared with that from hadron colliders; with 200 fb^{-1}, the constraint is ± 0.16 (95% conf.).

The F_1 associated with the $Wt\bar{b}$ vertex can be obtained from a measurement of the total width of the top quark, assuming that no exotic decay modes are seen. This quantity can be measured at the threshold for $t\bar{t}$ production by three techniques: the width of the 1S resonance or shoulder, the momentum distribution of reconstructed top quarks decaying to Wb, and the forward-backward asymmetry of top production due to the interference of overlapping S and P wave states. Fujii, Matsui, and Sumino [117] have found that, by combining these techniques, the width of the t can be determined with 100 fb^{-1} of data to an accuracy of 4%, plus an error from the uncertainty in α_s that boost this number to about 10%. We should also note that, in models with a relatively light Higgs boson, of mass about 100 GeV, the height of the 1S enhancement at the $t\bar{t}$ threshold measures the $t\bar{t}$ Higgs coupling constant, to about 25% accuracy.

In addition to magnetic dipole (F_2) couplings, the top quark may also have electric dipole couplings which violate CP. Direct measurements of the form factors give limits on the CP-violating couplings similar to those cited above. This is probably not adequate for those models of CP violation involving CP-violating Higgs bosons coupling strongly to the top quark. However, Schmidt and Peskin have suggested observing the energy asymmetry in e^+ vs. e^- decay products of $t\bar{t}$. Combined with the high statistics available at the LHC, this technique could reasonable reach the CP asymmetries of $10^{-3} - 10^{-4}$ predicted by the models [118].

10.3 Exotic Decays of the Top Quark

The top quark may also connect to exotic states by its decays. Many exotic decay modes of top have been considered in the literature. Here are the most important ones:

$t \to H^+b$: This mode is interesting because it is the best way to find H^+ at hadron colliders. In addition, the ratio of t-quark branching ratios to H^+b and W^+b depends on $\tan\beta$, providing one of the best ways to measure this parameter of an extended Higgs sector. The decay mode should be found at LHC with 10^{33} luminosity, or with a 500 GeV e^+e^- collider with 1 fb^{-1} of data.

$t \to \widetilde{\chi}^0\widetilde{t}$: This is a supersymmetric decay of top which occurs in a reasonable volume of parameter space, with a branching ratio of roughly 5% if $\widetilde{t}$ is much lighter than t [119].

$t \to ch^0$, $t \to cZ$, $t \to cg$, *etc.*: These flavor changing neutral current modes are extremely rare in the Standard Model, with branching ratios $< 10^{-10}$. If these modes are dramatically enhanced by new physics, they might be observed at the LHC.

$t \to sW^+$. This mode is expected to be at the 0.1% level in the Standard Model. If this mode is enhanced to the few-percent level, it should be observable at an e^+e^- collider [120].

11 Virtual Effects of New Physics

11.1 Overview

The study of virtual effects can open an important window on electroweak symmetry breaking and physics beyond the Standard Model (SM). The examination of indirect effects of new physics in higher order processes offers a complementary approach to the search for direct production of new particles at high energy accelerators. In fact, tests of loop induced couplings can provide a means of probing the detailed structure of the SM at the level of radiative corrections where Glashow-Iliopoulos-Maiani (GIM) cancellations are important. In some cases the constraints on new degrees of freedom via indirect effects surpass those obtainable from collider searches. In other cases, entire classes of models are found to be incompatible. Given the large amount of high luminosity data which will become available during the next decade, this approach to searching for physics beyond the SM will become a valuable tool.

11.2 Precision Electroweak Measurements

Virtual effects from new physics are easily detected in precision electroweak measurements. These include the measurements of the Z line-shape and asymmetries at SLD/LEP, deep inelastic muon–neutrino scattering, and atomic parity violation. Because of the accurate theoretical predictions in the electroweak sector of the SM, especially in purely leptonic processes that are free of QCD corrections, any observed deviations must be explained as due to new physics not encompassed within the SM.

Radiative corrections from new physics are often classified into two groups; the 'oblique' and the 'direct' corrections. 'Oblique' refers to the vacuum polarization corrections to the gauge boson propagators. They are independent of the external particles and thus are universal to all electroweak processes. 'Direct' refers to the vertex and box corrections which are specific to each process. Because of the universality of oblique corrections, it is possible to express them in a model independent fashion with a few phenomenological parameters. Under the two assumptions that the electroweak gauge group is the standard $SU(2)_L \times U(1)_Y$, and that the scale of new physics is large compared to the electroweak scale, the virtual effects of new physics can be expressed in terms of 3 parameters, often called S, T, and U [121]. Using a set of measurements to which the direct corrections from new physics are expected to be small, one can place tight constraints on the values of S and T. These include the purely leptonic processes at SLD/LEP, and atomic parity violation measurements. Currently, the strongest constraint on S and T comes from the SLD/LEP measurement of $\sin^2\theta_w^{\rm eff}$ and $\Gamma(Z \to \ell^+\ell^-)$.

If one relaxes the second assumption that the scale of new physics is large compared to the electroweak scale, then it becomes necessary to introduce three additional parameters called V, W, and X [122]. If one relaxes the first assumption and extends the electroweak gauge group to $SU(2)_L \times U(1)_Y \times U(1)_{Y'}$, $SU(2)_L \times SU(2)_R \times U(1)_{B-L}$, $SU(3) \times U(1)$, etc. then it becomes necessary to take into account the exchange of the additional gauge bosons as well as their virtual loop effects so these models cannot be treated within the S–T formalism and each case must be analyzed separately.

A process where the direct corrections from new physics is expected to be sizable is $Z \to b\bar{b}$. This is due to the fact that the b–quark is the isospin partner of the t–quark so the same mechanism which generates the large t–quark mass will also generate a large vertex correction to this process. This is true in the SM which has a vertex correction proportional to m_t^2, and also in Technicolor models and SUSY models. If such new vertex corrections exist, then one should be able to see a deviation from the SM prediction after the universal oblique correction effects have been folded out. Such a deviation at the 2σ-level is presently observed at LEP in the ratio $R_b = \Gamma_{b\bar{b}}/\Gamma_{\rm had}$.

11.3 Rare Processes in the Quark Sector

Rare K processes have played a strong and historical role in constraining new interactions. The box and penguin diagrams which typically mediate these processes receive contributions from the SM, as well as potentially sizeable contributions from new physics. For example, the strongest limit (albeit assumption dependent) on the mass of a right-handed W-boson is derived from its contributions to K^0–$\bar{K}^0$ mixing, and the requirement of near degeneracy of squark masses results from a super-GIM mechanism imposed in the K sector. Rare K decay experiments at Brookhaven and KTeV at the Tevatron will provide further probes in the near future. The decay $K^+ \to \pi^+\nu\bar{\nu}$ is an interesting mode as it is theoretically clean and experiment will reach the SM level during the next decade. $K_L^0 \to \mu^\pm e^\mp$ is forbidden in the SM and provides a probe of lepton number violation, leptoquark exchange, and family symmetries.

Short distance SM contributions to rare charm processes are small due to the effectiveness of the GIM mechanism (as there are no large quark mass splittings in the loops). Long distance effects usually dominate and are difficult to reliably calculate. Observation of $D \to \rho\gamma$ would provide an excellent test of the techniques used in these calculations, and could be used to reliably evaluate the related long distance contributions to $B \to K^*\gamma, \rho\gamma$. However, there is a window for the clean observation of possible new physics contributions in some processes, such as D^0–$\bar{D}^0$ mixing, $D \to X_u \ell^+ \ell^-$, and $D \to \mu\mu$. The types of non-standard scenarios which give the largest contributions to these reactions are

- two-Higgs-doublet models in the large $\tan\beta$ region,
- multi-Higgs doublet models with flavor changing couplings,
- new $Q = -1/3$ quarks,
- supersymmetric models with non-degenerate squarks, and
- variants of left-right symmetric models.

Several high volume charm experiments are planned for the future, with 10^{7-8} charm mesons expected to be reconstructed. D^0–$\bar{D}^0$ mixing can then be probed another 1–2 orders of magnitude below present sensitivities.

A large amount of data in the B-meson system will be acquired during the next decade at SLC/LEP, CESR, the Tevatron, and the SLAC and KEK B Factories. Here, one-loop processes occur with sizeable rates in the SM, due to large top-quark contributions. Many classes of new models can give significant and testable contributions to rare B processes. The inclusive process $B \to X_s\gamma$ and the related exclusive decay $B \to K^*\gamma$ has been observed by CLEO and is the first direct observation of a penguin mediated process. It has provided strong restrictions on parameters in several theories beyond the SM [123], with the bounds being competitive with those from direct collider searches. For example, it constrains the mass of a charged Higgs boson to be $> 260\,\text{GeV}$ for $m_t = 175$ GeV, and provides complementary bounds to those obtained at colliders on an anomalous $WW\gamma$ vertex. The inclusive processes $b \to s\ell\ell, s\nu\bar{\nu}$ are also excellent probes of new physics. They are sensitive to the form of possible new interactions since they allow measurements of various kinematic distributions as well as the total rate. Techni-GIM models predict $B(b \to s\mu^+\mu^-) \sim \mathcal{O}(10^{-4})$ and appear to already be in conflict with the present experimental limits.

CP violation in the B system will be examined at the SLAC and KEK B Factories. Signals for new sources of CP violation include,

- non-closure of the 3 generation unitarity triangle,
- non-vanishing CP asymmetries for the channels $B_d^0 \to \phi\pi^0, K_S^0 K_S^0$,
- inconsistency of separate measurements of the angles of the unitarity triangle, and
- a deviation of CP rates from SM predictions.

The large data sample and asymmetric configuration of the B Factories will provide a series of unique consistency tests of the quark sector and will challenge the SM in a new and quantitatively precise manner.

Loop-induced top-quark decays are small in the SM due to the effectiveness of the GIM mechanism. Branching fractions for $t \to c\gamma, cZ, cg, ch$ lie in the range 10^{-12}–10^{-8}. Long distance effects are expected to be negligible. Contributions from new physics can enhance these rates by 3-4 orders of magnitude. The tree-level decay $t \to ch$, if kinematically accessible, is possible in multi-higgs models without natural flavor conservation. $t \to \tilde{t}\tilde{\chi}^0$ can occur in SUSY if the stop-squark is light. These would provide clean signals for new physics and may be observable at a high luminosity Tevatron.

11.4 Electric Dipole Moments

Electric dipole moments (EDMs) of atoms, molecules, and the neutron are the most sensitive test of low energy flavor conserving CP violation. Such CP violation from the CKM angle is highly suppressed (by at least 3 loops), while other sources of CP violation are generally not so suppressed. Atoms with an unpaired electron (such as ^{133}Cs) are generally most sensitive to the electron EDM, while the neutron and atoms with paired electrons (such as ^{199}Hg) are sensitive to CP violation in the strongly interacting sector. Although no EDM has yet been observed, the current bounds already put stringent constraints on CP-violating extentions of the SM. For example in supersymmetric theories the bound of 1.3×10^{-27} e cm for the EDM of ^{199}Hg places a limit on new CP-violating phases that can arise in gaugino mass terms and in off-diagonal squark masses and squark interactions. These new phases must be less than a few tenths of a percent for $M_{\text{SUSY}} \simeq 100$ GeV. This bound also limits the CP-violating phases in the Higgs potential in multi-Higgs theories to be less than roughly 10^{-1} for $m_{\text{Higgs}} \simeq 100$ GeV. The bound from the neutron gives a slightly less stringent limit. Improvement in most experiments is anticipated over the next few years. New techniques, such as atomic traps, may allow improvement of up to a few orders of magnitude. If positive measurements are eventually made the QCD vacuum angle contribution can be distinguished from new physics by comparing atoms or molecules with paired and unpaired electrons.

11.5 Double Beta Decay

Neutrinoless double beta decay requires the violation of lepton number and is thus a clear signal for physics beyond the SM. Such a decay could arise from a nonzero Majorana mass for the electron neutrino. In theories with spontaneously broken lepton number, the decay could also be accompanied by the emission of a Majoron. The current half life limit for ^{76}Ge gives a bound on the effective Majorana neutrino mass of $m_\nu < 1$–2 eV. The neutrino-Majoron coupling is also constrained to be $g_{\nu,\chi} < 10^{-4}$. The current generation of enriched Ge experiments could reach the 0.3 eV level in m_ν. A large experiment, NEMO III, using ^{100}Mo, is being designed to reach the 0.1 eV level.

12 Experimental Issues at Hadron Colliders

The frontiers of particle physics have been significantly pushed forward by discoveries made at colliders with the highest available center-of-mass energies. Since the early 1980's, hadron colliders have become the tool of choice for extending our knowledge of the high energy frontier. The ability of collider detectors to cover a broad range of new phenomena over a wide range of mass scale allows many of the most important measurements to be made and limits to be set in an economical way. A mix of both "discovery" and precision measurements have been successfully carried out at such machines over the last ten years. Progress has depended upon a mix of both theoretical and experimental developments.

During the last ten years, the theoretical community has been able to refine our knowledge of hadron-hadron interactions to enable calculations with increasing precision. The soundness of the theoretical framework puts the hadron-hadron collider on a firm footing for reliable estimation of event rates and backgrounds for a variety of new phenomena.

The experimental community has likewise made important advances during this period. Operating experience at hadron colliders with precision tracking detectors has been gained. These devices open up the possibility of tagging jets containing b-quarks, a capability that considerably broadens the physics reach of collider detectors as the recent CDF top quark evidence has shown. In addition, the broad-based program of detector R&D for SSC and LHC detectors has deepened our understanding of how to handle the high occupancies and radiation fluxes at high luminosity hadron colliders. The knowledge gained from the detector R&D and design studies gives us confidence that the theoretically predicted physics capabilities of the hadron collider machines can be exploited in detectors that are practical to build and operate.

12.1 Machine Environment

The LHC program is designed to study physics beyond the Standard Model over a wide mass range for all types of hypothetical particles. To achieve a substantial reach for the production and observation of new particles, LHC plans to run at a luminosity of $10^{34}\text{cm}^{-2}\text{s}^{-1}$ with a 25 ns bunch spacing. There will be on average 25 interactions per crossing contributing considerable complexity to the detection of interesting phenomena. Future luminosity upgrades at Fermilab may also generate conditions comparable to these. The challenge met by the SSC and LHC R&D programs was to create detector designs that could survive in these harsh conditions of high occupancy and high radiation, while providing adequate resolution and robust triggering capability.

The general results of the R&D program showed that existing detector technologies could be successfully employed to meet the SSC/LHC requirements. The higher occupancies expected are accommodated in first order by more finely segmenting the detector. The main R&D challenges then lay in the area of electronics and DAQ design. The widespread use of custom integrated circuits in the detector front-end electronics allows practical designs to handle the high rates and fine segmentation at finite cost. Modest extension of existing multi-level trigger designs are able to handle the expected event rates.

The high radiation levels in high-luminosity hadron machines have been a cause for concern for both the SSC and LHC experimental groups. Again, the R&D program has shown that radiation resistant versions of the most sensitive detector elements can be fabricated with only modest extension of existing capability. In particular, many front end electronic circuits that would be exposed to the harsh radiation conditions have been fabricated in radiation-hard versions and shown to work. Also, the level of thermal neutron flux inside the inner tracking cavity and within the detector hall is brought within acceptable limits for detector operation by appropriate shielding design, especially in the region around the machine final focusing quads. Thus, while the problem of radiation resistance remains an ongoing area of further development for the experimenters, solutions to all the most critical problems appear to be in hand. For more specific details on proposed solutions, see the SSC and LHC detector design reports.

12.2 Detectors

Two large general purpose detectors are being designed for the LHC, namely ATLAS and CMS. Both experiments have been designed to detect all hypothesized particles accessible to the LHC. Many studies have been carried out by these collaborations, and are detailed in the ATLAS and CMS letters of intent. Generally, to fully exploit the potential physics reach of future hadron colliders, detectors must feature good charged particle tracking, precision electromagnetic calorimetry, full angular coverage for the hadronic calorimetry, and a robust muon detection system.

How physics processes set the requirements on the

detector design has been fairly thoroughly explored by the SSC and LHC experimental groups. As examples, detection of the signal $H \to \gamma\gamma$ for low mass Higgs requires shower counter resolution in the range $\Delta E/E = 10\%/\sqrt{E} \oplus .7\%$, measurement of photon position with accuracy in the range ~ 5 mm$/\sqrt{E}$, and timing resolution adequate for tagging from which bunch the photon originated. Measurement of high mass like-sign W production requires measurement of the sign of the charge of the W decay lepton out to $p_\perp \sim 500$ GeV$/c$. Angular coverage for both electrons and muons must extend to rapidities of ~ 2.5 for good angular acceptance for the $H \to 4\ell$ signals. Many other physics processes have been studied to set detailed design parameters for the detectors.

SSC and LHC R&D programs were required to show that detectors with the requisite capabilities are practical to construct. The SSC and LHC proposals have shown that detectors with resolutions comparable to existing detectors are adequate for the task. In large measure, the higher event rates expected at future machines are handled by finer segmentation of the detector elements, as mentioned earlier.

The existence of multiple interactions per crossing provides a further complication of this problem, mainly by degrading the pattern recognition capabilities of the detector. Each physics process must be studied with a detailed detector simulation to quantify the capability of the design to detect the signal of interest. Such studies have been carried out in detail, with full event reconstruction, for many of the important Standard Model Higgs detection modes. Numerous other physics processes have also been studied at this level of detail. However, the full extent of the capabilities of the designs as they currently stand has not been explored. In several areas, preliminary parton-level estimations of detector capability will be supplemented by more detailed simulations later. Areas of active investigation by the LHC collaborations include b-jet tagging capability at high luminosity, forward jet tagging capability, and alternate modes for MSSM Higgs detection.

12.3 Outlook

During the last five years, important progress has been made in understanding the experimental challenges of hadron colliders. This progress includes improved theoretical understanding of the basic Standard Model physics processes which occur at such machines, improved experimental technique through the operating experience gained mainly at Fermilab, and improved instrumentation mainly through the program of SSC and LHC R&D. Over the next five years, we expect to maintain a program of continual upgrade at FNAL, and continue preparations for LHC. New limits will be set at FNAL during this period on physics beyond the Standard Model. In addition, measurement of the top and W masses with improved precision will provide indirect information on the electroweak symmetry breaking sector. Within the next ten years, we hope to see the turn-on of the LHC with a first real chance for direct production of more massive Higgs particles, and considerable risk of discovery for physics beyond the Standard Model.

13 Experimental Issues at e^+e^- Linear Colliders

13.1 The Physics Environment

Experimental studies at e^+e^- colliders over the past two decades have provided key observations and insights to the nature of the fundamental particles and interactions of the Standard Model. Electron-positron collisions yield events with simple and transparent structures. Annihilation events carry the full energy of the beam, and produce final states with few partons. Searches for new phenomena give complete and unambiguous results, and precision studies of strong and electroweak interactions are made with a minimum of bias and background. The physics environment at TeV-scale colliders will continue to be as ideally suited to the exploration of particle physics [124].

The Standard Model final states that will predominate at a $\sqrt{s} = 500$ GeV e^+e^- collider are *udscb* pair production (9 units of R), W^+W^- (20 units of R), ZZ (1.2 units of R), and $t\bar{t}$ (1 unit of R), where $R = 87$ fb$/s$ if s is in TeV^2. These processes correspond to relatively simple final states, and, with a typical large solid-angle detector, event structures can be sorted to reduce the 50 or so pions, kaons, and photons to a few 4-vectors that accurately describe the underlying partons. With the partons reconstructed with high precision, topological and kinematic constraints can be utilized to take complete advantage of the simplicity of the physics environment at e^+e^- colliders.

Polarization of the electron beam in a linear collider is a unique tool that provides new and important views of particle physics. It can be utilized to isolate longitudinal W boson states, probe the helicity structure of the interactions of the top quark, reveal the underlying symmetry of SUSY, and determine the gauge structure of new interactions that might be found in high energy collisions - to name just several targets of opportunity. Sources of intense bunches of polarized electrons are presently used at the SLC where beam polarizations of 80% are routinely delivered to the collision point. Development of sources is continuing and it is reasonable to expect polarizations of 90-95% in the future. Instrumentation has been developed to measure beam polarization with accuracies of a per cent or better.

13.2 Detectors

The analysis of events at future e^+e^- colliders requires good tracking of high-momentum leptons and good reconstruction of quark jets. The isolation of heavy quarks with precision vertex detectors will continue to play an important role in the analysis of events. Particles that go unseen in the detector, such as neutrinos and beamstrahlung photons, can be fully reconstructed in many instances by applying energy-momentum and mass constraints.

Most studies of physics at future electron-positron colliders have used Monte Carlo simulations of detectors with capabilities that have already been achieved at LEP and SLC. The parameters of a "Standard Detector" are listed in Table 12. We see that the charged-particle tracking and electromagnetic and hadronic calorimetery does not extend any technology beyond that now in existence.

Table 12: "Standard Detector" Parameters. All momenta and energies are in GeV. The symbol $\oplus$ means quadratic sum.

Calorimetry	
σ_E/E (electromagnetic)	$8\%/\sqrt{E} \oplus 2\%$
σ_E/E (hadronic)	$60\%/\sqrt{E} \oplus 2\%$
Cell size (electromagnetic)	2°
Cell size (hadronic)	4°
Tracking	
σ_{P_t}/P_t	$10^{-3} \cdot P_t$
Vertex resolution (impact)	$20\mu\text{m} \oplus (\frac{100\mu\text{m}}{P})$
Hermiticity	
Calorimetery and tracking	$\theta_{e^\pm} > 10^\circ$

Table 13: "Improved Detector" Parameters. All momenta and energies are in GeV. The symbol $\oplus$ means quadratic sum.

Calorimetry	
σ_E/E (electromagnetic)	$8\%/\sqrt{E} \oplus 1\%$
σ_E/E (hadronic)	$35\%/\sqrt{E} \oplus 2\%$
Cell size (electromagnetic)	1°
Cell size (hadronic)	2°
Tracking	
σ_{P_t}/P_t	$2 \times 10^{-4} \cdot P_t$
Vertex resolution (impact)	$5\mu\text{m} \oplus (\frac{50\mu\text{m}}{P})$
Hermiticity	
Calorimetery and tracking	$\theta_{e^\pm} > 10^\circ$

We can anticipate that improvements in the techniques and technologies used in the design and construction of particle detectors will occur. At center of mass energies of several hundred GeV, most applications of electromagnetic and hadronic calorimetery do not suffer from lack of shower statistics, but it is increasingly important to minimize systematic errors in the reconstruction of the energies and positions of neutral hadrons and photons. The charged particles of greatest interest are similarly energetic, so that the scattering of tracks in the material of the detector becomes less problematic. The more aggressive set of goals shown in Table 13 might be realizable and certainly would enhance the physics output of the experiment. More systematic and quantitative study is needed to arrive at a complete set of design parameters, but it is clear that existing detectors are not far from ideal.

13.3 Event Rates and Time Structures

The total e^+e^- annihilation cross section at high energies is approximately 30 units of R, or 10^{-35}cm^2 at 500 GeV. Colliders at these energies are designed to generate luminosities typically $5 \times 10^{33}\text{cm}^{-2}\text{s}^{-1}$, with an interesting event produced every 10 seconds or so. Small-angle Bhabha scattering of electrons and positrons into the region covered by detector elements (above 10 degrees polar angle) occurs with a cross section 5-10 times the annihilation rate. This serves as a good monitor of the luminosity and energy spectra of collected data samples.

Cross sections for backgrounds produced by the coherent beam-beam interaction and low-energy peripheral two-photon reactions are also larger than those of annihilation events. For example, numerically the total cross section for the reaction $ee \to eeee$ is $\approx 10^{-26}\text{cm}^{-2}$ (at beam energy of 500 GeV). Most of this cross section corresponds to production of e^+e^- pairs near threshold, but the total rate can be sufficiently large that care must be taken in the design of the accelerator and detector to minimize the probability that soft particles from these interactions will overlap with the desired annihilation events. Accelerator and detector designs have evolved that account for these processes.

The luminosity per bunch crossing and the time interval between crossings varies considerably from one machine design to another. Those based on superconducting linacs yield the lowest luminosity per crossing and spacing between bunches of up to a microsecond. The duty cycle and time structure of these machines are similar to those found at LEP for example, and place similar demands on detector data acquisition systems. Colliders that utilize linacs with high frequency rf (at X-Band for example), create brighter collisions in short trains of typically 100 bunches pulsed at repetition rates of 100–200 Hz. Bunches within a train will be spaced by a few nanoseconds. Detector trigger and data acquisition systems will have long periods to analyze and format information from each repetition cycle, but it will be advantageous for detectors to be able to assign observed charged and neutral particles to their proper bunch-bunch interaction. Detectors have been built with this capability, and these issues have been considered in the evaluation of physics analyses at colliders with center of mass energies as high as 1.5 TeV.

13.4 Beamstrahlung

Another property of high-energy linear colliders is that of "beamstrahlung" — the radiation emitted by beam particles as they pass through the electromagnetic field of the opposing bunch. Corrections for the effects of initial-state radiation in e^+e^- collisions has long been a well-understood process, and similar techniques will be needed at future colliders to account for the smearing of annihilation center of mass energies by beamstrahlung. All proposed e^+e^- machines are designed so that the smearing of the center of mass energy by beamstahlung is never greater than, and often less than, the smearing by initial state bremsstrahlung.

The consequences of these effects on the analysis of data have been studied by many groups, and techniques to correct data, and even individual events, for radiation of energy by the incident particles prior to collision have been developed. It is extremely rare that more than one hard photon is emitted in any event, and those that are radiated travel along the incident beam direction, so it is often possible to simply include the radiation as an unknown parameter in a likelihood fit to the event topology. This works well for analyses of exclusive final states such as W^+W^-, ZH^0, and $t\bar{t}$ pairs.

13.5 Machine-Induced Backgrounds

While the physics environment created by electron-positron annihilation is remarkably neat and orderly, the single-pass nature of a linear collider poses a very special challenge to the experimenter and machine designer. Unlike a storage ring in which the circulating beam is quickly reduced to only those particles that are well-contained within the phase space of the physical and dynamic apertures of the machine, a pulsed collider will transmit particles that are very near to its aperture limit. Such particles will create backgrounds in experimental detectors either through the emission of synchrotron radiation as they pass through magnetic fields near the detector, or by the creation of secondary debris when they strike physical elements along the beamline. The detector must be properly masked, and the beam must be properly collimated to remove beam extremities, or "tails", well upstream of the interaction region. It is important that the design of the collider include accommodation for special optical sections to allow for complete collimation of the phase space of the beam.

The experience of the MARK II and SLD detectors [125] at the SLC has been an invaluable guide to the problems that will be encountered by experiments at future linear colliders. Successful operation of these experiments and the lessons that have been learned from the SLC provide a good basis for design and operation of experiments at the NLC.

14 Conclusions

Exploration of the TeV scale requires a new generation of colliders and detectors beyond those currently in operation, and is essential for the long-term vitality of particle physics. In this report, we describe an in-depth study of the phenomenology of electroweak symmetry breaking and quantify the "physics reach" of present and future colliders. Our investigations focus on the Standard Model (with one Higgs doublet) and beyond: models of low-energy supersymmetry, dynamical electroweak symmetry breaking, and other approaches containing new particles and interactions. Implications for future experiments at hadron and e^+e^- colliders are considered. On the basis of this work, we present the following conclusions and recommendations.

The participation of the United States in the LHC program should be vigorously supported. An extensive study of the electroweak symmetry breaking sector and the elucidation of physics beyond the Standard Model will require a robust experimental program at the LHC. The LHC possesses an impressive discovery potential for Higgs bosons, low-energy supersymmetry, and a large variety of beyond the Standard Model phenomena, with a mass reach in many cases five to ten times the corresponding mass reach at existing facilities. Of course, with the LHC approximately ten years from its initial run, one should not overlook the possibility that initial evidence for physics associated with electroweak symmetry breaking and/or physics beyond the Standard Model could be uncovered at upgrades of existing facilities.

Detailed simulations of high-energy particle physics processes at hadron colliders are often required to obtain an accurate assessment of the new physics discovery capabilities of future experiments. These include realistic treatments of hadronic jets (beyond parton-level Monte Carlo analyses) and full detector simulations. An active collaboration between experimentalists and theorists is crucial to the further development of strategies to extract evidence for electroweak symmetry breaking and beyond the Standard Model physics at future colliders.

There are many instances of the complementarity of future hadron colliders and future e^+e^- colliders. A high-energy e^+e^- collider will provide a unique low-background environment for the study of the production and decay of heavy gauge bosons and top quarks and of new physics beyond the Standard Model. A Next Linear Collider (NLC) at 0.5 TeV is an ideal laboratory for the study of intermediate-mass Higgs bosons, while at 1.5 TeV, it can begin to probe the physics of a strongly-coupled electroweak symmetry breaking sector. Experimental studies at the NLC would provide outstanding opportunities for discovery, while significantly enhancing the ability of the LHC to interpret evidence for new physics.

Other possible future collider scenarios are beginning to be examined. Physics studies at a number of hypothetical Tevatron upgrades beyond the Main Injector have been initiated. A full study of the scientific potential of such machines is necessary to provide input into any future cost/benefit evaluation that might be undertaken in the future.

In plotting out the course for future colliders, the US must work closely with its international partners. By avoiding the duplication of resources and taking into account the complementarity of various approaches, it should be possible to propose a set of future facilities that can fully explore the TeV energy scale and elucidate the physics of electroweak symmetry breaking. In doing so, we can push the frontiers of particle physics forward and keep our field intellectually vital and exciting.

References

[1] *Electroweak Symmetry Breaking and Physics Beyond the Standard Model*, eds. T. Barklow, S. Dawson, H.E. Haber and J. Siegrist, (World Scientific, Singapore) to be published.

[2] For convenient summaries of early material on the SM and MSSM Higgs sectors, see J.F. Gunion, H.E. Haber, G.L. Kane and S. Dawson, *The Higgs Hunters Guide* (Addison Wesley, Redwood City, CA, 1990). For more recent theoretical and phenomenological developments see *Perspectives on Higgs Physics*, ed. G. Kane (World Scientific, River Edge, N.J., 1992).

[3] H.E. Haber and R. Hempfling, *Phys. Rev. Lett.* **66**, 1815 (1991); *Phys. Rev. D* **48**, 4280 (1993); Y. Okada, M. Yamaguchi and T. Yanagida, *Prog. Theor. Phys.* **85**, 1 (1991); *Phys. Lett. B* **262**, 54 (1991); J. Ellis, G. Ridolfi and F. Zwirner, *Phys. Lett. B* **257**, 83 (1991); **262**, 477 (1991); R. Barbieri, M. Frigeni, and F. Caravaglios, *Phys. Lett. B* **258**, 167 (1991).

[4] For a review, see *e.g.*, P. Janot, in *Electroweak Interactions and Unified Theories*, 27th Rencontres de Moriond, Les Arcs, France, March 1992, ed. J. Tran Thanh Van (Gif-sur-Yvette, France, Editions Frontieres, 1992) Vol. 1, p. 317.

[5] J. Gunion, in *Properties of SUSY Particles*, Proceedings of the 23rd Workshop of the INFN Eloisatron Project, Erice, Italy, Sept. 28–Oct. 4, 1992, eds. L. Cifarelli and V. Khoze, (World Scientific, River Edge, N.J., 1993) p. 279.

[6] For a review and references, see J. Gunion, *Nucl. Phys. B* (Proc. Suppl.) **37**, 253 (1994), Proceedings of the Zeuthen Workshop on Elementary Particle Theory — *LEP 200 and Beyond*, Teupitz/Brandenburg, Germany, 10-15 April 1994, eds. T. Riemann and J. Blumlein.

[7] S. Mrenna and G. Kane, CALT-68-1938 (1994).

[8] ATLAS Technical Proposal, CERN/LHCC/94-43, LHCC/P2 (1994).

[9] CMS Technical Proposal, CERN/LHCC/94-38, LHCC/P1 (1994).

[10] H. Baer and X. Tata, in *Properties of SUSY Particles*, Proceedings of the 23rd Workshop of the INFN Eloisatron Project, Erice, Italy, Sept. 28–Oct. 4, 1992, eds. L. Cifarelli and V. Khoze, (World Scientific, River Edge, N.J., 1993) p. 244.

[11] F. Zwirner, *ibid.*, p. 509.

[12] For recent studies, see P. Grosse-Wiesemann, D. Haidt, and H.J. Schreiber, in *Workshop on e^+e^- Collisions at 500 GeV: the Physics Potential*, Munich, February; Annecy, June; Hamburg, September 1991, ed. P.M. Zerwas, DESY 92-123A (1992), p. 37; P. Janot, *ibid.*, p. 107; A. Djouadi, D. Haidt, B.A. Kniehl, B. Mele, and P.M. Zerwas, *ibid.*, p. 11.

[13] J.F. Gunion, in *Proceedings of the 2nd International Workshop on "Physics and Experiments with Linear e^+e^- Colliders"*, Waikoloa, HI, 1993, eds. F. Harris, S. Olsen, S. Pakvasa and X. Tata, Waikoloa, (World Scientific, Singapore, 1994) p. 166.

[14] P. Janot, *ibid.*, p. 192.

[15] A. Djouadi, J. Kalinowski, P.M. Zerwas, *Z. Phys.C* **57**, 569 (1993); A. Brignole, J. Ellis, J.F. Gunion, M. Guzzo, F. Olness, G. Ridolfi, L. Roszkowski and F. Zwirner, in *Workshop on e^+e^- Collisions at 500 GeV: The Physics Potential*, Munich, February; Annecy, June; Hamburg, September 1991, ed. P.M. Zerwas, DESY 92-123A/B, p. 613.

[16] V. Barger, K. Cheung, A. Djouadi, B.A. Kniehl, and P.M. Zerwas, *Phys. Rev. D* **49**, 79 (1994).

[17] B. Grzadkowski and J.F. Gunion, *Phys. Lett. B* **294**, 361 (1992), and preprint UCD-95-5 [hep-ph/9501339].

[18] M. D. Hildreth, T. L. Barklow, and D. L. Burke, *Phys. Rev. D* **49**, 3441 (1994).

[19] P. Langacker and M. Luo, *Phys. Rev. D* **44**, 817 (1991); J. Ellis, S. Kelley and D.V. Nanopoulos, *Phys. Lett. B* **260**, 131 (1991); U. Amaldi, W. de Boer and H. Fürstenau, *Phys. Lett. B* **260**, 447 (1991).

[20] L.E. Ibáñez and G.G. Ross, *Phys. Lett. B* **110**, 215 (1982); K. Inoue *et. al.*, Prog. Theor. Phys. **68**, 927 (1982); J. Ellis, D.V. Nanopoulos and K. Tamvakis, *Phys. Lett. B* **121**, 123 (1983); L. Alvarez-Gaumé, J. Polchinski and M.B. Wise, *Nucl. Phys. B* **221**, 495 (1983).

[21] L. Girardello and M.T. Grisaru, *Nucl. Phys. B* **194**, 65 (1982).

[22] For reviews, see H.P. Nilles, *Phys. Rep.* **110**, 1 (1984); P. Nath, R. Arnowitt and A. Chamseddine, *Applied N=1 Supergravity*, ITCP Series in Theoretical Physics (World Scientific, Singapore, 1984), Vol. 1.

[23] See *e.g.*, G.L. Kane, C. Kolda and J.D. Wells, *Phys. Rev. Lett.* **70**, 2686 (1993) and references therein.

[24] M. Drees, *Phys. Lett. B* **181**, 279 (1986); J.S. Hagelin and S. Kelley, *Nucl. Phys. B* **342**, 95 (1990).

[25] Y. Nir and N. Seiberg, *Phys. Lett. B* **309**, 337 (1993).

[26] T. Elliott, S.F. King and P.L. White, *Phys. Rev. D* **49**, 2435, 1994; U. Ellwanger, M. Rausch de Traubenberg and C.A. Savoy, *Phys. Lett. B* **315**, 331 (1993).

[27] L.E. Ibáñez and G.G. Ross, *Nucl. Phys. B* **368**, 3 (1992).

[28] H. Baer, C. H. Chen, C. Kao and X. Tata, FSU-HEP-950301 (1995); H. Baer, C.H. Chen, F. Paige and X. Tata, *Phys. Rev. D* **50**, 4508 (1994).

[29] H. Haber and G. Kane, *Phys. Rep.* **117**, 75 (1985); X. Tata, in *The Standard Model and Beyond*, p. 304, edited by J. Kim (World Scientific, Singapore, 1991); R. Arnowitt and P. Nath, Lectures presented at *VII J. A. Sweica Summer School, Campos do Jordao, Brazil, 1993*, CTP-TAMU-52/93.

[30] F. Paige and S. Protopopescu, in *Supercollider Physics*, p. 41, ed. D. Soper (World Scientific, Singapore, 1986); H. Baer, F. Paige, S. Protopopescu and X. Tata, in *Proceedings of the Workshop on Physics at Current Accelerators and Supercolliders*, ed. J. Hewett, A. White and D. Zeppenfeld, (Argonne National Laboratory, 1993).

[31] H. Baer, A. Bartl, D. Karatas, W. Majerotto and X. Tata, *Int. J. Mod. Phys.* A **4**, 4111 (1989).

[32] T. Kamon, J. Lopez, P. McIntyre and J. White, *Phys. Rev. D* **50**, 5676 (1994).

[33] H. Baer, C. Kao and X. Tata, *Phys. Rev. D* **48**, 2978 (1993).

[34] H. Baer, J. Sender and X. Tata, *Phys. Rev. D* **50**, 4517 (1994).

[35] V. Barger, M. Berger, P. Ohmann and R. Phillips, *Phys. Rev. D* **50**, 4299 (1994); H. Baer. C. Kao and X. Tata, *Phys. Rev. D* **51**, 2180 (1995).

[36] H. Baer, C. H. Chen, F. Paige and X. Tata, *Phys. Rev. D* **49**, 3283 (1994).

[37] H. Baer, C. H. Chen, R. Munroe. F. Paige and X. Tata, *Phys. Rev. D* **51**, 1046 (1995).

[38] H. Baer, J. Gunion, C. Kao and H. Pois, *Phys. Rev. D* **51**, 2159 (1995).

[39] For example, see, J.-F. Grivaz, in *Properties of SUSY Particles*, Proceedings of the 23rd Workshop of the INFN Eloisatron Project, Erice, Italy, Sept. 28–Oct. 4, 1992, eds. L. Cifarelli and V. Khoze, (World Scientific, River Edge, N.J., 1993).

[40] J.L. Feng and M.J. Strassler, SLAC-PUB-6497 [hep-ph/9408359].

[41] R. Becker and C. Van der Velde, in *Proceedings of European Meeting of the Working Groups on Physics and Experiments at Linear e^+e^- Colliders*, ed. by P.M. Zerwas, DESY-93-123C.

[42] J.-F. Grivaz, in *Proceedings of Physics and Experiments with Linear Colliders*, Saariselkä, Finland, September 1991, eds. R. Orava, P. Eerola, and M. Nordberg (World Scientific, Singapore, 1992).

[43] T. Tsukamoto, K. Fujii, H. Murayama, M. Yamaguchi, and Y. Okada, *Phys. Rev. D* **51**, 3153 (1995).

[44] JLC Group, "JLC-I", KEK-Report 92-16 (1992).

[45] A. Bartl, W. Majerotto and A. Sopczak, in preparation.

[46] R. Casalbuoni, P. Chiappetta, S. De Curtis, F. Feruglio, R. Gatto, B. Mele, and J. Terron, *Phys. Lett. B* **249**, 130 (1991); S. Dawson and G. Valencia, *Nucl. Phys. B* **352**, 27 (1991); A. Dobado, M. Herrero and J. Terron, *Z. Phys.C* **50**, 205 (1991).

[47] M. S. Chanowitz and W. Kilgore, *Phys. Lett. B* **322**, 147 (1994); **347**, 387 (1995).

[48] J. Bagger, V. Barger, K. Cheung, J.F. Gunion, T. Han, G. Ladinsky, R. Rosenfeld, and C.-P. Yuan, *Phys. Rev. D* **49**, 1246 (1994); new analyses in progress; J. Bagger, S. Dawson, and G. Valencia, *Nucl. Phys. B* **399**, 364 (1994) .

[49] M.E. Peskin, in *Proceedings of Physics and Experiments with Linear Colliders*, Saariselkä, Finland, September 1991, eds. R. Orava, P. Eerola, and M. Nordberg (World Scientific, Singapore, 1992).

[50] T. Barklow, talk presented at Albuquerque DPF 94, SLAC-PUB-6618.

[51] R. Casalbuoni *et al.*, in *Proceedings of the 2nd International Workshop on "Physics and Experiments with Linear e^+e^- Colliders"*, Waikoloa, HI, 1993, eds. F. Harris, S. Olsen, S. Pakvasa and X. Tata, Waikoloa, (World Scientific, Singapore, 1994) p. 887.

[52] V. Barger, K. Cheung, T. Han, and R. Phillips, MAD-PH-95-865 [hep-ph/9501379].

[53] R. S. Chivukula and M. Golden, *Phys. Lett. B* **267**, 233 (1991).

[54] M. Berger, contribution to this report.

[55] S. J. Brodsky, talk presented at the 2nd International Workshop on Physics and Experiments with Linear e^+e^- Colliders, Waikoloa, HI, April 1993.

[56] G. Jikia, IFVE-94-77 [hep-ph/9407393].

[57] K. Cheung, *Phys. Rev. D* **50**, 4290 (1994).

[58] L. Susskind, *Phys. Rev. D* **20**, 2619 (1979); S. Weinberg, *Phys. Rev. D* **13**, 974 (1976); **19**, 1277 (1979).

[59] GEM Technical Design Report, SSCL-SR-1219, 1993.

[60] A. Manohar and L. Randall, *Phys. Lett. B* **246**, 537 (1990); L. Randall and E. H. Simmons, *Nucl. Phys. B* **380**, 3 (1992); V. Lubicz, *Nucl. Phys. B* **404**, 559 (1993).

[61] K. Lane and M. V. Ramana, *Phys. Rev. D* **44**, 2678 (1991).

[62] V. Lubicz, work in progress.

[63] J.F. Gunion and H.E. Haber, *Phys. Rev. D* **48**, 5109 (1993); D.L. Borden, D.A. Bauer and D.O. Caldwell, *Phys. Rev. D* **48**, 4018 (1993).

[64] V. Barger, R.J.N. Phillips and D.P. Roy, *Phys. Lett. B* **324**, 236 (1994); J. F. Gunion, *Phys. Lett. B* **322**, 125 (1994).

[65] E. Eichten and K. Lane, *Phys. Lett. B* **327**, 129 (1994); K. Lane, BUHEP-94-12 (1994) [hep-ph/9406344].

[66] R.S. Chivukula, M. Golden and E.H. Simmons, *Nucl. Phys. B* **363**, 83 (1991).

[67] E. Eichten, I. Hinchliffe, K. Lane and C. Quigg, *Rev. Mod. Phys.* **56**, 579 (1984) [E: **58**, 1065 (1986)]; E. Eichten, I. Hinchliffe, K. Lane and C. Quigg, *Phys. Rev. D* **34**, 1547 (1986).

[68] J.L. Hewett and T.G. Rizzo, *Phys. Rep.* **183**, 193 (1989) and references therein.

[69] For a review and original references see R.N. Mohapatra, *Unification and Supersymmetry* (Springer, New York, 1986).

[70] P. Langacker and M. Luo, *Phys. Rev. D* **44**, 817 (1991).

[71] F. Abe *et al.* (CDF Collaboration), *Phys. Rev. Lett.* **74**, 2900 (1995).

[72] S. Godfrey, *Phys. Rev. D* **51**, 1402 (1995).

[73] F. del Aguila, M. Cvetič, and P. Langacker, in *Proceedings of the 2nd International Workshop on "Physics and Experiments with Linear e^+e^- Colliders"*, Waikoloa, HI, 1993, eds. F. Harris, S. Olsen, S. Pakvasa and X. Tata, Waikoloa, (World Scientific, Singapore, 1994) p. 490 and p. 929.

[74] For reviews on exotic fermions see, A. Djouadi, D. Schaile and C. Verzegnassi [conv.] *et al.*, *Proceedings of the Workshop on e^+e^- Collisions at 500 GeV: the Physics Potential*, Report DESY 92-123A/B, P. Zerwas, ed; P. Langacker and D. London, *Phys. Rev. D* **38**, 244 (1988).

[75] J. Maalampi and M. Roos, *Phys. Rep.* **186**, 53 (1990).

[76] R. Robinett and J. Rosner, *Phys. Rev. D* **25**, 3036 (1982); P. Langacker, R. W. Robinett and J. L. Rosner, *Phys. Rev. D* **30**, 1470 (1984); V. Barger *et al.*, *Phys. Lett. B* **118**, 68 (1982); W. Buchmüller and C. Greub, *Nucl. Phys. B* B363 345 1991 and *Nucl. Phys. B* B381 109 1992 .

[77] For a recent discussion on exotic fermions properties, decays and production at e^+e^- colliders, and for references on previous work, see A. Djouadi, *Z. Phys.C* **63**, 317 (1994); G. Azuelos and A. Djouadi, *Z. Phys.C* **63**, 327 (1994).

[78] W. Buchmüller, R. Rückl, and D. Wyler, *Phys. Lett. B* **191**, 442 (1987); W. Buchmüller, and D. Wyler, *Phys. Lett. B* **177**, 377 (1986).

[79] P.H. Frampton and D. Ng, *Phys. Rev. D* 45 4240 1992 ; P.H. Frampton *et al.*, *Phys. Lett. B* B317 369 1993 ; J. Agrawal, P.H. Frampton, and D. Ng, *Nucl. Phys. B* B386 267 1992 ; T.G. Rizzo, *Phys. Rev. D* **45**, 42 (1992); **46**, 910 (1992).

[80] For reviews on fermion substructure, see: M. Peskin, *Proceedings of the 10th International Symposium on Lepton–Photon Interactions at High Energies*, Bonn FRG, September, 1981, ed. W. Pfeil (Bonn University, Physik Inst., 1981); H. Terazawa, *Proceedings of the Europhysics Conference on Flavor Mixing in Weak Interactions*, Erice Italy, March 1984, ed. Ling-Lie Chau (Plenum Press, New York, 1984); W. Buchmüller, Lectures at the *24th Int. Wocheneuniversität für Kernphysik*, Schladming, Austria, 1985, eds. H. Mitter and W. Plessas (Springer-Verlag, Vienna, 1985); R. Peccei, *Proceedings of the Lake Louise Winter Institute, "Selected Topics in Electroweak Interactions"* Lake Louise, Canada, February 1987, eds. J.M. Cameron *et al.*, (World Scientific, Singapore, 1987); F. Boudjema, *Int. J. Mod. Phys.* A **6** (1991) 1.

[81] For a detailed analysis of excited fermions properties and production at e^+e^- and ep colliders, and a complete set of references on previous work, see F. Boudjema, A. Djouadi and J. L. Kneur, *Z. Phys.C* **57**, 425 (1992); K. Hagiwara, S. Komamiya and D. Zeppenfeld, *Z. Phys.C* **29**,115 (1985).

[82] Excited fermions at pp colliders are discussed in: U. Baur, M. Spira, P.M. Zerwas, *Phys. Rev. D* **42**, 815 (1990); U. Baur, I. Hinchliffe and D. Zeppenfeld, Proceedings of the Workshop *From Colliders to Supercolliders*, Madison, May 1987, eds. V. Barger and F. Halzen (World Scientific, Singapore, 1987).

[83] A. Djouadi, contribution to this report.

[84] M. Drees, *Nucl. Phys. B* **298**, 333 (1988); W. Buchmüller, C. Greub and P. Minkowski, *Phys. Lett. B* **267**, 395 (1991).

[85] M. Chanowitz, M. Furman and I. Hinchliffe, *Phys. Lett. B* **78**, 285 (1978); F. Csikor and I. Montvay, *Phys. Lett. B* **231**, 503 (1990).

[86] E. Nardi, E. Roulet and D. Tommasini, contribution to this report.

[87] M. Davier, Rapporteur talk at the Lepton–Photon Conference, Geneva 1991.

[88] G. Bhattacharya, P. Kalyniak and K. Peterson, contribution to this report.

[89] M. Spira, contribution to this report.

[90] C. Greub, contribution to this report.

[91] T. G. Rizzo, contribution to this report.

[92] J. Blümlein, contribution to this report.

[93] G. Bélanger, M. Doncheski, S. Godfrey, D. London, H. Nadeau, contribution to this report.

[94] J. Hewett, T. Rizzo, S. Pakvasa, A. Pomarol, and H.E. Haber, contribution to this report.

[95] P. Chiappetta, contribution to this report.

[96] N. Leporé, B. Thorndyke, H. Nadeau and D. London, contribution to this report.

[97] J. L. Kneur, contribution to this report.

[98] E. Eichten, K. Lane and M.E. Peskin, *Phys. Rev. Lett.* **50**, 811 (1983).

[99] K. Hagiwara *et al.*, *Nucl. Phys. B* **282**, 253 (1987).

[100] A. de Rujula *et al.*, *Nucl. Phys. B* **384**, 3 (1992); C. Arzt, M.B. Einhorn, and J. Wudka, *Phys. Rev. D* **49**, 1370 (1994); *Nucl. Phys. B* **433**, 41 (1994); C.P. Burgess, M. Frank and C. Hamzaoui, MCGILL-93-40 [hep-ph/9312238] and references therein.

[101] K. Hagiwara *et al.*, *Phys. Rev. D* **48**, 2182 (1993).

[102] F. Abe *et al.* (CDF Collaboration), *Phys. Rev. Lett.* **74**, 1936 (1995); 1941 (1995); S. Errede, FERMILAB-Pub-94/306-E preprint (September 1994), to appear in the *Proceedings of the 27th International Conference on High Energy Physics*, Glasgow, Scotland, July 20–27, 1994; J. Ellison

(DØ Collaboration), FERMILAB-Conf-94/329-E preprint (November 1994) to appear in the *Proceedings of the DPF 1994 Conference*, Albuquerque, NM, August 1994; H. Aihara, talk given at the International Symposium on Vector Boson Self-Interactions, Los Angeles, CA, February 1995.

[103] M. Acciarri *et al.* (L3 Collaboration), *Phys. Lett. B* **346**, 190 (1995).

[104] A. Miyamoto, in *Proceedings of the 2nd International Workshop on "Physics and Experiments with Linear e^+e^- Colliders"*, Waikoloa, HI, 1993, eds. F. Harris, S. Olsen, S. Pakvasa and X. Tata, Waikoloa, (World Scientific, Singapore, 1994) p. 141.

[105] F. Abe *et al.*(CDF Collaboration), *Phys. Rev. D* **50**, 2966 (1994).

[106] F. Abe *et al.* (CDF Collaboration), *Phys. Rev. Lett.* **73**, 225 (1994).

[107] F. Abe *et al.* (CDF Collaboration), FERMILAB-PUB-95/022-E (1995).

[108] S. Abachi *et al.* (DØ Collaboration), FERMILAB-PUB-95/028-E (1995).

[109] C. T. Hill and S. J. Parke, *Phys. Rev. D* **49**, 4454 (1994).

[110] G. Kane, G. A. Ladinsky, and C.-P. Yuan, *Phys. Rev. D* **45**, 124 (1992); D. Atwood and A. Soni, *Phys. Rev. D* **45**, 2405 (1992); G. A. Ladinsky and C.-P. Yuan, *Phys. Rev. D* **49**, 4415 (1994); D. O. Carlson, E. Malkawi, and C.-P. Yuan, *Phys. Lett. B* **337**, 145 (1994); E. Malkawi and C.-P. Yuan, *Phys. Rev. D* **50**, 4462 (1994); T. Barklow and C. Schmidt, contribution to the 1994 DPF conference.

[111] R. S. Chivukula, S. B. Selipsky, and E. H. Simmons, *Phys. Rev. Lett.* **69**, 575 (1992); R. S. Chivukula, E. H. Simmons, and J. Terning, *Phys. Lett. B* **331**, 383 (1994).

[112] T.G. Rizzo, *Phys. Rev. D* **50**, 4478 (1994).

[113] D. Atwood, A. Kagan, and T. G. Rizzo, SLAC-PUB-6580 (1994).

[114] C-P. Yuan *et al.*, in *Proceedings of the Workshop on Physics at Current Accelerators and Supercolliders*, eds. J. Hewett, A. White, and D. Zeppenfeld (Argonne, 1993).

[115] D. Amidei, in preparation.

[116] T.G. Rizzo, *Phys. Rev. D* **51**, 1064 (1995).

[117] K. Fujii, T. Matsui, and Y. Sumino, *Phys. Rev. D* **50**, 4341 (1994).

[118] C. Schmidt and M. E. Peskin, *Phys. Rev. Lett.* **69**, 410 (1992).

[119] G. Bagliese *et al.*, in *Proceedings of the Workshop on e^+e^- Collisions at 500 GeV*, ed. P. M. Zerwas (DESY, 1992).

[120] M. Hildreth, in preparation.

[121] M. E. Peskin and T. Takeuchi, *Phys. Rev. Lett.* 65 964 1990 ; *Phys. Rev. D* 46 381 1992 .

[122] I. Maksymyk, C. P. Burgess, and D. London, *Phys. Rev. D* 50 529 1994 .

[123] For a review of implications of non-SM physics in $b \to s\gamma$, see J.L. Hewett, *21st Annual SLAC Summer Institute on Particle Physics*, Stanford, CA, July 1993.

[124] For a review, see: D.L. Burke, in *Workshop on Physics and Experiments with Linear e^+e^- Colliders*, Saariselkä, Finland, September 1991; and W. Ash, *Proceedings of the 2nd International Workshop on "Physics and Experiments with Linear e^+e^- Colliders"*, Waikoloa, HI, 1993, eds. F. Harris, S. Olsen, S. Pakvasa and X. Tata, Waikoloa, (World Scientific, Singapore, 1994).

[125] R. Jacobsen *et al.*, SLAC-PUB-5205, April, 1990.

Astrophysics, Cosmology, and Unification of Forces

Barry C. Barish

California Institute of Technology, Pasadena, CA 91125

Michael S. Turner

The University of Chicago, Chicago, IL 60637
and
Fermi National Accelerator Laboratory, Batavia, IL 60510

Frank Wilczek

Institute for Advanced Study, Princeton, NJ 08540

1 Executive Summary

The fields of astrophysics, cosmology, and unification are growing areas of research that are at the intellectual frontier of particle physics. Because the "energy reach" of theory has exceeded that of terrestrial experiment, the pursuit of fundamental physics has more and more come to involve astrophysics and cosmology. Today many particle physicists, both theorists and experimentalists, are working at the interface of particle physics with astrophysics and cosmology, and this trend is likely to increase.

The boundaries of particle physics now touch on cosmology, astrophysics and gravity. Our best attempts to understand the origin of the cosmic asymmetry between matter and antimatter, the tiny primordial inhomogeneities that seed the growth of all structure in the Universe, and the nature of the mysterious dark matter involve fundamental physics. Studies of the highest-energy cosmic rays or of the low-energy solar neutrinos already hint at physics beyond the Standard Model. The unification of the strong, weak and electromagnetic forces seems to implicate gravity and to force the question of its quantization.

Large-scale experiments addressing these fundamental problems have already been undertaken and many more are planned. They include the search for proton decay and particle dark matter, measurements of the solar-neutrino flux, large redshift surveys, programs to observe gravity waves, map out the anisotropy of the cosmic background radiation, and study the highest-energy cosmic-ray particles.

Increasingly these experimental efforts have come to involve scientists from diverse disciplines (*e.g.*, astronomy, space science, nuclear physics, and high-energy physics). The growing scope and importance of efforts in particle astrophysics requires that difficult structural issues facing an inter-disciplinary field be addressed. Some of these include determining the role of the national laboratories, developing methods for planning and setting priorities, and coordinating with other disciplines and several funding agencies (*e.g.*, Divisions of Physics and of Astronomy at NSF, Divisions of High Energy and of Nuclear Physics at DOE, and various programs at NASA).

2 Cosmology

2.1 Status

The standard model of cosmology, the hot big-bang model, provides a remarkably successful account of the history of the Universe from about 0.01 sec (temperature $\sim 10\,\text{MeV}$) until the present (about 15 billion years later and temperature 2.726 K). The standard cosmology supplemented by modern ideas in particle physics has allowed sensible speculations about the earliest history of the Universe which address a deeper set of cosmological questions. Conversely, the early Universe provides a unique "laboratory" in which physics at unification energies (say, greater than $10^{10}\,\text{GeV}$) can be studied.

Four pillars provide the observational support on which the standard cosmology rests: (1) The uniform distribution of matter on large scales and the isotropic expansion that maintains this uniformity; (2) The existence of a nearly uniform and accurately thermal cosmic background radiation (CBR); (3) The abundance (relative to hydrogen) of the light elements D, ^{3}He, ^{4}He and ^{7}Li; and (4) The existence of small fluctuations in the temperature of the CBR across the sky at the level of about 10^{-5}.

The nearly linear relationship between velocity and distance (Hubble's expansion law) has been established out to distances of about 300 Mpc (5% of the distance to the edge of the observable Universe). Except for a handful of nearby galaxies, all galaxies are observed to have redshifted, as opposed to blueshifted, spectral features, indicating a universal expansion. [The shift of spectral lines toward longer wavelength, $\lambda_{\text{detected}} \equiv (1+z)\lambda_{\text{emitted}}$, is related to the size of the Universe when the radiation was emitted: $R_{\text{today}}/R_{\text{emission}} = (1+z)$.] Hundreds of objects with redshifts in excess of unity have been seen, and the most distant of these is a quasar with redshift $z = 4.9$. There is no doubt that the Universe is expanding in accord with the big-bang model. Unfortunately, the very important proportionality constant in Hubble's law, the present expansion rate $H_0 = 100h\,\text{km/sec/Mpc}$ which sets the scale for cosmic distances as well as cosmic times, has yet to be determined accurately.

Measurements made by the Far Infrared Absolute Spectrometer (FIRAS) on the Cosmic Background Explorer (COBE) spacecraft have established the blackbody nature of the CBR spectrum to better than 0.03%, leaving little room to question that the radiation is a relic of an early, hotter and denser epoch. The dipolar variation of the CBR temperature on the sky indicates that our galaxy is moving with respect to the cosmic rest frame at a speed of 620 km/s. Remarkably, the yearly modulation of the CBR temperature due to orbital motion about the sun at 30 km/s has also been measured by COBE (the earth really does move!).

According to the big-bang model the temperature of the CBR decreases as the Universe expands, and a recent measurement has confirmed this. The relative populations of hyperfine states in neutral Carbon atoms seen in a gas cloud at redshift $z = 1.776$ indicated a thermodynamic temperature, $7.4 \pm 0.8\,\text{K}$, which is consistent with the big-bang prediction for the CBR temperature at this earlier time $T(z) = (1+z)2.726\,\text{K} = 7.58\,\text{K}$.

The earliest and perhaps most impressive test of the standard cosmology involves the light elements D, ^{3}He, ^{4}He and ^{7}Li which were synthesized when the Universe was seconds old. The predictions for the "primeval" (just after big-bang nucleosynthesis) abundances of these elements depend upon the number of light neutrino species (now known to be three) and a handful of nuclear matrix elements (which for the most part are relatively well determined). The comparison between the predicted abundances and the light-element abundances measured today is not a simple matter; it is complicated by 15 Gyr of "chemical evolution" (astrophysical processes destroy D, produce ^{4}He, and destroy or produce ^{3}He and ^{7}Li). However, three decades of careful theoretical and observational work has put the comparison on a firm footing, and there is excellent agreement provided that the ratio of baryons to photons is between 2.5×10^{-10} and 6×10^{-10} (see Fig. 1).

Accepting the validity of the standard cosmology, this then provides the most accurate determination of

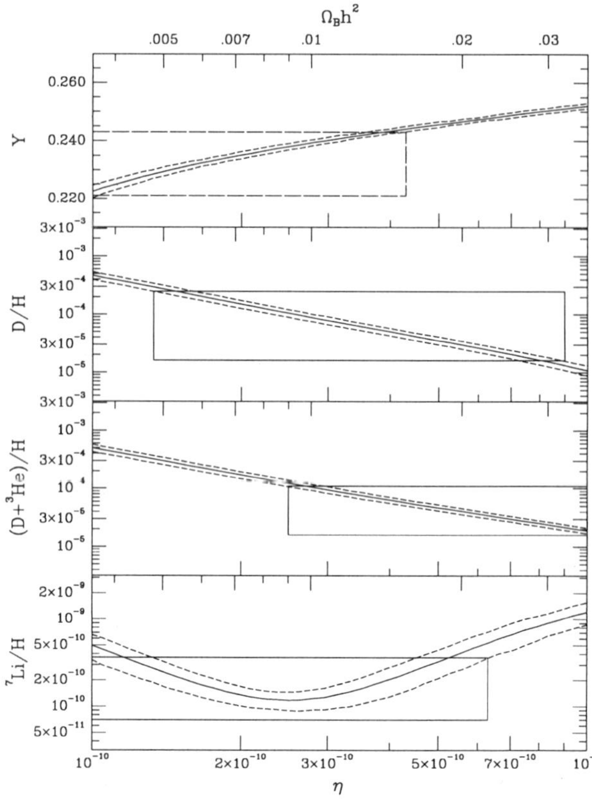

Figure 1: The predicted light-element abundances as a function of the baryon-to-photon ratio η and $\Omega_B h^2$. The broken curves delineate the two-sigma theoretical uncertainties and the boxes delineate the acceptable range for η as allowed by measured abundances. The predictions for all four light elements are consistent with the observations for $\eta \simeq 2.5\times10^{-10} - 6\times10^{-10}$. (Figure courtesy of C. Copi.)

the density of ordinary matter, between about $1.5\times10^{-31}\,\text{g/cm}^3$ and $4.1\times10^{-31}\,\text{g/cm}^3$. However, because the value of the critical density depends upon the Hubble constant, $\rho_{\text{crit}} = 3H_0^2/8\pi G$, the fraction of critical density contributed by baryons is not as well determined, between about $0.01h^{-2}$ and $0.02h^{-2}$ or 1% and 15% for a generous range of the Hubble constant ($h = 0.35 - 1.0$).

The final pillar involves the existence of small variations in the CBR temperature (around 30μK) at points on the sky separated by angles from about 0.5° to 90° (see Fig. 2). These fluctuations imply the presence of inhomogeneity (of about the same magnitude) at early times. These tiny seed inhomogeneities, amplified by gravity over the life of the Universe, develop into the rich variety of structures we observe today—galaxies, clusters of galaxies, superclusters, voids, and great walls. This picture for how structure evolved is an essential aspect of the standard cosmology as it links the smooth early Universe with the markedly inhomogeneous Universe today.

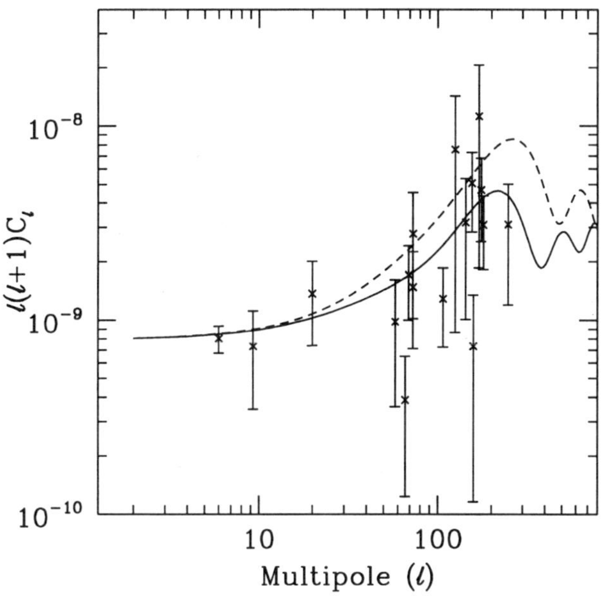

Figure 2: Summary of current measurements of CBR anisotropy in terms of a spherical-harmonic decomposition, $C_l \equiv \langle |a_{lm}|^2 \rangle$. The rms temperature fluctuation measured between two points separated by an angle θ is roughly given by: $(\delta T/T)_\theta \simeq \sqrt{l(l+1)C_l}$ with $l \simeq 200°/\theta$. The curves are the cold dark matter predictions, normalized to the COBE detection, for Hubble constants of 50 km/s/Mpc (solid) and 35 km/s/Mpc (broken). (Figure courtesy of M. White.)

2.2 Open problems

Building upon the success of the standard cosmology we can now address questions that involve the most fundamental aspects of the Universe. Our best efforts to answer these questions involve events occurring during the earliest moments and physics at the highest energies. Today, cosmologists and particle physicists are involved in a common quest to understand physics beyond the Standard Model.

Here we highlight some pressing issues. An exciting common theme is that data are pouring in from a variety of efforts—observations made by telescopes to particle searches at the highest-energy accelerators—and are testing both early-Universe cosmology and fundamental physics at the highest energies. We are poised for major advances in our understanding of both cosmology and fundamental physics.

2.2.1 Nature of the Dark Matter

The mean density of the Universe, expressed as a fraction of the critical density ($\bar{\rho}/\rho_{\rm crit} \equiv \Omega_0$), is a very important cosmological parameter which, like the Hubble parameter, is still not known with good precision. With confidence it is only possible to say that Ω_0 is between 0.1 and 2. The value of Ω_0 takes on additional importance because nucleosynthesis indicates that ordinary matter can account for between about 1% and 15% of critical density. Thus, if the total density is significantly greater than 15% of critical, a strong case exists for an additional, exotic component. There is a ready candidate for this exotic component: elementary particles that remain today from the earliest moments.

To be more specific, it is known that: (i) luminous matter (that closely associated with stars) contributes much less than 1% of the critical density; (ii) based upon nucleosynthesis, ordinary matter contributes between about 1% and 15% of critical density; (iii) dynamical measurements of the mean mass density indicate that it is certainly greater than 10% of the critical density, probably at least as large as 20% to 30% of the critical density, and a few measurements even suggest that it is close to the critical density. These dynamical measurements include: determinations of the masses of rich clusters from x-ray measurements, galactic motions (virial theorem), and gravitational lensing; estimates of the mean density based upon relating the peculiar motions of galaxies to the inhomogeneous distribution of matter; and measurements of the masses of spiral galaxies based upon rotation curves and gravitational lensing by their halos. It is important to state again, no method has yet provided a definitive answer for the mean density.

A number of indirect arguments favor the spatially flat, critical Universe; among them, structure formation—the most successful models involve a flat Universe—and theoretical notions—inflation and the Eddington-Dicke-Peebles naturalness argument (regardless of the value of Ω today, at early times Ω approaches unity, and if $\Omega \neq 0$, we are living at the special epoch when it is first beginning to deviate from unity).

Two ark-matter problems are indicated: the form of the dark baryons, and the form of the exotic dark matter—provided the density of the Universe is greater than that which ordinary matter can contribute. If the density of the Universe is equal to the critical density, then exotic matter is the dominant form of matter in the Universe.

The "dark" baryons are likely to exist in different forms in different places. In rich clusters the majority of baryons exist in the form of hot, x-ray emitting gas, and thus by "optical standards" are dark. The abundance of hot gas in clusters has measured by satellite-borne x-ray instruments (Einstein, ROSAT and ASCA) and is only around 5% to 10% of the total cluster mass, supporting the idea of nonbaryonic dark matter.

In spiral galaxies like own Milky Way, the bulk of the baryons are certainly not hot gas and are likely to be dark stars, objects not massive enough to have ignited their nuclear fuels (mass less than about $0.08M_\odot$) or stars that have exhausted their nuclear fuels (old neutron stars, white dwarfs or black holes). Three large-scale efforts are underway (MACHO, EROS, and OGLE collaborations) to search for dark stars in our galaxy through the effects of microlensing, and more than fifty microlensing events have now been seen. The event rates suggest that dark stars can only account for about 10% of the mass of the dark halo of our galaxy, further strengthening the case for nonbaryonic dark matter.

The second dark matter problem seems to directly involves fundamental physics, and the idea that the dark matter is something exotic receives support from such considerations. For three particles, the axion, the lightest supersymmetric partner (usually the neutralino), and the neutrino, the relic mass density contributed by these particles can be about equal to the critical density.

The axion and neutralino are hypothetical particles that arise from attempts to solve problems in particle physics (strong-CP problem for the axion and mass hierarchy problem for the neutralino). In the case of the neutrino, three species are known to exist and their relic abundance is accurately known ($113\,\mathrm{cm}^{-3}$); the issue is mass. To be cosmologically interesting, a neutrino mass of order 3 eV to 30 eV is required. Virtually all attempts to unify the forces and particles lead to the prediction that neutrinos have mass; while the predictions vary wildly, in some instances they fall into this range.

2.2.2 Structure Formation

The basic picture of how structure developed has been established: Small inhomogeneities in the density of the Universe were amplified by gravity and ultimately grew into the structure seen today. The challenge is to formulate a coherent and detailed picture. The two basic elements of such a description are a specification of the quantity and composition of the dark matter and a quantitative description of the density perturbations. We have just discussed the relationship of the dark matter problem to fundamental physics; the most attractive models for the origin of the density perturbations also involve fundamental physics.

The first and probably most attractive possibility is that the density perturbations arose from quantum-mechanical fluctuations during inflation (see below). Inflation-produced density perturbations and a critical-density Universe comprised mainly of "cold dark matter" (slowly moving particle dark matter—axions or neutralinos) leads to the most successful and well studied picture for structure formation, known as the cold dark matter theory. (The other alternative with inflationary perturbations, hot dark matter, where the bulk of the mass density is in fast-moving neutrinos does not lead to a viable picture of structure formation because galaxies and even smaller structures form far too late.)

The measured CBR anisotropy and the distribution of galaxies in the Universe today (as determined by redshift surveys) generally agree with the predictions of cold dark matter (see Fig. 3). However, there is evidence for a discrepancy: when the spectrum of density perturbations is normalized to agree with the CBR anisotropy as measured by COBE, the level of inhomogeneity predicted on small scales seems to be too large. This discrepancy, which is supported by several other observations (including the abundance of clusters and their clustering properties), suggests that the cold dark matter theory needs to be modified slightly. A number of "fixes" which lead to a better fit to the observations have been proposed: hot + cold dark matter, cosmological constant + cold dark matter, unstable tau neutrino + cold dark matter. It is interesting to note that the discrepancy would vanish if the Hubble constant were around 35 km/s/Mpc; however, such a value is significantly lower than the current range of measurements.

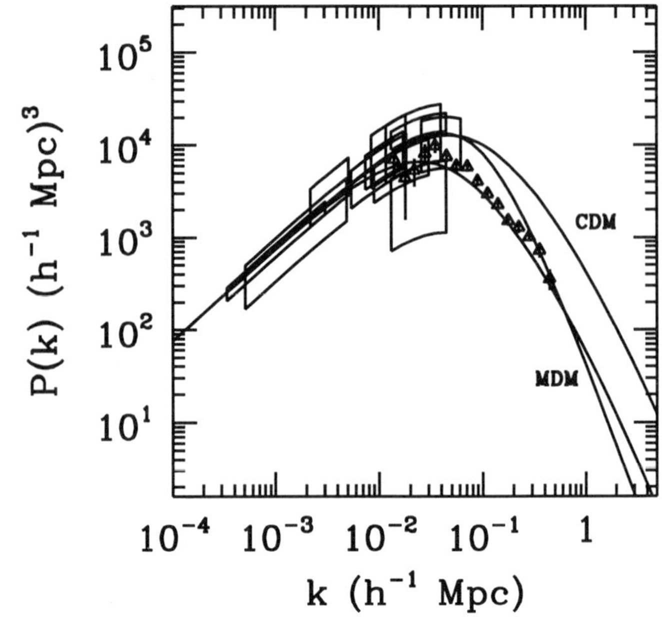

Figure 3: Comparison of the cold dark matter perturbation spectrum, $P(k) \equiv |\delta_k|^2$, with CBR anisotropy measurements (boxes) and the distribution of galaxies today (triangles). Wavenumber k is related to length scale by: $k = 2\pi/\lambda$; error flags are not shown for the galaxy distribution (since it is indicative of the distribution of "light" as opposed to mass there could well be a systematic offset). The curve labeled MDM is hot + cold dark matter ("5 eV" worth of neutrinos); the other two curves are cold dark matter models with Hubble constants of 50 km/s/Mpc (labeled CDM) and 35 km/s/Mpc. (Figure courtesy of M. White.)

The second possibility is that the seeds are topological defects (cosmic string, textures, or global monopoles) produced in an early-Universe phase transition. An attractive feature of this idea is that the energy scale desired for these defects, around 10^{16} GeV, is close to estimates for the unification scale. In these models the dark matter could either be neutrinos (in the case of cosmic string) or cold dark matter (in the case of textures). It is technically much more challenging to make sharp predictions for these models because density perturbations arise due to the defect network which constantly evolves with time. However, based upon the work done to date, defect models appear to be less successful in accounting for the observed structure and measured CBR anisotropy.

The first two pictures involve fundamental physics; a third, more conservative picture of structure formation has been advocated by Peebles and is known as the primeval baryon isocurvature scenario (PBI). This model begins with a sub-critical Universe comprised of baryons only and an ad-hoc spectrum of perturbations in the distribution of baryons. At the very least PBI provides a "standard" against which the two early-Universe scenarios can be judged. Thus far it has been less successful, requiring a baryon density that is higher than the nucleosynthesis limit and having difficulty conforming to CBR anisotropy measurements.

In sum, because the density perturbations that seed structure may well trace their origins to the earliest moments of the Universe, the study of large-scale structure has become an important window on physics beyond the Standard Model.

2.2.3 Inflation

Guth proposed the idea of cosmic inflation to address three vexing problems associated with the standard cosmology:

- Why is the observable Universe so smooth (horizon problem)?
- Why has the Universe not recollapsed or gone into free expansion long ago (flatness problem)?
- The extreme overproduction of magnetic monopoles in the early Universe (monopole problem).

Inflation not only squarely addresses these three problems, but it also provides an explanation for the origin of the tiny primeval density perturbations.

The key to inflation's beneficial effects is a very rapid period of expansion driven by the false-vacuum energy associated with a scalar field that is displaced from the minimum of its potential-energy curve. During the inflationary period the Universe grows in size by a larger factor than it has since. This tremendous growth allows a small smooth part of the Universe to become large enough to encompass all that we see today; it also allows quantum-mechanical fluctuations on very small scales ($< 10^{-23}$ cm) to become density fluctuations on astrophysical scales ($> 10^{25}$ cm). (Quantum-mechanical fluctuations also lead to a spectrum of relic gravitational waves.)

Inflation ends with the enormous entropy production associated with the conversion of the false-vacuum energy into a thermal bath of radiation (that ultimately becomes the CBR) and exponentially dilutes the abundance of any unwanted relic such as magnetic monopoles. Any baryon number the Universe possessed before inflation is also exponentially diluted and baryogenesis (see below) is a necessity.

Inflation has revolutionized the way cosmologists think about the early Universe as well as the Universe today. It has been one of, if not the most, important ideas in cosmology in the past two decades. At the moment, there is no standard model of inflation; there are many viable models, most of which involve energy scales of 10^{14} GeV and greater (although there is at least one model where inflation occurs at the electroweak energy scale). Great effort is being given to "connecting" the inflationary scalar-field potential which drives inflation to models for unification or to superstring theory.

One of the best prospects for testing inflation and learning about the underlying inflationary model is by testing the cold dark matter theory of structure formation. Observations of large-scale structure and the anisotropy of the CBR can be used to determine the spectra of density and gravity-wave perturbations (amplitudes and power-law indices). They in turn can be used to determine the shape and scale of the scalar-field potential. Such information would surely provide valuable insights about physics at energies well beyond those accessible in terrestrial laboratories.

Inflation not only holds the promise of extending the standard cosmology to very early times but also of providing a "bridge" from cosmological observations to unification-scale physics.

2.2.4 Baryogenesis

There is good evidence that the Universe contains only matter and thus has a net baryon number, about 10^{-10} relative to the photon number. The matter-antimatter asymmetry is very small, very mysterious, and very important—without it, matter and antimatter would have annihilated to negligibly small levels when the Universe was a fraction of a second old. In 1967 Sakharov suggested that this asymmetry could have evolved dynamically in the early Universe due to interactions that do not conserve baryon number, do not respect C and CP symmetry (matter/antimatter symmetry) and occurred out of thermal equilibrium. For a decade the paper went unnoticed, in part because there was little motivation for the nonconservation of baryon number. The advent of grand unification and its prediction of baryon-number violation led to the first working model for the origin of the baryon asymmetry (baryogenesis). It involves the out of equilibrium decay of a superheavy Higgs boson which mediates baryon-number violation (such bosons are predicted to exist in virtually all grand unified theories).

Almost two decades after Sakharov's paper, it was suggested that the required baryon-number violation could arise due to nonperturbative effects in the Standard Model. Since, a great deal of activity has focussed on whether or not physics at the electroweak scale—perhaps even within the Standard Model itself—can account for the baryon asymmetry of the Universe. The issue, which is a pivotal one, has yet to be resolved.

The baryon asymmetry of the Universe appears to be deeply connected to the physics of baryon-number violation and C, CP violation. Conversely, the existence of a cosmic baryon asymmetry provides strong circumstantial evidence for baryon-number violation and C, CP violation beyond the K meson system. It seems certain that advances in the understanding of both baryon-number violation and CP violation and the origin of the cosmic baryon asymmetry will be closely linked.

2.2.5 Relic neutrinos

Neutrinos have provided and continue to provide a pivotal connection between particle physics, astrophysics and cosmology. The success of primordial nucleosynthesis (as well as our understanding of the weak interactions) gives us confidence that the relic abundance of neutrinos is $113\,\mathrm{cm}^{-3}$ with temperature equal to $(4/11)^{1/3}$ that of the CBR. The fraction of critical density contributed by neutrinos is related to the sum of their masses, $\Omega_\nu h^2 = \sum m_\nu/93\,\mathrm{eV}$, and if the sum of neutrino masses is a few eV or greater neutrinos contribute at least as much mass density as baryons do. If they are unstable, then they can have variety of interesting effects for a much wider range of masses (*e.g.*, their decays can reionize the Universe after decoupling or raise the level of relativistic particles and improve the agreement between the cold dark matter theory and observations).

In any case, neutrinos are a cosmological relic that we are confident must exist, and, if detected, they would provide a window on the Universe at an age of around 1 sec (the surface of last scattering for neutrinos). Unfortunately, there are no practical ideas for their detection (yet!).

2.2.6 Phase transitions

It is very likely that the Universe has undergone two phase transitions (quark/hadron and electroweak symmetry breaking) and perhaps several others (*e.g.*, symmetry breaking associated with grand unification and compactification of extra dimensions). There may be remnants left over from these phase transitions (*e.g.*, topological defects) or other important consequences (production of the baryon asymmetry or inflation). The early Universe provides one of the few "laboratories" where the full nonperturbative effects of this physics can be studied.

2.2.7 Planck Epoch

General relativity together with current ideas in particle physics allows us to self-consistently extrapolate to times as early as 10^{-43} sec. Earlier than about 10^{-43} sec (temperatures greater than 10^{19} GeV) quantum corrections to general relativity should be very significant and must be considered. Thus, the very early Universe offers one of the few venues for studying the quantum effects of gravity. Superstring theory and other attempts to unify/quantize gravity make some notable predictions, *e.g.*, the existence of extra spatial dimensions or that gravity at high energies differs from general relativity. The hope exists that through extra dimensions or connections to inflationary theory or something else cosmology will allow the exploration of the quantum gravity regime.

2.2.8 Cosmological Parameters

From the discussion of open problems it is clear that more precise knowledge of the cosmological parameters, H_0, Ω_0, the deceleration parameter q_0, the age of the Universe t_0, and the cosmological constant Λ, is crucial to testing many of our most interesting ideas about the earliest history of the Universe, *e.g.*, nonbaryonic dark matter, the cold dark matter theory, and inflation. The need to know these parameters is so pressing that physicists have become involved in trying to determine some of them, *e.g.*, the use of distant (redshifts $z \sim 0.3$) type Ia supernovae to determine q_0. It is likely that there will be even more interest in using new techniques to measure these important parameters.

3 Unification

3.1 The Case for Unification

One primary goal of physics is to achieve an integrated description of Nature. History teaches us that progress toward this goal is not continuous. Rather, long periods of piecemeal accumulation of insight within limited domains of experience have been punctuated by a few bold syntheses which suddenly connect what had seemed quite separate. Historical examples include Newton's synthesis of terrestrial mechanics and celestial dynamics, and Maxwell's synthesis of electricity, magnetism, and optics. More recent examples include Einstein's synthesis of space, time, and gravitation; and the profound synthesis, made possible by quantum theory, of chemistry, material science and fundamental physical law.

Each of these great syntheses not only formed connections between what was previously known, but produced essentially new, unanticipated consequences, which often did not become fully apparent for many years. Thus Newtonian mechanics leads directly to modern structural engineering on the one hand and inertial guidance on the other; Maxwellian electrodynamics underlies every aspect of modern telecommunications; and the insights provided by quantum theory provide the foundation for lasers, transistors, and a host of other innovations. Even Einstein's general relativity theory, which was deduced by "thought experiments" and for many years was notorious for being difficult to test, has become the foundation of modern cosmology and a workhorse of contemporary astrophysics, describing in exquisite detail beautiful observations regarding pulsar timing and gravitational lensing, among other things.

As documented in an earlier Chapter, a remarkable new synthesis between what previously appeared to be widely disparate areas of physics—electrodynamics and the weak nuclear force—has evolved, over the last thirty years or so, from a tentative, schematic suggestion into a fully predictive and experimentally verified theory. This synthesis, together with the modern theory of the strong nuclear force, quantum chromodynamics or QCD, forms the basis for the Standard Model of particle physics. The Standard Model accurately describes an astonishing range of phenomena in terms of a few deep principles, and by any measure represents one of the greatest intellectual achievements of all time.

The guiding principle of the Standard Model is the symmetry of physical law. To be more precise, the principle of gauge invariance dictates how many gauge bosons exist and how they interact. And precisely these interactions are responsible for the structure of matter. It is impossible to do justice to the elegance and power of these mathematical principles in a small number of ordinary words; but one can fairly say that the fundamental laws governing (in Dirac's phrase) "all of chemistry and most of physics" are embodied, within their own appropriate language, in a few lines of poetry.

Yet there are important clues leading us to suspect that the Standard Model is but a way-station on the road to a still grander and more complete synthesis, some of whose far-reaching implications—for cosmology, in particular—we can already begin to discern.

To put these clues in context, we must now mention another central feature of the Standard Model, a feature that is perhaps unique in the history of physics. Very roughly speaking, in the past one has arrived at successively deeper and more complete levels of understanding matter by discovering and analyzing ever smaller and more basic building blocks. Thus ordinary matter was analyzed into molecules, molecules into atoms, atoms into electrons and nuclei, nuclei into protons and neutrons, protons and neutrons into quarks and color gluons. Thinking along these lines, one might anticipate that the next thing that will happen would be that electrons, quarks, gluons, and the other "fundamental particles" of the Standard Model will be further analyzed into smaller, more basic entities. There are important reasons to believe, however, that this is not the case—that electrons, quarks, and gluons are in a real sense truly fundamental, and that the next step beyond the Standard Model will be profoundly different from those that went before.

What suggests this extraordinary conclusion? First, of course, there is the fact that experiments attempting to find substructure have uniformly failed. Second, within the Standard Model itself one finds that all interactions turn off at short distances or high energies. This asymptotic freedom property is central to the modern theory of the strong interaction, and has been very extensively checked, as is discussed elsewhere in this Report. This property implies that it is internally consistent to assume that the Standard Model is valid down to arbitrarily short distances. Indeed not only does the assumption of simplicity and closure avoid contradictions—it becomes ever more accurate, as the interactions turn off completely.

Just because the Standard Model is potentially so accurate and complete, we must judge it by the highest standards. Every imperfection, in a theory which presents itself as potentially a nearly complete description of the fundamental laws of the material world, must be cause for pain and puzzlement.

Upon critically scrutinizing the Standard Model, one does readily discern imperfections. The gauge bosons fall into three unrelated sets, associated with the three different gauge groups $SU(3) \times SU(2) \times U(1)$ that appear in the mathematical formulation of the theory. Similarly, the fundamental fermions fall into fifteen different unrelated sets—three identical repeats of five essentially different structures. One cannot be satisfied with such a proliferation of structures. The situation becomes even more disturbing when one considers the electric charges of the different particles, which likewise appear rather helter-skelter.

A considerable effort of focussed imagination is required to discern any pattern implicit in this apparent jumble. Fortunately, an extremely attractive and convincing one was found for us by Georgi and Glashow. They showed that the various gauge bosons could be grouped into a single larger structure, $SU(5)$, and that when this was done the fundamental fermions fell into just six groups (three identical families each with two different groups), and furthermore that when this was done the electric charges were determined uniquely and correctly. Many variants and extensions of this idea have been proposed since, but all successful ones employ the same core pattern. The idea, then, is that at some very short distance (or, equivalently, in interactions at very high energies) the full symmetry of the world is larger than that of the Standard Model. The Standard Model as we see it at comparatively large distances, or in interactions at low energies, descends from the larger, more symmetric model by a process of spontaneous symmetry breaking. The sub-idea of spontaneous symmetry breaking, of course, is partially inspired by the similar mechanism known to be at to work in the electroweak sector of the Standard Model.

Most remarkably, these bold speculations involve a concrete, testable numerical consequence. For the gauge bosons of $SU(3) \times SU(2) \times U(1)$ to unify into the single structure $SU(5)$ (or any of its variants), it is necessary that their coupling strengths become equal. Onc can compute how the actual coupling strengths change with distance, following the procedures known and tested in QCD, to see if they do indeed become equal. When this is done a stunning—and most encouraging—result emerges.

It appears that straightforward extrapolation of the couplings measured at low energies, incorporating the effect of known particles only, comes very close to insuring the unification of coupling strengths at high energies. This gives us confidence that our ideas about unification are on the right track. Recent extremely accurate measurements of the couplings allow us to discriminate further among unified models. It now appears that the minimal extension of the Standard Model does not quite allow unification, but that its minimal extension supporting supersymmetry does. Supersymmetry is a very attractive idea, that was proposed for other independent reasons, and is discussed in an earlier Chapter. Taken at face value, then, the unification of couplings seems to encourage not only the quest for a unification of gauge forces, but also the quest for a further unification of gauge bosons, fundamental fermions, and Higgs particles all together in the framework of supersymmetry.

3.2 Open Questions

The main open question regarding these ambitious ideas of unification is simply: Are they true? Most physicists regard the clues discussed above: that the gauge bosons and fermions fit beautifully into a unified framework, and that the couplings do appear to approach a common value at short distances, as extremely encouraging—but perhaps not yet compelling. Will we be able to elevate our present tentative, schematic suggestions into the scientific realm of predictions and verifications? There are good reasons to hope that, given new tools and some hard work, we can.

As was mentioned above, the unification of couplings works best quantitatively if one assumes that supersymmetry is only mildly broken. More precisely, one requires that the supersymmetric partners of the known particles, as yet undetected, should not be much heavier than the W bosons. This is also desirable from other points of view, slightly too technical for discussion here, having to do with subtler details of unification. It is an exciting prospect, because it implies that the discovery of these partners, so pregnant with meaning for the unification of fundamental forces, is almost within our grasp.

Unified theories also suggest the possibility of certain small but very significant effects that are altogether forbidden within the Standard Model. One is the spontaneous decay of protons. This process is mediated, at a small but potentially detectable level, by the extra gauge bosons and Higgs particles needed to consummate the unification. Supersymmetric theories support additional decay mechanisms, which specifically facilitate decay into strange particles. Another is the existence of neutrino masses and the concomitant possibility of neutrino oscillations. There may already be evidence for this effect, from the long-standing but still not completely elucidated discrepancy between the predicted rate of arrival of neutrinos from the sun and the rate actually observed. A third prediction of unified theories is the existence of superheavy magnetic monopoles; a large-scale search for such particles is being carried out in an underground laboratory in the Gran Sasso tunnel.

There is a remarkable connection between all these ideas and cosmology. The production of supersymmetric particles during the Big Bang can be estimated with some confidence. Indeed, the methods used for these estimates are not very different from the methods used to estimate—with great success—the production of different light nuclear species and of the microwave background radiation. When this is done, it is found that for a wide range of parameters the lightest supersymmetric partner is a stable particle (usually the neutralino, a combination of the supersymmetric partners of the gauge and Higgs particles), and that it would have been produced during the Big Bang in such abundance as to presently contribute a significant fraction of the total mass density of the Universe. In this way, theoretical considerations concerning unified theories have produced a plausible candidate to solve the "dark matter" or "missing mass" problem of cosmology. The relic particles, though they are predicted to be omnipresent, are also predicted to be so feebly interacting that they are very difficult to detect. Several heroic, technologically innovative experiments are being mounted in response to this challenge. It is also possible that some or all of the missing mass is provided by relic massive neutrinos, or by axions, another particle species whose existence is suggested by unified theories.

The ideas we have just sketched regarding unification of gauge forces seem to us—given their scope—both astonishingly specific, and in their broad outline persuasive. They are not complete, however, and in concluding this brief survey of open problems it seems appropriate to mention two of their most obvious omissions.

First, although they succeed in organizing fundamental fermions into large families, existing unified gauge theories leave unexplained the fact that three such families of particles are known to exist, identical except for their masses. Are there additional families yet to be discovered? Or additional symmetries and interactions that group the known families, together perhaps with other particles, into a larger, more coherent whole? Experimental searches for the tell-tale signs of such interactions, including especially hypothetical rare processes in which members of one family spontaneously decay into members of another, could help us decipher the meaning of this cryptic but doubtless profound message from Nature.

Second, the unified gauge theories do not include gravity. Some promising but as yet highly speculative attempts to construct fully unified theories, including both the interactions of the Standard Model and gravity, are discussed in the previous Chapter.

4 Gravity

4.1 Status

Gravity is central to the interface of astrophysics, cosmology, and particle physics. It is fundamental to cosmology and to the understanding of many objects in the Universe. Further, the unification of gravity with the other forces is a major focus in theoretical particle physics.

Our current understanding of gravity is embodied in general relativity. It provides the theoretical foundation for the big-bang cosmology as well as the means for understanding many of the most interesting objects and phenomena in the Universe—neutron stars, black holes, gravitational waves, and gravitational lenses.

While general relativity has passed a long succession of tests, starting with Eddington's measuring the bending of starlight in 1919 to a host of more recent tests involving the binary pulsar (PSR 1913+16), most theorists are confident that it is at best the classical limit of a more complete theory that brings together quantum mechanics and gravity. Superstring theory provides the first example of a theory that does this self consistently, and at the same time unifies gravity with the other forces of nature.

4.2 Open Problems

Relativity is a vibrant area of research in its own right. Here we only mention the topics that overlap significantly with particle astrophysics, cosmology, and unification. To be sure, there is much interesting work going on beyond this, from the formal aspects of relativity (cosmic censorship hypothesis, entropy of gravitating systems, possibility of time machines, and the application of supercomputers to address numerically important problems in relativity) to the application of general relativity to astrophysical systems (massive accreting black holes that power quasars, colliding black holes).

4.2.1 Fundamental Aspects of Gravitation

It could well be that the classical description of gravity is a theory that closely resembles general relativity, but is not general relativity. This is because none of the tests of general relativity have yet probed the strong-field regime (*e.g.*, near black holes) or its limits of applicability (*e.g.*, in the very early Universe when quantum corrections to general relativity are important). There is a hint from superstring theory that the classical gravitation theory that emerges might be more similar to Brans-Dicke theory (where the gravitational constant is determined by the expectation value of a scalar field), and a class of very interesting models of inflation are based upon the idea that at the time of inflation gravity is not described by general relativity. The formulation, study, and testing of alternative theories of gravity continues to be an important area of research.

Black holes are a possible key to understanding the quantum nature of gravity. Hawking showed that a black hole should radiate a thermal spectrum of particles due to quantum processes near its horizon and in the process reduce its mass. This semi-classical calculation (quantum fields are treated on the classical background of a black hole) provided a crucial first link between gravity, quantum mechanics and thermodynamics, and in the process has raised many issues. What is left when a black hole evaporates? What happens to information (*e.g.*, coherent radiation or an encyclopedia) that falls into a black hole; is it lost? As a result of ideas from string theory two-dimensional black-hole models have been developed to describe the formation of a black hole, the "turning on" of Hawking radiation, and the evaporation process itself. This has allowed fundamental questions about black-hole physics to be addressed quantitatively.

The interest in the unification of gravity with the other forces has stirred theorists to address other fundamental issues, *e.g.*, topology change and wormholes. (Many of these issues are discussed in more detail in the previous Chapter.)

4.2.2 Phenomenological Aspects

Gravitational waves have yet to be detected directly, though there is strong evidence for their existence: The measured decrease in the orbital period of the binary pulsar agrees to a precision of better than 1% with the prediction of general relativity based upon energy loss due to the emission of gravitational radiation. When the LIGO and VIRGO facilities that are currently under construction become operational, the chances for direct detection of gravitational waves will improve dramatically. Not only can the detection of gravitational radiation probe the behaviour of gravity in a new regime, but it can also open a new window to the Universe.

The "bread and butter" source for these broad-band laser interferometric detectors are coalescing compact objects: neutron-star binaries, neutron star–black hole binaries, and black hole-black hole binaries. These sources can all be detected at cosmological distances and have potential for probing the Universe (*e.g.*, a measurement of the Hubble constant or even the deceleration parameter). In addition to a variety of astrophysical sources (pulsars and supernovae), there are interesting early-Universe sources, *e.g.*, gravity waves from first-order phase transitions, inflation, or cosmic strings. The detection of gravity waves from the early Universe would reveal a wealth of information about the earliest moments of the Universe. While these early-Universe sources are probably too weak and in the wrong frequency band to be detected by earth-based interferometers, they might be detectable with future space-based interferometers (*e.g.*, LISA).

5 Interface with Cosmic Rays, Nuclear Physics, and Astrophysics

The interface between these fields with particle physics is both scientifically rich and very complex. We highlight some of the connections.

5.1 Cosmic rays

Cosmic rays in our galaxy have energies from less than 10^6 eV to more than 10^{20} eV, and include the highest energy particles we have access to. The particles in the cosmic rays include photons, neutrinos, electrons, positrons, protons, antiprotons, nuclei from the lightest to the heaviest elements, and perhaps exotic particles yet to be identified. There is good evidence that cosmic-ray particles of energy up to 10^{13} eV (or even 10^{15} eV) were produced and accelerated by supernovae. Likewise, the propagation of these particles throughout the galaxy is relatively well understood.

The change in the shape of the spectrum of cosmic rays around 10^{16} eV (the knee region) is still not understood (it could involve the escape of cosmic rays from the galaxy or a change in composition or something else). The origin, composition, and acceleration of the highest-energy cosmic rays remain a mystery. They could be produced by topological defects (such as cosmic string) remaining from the early Universe, or by astrophysical accelerators whose interworkings are still not understood at the crudest level. The gamma-ray spectrum above 10^{12} eV remains a mystery: Are there ultra-high energy point sources? Is there a diffuse background of ultra-high energy gamma rays; if so, what is its origin?

Even the low-energy cosmic rays could provide important insights on new physics. For example they could include the annihilation products from neutralinos that may comprise the dark halo of our galaxy. Low-energy cosmic rays offer us the best opportunity for searching for antimatter in the Universe (antiprotons and antinuclei).

5.2 Astrophysical phenomena

Aspects of fundamental physics (*e.g.*, ordinary weak interactions, a variety of nuclear cross sections, and new phenomena such as neutrino oscillations or the existence of particles such as axions or righthanded neutrino states) provide important "input microphysics" for astrophysical situations, *e.g.*, structure of neutron stars, cooling of white dwarfs, red giant stars or newly born neutron stars, explosion of massive stars (supernovae) and explosive synthesis of heavy elements. In some instances the input microphysics might better be described as nuclear physics, *e.g.*, in neutron stars, in the synthesis of the light elements in the early Universe, or in the synthesis of heavy elements in supernovae. Likewise, some of the speculative physics is more closely related to nuclear physics, such as pion or Kaon condensates at high density, or strange quark matter.

Because astrophysical and cosmological phenomena in a wide range of circumstances—from the cosmic rays to the early Universe—depend upon input microphysics they can be exploited to study fundamental physics in regimes beyond the reach of terrestrial laboratories. The knowledge derived in the past has been very useful: hadronic cross sections at very high energies; nucleosynthesis limit to the number of light neutrino species (and other light particle species); cosmological limits to neutrino masses and other properties; constraints to the mass of the axion based upon the cooling of hot neutron stars, white dwarfs, and red giant stars; limits to the flux of superheavy magnetic monopoles based upon the existence of magnetic fields in neutron stars, galaxies and clusters of galaxies; a host of limits to neutrino properties based upon SN 1987A.

Since the energy reach of theorists continues to exceed that of accelerators and other terrestrial-based experiments, the "heavenly laboratory" is certain to continue to play an important role in the quest for our understanding of the most fundamental aspects of physics.

6 Experiment

Our focus here is on experiments not involving high-energy accelerators. That is not to say that accelerator-based experiments are of no interest for astroparticle physics, cosmology, and unification. They are of great interest; accelerator-based experiments have the potential to discover new phenomena that bear on important issues for this discipline, *e.g.*, neutrino oscillations, supersymmetric particles (including the neutralino), Higgs particles, and evidence for physics beyond the Standard Model.

Experiments addressing a broad range of questions involving astroparticle physics, cosmology and unification have been implemented in the past decade, more are underway, and still more are planned for the future. Since the experiments address such a wide range of questions they are very diverse in approach, techniques and scale. However, some trends stand out. Many such experiments in this general area require a quiet environment, and, as a consequence, we have seen major facilities develop underground at the Gran Sasso Laboratory in Italy, at the Baksan Laboratory in Russia, at Kamiokande in Japan, and at the Soudan and IMB facilities in the U.S.

The experiments often employ techniques from high-energy physics, such as the use of tracking chambers and scintillators. However, in addition, very important and successful new techniques were pioneered for this research, the most visible being large-scale, imaging H_2O Cherenkov detectors. Also, detector development is crucial to efforts to directly detect particle dark matter. Experiments addressing the questions discussed here are by no means limited to underground facilities and very challenging experiments that require advanced instrumentation are being done on the earth's surface.

There are a number of exciting new projects under construction, such as a next generation H_2O Cherenkov detector (SuperKamiokande), a heavy-water detector (SNO), and an interferometer facility to detect gravitational waves (LIGO/VIRGO). In the planning stage are very-large neutrino detectors for deep underwater, DUMAND and NESTOR, or under ice, AMANDA. In addition to these very-large facilities, special efforts to build sensitive and massive dark-matter detectors are underway. Space-based experiments (high-resolution mapping of the CBR anisotropy, search for antimatter using a large magnet in space, a small telescope in space to study microlensing by dark stars in our galaxy) and large-scale astronomical observational efforts (large red-shift surveys, a dedicated cosmology telescope) are beginning to emerge.

Below, we highlight some of these experimental efforts.

6.1 High-Energy Cosmic-ray Physics

High-energy cosmic rays have been of common interest to particle physics and astrophysics for many years. Many important questions remain to be answered. Moreover, their answers could well shed light on fundamental physics.

At the lower end of the energy scale for cosmic-ray studies, satellite experiments have been very successful. The Gamma Ray Observatory (GRO) for instance has detected γ-rays from pulsars and active galactic nuclei, as well as from unidentified objects, most notably, the mysterious gamma-ray bursters. Are they neutron-star collisions seen out to the edge of the Universe or accretion instabilities associated with neutron stars in the halo of our own galaxy?

At energies above 10 GeV or so, the rates are too low for satellites and ground-based instruments are used. Large electromagnetic shower arrays only become useful well above 1 TeV. To partially fill the gap, advanced imaging Cherenkov techniques have been developed and are quite successful. The imaging Cherenkov detector at Whipple Observatory has detected the Crab Nebula, a galactic source, and an extragalactic source, Mrk 421, in the 200 GeV to 1 TeV energy range. However, an important gap—between about 20 GeV and 200 GeV—still remains.

At higher energies, above 10 TeV, searches using large-scale surface shower arrays like CASA-MIA and the MILAGRO H_20 water-Cherenkov detector have thus far failed to find gamma-ray point sources or evidence for a diffuse gamma-ray background. It is interesting to note that the origin of gamma rays at such very high energies may be hadronic, rather then electromagnetic. Such sources would provide a window on particle acceleration at very high energies. These large-scale devices are also very useful in studying another important problem: the composition in the "knee" region of the cosmic-ray spectrum near 10^{16} eV. The origin of the knee is unknown and these experiments are aimed at resolving whether it is due to particles escaping the galaxy or another mechanism.

Finally, the very-highest energy cosmic rays are detected using very-large air-shower arrays (Akeno in Japan) and a different technique (fluorescence in the atmosphere) with a Cherenkov array (Fly's Eye). These arrays have detected many more cosmic rays with energies in excess of 10^{19} eV than expected (hadrons at these energies have very short mean-free paths for inelastic scattering with CBR photons). Future improvements (Hi Res at Fly Eye's) will investigate this interesting puzzle with improved sensitivity that hopefully will confirm previous results and shed light on the origin of the highest-energy cosmic rays.

Complementary investigations are performed deep

underground by studies of penetrating multiple muons in large detectors like MACRO, and these also yield primary composition information at very-high energies through muon multiplicity and the transverse distributions. Such studies are now beginning to include hybrid experiments that use coincidences of muons in a large underground detector with the signal from a large area surface electromagnetic shower array. This added information is very promising for composition studies.

Future plans in high-energy cosmic-ray physics are focussed around the development of a very-large hybrid (air shower/fluorescence) array (about $10^4\,\mathrm{km}^2$), and presently involves an international design effort. Such a facility will provide greatly improved sensitivity directed at searches for point-like sources of high energy γ-rays at levels where detection is expected as well as studies of the cosmic ray cut-off.

6.2 Neutrinos

The range of physics and large number of different facilities developed to explore neutrino physics make this area a major focus of astroparticle physics. The implementation of large-scale underground detectors has opened up a large range of studies involving neutrinos. These studies cover a very-large dynamic range of energies ranging from an MeV to several TeV, presenting a large experimental challenge.

At the lowest energies (less than one MeV), solar neutrinos have been detected by radiochemical techniques (using Chlorine and Gallium) and at somewhat higher energy by water Cherenkov techniques. The experiments indicate a deficit of neutrinos whose best explanation involves neutrino oscillations. The next generation experiments at SuperKamiokande, using a much larger volume of H_2O, at Sudbury, Canada (SNO), using D_2O, which allows the observation of the neutral-current channel, and at Gran Sasso (Borexino and ICARUS) should shed important new light on this question. It seems likely that a resolution of the solar-neutrino problem is near and will advance our understanding of both neutrino physics and the detailed interworkings of the sun.

At somewhat higher energies ($> 5\,\mathrm{MeV}$), large-volume underground detectors (*e.g.*, SuperKamiokande and MACRO) are sensitive to a supernova collapse in our galaxy. Observation of SN 1987A validated the idea that the collapse of a massive star to a neutron star liberates most of its energy in neutrinos, but the source was too distant and therefore the rate too small to enable detailed studies. A future stellar collapse will be observed by several complimentary detectors and will yield hundreds to thousands of events if it is in our own galaxy. Detection of neutrinos from a galactic supernova would greatly improve our understanding of type II supernovae and perhaps the equation of state of matter at supernuclear density, as well as provide useful information about the fundamental properties of neutrinos.

The primary source of neutrinos in the GeV range comes from atmospheric showers created when a high-energy primary cosmic ray interacts at the top of the atmosphere producing a shower of secondary, tertiary, etc. particles. Such a cascade should yield ν_μ's and ν_e's in an approximate ratio of two to one. At present, the observations of stopping muons from a number of underground detectors indicate a ratio closer to unity. This is yet another hint of neutrino oscillations. Further investigations of this anomaly appear best done through a hybrid experiment, where neutrinos are produced at a high-energy accelerator, and are detected at a distant detector. The distance and neutrino energy can be chosen to explore the same region as available with atmospheric neutrinos. Experiments are presently being considered for Fermilab/Soudan or BNL, CERN/Gran Sasso, and KEK/SuperKamiokande. These are all in the proposal stage and involve intense new beams and sophisticated new detectors.

For many years there has been an ongoing program to develop a very-large scale (*e.g.*, $1\ \mathrm{km}^3$) underwater or under-ice detector that could sensitively search for astrophysical point sources of ultra high-energy neutrinos. Prototypes are under study, and sites and detectors under ice at the South Pole (AMANDA) or under water near Hawaii (DUMAND) or Greece (NESTOR) are being considered.

The window to the Universe provided by ultra-low and ultra-high energy neutrinos has yet to be opened. It is interesting to note that ultra-high energy neutrinos offer one of the only plausible means of detecting the cosmic background of ultra-low energy neutrinos: at the Z-pole the neutrino seas provide about a mean-free path for absorption and the dip in the spectrum of an ultra-high neutrino source could reveal their presence.

6.3 Particle/Nuclear Physics

Another important interface is between particle and nuclear physics where experiments are directed at fundamental-physics questions, but involve nuclear physics or nuclear physics techniques. For example, direct studies of the electron-neutrino mass involve precision measurements of the nuclear beta-decay spectrum. These have yielded a sensitivity for the electron neutrino with a current best limit of about $7\mathrm{eV/c}^2$. These measurements are limited by our knowledge of atomic and molecular physics effects.

In addition to experiments directly sensitive to neutrino mass, effects are expected that involve the mass differences between different neutrino types. The solar-neutrino and atmospheric-neutrino experiments are important examples, and both show hints of evidence for neutrino oscillations. However, systematic uncertainties—both theoretical and instrumental in character—preclude a definite statement at present.

Another important class of nuclear physics experiments are the double beta-decay experiments. In particular, new physics beyond the Standard Model can involve massive Majorana-type neutrinos which do not distinguish between neutrino and antineutrino. One consequence of this that would be observable are neutrinoless double-beta decays that yield two electrons but no neutrinos in the final state. Sensitive searches have been performed setting the present limits (less than around $1\,\mathrm{eV/c^2}$) and future improvements are expected.

Finally, there are experiments done at reactors and small accelerator facilities (some involving radioactive beams) that make crucial measurements for particle astrophysics and cosmology. These measurements include: cross sections for reactions that are key to the solar-neutrino problem (*e.g.*, $^7\mathrm{Be(p,\gamma)^8B}$), to stellar physics (*e.g.*, $^{12}\mathrm{C}(\alpha,\gamma)^{16}\mathrm{O}$), to explosive nucleosynthesis (*e.g.*, r-process reactions and neutrino cross sections); and to big-bang nucleosynthesis (*e.g.*, neutron mean lifetime, $^4\mathrm{He}(\alpha,\gamma)^7\mathrm{Be}$, $^3\mathrm{H}(\alpha,\gamma)^7\mathrm{Li}$, and $^7\mathrm{Li}\ (\mathrm{p},\alpha)^4\mathrm{He}$).

6.4 Particle Dark-Matter Detectors

The search for dark matter is a central theme of particle astrophysics. The evidence for dark matter based on cosmological and astronomical observations is very strong. In addition, there are compelling cosmological and astrophysical arguments that suggest that the bulk of the dark matter is nonbaryonic. Searches for baryonic dark matter have been performed, including the search for Massive Astrophysical Compact Halo Objects (MACHOs) using the gravitational microlensing technique.

Of most interest to particle physics, however, are the class of candidates that involve new elementary particles or massive neutrinos. The simplest solution to the dark matter problem involving particle physics is finite-mass neutrinos. Light neutrinos of mass in the few to 20 eV could represent some or all of the dark matter. Neutrino-oscillation experiments, involving accelerators with short and long baselines and nonaccelerator experiments are addressing the possibility of such neutrino oscillations, both in this mass range and in those indicated by the atmospheric neutrino data.

Cold dark matter is favored from the point of view of structure formation. Weakly Interacting Massive Particles (WIMPs) are an attractive solution, especially the lightest supersymmetric particle (usually the neutralino). Direct detection of such particles requires technological detector developments before a large-scale direct searches with sufficient sensitivity are possible. Such developmental work is underway and the first generation detectors of small scale have already been used to set limits. This program is planned to evolve in stages to larger and more sophisticated detectors to be installed deep underground with improved sensitivity. A definitive search at the expected sensitivity level will require a major initiative.

There are specialized experiments for other dark-matter candidates, *e.g.*, $10^{-6} - 10^{-5}\,\mathrm{eV}$ axions (Livermore and Kyoto) and superheavy magnetic monopoles (MACRO), and improved results are expected in the next few years.

Complementary to these direct searches, indirect searches, through WIMP-antiWIMP annihilation in the earth and sun that yield high-energy neutrinos, are being performed using data from the large underground detectors (*e.g.*, Kamiokande and MACRO). These searches have already ruled out large regions of neutralino parameter space, but eventually are limited by background. The best prospect for indirect detection in the future probably involves the search for annihilation products (positrons, antiprotons, and gamma rays) from WIMP-antiWIMP annihilations in the halo.

6.5 Gravitational Waves

Two projects (LIGO, VIRGO) are under construction with the goal to observe cosmic gravitational waves. These novel scientific facilities are directed first at direct detection and then at becoming gravitational-wave observatories that will open a new window on the Universe. The method is to detect small displacements between widely separated suspended test masses that result from the passage of a gravitational wave. The LIGO and VIRGO projects use laser interferometers with arm lengths of 3-4 kilometers which are capable of resolving 10^{-18} meter change in arm length due to the passage of a gravitational wave.

The best understood source of gravitational waves is the inspiral of binary neutron-star systems. The design sensitivity of VIRGO and LIGO are such as to produce a few events per year from these sources by the initial interferometers with substantial long-term improvements envisioned. The LIGO detectors are located in Hanford, Washington and Livingston, Louisiana, and are used in coincidence. These will also be combined with the VIRGO detector near Pisa, Italy to decompose the polarizations of the gravitational waves and for pointing to the sources. The time scale for initial physics with these detectors is about 2002.

6.6 CBR Studies, Redshift Surveys, Gravitational Lensing, Etc.

Some of the most decisive tests of the predictions of new physics beyond the Standard Model involve observations that are more astronomical in character. However, even here the boundaries between fields have become blurred, and techniques and people move freely across traditional boundaries in the quest to answer fundamental questions. We briefly mention some of the most exciting work.

The anisotropy of the CBR provides the best record of density perturbations on very-large scales and thereby probably the most powerful test of inflation and topological-defect theories. Measuring the anisotropy of the CBR is a rapidly developing field (see Fig. 2) and more experiments, including high-resolution, balloon-borne and space-based experiments to map the CBR sky, are being proposed.

On small scales the spectrum of inhomogeneity is probably best probed by directly "mapping the Universe," *i.e.*, determining the redshifts of galaxies (and through Hubble's law their distances) in a representative sample of the Universe. An effort to obtain a million redshifts and involving astronomers at a number of universities as well as high-energy physicists at Fermilab is well underway (Sloan Digital Sky Survey).

The MACHO and Sloan projects represent a new style. For the MACHO project a special CCD camera was built for a telescope dedicated to searching for microlensing; in the case of the Sloan Survey a special purpose 2.5 meter telescope and instruments are being built to map the Universe. Similar initiatives on the horizon. For example, a diverse group of scientists (astronomers, high-energy physicists, experimental physicists) has discussed a dedicated cosmology telescope to probe the Universe using gravitational-lensing techniques. Such an instrument could map dark matter in clusters and the halos of individual galaxies as well as determining the clustering properties of dark matter.

7 Structural Issues

High-energy physics research is built around strong partnerships between the national laboratories devoted to HEP (Fermilab, SLAC, and BNL) and university-based research groups. In recent years, we have seen a growing centralization of major facilities for research in particle physics resulting from the development of high-energy accelerators and colliders, coupled with the construction of very-large detector facilities. As a result, the laboratories now both provide the location and the focus of activity, including much of the resources, technical expertise, and infrastructure required. A system of peer review through program advisory committees, long-range planning, and the necessary management has developed in each laboratory.

The increased importance of particle astrophysics and other nonaccelerator experiments, coupled with the growing scale and technical challenges, presents new challenges. In terms of planning, a method for setting priorities and the ability to monitor and track the execution of complicated nonaccelerator projects does not naturally fit into the current mode of operation where this is done through the national laboratories. In particular, many of the current projects have no direct national laboratory involvement and are located elsewhere.

Presently, about 20% of the experimental program in HEP involves experiments not using the major accelerators at the national laboratories. This research is conducted by individual groups and by large collaborations. Some centers for these studies (*e.g.*, NASA/Fermilab Astrophysics Center, Experimental Astrophysics Group at Fermilab, Center for Particle Astrophysics in Berkeley, Institute for Nuclear/Particle Astrophysics at LBL) have emerged in recent years, responding to the need to provide coherence to these efforts and, also, to increased interest in pursuing this research. The Baksan Laboratory in Russia, and most notably, the Gran Sasso Laboratory in Italy, have developed as major laboratories providing facilities and infrastructure for such experiments. Most of the major projects are conducted through large international collaborations (*e.g.*, SuperKamiokande is U.S./Japan; MACRO is Italian/U.S.; GALLEX is German, French, Italian and U.S.; SAGE/RAGE is Russian and U.S.). It is also worth noting that these large experiments often involve funding by multiple agencies, either in the U.S. or in different countries, which presents additional logistical problems.

Conducting this research which has no host national laboratory presents a number of new problems. Perhaps the most important is to find a satisfactory mechanism for long-range planning and to set priorities. Given the diversity of the field and the number of funding agencies involved this is a difficult challenge. In high-energy physics such planning is generally provided by laboratory long-range planning and program advisory committees, with HEPAP and DPF Workshops used for global studies and to set global priorities. There are no standing committees to help set priorities and provide oversight for particle-astrophysics experiments and especially with limited resources this might be of value. A related problem is how best to make available or provide the technical and project expertise that is centralized in our national laboratories for this research.

An emerging problem involves the fact that the collaborations that are developing are becoming even more diverse. For example, the Sloan Digital Sky Survey involves astronomers, astrophysicists, high-energy physicists, and computer scientists at universities, Fermilab, the Naval Observatory, and the National Astronomical Observatory in Japan. The funding comes from NSF, the Sloan Foundation, the DOE (through Fermilab), the universities, and Japan. The challenges include the merging of scientific cultures, the usual ones associated with international collaboration, and dealing with several funding agencies.

The research opportunities in particle astrophysics,

cosmology, and unification address some of the most important and exciting problems in our field. The future will involve implementation of the next generation of facilities (*e.g.*, SNO, SuperKamiokande, LIGO/VIRGO, etc.) as well as choosing and implementing new large projects now under design (long-baseline neutrino oscillation experiments, large-scale cosmic-ray arrays, large underwater or under ice neutrino detectors, large-scale dark-matter detectors, CBR anisotropy). To select the best projects, obtain funding, and successfully implement them is an increasing challenge. The creation of infrastructure for planning and technical support may well be needed.

Selected Bibliography

1. *Proceedings of Snowmass 1994: Nuclear and Particle Astrophysics and Cosmology in the Next Millennium*, eds. E.W. Kolb and R.D. Peccei (World Scientific, Singapore, 1995).

2. E.W. Kolb and M.S. Turner, *The Early Universe* (Addison-Wesley, Redwood City, CA, 1990).

3. P.J.E. Peebles, *Principles of Physical Cosmology* (Princeton University Press, Princeton, 1993).

4. *The Birth and Early Evolution of Our Universe (Nobel Symposium 79), Physica Scripta* **T36** (1991).

5. C. Copi, D.N. Schramm, and M.S. Turner, "Big-bang Nucleosynthesis and the Baryon Density of the Universe," *Science* **267**, 192 (1995).

6. J.C. Mather *et al.*, "Measurement of the Cosmic Microwave Background Spectrum by the COBE FIRAS Instrument," *Astrophys. J.* **420**, 439 (1994).

7. M. White, D. Scott, and J. Silk, "Anisotropies in the Cosmic Microwave Background," *Annu. Rev. Astron. Astrophys.* **32**, 319 (1994).

8. A. Cohen, D. Kaplan, and A. Nelson, "Progress in Electroweak Baryogenesis," *Annu. Rev. Nucl. Part. Sci.* **43**, 27 (1992).

9. G.G. Ross, *Grand Unified Theories* (Addison-Wesley, Reading, MA, 1984).

10. C.W. Misner, K.S. Thorne, and J.A. Wheeler, *Gravitation* (Freeman, San Francisco, 1973).

11. T.K. Gaisser, *Cosmic Rays and Particle Physics* (Cambridge University Press, Cambridge, 1990).

12. J.W. Cronin, K.G. Gibbs, and T.C. Weekes, "The Search for Discrete Astrophysical Sources of Energetic Gamma Radiation," *Annu. Rev. Nucl. Part. Sci.* **43**, 883 (1993).

13. J. Wdowczyk and A.W. Wolfendale, "The Highest Energy Cosmic Rays," *Annu. Rev. Nucl. Part. Sci.* **39**, 43 (1989).

14. G.G. Raffelt, "Astrophysical Methods to Constrain Axions and Other Novel Particle Phenomena," *Phys. Repts.* **198**, 1 (1990).

THE STATE OF EXPLORATORY THEORY BEYOND THE STANDARD MODEL

DAVID J. GROSS

Princeton University, Princeton, NJ 08544

GORDON KANE

University of Michigan, Ann Arbor, MI 48109

EDWARD WITTEN

Institute for Advanced Study, Princeton, NJ 08540

1 Introduction

One of the most fundamental unsolved problems in physics is how the electroweak symmetry is broken and why the symmetry breaking scale is so tiny compared to the natural scale of unification or gravitational phenomena. The latter question is a modern version of the "large numbers problem" of Dirac, who wondered why the proton mass was so much less than the Planck mass.

One of the most attractive frameworks for addressing this question is supersymmetry. The standard model has a fascinating supersymmetric extension which has remained viable over the years while other alternatives such as composite models have encountered difficulties. The supersymmetric standard model resolves the large numbers problem and is also compatible with our limited experimental constraints on TeV physics. It hints that a whole new world of physics is just around the experimental corner, just out of reach of present day experimental tools. But is it right? Only experiment can tell; the supersymmetric standard model can be vindicated only by discovery of the Higgs boson and the other new particles that it predicts at energies not much greater than that of the Higgs.

Discoveries of the Higgs boson and other possible TeV scale particles might also give clues about some of the other mysteries left open by the standard model of particle interactions:

- What determines the gauge group of the world at ordinary energies to be $SU(3) \times SU(2) \times U(1)$?
- How does the other force in nature, gravity, enter the picture?
- How can the gauge forces be unified with each other and with gravity?
- What determines the quantum numbers of the quarks and leptons?
- Why does nature seem to repeat itself, with several "families" of quarks and leptons that appear to be identical except for the masses?
- What determines the bare masses of the quarks and leptons and the mixing angles that control their lifetimes?
- Why, in particular, are some of the quarks and leptons – like the electron and the up and down quarks – so much lighter than what would appear to be the natural scale, which is the mass scale of the W and Z bosons, the top quark, and presumably the Higgs boson?
- Why does the cosmological constant vanish?

One of the first significant approaches to these questions, by now twenty years old, was the idea of a "grand unified" gauge theory (GUT) with $SU(3) \times SU(2) \times U(1)$ unified in a simple group such as $SU(5)$. The $SU(5)$ model, and its variants, gave such a beautiful explanation of the strange fractional quantum numbers of the quarks and leptons that it is hard to believe that there is not some truth in it. These theories also made a spectacular prediction of proton decay and gave a strong hint of neutrino masses and mixing. These predictions have not yet been confirmed (though the solar neutrino problem may be giving us an emerging discovery of neutrino masses and mixing); on the contrary the original $SU(5)$ model in its simplest form was eventually excluded by failure to observe proton decay at the predicted rate.

Another early success of the $SU(5)$ model was the rather good prediction it made for the mixing angle of the electroweak theory – the parameter called $\sin^2\theta_W$ that measures the ratio of the $U(1)$ and $SU(2)$ gauge couplings. In more recent years, however, experiments – such as the precision measurements of the strong, weak,and electromagnetic couplings at LEP – have shown that while the $SU(5)$ prediction for $\sin^2\theta_W$ is quite close, it is not close enough.

Just as the standard model of TeV physics has an attractive alternative in the supersymmetric standard model, grand unified theories have viable supersymmetric extensions. Indeed, the effect of supersymmetry is quite fascinating – the small error in the original $SU(5)$

prediction of $\sin^2\theta_W$ is corrected, bringing the prediction right on top of the best experimental values (if supersymmetry survives to energies close to a TeV); and the proton lifetime is extended beyond the experimental limits. (Of course the supersymmetric grand unified theories also solve the large numbers problem.) Are these fundamental facts about nature or frustrating coincidences? The best way to find out is to find– or not find – superpartners at the TeV scale.

Another fascinating property of GUT's is that the mass scale of unification can be computed using the renormalization group and the values of the couplings at accelerator energies, and,especially in the supersymmetric version, the answer turns out to be rather close to the Planck mass. Though we certainly don't fully know why this is so, it is a promising sign since we do know that the Planck mass is an energy at which new things must be happening in physics.

Yet GUT's, with or without supersymmetry, leave open many of the questions on our above list. They suggested some new perspectives but in fact led to limited progress in explaining the multiplicity of fermions and the quark and lepton masses and mixing. Their potential to constrain the low energy gauge group appears to be limited. They of course leave gravity and the cosmological constant out of reach.

To incorporate gravity is perhaps the thorniest knot in theoretical physics because – as has long been perceived – the nonlinear structure of general relativity clashes with the requirements of renormalizable quantum field theory. To make sense of gravity in the light of quantum physics would appear to require a new framework, and luckily a new framework did appear in the form of string theory, which originally was developed in the late 1960's and early 1970's in the context of hadron physics.

At first, when it was perceived that string theory could be interpreted as a new framework for physics including gravity, it appeared that the theories that arose, though they were unified theories of gravity and matter, were unrealistic (the weak interactions had to conserve parity, for instance). By 1984, however, new developments involving anomaly cancellation and the heterotic string had made it possible to make more or less realistic models of particle physics from string theory. New insights appeared about why the gauge groups that we actually see might be natural and why nature might naturally generate several families of fermions.

The whole framework of string theory has proved to be incredibly rich. It is amazing to see how phenomenologically interesting gauge groups and fermion representations, not to mention general relativity, appear out of elegant and simple world-sheet constructions. Many new ideas have come from string theory – just to give one example, supersymmetry originated nearly twenty-five years ago in attempts to incorporate fermions in string theory. It is hard to imagine what theoretical physics would be like today without this enriching influence.

On the other hand, there is so much that is not understood. We not only do not know how to determine the vacuum state – which underlies observed particle physics – but we do not even have a sensible framework for asking the question. All we really know is how to construct perturbation theory around classical solutions. TeV scale supersymmetry is very natural from the point of view of string theory, but it has not been deduced as a true prediction. Almost all more or less phenomenological work in string theory has assumed TeV scale supersymmetry, but – pending experiments at the necessary energies – it is possible for all we know that this assumption is completely on the wrong track.

Assuming that TeV supersymmetry is correct, one of the most compelling questions is to find a sensible mechanism of supersymmetry breaking. This has not emerged yet. The problem is not to break supersymmetry– this occurs naturally through chiral symmetry breaking in one or another piece of the gauge group – but to break it in a way that leads to a stable vacuum and vanishing cosmological constant.However, the absence of an understanding of supersymmetry breaking does not prevent proceeding with useful phenomenology and studies, because the non- renormalization theorems and the knowledge of the form of the general soft supersymmetry breaking terms means one can write the most general effective Lagrangian at a high scale and study its implications for experiments and measure its parameters.

Studies of dynamics of supersymmetric models beyond perturbation theory have in fact been carried out most intensively in field theory – where better nonperturbative formulations are known and more nonperturbative methods are available. These studies, pursued intensively at different times since the early 1980's, have uncovered many fascinating phenomena and shed much light on four-dimensional quantum field theory in general. But phenomenologically useful mechanisms of supersymmetry breaking have not yet emerged.

Should supersymmetry indeed be discovered at the TeV scale, many other discoveries will inevitably come with it. There should be a whole garden of new particles. Which are lighter and which are heavier? Is the mechanism of supersymmetry breaking visible at the TeV scale, or does it come from higher energies? And in the latter case, through what effective interactions in the low energy theory is the supersymmetry breaking manifested? And what clues can one draw from that about the energy scale at which the supersymmetry breaking originates?

It is impossible to foresee all the ramifications of a discovery of supersymmetry at TeV energies, except that it would open up one of the golden ages in experimental physics as answers appear to some of the questions just raised. And it would give a tremendous, though not fully predictable, boost to the efforts of theorists, as well as experimenters.

2 Experimental Opportunities to Detect Supersymmetric Physics

As we have already suggested, there are stronger reasons than ever to think that nature may be supersymmetric at the TeV scale. The arguments of the early 1980's have held up: a supersymmetric world could maintain a hierarchy of scales and thus allow the particles and forces to unify; the supersymmetric standard model has candidates for cold dark matter, unlike the ordinary standard model; and the Higgs mechanism can be derived more naturally than in the ordinary standard model. Meanwhile, new arguments have been added, both theoretical and phenomenological ones. On the theoretical side, the emergence of phenomenologically attractive superstring models has focussed renewed attention on the possibility that supersymmetry survives down to electroweak energies. Phenomenologically, a major development hinting at supersymmetry – already mentioned above – is the beautiful quantitative agreement of precision measurements of the low energy coupling constants with predictions of supersymmetric unified theories; the non-supersymmetric counterparts of those theories fail by many standard deviations given the precise data now available. Another important phenomenological result, unexpected by most physicists ten years ago,is that the top quark is "heavy" – that is, unlike the other quarks and leptons, its mass is comparable to the electroweak scale. It was already clear in the early 1980's that gauge symmetry breaking would work out correctly for a much larger range of the parameters if the top quark was heavier than about M_W; that prediction, which was not considered terribly attractive at the time, has indeed been satisfied. Further, in recent years very detailed supersymmetric models that impose all known theoretical and phenomenological constraints have been successfully constructed; it could easily have happened that once one went beyond general arguments it would turn out that it was not possible to satisfy all the constraints simultaneously. Among other things, the lightest superpartner of such models (the LSP), indeed has the properties to give about the right amount of cold dark matter.

Of course, the above indications for supersymmetry could be coincidental and misleading. Until superpartners or the supersymmetric Higgs spectrum are discovered experimentally, the situation will not be settled. Nevertheless, the indications sketched above are suggestive enough that, in our opinion, testing supersymmetry and discovering superparticles should be one of the prime goals of future experiments. Discovery of supersymmetry would be one of the great event in physics; it would profoundly affect experimental priorities and decisions about new facilities and refocus the efforts of many theorists. If electroweak physics is not supersymmetric,many theorists are working in the wrong direction, and it would be useful to know that as soon as possible.

How can we get information about the masses of the superpartners and Susy-Higgs bosons, in order to tell us what facilities are needed to detect them? How can we get information on cross sections and signatures so we can know what luminosities are needed to produce and detect them? There has been progress here too in the past couple of years. Detailed quantitative models have been constructed that give a complete set of predictions for masses and couplings, so that all observables can be calculated. There are parameters in the models, and some assumptions underlying them, mainly reflecting our lack of understanding of how supersymmetry is broken, so that most predictions are not unique. Nevertheless, the constraints are strong enough that many useful predictions can be made.

First, the models imply that we would have been very lucky if superpartners or a Higgs boson were already detected; only a few percent of the parameter space has been covered in existing experiments. Further, the masses of the charginos and neutralinos are always expected to be rather light, within a factor of two or so of the W and Z masses, and therefore detectable at LEP2 and at FNAL (particularly if its luminosity or energy or both are upgraded) relatively soon. Other partners are often very light, *e.g.* the stop or superpartner of the top quark. If very recent indications from LEP, particularly concerning the branching ration for $Z \to b + \bar{b}$, persist as the data become more precise, then upper limits of order 100 GeV can be set on chargino and stop masses; they have large production cross sections at LEP and FNAL, and good signatures. Further, the models indicate that the LHC will produce large quantities of most (or all) of the superpartners and SUSY-Higgs bosons, so that with sufficiently good detectors it will be possible to study many of them and learn the basic parameters of the theory. A high-energy electron-positron collider (NLC) will also be valuable to study many superpartners, and it is complementary to the LHC to a large extent; for example, gluinos will be produced copiously at LHC but little at NLC, while charginos and sleptons will be easily studied at LEP or NLC, particularly with polarized beams that can play a major role in separating different states and measuring their masses and couplings. If indeed nature is supersymmetric on the electroweak scale, and a perturbative unification occurs at a scale of order 10^{16} GeV, a remarkable consequence is that it is possible to calculate predictions for collider data from the theory at the unification scale, and conversely to learn the form of the effective Lagrangian at the unification scale from collider data. Effectively we are able to make certain measurements at the unification scale. Connecting the Planck scale to the electroweak scale may seem difficult, but if we can probe the unification scale then it does not seem so implausible to go the remaining two orders of magnitude to the Planck scale.

Another major consequence of finding superpartners is that automatically the LSP is produced in the decays of all of the others. Thus its mass and couplings will be measurable at whatever collider first detects a superpartner. That will allow determination of whether it has the right properties to be the cold dark matter of the universe. Furthermore, knowing the characteristics of a candidate particle would also help focus efforts at direct detection of cosmic dark matter.

If TeV scale supersymmetry is wrong, one may ask how can it best be disproved experimentally? One particularly efficient and rigorous way to do this is known. In the supersymmetric standard model, there is an upper limit on the mass of the lightest SUSY-Higgs boson, independent of detailed assumptions. The numerical value of the limit (but not its existence) depends a little on the gauge theory. For the simplest gauge theories the limit is below about 150 GeV; for more complicated ones it can go up of order $10 - 15\%$. Upper limits on superpartner masses also exist in all models that can satisfy the existing theoretical and phenomenological conditions. These limits, presently of order a TeV for the heaviest partners, are still dependent on parameters and assumptions, but data are becoming strong enough to set lower and tighter limits. These limits depend less on the parameters of the theory and on other assumptions. Thus if experiments do not find superpartners at LHC (assuming sufficiently good detectors) it will be possible to conclude that supersymmetry is not relevant to understanding the physics of the electroweak scale and that the values of the measured gauge couplings are not really results of TeV supersymmetry. Indeed, if the recently reported branching ratio $\Gamma(Z \to b\bar{b})$ at LEP does not change by more than about one standard deviation, it maybe possible in a few years to exclude supersymmetry very generally if charginos and stops and the LSP are not discovered at LEP2 or an upgraded FNAL.

There may be further indirect indications of supersymmetry. If protons decay it will be a strong signal of unification in general. In a supersymmetric theory these decay modes are generally different than in other theories (usually $\nu + K^+$ dominates in supersymmetry). The branching ratio for $b \to s + \gamma$ can be greatly affected by superpartners since it must vanish in the supersymmetric limit. Thus depending on what its measured value is and on the errors, this process may be able to distinguish supersymmetry from the standard model. Other rare decay modes and loop contributions may also be important, particularly in conjunction with other measurements for testing supersymmetry.

Because of the existing indirect indications for supersymmetry, and the extraordinary consequences for physics if it is detected, we urge that every opportunity to get evidence of supersymmetry be exploited as quickly and thoroughly as possible, and that planning for detectors, upgrades, facilities, and funding priorities include consideration of opportunities to learn about whether nature is supersymmetric.

3 Recent Theoretical Progress

Even in the absence of direct experimental stimulus exploratory theory remains vital and exciting. New ideas have been developed over the last few years that could shed light on some of the deep issues in string theory, quantum gravity and in supersymmetric theories. In addition many of the methods and techniques that have been developed to explore these questions have applications in other areas of physics and mathematics.

We comment below on four areas of research that have been active over the few years:

- the study of classical solutions of string theory—two dimensional conformal field theory,
- black hole physics,
- matrix models and large N techniques, and
- supersymmetric field theory.

One major focus in the last few years has been on understanding classical solutions of string theory. A basic paradigm here relates classical solutions of string theory to conformally invariant field theories in two dimensions; this paradigm, many of whose implications are still mysterious, has no real analog in pre-string physics. Because of the paradigm, work on classical solutions of string theory often interacts with studies of critical phenomena and many-body physics, such as the Kondo problem and the fractional quantum Hall effect, which can be described by two-dimensional field theory.

Working within this paradigm, string theorists have discovered many unusual physical phenomena that occur in string theory under extreme conditions. One important observation is that there seems to be in string theory a minimum size of any dimension of space-time; if one tries to contract space-time below that scale, it re-expands in a new "direction" – the phenomenon cannot be described in pre-string terms. It seems that the distance at which this occurs is actually in some sense the smallest measurable space-time distance. For instance, the Heisenberg microscope fails to work at very short distances: the ability to probe shorter distances by accelerating particles to ever-higher energies is eventually limited by the fact that at high enough energies strings begin to *grow*. Other strange things occur when one probes the limiting distance scale. One example is that the topology of space-time can change purely at the classical level, with no reference to quantum corrections:

one can continuously interpolate from one classical space-time to another, passing through an intermediate stage in which stringy effects are big.

Closely related to this are new, non-classical symmetries found in string theories, such as mirror symmetry; our knowledge of such symmetries has greatly increased in the last few years. One of the main motivations in this work has been to make more computable some of the models of elementary particle physics that can be derived from string theory. For instance, mirror symmetry was used—and in large part was developed—to make it possible to calculate Yukawa couplings that were otherwise out of reach.

The strange new phenomena found in classical solutions of string theory hint at a new, not yet understood, formulation of the theory in which space-time would be a purely derived and approximate concept, just as traditional concepts like the position and momentum of a particle became approximate concepts in quantum mechanics. These investigations might perhaps be compared to the discovery of the Klein paradox in the early days of quantum electrodynamics—a strange phenomenon in classical solutions of the Dirac equation that pointed the way to a better understanding of the role of the equation in physics.

Some classical solutions found in string theory are closely related to black holes. Thus, one has been led to ask what light string theory—which after all is claimed to be a consistent theory of quantum gravity—sheds on the longstanding paradoxes that arise when black holes are considered in of quantum theory. In the last few years, there has been intensive study of two dimensional black hole models that can be treated quantum mechanically. The analyses are substantially more precise than previous treatments of quantum black holes, and various attempts have been made to extract lessons that could apply to more realistic models. Partly based on some of these analyses, arguments have been made that the details of black hole evaporation – even at an early stage – depend on Planck scale physics, which would then play an essential role in the resolution of the black hole paradoxes. Indeed, one interesting line of argument has claimed that some of the peculiar properties of string theory – like the tendency of strings to grow when probed over short times – are just what is needed to resolve the paradoxes of quantum black holes.

In addition to the study of purely classical solutions of string theory, examples have been found (with low dimensional target spaces) in which the *quantum* theory is soluble. The *matrix model* techniques for finding these solutions grew in part from methods that were originally developed to understand the $1/N$ expansion of QCD. The $1/N$ expansion is of interest because it is widely suspected to offer the best hope for better understanding of those important aspects of QCD (quark confinement, mass generation and chiral symmetry breaking) for which there is only very limited analytic understanding. Techniques of the $1/N$ expansion have been used to make possible exact quantum treatments of special string models. Perhaps these will one day play a role similar to the Schwinger model of two dimensional quantum electrodynamics – a soluble model which despite its special nature is now understood to illustrate many mechanisms of field theory that are important in four dimensions. The success in applying $1/N$ methods to special string models also helped revive interest in the analogy between the $1/N$ expansion and string theory; this analogy, and the possible interpretation of some gauge theories – at least in two dimensions – as string theories, have been much investigated in the last few years.

Dramatic advances in understanding four dimensional strongly interacting field theories have come from another front: the study of supersymmetric models. Supersymmetric field theory may very well turn out to be the real thing at the TeV scale (though if so the parts of the theory that survive down to that energy scale may well be weakly coupled, except of course for QCD). Whether that is so or not, use of the supersymmetric case as a test case has led to much greater progress in understanding four dimensional gauge theory dynamics than has been obtained (from any point of view) at any time since the late 1970's. The phenomena that are now under control – such as composite massless fermions, confinement and chiral symmetry breaking via condensation of magnetic monopoles, baryons as solitons, the dynamical role of electric-magnetic duality in four-dimensional field theory – are not special to supersymmetry, but at present supersymmetry is needed to bring them under control. Some of the ideas in this work are known to have interesting extensions to or analogs in string theory, but the recent developments have mainly involved analyses of the field theories.

Partly stimulated by the progress with gauge theory dynamics but in equal part because of the potential phenomenological importance, there have also been renewed and reinvigorated studies of dynamical supersymmetry breaking, which might be relevant to physics at the next generation of accelerators either directly (if the mechanism of spontaneous supersymmetry breaking is manifest at a TeV or so) or indirectly (if supersymmetry is dynamically broken in a hidden sector). Both new mechanisms of supersymmetry breaking and new ways to use old mechanisms have been developed.

4 Theoretical Challenges

Theoretical particle physics is at the present time in a very unusual state. On the one hand, the standard model, developed in the 1960's and 1970's, has provided an extraordinarily successful theory of the electro-weak and strong interactions that has passed many precise tests over the last twenty years. This is a great theoretical triumph. On the other hand, the very success of this theory has made it clear that its scope is limited. It cannot address nor answer most of the questions posed above. Experiments have not yet come to the aid. Thresholds for new and relevant data seem to be beyond reach of existing accelerators, though perhaps very close.

As a consequence of these developments, many theorists have proceeded to study speculative new theories which do offer the potential to address and answer the questions posed by the success of the standard model, yet are too undeveloped to provide precise predictions. Some of these new ideas, in particular string theory, offer a tantalizing vision of a fundamental revolution in the conceptual framework of physics and the possibility of a truly unified and totally predictive theory of gravity and matter.

Under these circumstances a rather unhealthy rift has developed in the community of theoretical physicists, between the more phenomenologically oriented theorists and those who are engaged in more theoretical, formal, speculation. This rift threatens the vitality of theoretical particle physics. This danger is made more severe by the fact that new clues from experiment may be some years off and that the new thresholds suggested by current speculation are difficult to reach. It is imperative that in these difficult times, as experimental particle physics is being subjected to increasing fiscal constraints, that theoretical particle physics remain as healthy as possible. Therefore we feel it important to address this issue.

To begin with, we feel that it is crucial that the community of particle theorists show more tolerance towards the attitudes and approaches of others. More communication and interaction is required between the phenomenologists and the speculators. One can aim to bridge the gap through conferences that attract a mixed audience, but most important is a change of attitude.

We urge our formal colleagues not to ignore experiment and phenomenology. Present day experiments do not provide much help in solving string theory–but that can change and it can change overnight. Unless you continue to follow current phenomenology and experiment you will not be in a position to take advantage of the new experimental discoveries when they do take place. One piece of experimental data is often more valuable than dozens of technical and formal advances.

We also urge them not to ignore the many interesting and difficult theoretical problems that remain within the standard model. These are not only interesting and important problems, but the ideas and methods that are developed to solve them are often useful in other areas of physics, even in string theory. For example, the development of soluble toy string theories based on matrix model methods was a direct outgrowth of old attempts to solve QCD.

And we urge our more phenomenological friends not to ignore the more formal and speculative parts of our field. They should remember that what appeared to be pure speculation and formal theory in the past has now become enshrined in the standard model. It is a mistake for them to ignore the hints that speculative, yet rich and evocative, theories can provide for those interested in inventing new phenomenology.

Contrary to claims that are sometimes heard, speculative theories such as string theory can be disproved. String theory, for instance, risked being eliminated in the early 1980's because it appeared to predict that the weak interactions would conserve parity; it survived only because of unexpected theoretical advances. While string theory survived in this perilous way, other ambitious theories have gone by the wayside. For example, the more ambitious versions of supergravity and Kaluza-Klein theory have been all but excluded by combinations of theoretical and experimental difficulties. The very fact that string theory and the supersymmetric standard model have survived while other theories have been excluded is a hint of their vitality. The history of physics is filled with experimental and theoretical surprises and it is foolish to claim that a theory with the richness of string theory cannot be tested.

Talented young people should not be discouraged by gloomy claims occasionally made about the state of theoretical physics. The problems are great, but the opportunities are great also. Particle physicists are grappling with wonderful questions and marvelous, mysterious ideas. The best and the brightest continue to be attracted to particle physics because, even in these difficult times, it addresses the most fundamental questions. There is every possibility of new synthesis in coming years and what looks like calm may be the eye of the storm.

Accelerator Physics, Technologies, and Facilities

Donald L. Hartill

Cornell University, Ithaca, NY 14853

Stephen D. Holmes

Fermi National Accelerator Laboratory, Batavia, IL 60510

Robert B. Palmer

Brookhaven National Laboratory, Upton, NY 11973

Ronald D. Ruth

Stanford Linear Accelerator Center, Stanford, CA 94309

Working Group Members

R. Brinkman,	*DESY*	N. Phinney,	*Stanford Linear Accelerator Center*
A. Chao,	*Stanford Linear Accelerator Center*	A. Sessler,	*Lawrence Berkeley Laboratory*
B. Danly,	*Massachusetts Institute of Technology*	R. Siemann,	*Stanford Linear Accelerator Center*
G. Dugan,	*Cornell University*	K. Takata,	*KEK*
M. Harrison,	*Brookhaven National Laboratory*	P. Tenenbaum,	*SLAC and UC Santa Cruz*
M. Holtkamp,	*DESY*	M. Tigner,	*Cornell University*
G. Jackson,	*Fermi National Accelerator Laboratory*	T. Weiland,	*TH-Darmstadt*
T. Katsouleas,	*University of Southern California*	I. Wilson,	*CERN*
D. Neufer,	*CEBAF*		

1 Introduction

Since the invention of the cyclotron in the early 1930's, the particle accelerator and storage ring have been the instruments of choice for scientists engaged in particle physics research. Historic advances in high-energy and particle physics have been, and are expected to continue to be, dependent upon advances in accelerator facilities and the underlying technologies upon which they are based. Over the past sixty years these technologies have provided nearly a million-fold increase in beam energy, accompanied by facilities whose size is measured in kilometers rather than meters, and whose costs are now measured in hundreds of millions, or even billions, of dollars.

Accelerators in operation today generally fall into two classes: hadron accelerators and electron accelerators. The majority of contemporary high-energy physics experiments use accelerators in which counter-circulating beams are brought into collision. This mode of operation provides the most efficient utilization of beam energy for producing high-mass particles. The highest energies are attained in hadron colliders while electron-positron colliders are used in applications in which a premium is placed upon having all available energy carried by the interacting constituents.

The highest energy accelerator facility in the world today is the Tevatron at Fermilab. This facility, the first high-energy accelerator ever constructed utilizing superconducting magnets, provides proton-antiproton collisions at 1.8 TeV in the center-of-mass, with a luminosity currently exceeding 1×10^{31} cm^{-2} s^{-1}. The Tevatron collider is the only currently operational hadron collider in the world and is the only facility of any type at which the top quark can be produced and its properties measured. The construction and operation of the evatron made possible development f the concepts of the Superconducting Super Collider and the Large Hadron Collider (LHC).

Hadron accelerators are often operated to provide opportunities for fixed target experiments. The Alternating Gradient Synchrotron (AGS) at Brookhaven National Laboratory currently provides 30 GeV proton and lower energy secondary beams for use in neutrino physics, searches for new particles, a measurement of the muon's anomalous magnetic moment, and searches for and studies of rare kaon decays. This accelerator is planned to serve as the injector to the new Nuclear Physics facility RHIC later in this decade. Proton beams at 800 GeV, accompanied by a variety of secondary beams, are also available at the Fermilab Tevatron for use in fixed target experiments during periods in which the collider is not running. The focus here is on deep inelastic neutrino scattering, CP violation in rare K decays, and charm spectroscopy.

Electron colliders operational in the world today come in two types and operate in two distinct energy regimes. Colliders in the lower energy regime are exclusively based on circular accelerators. These facilities include the VEPP–2M collider in Novosibirsk, Russia, the Beijing Electron Positron Collider (BEPC), the Cornell Electron Storage Ring (CESR), and the Tristan collider at KEK in Tsukuba, Japan. These facilities operate at center-of-mass energies that range between 1.0 GeV and 60 GeV. Typical operations are tuned to the resonance associated with a particular heavy quark-antiquark bound state, *e.g.*, the ϕ at VEPP-4m, the J/Ψ at BEPC, and the $\Upsilon(4S)$ at CESR. The highest luminosity attained in any of these colliders is 3×10^{32} cm^{-2} s^{-1} at the CESR facility.

Two electron collider facilities currently operate at the energy of the Z (90 GeV in the center of mass)—the Stanford Linear Collider (SLC) and the Large Electron Positron Collider (LEP) at CERN. The SLC is the first linear collider ever constructed and demonstrates the technology that is required to extend further the energy reach of electron-positron colliders. The LEP facility is a circular storage ring 27 km in circumference and most likely represents the largest circular electron accelerator that will ever be built. LEP currently operates with a luminosity in excess of 2×10^{31} cm^{-2} s^{-1} and SLC at about 5×10^{29} cm^{-2} s^{-1}.

Finally, HERA, a hybrid facility currently operational at the DESY laboratory in Hamburg, Germany, provides electron-proton collisions at approximately 300 GeV in the center-of-mass. The facility is based on an electron accelerator operating at 30 GeV and a superconducting proton accelerator operating at 820 GeV both located in a common, 6.3 km, tunnel.

Two major new facilities, and one major upgrade, are currently under construction in the United States. The Fermilab Main Injector Project will replace the existing Main Ring at Fermilab and boost luminosity in the Tevatron Collider by a factor of five, while simultaneously providing beams for a 120 GeV fixed target program. The PEP–II project at SLAC will construct an asymmetric electron-positron collider operating at the $\Upsilon(4S)$ resonance for use in CP violation studies in the B system. The luminosity goal for this facility is 3×10^{33} cm^{-2} s^{-1}. Finally, an upgrade of the existing CESR facility is aimed at providing a luminosity of 1×10^{33} cm^{-2} s^{-1}, again at the $\Upsilon(4S)$ but in a symmetric collider mode. Each of these projects is expected to be completed around 1998. World-wide facilities under construction today include an asymmetric B Factory at KEK and ϕ Factories in Frascatti, Italy (DAΦNE), and in Novosibirsk, Russia (VEPP–2M).

It currently appears likely that CERN will proceed with construction of the LHC sometime in the near future. This facility, to be located in the existing LEP-tunnel, is proposed to provide proton-proton collisions at 14 TeV in the center-of-mass with a luminosity of 1×10^{34} cm^{-2} s^{-1}. It is expected that this project will be completed in the middle of the coming decade.

Further new facilities and upgrades that could be brought to a state of reviewable proposals over the next several years include further possible luminosity (to 1×10^{33} cm^{-2} s^{-1}) and/or energy (to 4 TeV in the center-of-mass) upgrades of the Tevatron collider, and a next generation of electron-positron linear colliders. Research and development aimed at development of technologies that could support a 500 GeV linear collider, upgradable to 1500 GeV, is underway at a variety of centers world-wide.

Despite the existence of a plan for continued extension of research opportunities over the next decade, it should be recognized that the long-term future of high-energy physics requires a continuing investment in development of accelerator technologies. Technologies currently or nearly in hand are sufficient to allow exploration of physics at the TeV scale. Moving beyond, within the confines of facilities of the current physical size and cost, will require either the development of new technologies or achievement of significant cost reduction in existing technologies. It is not too early to start thinking about a hadron collider that could be built after the LHC with 4–5 times the energy reach. Cost would be a major consideration in the construction of such a facility and so any development program would have to focus on cost reduction as well as technological goals. One may also find that departures from the standard way of thinking, *e.g.*, muon or gamma colliders, represent a highly effective means of proceeding.

As seen in the Table of Contents, the results of this working group have been divided into six primary sections. Section 2 covers Hadron Facilities, from the current state of the art all the way to a possible 30×30 TeV collider. Section 3 treats Electron Linear Collider Facilities, beginning with a summary of the status of the first linear collider, the SLC, and moving on to cover various approaches and test facilities for validating the enabling technologies. Section 4 covers Electron-Positron Circular Collider Facilities. In Section 5, we look further to the future in our review of Advanced Accelerators. Finally, in the Conclusions, we discuss several key points that have been distilled from the contributions to this report.

2 Hadron Facilities

Hadron accelerators and storage rings have provided the US HEP program with research opportunities at the high-energy frontier for nearly forty years. Following the invention of the strong focusing concept in the early 1950's a series of proton accelerators of ever increasing energy has been constructed in the United States and Europe. Included in this group are the Brookhaven Alternating Gradient Synchrotron (AGS); the Proton Synchrotron (PS), the Intersecting Storage Rings (ISR), and Super Proton Synchrotron (SPS) at CERN; the proton ring in the HERA facility at DESY; and the Main Ring and Tevatron accelerators at Fermilab. While the basic accelerator design has largely changed only in scale, advances in the underlying technology have allowed a nearly 250-fold increase in the center-of-mass energy achieved in these facilities. The advent of superconducting magnets and the invention of stochastic cooling have spurred the latest advances.

The highest energy accelerator facility in the world today is the Tevatron at Fermilab. This facility, the first high-energy accelerator ever constructed utilizing superconducting magnets, provides proton-antiproton collisions at 1.8 TeV in the center-of-mass, with a luminosity currently exceeding 1×10^{31} cm^{-2} s^{-1}. The construction and operation of the Tevatron made possible development of the concepts of the Superconducting Super Collider (SSC) and the Large Hadron Collider (LHC).

The construction of the SSC has been the highest priority construction project in the US HEP program over the preceding decade. The construction in Waxahachie, Texas of this project, a superconducting proton-proton collider operating at a center-of-mass energy of 40 TeV, was canceled by the United States government in the fall of 1993. The cancellation of the SSC has prompted a complete reexamination of future options in both the United States and abroad, most notably in Europe. It now appears likely that the construction of the LHC will proceed in Europe over the next ten or so years. This facility, when completed, will provide proton-proton collisions at 14 TeV in the center-of-mass, eclipsing the energy reach of the Fermilab Tevatron early in the next century.

This report describes a study of options for future hadron facilities in the United States and identification of the R&D programs that need to be supported if such facilities are to be realized. The material presented here was prepared at a workshop held at the Indiana University Cyclotron Facility over the period July 6–10, 1994 [1].

2.1 Current State of the Art in Hadron Colliders

The Fermilab Tevatron is the highest energy collider operational in the world today. Protons and antiprotons counter-circulate within the Tevatron superconducting accelerator, colliding in two interaction regions. Collisions at 1.8 TeV in the center-of-mass are observed by the "CDF" and "DØ" detectors. The Tevatron is currently operating with a luminosity in the range 1–1.5×10^{31} cm^{-2} s^{-1}. Completion of the Fermilab Main Injector (FMI) project, currently under construction, is expected to yield a luminosity in the range 5–10×10^{31} cm^{-2} s^{-1} in the year 1999. The increase in luminosity is primarily related to an increase in the antiproton production rate achievable with the newly constructed Main Injector accelerator. It is anticipated that improvements to the Tevatron refrigeration system will allow operation at 2.0 TeV in the center-of-mass during this period. However, further energy enhancements are dependent on construction of either higher field dipole magnets (presently 4.4 T), or of a larger circumference tunnel (presently 6280 m), or both. Completion of the FMI will also afford the opportunity for a 120 GeV fixed target program concurrent with collider operations at Fermilab.

The design of the SSC was based on the technology developed in the Fermilab Tevatron. Proton-proton collisions were to be produced by colliding protons counter-circulating in separate rings utilizing dipole magnets operating at 6.6 T. The total circumference was to have been 87 km. A total energy of 40 TeV in the center-of-mass with a luminosity of 1×10^{33} cm^{-2} s^{-1} was anticipated. The magnetic dipole field of 6.6 T represented a 50% improvement over the Tevatron design. This improvement was primarily derived from improved performance in superconducting cable over the period 1978–1988. The design of the magnets also showed significant advances over the Tevatron in the area of reduced heat leak through advanced cryostat design. The new significant design issues in the SSC were largely related to the enhanced level of synchrotron radiation and to the achievement of significantly smaller transverse emittances in the SSC relative to the Tevatron. The superconducting magnet design for the SSC was fully developed with technology transfer to industry underway at the time of cancellation of the project.

The LHC complex, as currently proposed, is very similar to the SSC except for the smaller scale. The design energy is 14 TeV in the center-of-mass with a luminosity of 1×10^{34} cm^{-2} s^{-1}. The LHC is proposed to be constructed in the existing, 27 km, LEP tunnel. The magnet, a two-in-one design operating at 8.7 T, represents a significant extension of the SSC design. Models of this magnet have been successfully built and tested. If this project proceeds as currently proposed it is anticipated that operations will commence sometime around the year 2005.

2.2 Future Hadron Facilities

A workshop on "Future Hadron Facilities in the US" was held at the Indiana University Cyclotron Facility over the period July 6–10, 1994. Workshop participation included 52 registrants from 17 institutions in the United States and from CERN in Europe. This workshop was held under the auspices of the Accelerator Physics, Technologies and Facilities Working Group of the DPF Long Term Planning Study.

The specific goals of the Indiana workshop were three-fold:

- Development of a defensible parameter list for a 1×10^{33} cm^{-2} s^{-1} luminosity upgrade of the Tevatron $\bar{p}$–p collider at Fermilab, with an optional 2×2 TeV energy upgrade within the existing enclosure.
- Development of a defensible parameter list for a 1×10^{34} cm^{-2} s^{-1}, 30×30 TeV p–p collider.
- Identification of R&D requirements for achieving the stated parameters.

The two facilities examined were chosen because they were felt to span the potential needs of the US HEP community following the completion of the Main Injector upgrade at Fermilab, and through and beyond the period of utilization of the LHC in Europe. It was felt that the necessary R&D required to realize either of these facilities would likely provide a strong basis for nearly any direction that the community wished to move in the realm of hadron facilities over the next twenty years or so. The work of this meeting rests upon and extends previous studies of higher energy options in the Tevatron tunnel, as chronicled in the 1988 Snowmass Proceedings, and the extensive design and development work completed at the SSC Laboratory.

Six working groups were charged with identifying the technology issues related to each of the potential facilities. The working groups and their leaders were:

Magnets:	J. Tompkins, Fermilab
Cryogenics & Vacuum:	T. Peterson, Fermilab W. Turner, LBL
Antiproton Sources:	J. Marriner, Fermilab
Injectors:	R. York, Michigan State
Interaction Regions:	S. Peggs, Brookhaven
Lattice & Beam Dynamics:	M. Syphers, Brookhaven

These groups were assisted by two teams with overall responsibility for coordinating study and evaluating parameters for each of the two facilities mentioned above. These two teams were:

- 2×2 TeV $\bar{p}$–p collider at Fermilab:
  G. Jackson (Fermilab)
  R. Siemann (Stanford Linear Accelerator Center)
- 30×30 TeV p–p collider:
  A. Chao (Stanford Linear Accelerator Center)
  G. Dugan (SSC Laboratory)
  M. Harrison (Brookhaven National Laboratory)

The main body of this report is composed of the reports from each of the coordinating teams; reports from the working groups may be found in the workshop proceedings.

2.2.1 Design Issues

Luminosity and/or Energy Upgrades of the $\bar{p}$–p Collider at Fermilab. Design issues for an improvement program aimed at achieving a luminosity in excess of 1×10^{33} cm^{-2} s^{-1} in the Tevatron $\bar{p}$–p collider at Fermilab, as well as possibilities for creating a facility operating at 2×2 TeV within the existing enclosure were studied. Such a facility represents a factor of twenty improvement in luminosity over performance expected following the Main Injector upgrade. Performance parameters are listed in Table 1, where Tev* refers to the luminosity upgrade alone, and DiTevatron to the combined luminosity and energy upgrade. As noted in the table energy and luminosity upgrades can be considered separately with an additional factor of two in luminosity achievable through the energy upgrade.

The primary design issue for this facility is the antiproton production/accumulation rate. An accumulation rate of $9\times10^{11}\bar{p}$/hour is required to support a luminosity of 2×10^{33} cm^{-2} s^{-1}. This represents nearly a factor of twenty beyond the present performance of the Antiproton Source at Fermilab, and a factor of six beyond the performance expected in the Main Injector era. In addition a new facility is required in which up to 10^{13} antiprotons can be stored prior to injection into the collider. The current accumulation capability of the Antiproton Accumulator is about 2–3$\times10^{12}$ antiprotons. Potential solutions to these issues are related to an increase in the number of protons targeted per hour and increased cooling bandwidth. At least one, and perhaps two, new rings are required. Details may be found in Section 2.3 and in Ref. 1.

Secondary design issues are related to providing the required magnetic fields and gradients, and understanding beam stability and dynamics with large numbers of bunches in each beam. The achievement of 2 TeV in the existing Tevatron tunnel requires dipole magnets operating at 8.8 T and quadrupoles in the interaction regions with gradients of 210 T/m. The dipole field required is a factor of 33% beyond that achieved in the SSC development program and is comparable to current achievements in the LHC program. The achievement of this field is not considered a major challenge. Development of the large aperture, 210 T/m, quadrupole represents a greater challenge. The development of such a magnet would probably

Table 1: Tevatron Upgrade Parameters.

Parameter	Main Injector	TeV*	DiTevatron
Beam energy (GeV)	1000	1000	2000
Peak luminosity ($cm^{-2}s^{-1}$)	8.3×10^{31}	1.0×10^{33}	2.0×10^{33}
Protons per bunch	3.3×10^{11}	2.4×10^{11}	2.4×10^{11}
Antiprotons per bunch	3.6×10^{10}	1.0×10^{11}	0.9×10^{11}
Proton emittance 95%, normalized (mm-mr)	$30\ \pi$	$18\ \pi$	$18\ \pi$
Antiproton emittance, 95%, normalized (mm-mr)	$20\ \pi$	$18\ \pi$	$18\ \pi$
β^* (cm)	35	25	25
Bunches/beam	36	108	108
Bunch spacing (ns)	395	132	132
Total number of protons	1.2×10^{13}	2.6×10^{13}	2.6×10^{13}
Total number of antiprotons	1.3×10^{12}	1.1×10^{13}	1.0×10^{13}
Interaction regions	2	2	2
Bunch length, rms (cm)	43	26	22
Integrated luminosity (pb^{-1}/wk)	17	200	400
Interactions per crossing at 45 m	2.2	9	17
Total antiproton tune shift (2 IR's)	0.016	0.019	0.019
Total proton tune shift	0.003	0.008	0.008
Peak antiproton loss rate at 90 mb ($\bar{p}/hr^{-1}$)	5.4×10^{10}	7.0×10^{11}	13×10^{11}
Antiproton stacking rate ($\bar{p}/hr^{-1}$)	1.5×10^{11}	1.0×10^{12}	1.0×10^{12}

closely parallel the development of the interaction region quadrupoles required for the LHC program. It should be noted that the magnet design is strongly impacted by the potential need to support a slow extracted beam program at 2 TeV. The decision of whether to include such a performance specification would have to be made fairly early in the development program

A number of potential issues relating to operations with a large number of bunches deserve further study. These include the long range beam-beam interaction, intrabeam scattering, issues related to the production and preservation of low emittance, and single beam instabilities. Many of these issues could be addressed through beam studies in the Tevatron.

Many of the beam dynamics issues referred to above could be ameliorated through reduction of the number of bunches. The motivation for operating with large numbers of bunches is reduction of the number of interactions per crossing at the detectors. For example, with a luminosity of 2×10^{33} cm^{-2} s^{-1} the number of interactions/crossing seen in the detectors ranges from 2.5 to 17 as the number of bunches is varied between 750 and 108. It is obvious that a dialog between accelerator and detector designers will have to take place early within the developmental period of such a facility in order to select the appropriate number of bunches.

For a luminosity upgrade of the Tevatron unaccompanied by an energy upgrade essentially all the above issues remain with the exception of those related to magnet development.

The 30×30 TeV p-p Collider. Design concepts for a collider operating at an energy a factor of four beyond LHC and a factor of one-and-a-half beyond the SSC were investigated. Comparison of the LHC and 30×30 TeV parameters is given in Table 2. While the energy of such a facility is only 50% higher than that planned for the SSC, it was recognized fairly quickly that the design and operational issues of such a machine would be quite different due to the enhanced role of synchrotron radiation. A proton collider operating at 30 TeV per beam, with dipole fields of 10 T or greater would represent the first hadron facility in which the role of synchrotron radiation went beyond being irrelevant, as at the Tevatron, or a nuisance, as at LHC and SSC. Radiation damping in a 30×30 TeV collider has a significant impact on the operating characteristics. Every effort must be taken to understand how best to utilize synchrotron radiation as an aid for simplifying the design of the facility.

The number one design issue in such a facility is clearly the dipole magnets. Fields of 10–15 T are required to keep the size of the ring manageable (where manageable means equal to the SSC). Quadrupoles with gradients of 250 T/m are required for the interaction regions. Achieving fields above 10 T will not be

Table 2: LHC and 30×30 TeV Parameters. Peak luminosity for the 30 × 30 TeV machine is achieved after several hours of storage.

Parameter	LHC	30×30 TeV
Beam energy (TeV)	7	30
Peak luminosity ($cm^{-2}s^{-1}$)	1.0×10^{34}	1.0×10^{34}
Protons per bunch	1.0×10^{11}	2.4×10^{10}
Proton emittance, 95%, normalized (mm-mr)	22.5 π	9 π
β^* (cm)	50	20
Bunches	2835	2030
Bunch length, rms (cm)	7.5	4
Arc dipole field (T)	8.7	12.5
Circumference (km)	26.7	60
Rev. frequency (kHz)	11.25	5
Synchrotron radiation power (kW/ring)	3.5	29
Linear power density (W/m)	0.2	0.6
Crossing half-angle (μrad)	0	50
Integrated luminosity (pb^{-1}/wk)	2000	2000
Bunch spacing (ns)	25	100
Interactions per crossing at 60 mb	19	60

easy. It is felt that 10 T is achievable but probably represents the maximum reach of the currently employed NbTi/cosθ technology. Going past 10 T will require new technology. It is difficult to predict at the moment a rate of development for such technologies.

Synchrotron radiation damping, with damping times of 4–5 hours, is shown to have a substantial positive impact on the performance of a 30×30 TeV collider. However, one must remove the heat generated within the cold magnets. The linear heat load for the parameters described below is three times that being planned for in the LHC. As in the case of the Tevatron upgrade there is a strong coupling between relaxing the challenges to the accelerator builders at the expense of interactions per crossing seen by the experimenters. In this case fewer bunches reduces the magnitude of the synchrotron radiation heat load problems. For the example given in Section 2.4, a luminosity of 1×10^{34} cm^{-2} s^{-1} is obtained with a bunch spacing of 100 ns accompanied by 60 interactions/crossing.

Options for staging such a facility, *e.g.*, starting out with lower energy or with a single ring and proton-antiproton collisions, were not examined in this study. In general one would expect the luminosity achievable to scale with energy as the energy is lowered (at a fixed circumference), and the proton-antiproton luminosity to be approximately a factor of ten lower than the proton-proton case.

Conclusions. Upgrading the Fermilab $\bar{p}$–p collider complex to either 1×10^{33} cm^{-2} s^{-1} at 1 × 1 TeV, or to 2×10^{33} cm^{-2} s^{-1} at 2×2 TeV is an aggressive goal. Increasing the antiproton production rate by a factor of 6 beyond that anticipated following completion of the Main Injector is the key. Several ideas exist for accomplishing this, but none are mature at this stage. Further design and development work are required to generate assurance of achieving this performance. Long range beam-beam effects are likely to be important if the Tevatron were configured to run with many (> 100) bunches at high (>1×10^{33}) luminosity. Machine studies in the Tevatron could shed much light on this issue.

A proton-proton collider operating at 60 TeV in the center-of-mass, and with a luminosity of 1×10^{34} cm^{-2} s^{-1}, is a reasonable goal for the next hadron facility following the LHC. The possibility of utilizing synchrotron radiation emittance damping to enhance the performance, and simplify the construction, of a next generation hadron collider looks promising enough to warrant further attention. An operating energy in the range of 25–40 TeV/beam would require 12.5 ± 3 T dipoles. As in all high-energy hadron facilities, magnets are the key. A reinvigorated US superconducting magnet R&D program will be needed to support these aims. Facility cost will clearly be a significant design consideration and needs to be integrated into thinking on even the most preliminary designs. Cost minimization through simplification of injector performance specifications, utilization of existing facilities and infrastructure, optimization of fabrication and procurement strategies, and staging all deserve considerable thought.

There is a real trade off between relaxation of accelerator parameters and increasing the number of interactions per crossing in each of these facilities. There must be a close interaction between accelerator and detector designers to identify an optimum bunch spacing.

2.3 Luminosity and/or Energy Upgrades to the Tevatron $\bar{p}$–p Collider

A preliminary evaluation of a high-luminosity 2×2 TeV collider in the Tevatron tunnel was undertaken at the workshop. The parameters that were distributed at the beginning of the workshop and were the focus of much of the discussion are given in Table 3 in the "Original DiTevatron" column. The peak luminosity would be $L\sim 2\times10^{33}$ cm^{-2} s^{-1} achieved with 750 bunches spaced 19 ns apart. There would be 2.5 interactions per crossing in

Table 3: Tevatron Upgrade Workshop Parameters.

Parameter	1994 Actual	Main Injector	DiTevatron Original	DiTevatron Modified
Beam energy (GeV)	900	1000	2000	2000
Peak luminosity ($\times 10^{30}$ cm^{-2} s^{-1})	12.0	123	1980	1980
Protons/bunch ($\times 10^{9}$)	185	380	60	238
$\bar{p}$s/bunch	50	36	16.5	90
Proton emittance (95%,πmm-mrad)	22	30	5	18
$\bar{p}$ emittance (95%,π mm-mrad)	14	15	5	18
β^* (cm)	35	25	25	25
Bunches/beam	6	36	750	108
Bunch spacing (ns)	3493	395	18.9	132
Total protons ($\times 10^{12}$)	1.11	13.7	45.0	25.7
Total $\bar{p}$s ($\times 10^{12}$)	0.30	1.30	12.38	9.8
Interaction regions	2	2	2	2
Fract momentum spread ($\times 10^{-4}$)	6.0	5.0	0.9	2.4
rms bunch length (cm)	55	45	8	22
Crossing half angle (mr)	0	0	0.12	0
Luminosity form factor due to hourglass and crossing angle	0.62	0.58	0.69	0.79
Integrated luminosity (pb^{-1}/wk, 33% duty factor)	2.4	25	400	400
Interactions/crossing (45 mb)	1.9	3.2	2.5	17
Total $\bar{p}$ tune shift (2 IRs)	0.012	0.020	0.019	0.019
Total proton tune shift	0.005	0.004	0.005	0.005
Peak$\bar{p}$ loss rate (10^{10}/hr, 90 mb)	0.78	8.0	130	130
Stacking rate (10^{10}/hr)	4.5	15	100	100

the particle physics detectors assuming a 45 mb cross section. As discussed below some of the problems with the original parameters identified during the workshop could be solved by reducing the number of bunches at the cost of increasing the interactions per crossing to 17. These are given in the "Modified DiTevatron" column. The parameters for the current (1994) Tevatron run and the first run planned with the Main Injector are given in the table for comparison.

The major elements of either upgrade are additional accelerator(s) to increase substantially the antiproton production rate and, for the energy upgrade, a new superconducting ring replacing the existing Tevatron.

Implementation. *Lattice:* A crossing angle of 120 μrad is used to separate the beams at the interaction regions. After that, electrostatic separators are needed to establish spiral orbits in the arcs to avoid other unwanted conditions. These separators must be placed where the appropriate β-function is large and points are needed with 90° phase advance between horizontal and vertical. A lattice needs to be developed that satisfies these criteria with a footprint consistent with the Tevatron tunnel, possibly with minor modifications.

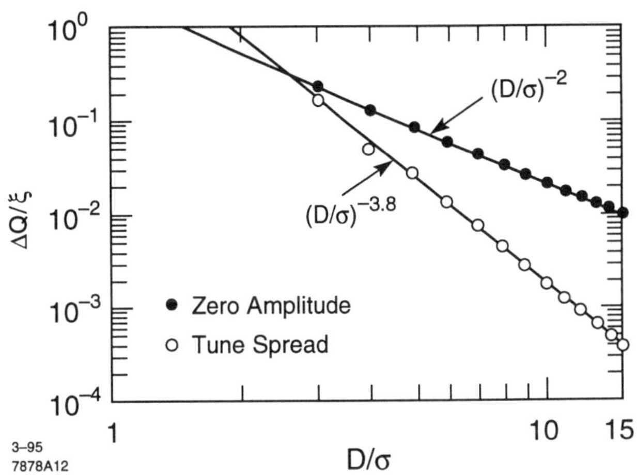

Figure 1: Parasitic beam-beam interaction. The tune shift of a zero amplitude particle and the tune spread for particles with $2\,\sigma$ amplitudes of oscillation. These are normalized to the head-on beam-beam tune shift. D is the center-to-center separation and σ is the rms beam size.

Beam Intensity: There are 1500 parasitic collisions. The tune shift of a zero amplitude particle and the tune spread for particles with oscillation amplitudes up to 2σ are plotted in Fig. 1. These are normalized to the head-on tune-shift, and the approximation has been made that the beams are round at the parasitic crossing point. Bunches are separated by D/σ = 4–5 in present Tevatron operation where there are ten parasitic collisions and no significant problems observed from them. Keeping the same ratio of parasitic to head-on tune spread, the separation should become $D/\sigma \sim 15$–20. Since the emittance and energy spread are smaller in the DiTevatron, the beam size is smaller, and roughly the same center-to-center separation as used at present should be adequate for this consideration.

The average tune shift is more of a problem because there are a number of effects that cannot be compensated by changing the lattice tune. These include the difference in proton and $\bar{p}$ tunes, approximately $0.014\sigma^2/D^2$ per parasitic collision, and rapidly changing tunes during injection of the opposing beam. A theoretical analysis is needed, and the effects of parasitic collisions should be studied experimentally by looking at the behavior of an antiproton bunch in the presence of a fixed target proton beam.

The average proton current is comparable to the beam used in fixed target operation. Coupled bunch instabilities have been observed there, and they have been cured by tuning of cavity higher-mode frequencies. In addition, reducing the higher mode Q's could be used as a cure also, and the average current does not present a problem.

The single bunch longitudinal and transverse instability limits have been estimated, and the longitudinal microwave instability sets the impedance limit. The high-frequency impedance limit is $Z/n \leq 2\Omega$ which is be compared with the estimated Tevatron impedance, $Z/n \approx 5\Omega$. The new collider ring will have to have a lower impedance, and the microwave instability seems to be a performance limit for upgrades that do not replace the Tevatron. Of course, this can be studied experimentally.

The emittance growth rate due to intrabeam scattering has been estimated in two ways. A two hour horizontal emittance growth time was calculated by scaling from the SSC [2], and a growth time an order of magnitude longer was found using approximations in the original Piwinski reference [3]. If the first result was correct, intrabeam scattering would dominate the luminosity lifetime and it would be unacceptably short. The disagreement between the calculations is due to strong dependence on lattice functions which are not known, and a better calculation must be performed when they are.

Magnets: The 8.8 T dipole magnet technology that is needed for the DiTevatron remains to be proved, but this should be done soon as part of LHC development. The good field aperture may have to be comparable to that of the present Tevatron, 7.5-cm diam., to allow adequate separation at parasitic crossing points. SSC magnets could be run at low temperature to test the feasibility of those magnets for producing 8.8 T, but they do not have sufficient good field aperture. A design iteration addressing field quality had been planned as part of SSC development.

A gradient of 150 T/m, as required for the lattice quadrupoles, is comparable to the present Fermilab IR quad gradient of 140 T/m and the SSC lattice quadrupole gradient of 220 T/m. Scaling of present Tevatron lattice would require IR quadrupole gradients of 280 T/m. This is not feasible, but this is not a problem since the design $\beta^* = 25$ cm has been reached in lattice designs with a lower gradient of 210 T/m. Accommodating the electrostatic separators is a more serious issue.

Vacuum: The critical energy for a 2 TeV proton beam in an 8.8 T magnetic field is 3.8 eV which is close to the work function of many metals. Desorption rates have not been measured for these energies, so the impact on the vacuum system is not clear.

Low Emittance Proton Source: The present performance of the Fermilab Booster is that it accelerates batches of 84 bunches with a single bunch intensity of 4×10^{10} and longitudinal and transverse emittances of 0.08 eV-s and 10 π mm-mrad, respectively. The intensity must be higher and the transverse emittance lower for the DiTevatron, and the space charge tune shift in the Booster would be a factor of 2–3 times higher. This cannot be achieved with the present linac, and the linac energy would have to be raised to roughly 0.8–1 GeV to avoid emittance degradation from space charge. The space charge tune shift at injection to the Main Injector is low, and that should not present a problem.

Modifications to the Fermilab ion source and low energy beam optics are needed for the low transverse emittance. The requirements are within the ranges of developments for the SSC.

Antiproton Production: The $\bar{p}$ stacking rate is 5×10^{10}/hour at the present time, and this must be increased to 10^{12}/hour for high luminosity. The consensus of the antiproton source group was that at least two additional accelerators for manipulating the $\bar{p}$ phase space, cooling and accumulating $\bar{p}$'s would be required [4]. Common features of the sources considered are:

- stochastic cooling in the present Accumulator ring;
- followed by additional stacking in a new, high-energy accumulator ring;
- $\bar{p}$ recovery from the previous collider store which gives some flexibility in store length; and
- an 8–16 GHz cooling system.

The differences between approaches were:

- the role of the present Debuncher ring in a new $\bar{p}$ source,
- whether one or six 84-bunch Booster batches of Main Injector beam were targeted,
- proton beam manipulations before targeting,
- targeting, and
- antiproton manipulations after targeting.

The different approaches are detailed in the report from the antiproton source group. All of them have advantages and drawbacks, and none is developed sufficiently to judge feasibility, reliability, and cost. The viability of a high-luminosity collider in the Tevatron tunnel depends on developing an adequate $\bar{p}$ source, and the research and development associated with that is crucial.

Number of Bunches: The original parameters had 750 bunches to get a small number of interactions per crossing. With a limit on $\bar{p}$ production, the transverse emittance must be small for high luminosity and the longitudinal emittance must be small for the short bunch needed for crossing at an angle. These considerations lead to a high proton phase space density and the dominant problems of the original parameter list: space charge tune shift at injection into the Booster and intrabeam scattering (which needs to be checked as discussed above).

The modified parameters are for 108 bunches. Longitudinal coalescing could be used, as it is now, to reduce the space charge tune shift in the Booster, and the larger emittances make intrabeam scattering unimportant.

The situation with parasitic beam-beam collisions does not change substantially. While the number of parasitic collisions is reduced by a factor of 7, the beam sizes are larger, and the required separation could be larger although D/σ is smaller.

Conclusions. This workshop was helpful for identifying potential problems and suggesting areas of research. The most important ones are:

- Antiproton production is the key to high luminosity in the Tevatron. No fatal flaws were found in the approaches considered, but because of their preliminary nature, the trade-offs, costs, and potential impact on the Main Injector fixed target program, they need to be evaluated carefully.
- Parasitic beam-beam interactions would produce substantial tune shifts and tune spreads because of the large number of bunches in either case. The parasitic beam-beam interaction needs to be studied theoretically and experimentally to evaluate the impact. The required separation will determine the required good field aperture and magnet costs.
- The two DiTevatron parameter lists are contrasted by the number of interactions per crossing and phase space density of the proton beam. If the intrabeam scattering lifetime is as short as calculated by scaling from the SSC, the original list is flawed. If that is not the case, a linac energy increase and modifications to the low energy portion of the linac would be required for the phase space density in the original list. The modified list does not have these accelerator physics problems, but does have 17 interactions per crossing, which needs to be considered by experimenters.

Finally, a comparison of the DiTevatron parameters with the actual current performance shows that a 2×2 TeV high-luminosity collider in the Tevatron tunnel would represent a substantial extension. Fortunately, experiments are possible to test many aspects of the design, clarify the issues we have raised, and provide an opportunity to see unexpected ones. A concerted experimental effort would be a wise investment.

2.4 A High-Luminosity 30×30 TeV Proton Collider

This section summarizes the basic conceptual ideas that were developed during the workshop. Using 30 TeV as an example we looked at the potential consequences of a hadron collider where synchrotron radiation was sufficient to produce significant emittance damping. The machine footprint was "SSC–like" and consisted of two interaction regions and two arc sections. The desired luminosity regime was $\sim 10^{34}\,\mathrm{cm}^{-2}\,\mathrm{s}^{-1}$ in a proton-proton, 2-in-1 magnet environment. No detailed parameter adjustment was made to achieve any exact performance goals however, interest was focused more on the consequences of parametric variations.

Optimizations. Synchrotron radiation, broadly speaking, produces two main issues. Emittance damping which results in enhanced beam phase space density and is a positive attribute, and radiated power into the bore tube producing vacuum and cooling complications. The conceptual goal of this study involved utilizing the former while mitigating the latter in what we perceive to be a reasonable fashion.

The synchrotron radiation damping rate is determined by the collision energy and the dipole field, thus for a fixed set of these parameters the factor linking the damping rate to the radiated power is the number of circulating protons in the machine. Minimizing the circulating current for a fixed luminosity argues for lower beta-star at the IP and increased bunch spacing/intensities. The maximum luminosity for a fixed circulating current is obtained by increasing the bunch intensity and bunch spacing. This can be accomplished by a bunch coalescing in the injector chain. The practical limits to this exercise are determined by the experimental consequences of the increased number of events per bunch crossing. Another issue which becomes important for low beam currents, is particle burn-off at the IRs. Significant particle burn-off results in short luminosity lifetimes which pushes the machine operation in the direction of short store lengths. This in turn requires a relatively rapid cycling injector.

On the positive side, significant emittance damping provides luminosity enhancement and may also result in less sensitivity to emittance growth arising from the beam-beam interaction, *i.e.*, increase the tolerable tune shift limit. Insensitivity to many of the traditional sources of emittance growth in hadron colliders can be expected to result in slackening of tolerances in many areas and thus a simpler, cheaper and more robust machine. Integrated luminosity is not strongly correlated to the initial beam size which simplifies the injector complex. Since significant emittance growth can be accepted during the injection cycle, field quality requirements on the arc magnets (invariably defined by low energy criteria) can be relaxed. Likewise, less sensitivity to persistent and eddy current phenomena, can be used to increase the dynamic range of the machine energy to simplify the injector specifications. Similarly, beam transfer issues are less complicated with a lack of stringent emittance growth requirements. Matching, kicker specifications and damping systems will all benefit. Instrumentation is helped by (relatively speaking) large beam sizes at injection and significant photon emission at high energies. Emittance growth due to vibration effects and power supply noise in the arcs will be ameliorated by the emittance damping, maintaining beam collisions must still be assured however. Reduced circulating beam currents also result in less stored energy in the beams, which, while still a very difficult issue, is less of a challenge than might have been expected from a straightforward scaling of LHC/SSC parameters.

Parameters. A partial parameter list for a 30×30 TeV machine is shown in Table 4. The column labeled "Original" represents the starting point for the study; "Modified" represents the developed set described here. The collision energy of 30 TeV was chosen purely to illustrate the potential consequences of operating a hadron machine in a radiation damping regime. It should be emphasized that there is an energy window where this behavior applies and is correlated to the assumed dipole field. If the machine energy is too low then the damping is insufficient to be useful. If the energy is too high then the radiated power becomes prohibitive for the cryogenic magnets. The injection energy is defined by the somewhat arbitrary choice of collision energy together with a choice of 30:1 for the dynamic range which seems reasonable with an allowance for 50% emittance growth at flat bottom. The dipole field of 12.5 T was chosen as a representation of next generation superconducting magnets though, as we demonstrate, a range of dipole fields can be accommodated in this approach. A marginal parameter remix can produce similar machine characteristics with lower dipole fields as can non-FODO lattices. The desired injector beam emittance of 1 π mm-mrad rms was taken from the SSC injector chain specifications somewhat arbitrarily. As alluded to in the next section, the luminosity evolution is relatively insensitive to initial emittance values. The criteria for injected beam parameters would appear to be set by the somewhat less demanding requirement that beam should stay in the machine for the $\sim$ 30 minute filling time rather than any emittance specifications. Since this is difficult to quantify in the short time period of a workshop we chose the SSC-like parameters. The 20 cm beta-star was picked to be the lowest value that the IR working group deemed realistic. The rf parameters were picked to deliver a bunch length consistent with the beta-star value.

The total beam current was loosely defined by the desire to produce an $\sim$ 8 hour optimum store length. The bunch intensity, bunch spacing, and hence number of bunches, was influenced by stability considerations and the projected 90 events per bunch crossing at maximum luminosity. It would appear that the LHC with 16 events per crossing has removed the option of event by event vertex reconstruction from the detectors. We have pushed this harder. Since this is one of the more controversial aspects of this design we expect further discussion on this point. Every cloud is reputed to have a silver lining, and the increased beam granularity does provide much more time between crossings for trigger processing than the LHC, for example.

The initial beam-beam tune shift is relatively benign but does increase during the store. We assumed a tolerable tune shift limit of $\sim$ 0.03 which is a modest increase on the 0.02–0.025 achieved today.

A crucial parameter in this conceptual exercise is the radiated beam power per unit length. We chose a value of 0.6 W/m which is approximately a threefold increase over the LHC. At this level a bore tube liner is necessary but vacuum issues arising from the desorbed gases are certainly tractable as are cryogenic problems associated with the heat removal. As discussed in the vacuum/cryogenic section of this workshop report power densities as high as 3 W/m could be handled with increased complexity in the various systems and beam conditioning scenarios. Since there is almost a straight trade-off between events per crossing and radiated power, and the situation with respect to either issue is complicated we imagine further discussions on this point.

The injector cycle time and the machine acceleration time are determined by the short store regime and correspond to the present Tevatron cycle time and a ramp rate of about 20% that of the Tevatron.

Operating Characteristics. The operating characteristics of this machine show, not surprisingly, more dynamic variation than present day hadron colliders, *i.e.*, more electron like behavior. Figures 2(a)–(d) show an analytic estimate of the time evolution of the transverse emittance, bunch intensity, luminosity, and head-on beam-beam tune shift. In obtaining these results we have assumed no emittance growth mechanisms. The

Table 4: 30×30 TeV Workshop Parameters.

Parameter	Original	Modified
Beam Energy (TeV)	30	30
Initial Luminosity ($\times 10^{34}$ cm^{-2} s^{-1})	1.0	0.4
Injection Energy (TeV)	—	1
Number IRs	—	2
Dipole field (Tesla)	12.5	12.5
Circumference (km)	60	60
Number dipoles	—	3335
Initial transverse emittance, rms (mm-mr)		
Collisions	1.0 π	1.5 π
Injection	—	1.0 π
Initial longitudinal emittance, rms (eV-s)	—	0.5
Radio frequency (MHz)	—	360
rf voltage(MV)	—	25
Bunch length, rms (cm)	—	4
Bunch intensity ($\times 10^{10}$)	2.0	2.4
Bunches	12000	2030
Bunch spacing (ns)	25	100
Number of events/crossing (@60 mb, 1×10^{34})	11	60
Revolution frequency (kHz)	5	5
Initial beam lifetime burn off (hr)	75	16
Initial $\Delta\nu$ beam-beam head-on/IR	.002	0.002
Crossing angle (μrad)	50	50
Initial $\Delta\nu$ long-range/IR	—	0.0039
β^* (cm)	50	20
Synchrotron radiation power/length/ring (W/m)	3.0	0.56
Total synchrotron radiation power/ring (kW)	151	29
Radiation damping time, FODO lattice (hours)		
Transverse	4.7	4.7
Longitudinal	2.3	2.3
Stored energy/beam (MJ)	1130	230
Filling time/ring (min)	—	10
Injector cycle time (min)	—	1
Acceleration time (min)	—	5
Injection dipole field (T)	—	4

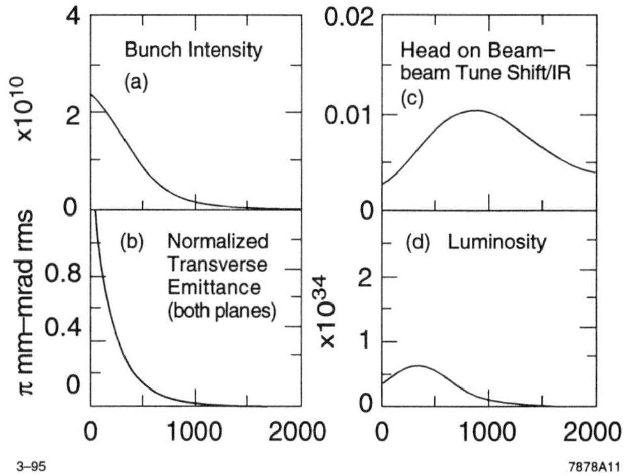

Figure 2: Time evolution of the store parameters for a dipole field of 10 T.

luminosity rises from an initial value of $\sim 5\times 10^{33}$ to $\sim 10^{34}$ over a period of $\sim$ 4 hours and then slowly falls again in a more or less symmetric fashion. The peak to average luminosity is a healthy 70% across this time span. During the 8 hour storage period the bunch intensity falls by a factor of 5 and the transverse emittance shrinks by an order of magnitude. The head-on tune shift increases with bunch density to $\sim$0.01/IR, but this is partially compensated by the reduction in long range tune shift which is not shown. The quite rapid fall in initial emittance intuitively leads one to suspect that the integrated luminosity is not strongly correlated to the assumed value for initial transverse emittance and other runs, not shown here, confirmed this. One interesting fact did emerge from this superficial exercise; initial emittances significantly less the 1 π result in a rapid saturation of the relaxed beam-beam tune shift limit *i.e.*this machine does not like very dense beams!

Other runs were performed with the same parameters using a 10 T dipole field (80 km circumference). With the reduced damping rate the luminosity peaks at a lower value and the optimum store length is longer but the same basic features are evident albeit at a $\sim$ 40% loss of performance. The lack of sensitivity to the dipole operating field is a healthy attribute given that we chose to use a "dipole" that is beyond the state of the art today. While these simulations were performed using a straight forward FODO lattice in the arc sections, the lattice working group demonstrated during the workshop enhanced damping lattices by manipulating the partition functions through the use of combined function magnets. Changing the partition functions from a FODO-like $J_x = J_y = 1.0$, $J_z = 2.0$, to a more desirable $J_x = J_y = 1.5$, $J_z = 1.0$, provides performance in the 10 T machine that is virtually identical to 12 T. Other techniques to enhance the damping rate such as operating with a small momentum offset are also potentially attractive from this point of view.

Since we are relying for luminosity on the increased phase space density to counter balance the loss of bunch intensity, it is natural to be concerned about the effect of emittance dilution mechanisms. It is evident from the figures that a beam emittance which asymptotically approaches zero is nonphysical in the limit, and a more rational simulation would include beam heating mechanisms where the equilibrium beam emittance is defined by the balance between heating and cooling. Intrabeam scattering (IBS) is such a heating term which increases with increasing beam density. A first estimate of the impact of IBS was made by including a paramaterized IBS heating term in a self-consistent way. These results can be compared to those of Fig. 2 (a different calculation). As expected, there is little effect at the beginning of the store when the beam size is large, but it becomes progressively more important as the beam shrinks. The peak luminosity is reduced by $\sim 20\%$ and instantaneous luminosity falls off faster after the peak value with the transverse emittance approaching an asymptotic value of $\sim 0.15\pi$. This results in a 20% decrease in integrated luminosity over a 10 hour period. Maintaining the longitudinal emittance at its initial value of 0.5 eV-s by beam heating halves this decrease. While the effect of IBS is significant the overall features of the store evolution are similar.

Observations. Since magnets are a crucial and dominant feature of any accelerator in this energy regime it is appropriate to include some observations on this topic which are not strictly related to machine performance. While our basic machine model assumed a 12.5 T dipole it became apparent during the workshop that there is no threshold field necessary to enter the regime of usable emittance damping. All things being equal, the biggest impact on synchrotron radiation emission is operating energy, which produces the somewhat interesting observation that for a constant damping rate, the lower the dipole field the higher the operating energy, and vice versa. Alternatively, for a desired operating energy there is a maximum dipole field beyond which radiated power becomes prohibitive. These factors together with fiscal realities would indicate that optimistic but otherwise reasonable people would conclude that dipoles in the range of 9–15 T and a machine energy of 25–40 TeV can be made to fit into the basic conceptual framework outlined here. Beyond this regime a different approach is called for.

Magnet aperture and field quality are related quantities. As discussed in the other workshop reports, at the back-of-an-envelope level, a 5 cm magnet aperture would appear to suffice, including the needs of bore tube liners. Since field quality in the arc regions is invariably defined by the low energy performance of the magnets, the short dwell time (20 min) and 50% allowance for emittance growth in this model should permit a relaxation of SSC–like field quality specifications (factor of 3?) which in turn will reduce magnet costs. A 2-in-1 magnet style is also preferred for both cost implications and simplifications in the IR regions when the beams are brought into collision.

The use of high-T_c superconductors for the magnets does not appear to be necessary to achieve the magnet performance alluded to here. An elevated operating temperature (say, 10 to 15° K), while not necessary, is still highly desirable. The bore tube liner could become passive with no dedicated cooling circuits, a greater allowable temperature fluctuation across the magnet strings greatly simplifies the cryogenic system, as does the removal of radiated power at the higher temperature. Cryogenic operating costs are also reduced. High-T_c materials also raise the possibility of permitting an increase in the tolerable synchrotron radiation power density in the magnets, *i.e.*, a potential increase in luminosity. Since these items are, in principle, quantifiable, the basic elements for a cost/benefit study of a high-T_c magnet R&D program are available.

Conclusions. The possibility of utilizing synchrotron radiation emittance damping to enhance the performance, and simplify the construction, of a next generation hadron collider looks promising enough to warrant further attention. An operating energy in the range of 25–40 TeV would require 12.5 ± 3 T dipoles. Some form of magnet R&D program will be needed to support these aims.

3 Electron Linear Collider Facilities

3.1 Introduction

Over the past 6–7 years there has been a concerted effort to evaluate the physics which could be addressed with a future e^+e^- linear collider. This work was conducted in a series of workshops shown in Table 5. During the same period a series of Linear Collider Workshops was held to address the accelerator physics and technology of such a collider (see Table 6). These workshops served to disseminate information and provided an informal collaboration of R&D efforts at laboratories around the world. The result of these studies was a broad international consensus that the next e^+e^- collider should have an initial energy of about 500 GeV and should be expandable to 1 TeV and beyond. This conclusion evolved both from considerations of the technical difficulty of the linear collider and studies of the physics opportunities in this energy range.

Table 5: Physics Studies.

Date	Study	Location
1987	SLAC Study Group	US
—	La Thuile CLIC Study	Europe
1988	SNOWMASS	US
1989	First JLC Workshop	Japan
1990	SNOWMASS	US
—	Second JLC Workshop	Japan
1991	EE500 Workshops	Europe
—	Saariselka, Finland	International
1992	Colliding Beams Workshops (*ee*, *pp*, *ep*)	US
1993	Waikoloa, Hawaii	International
1995	Japan	Japan

A typical layout for such a collider is shown in Fig. 3. The key elements of the collider are highlighted:

- Electron and positron sources produce the beams
- Damping Rings shrink the beam size.
- Compressors and Pre-accelerators prepare the bunch for acceleration.
- The main linac accelerates the beam to high energy.
- The Final Focus demagnifies the beam size for collision.
- The detector must have a clean environment with low enough background for physics.

Although the basic features of the linear collider shown above are generic, the linear collider work around the world has led to several possible approaches.

To illustrate the range of the designs, a subset of parameters for colliders is shown in Table 7. (Parameter lists for all approaches are given later in this report). They illustrate the primary trade offs which can be made in order to obtain high luminosity at high energy. When beam-beam backgrounds are held roughly constant there are two ways to get high luminosity: (1) increase the power in the beam, or (2) decrease the vertical spot size.

As we move from TESLA to NLC/JLC, the beam power drops from 16.5 MW to 4.2 MW in the 0.5 TeV design and from 15.2 MW to 9.4 MW in the 1 TeV design. The vertical spot size in the 0.5 TeV designs also drops as we move to the right in Table 7 to keep the luminosity roughly constant. The situation for 1 TeV is somewhat different. Both TESLA and SBLC must drop the vertical beam size to the few nanometer range in order to keep wall power and beam-beam effects under control. The NLC/JLC design has a smoother scaling to 1 TeV, and can even be scaled to 1.5 TeV, as shown in Table 7.

The final energy of the linear collider can be obtained with two primary technological choices: superconducting or warm radio-frequency (rf) acceleration. The TESLA

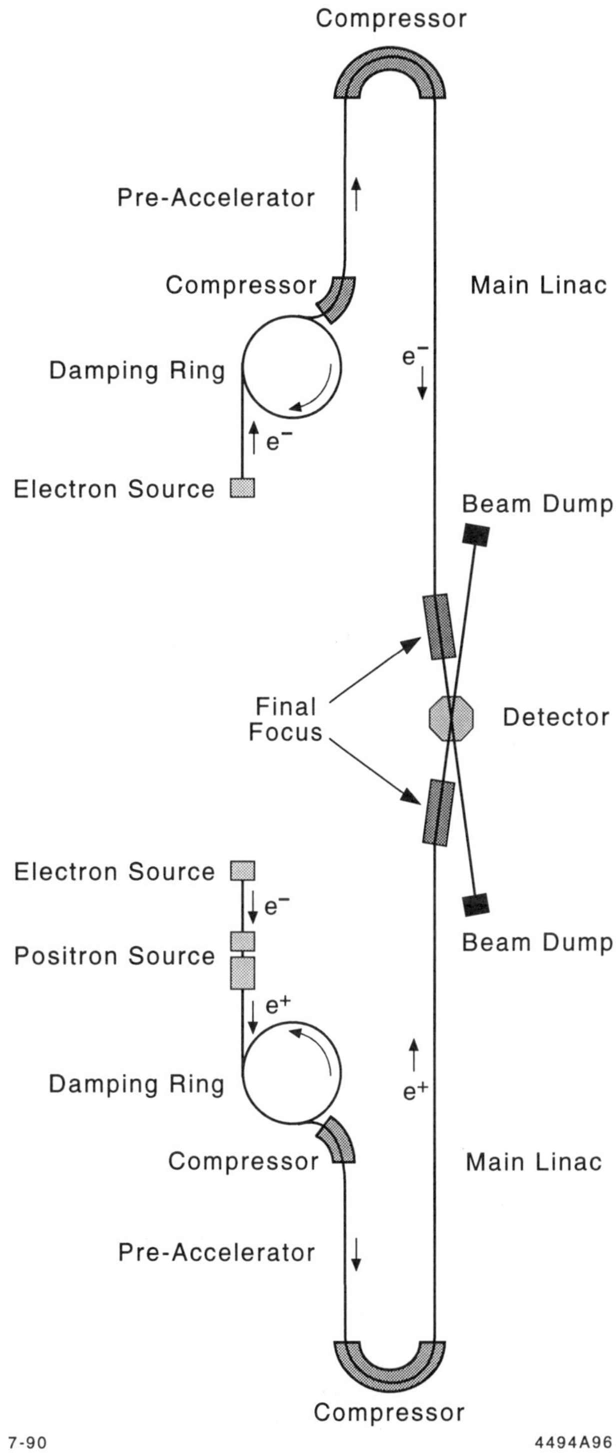

Figure 3: Schematic layout of a future e^+e^- Linear Collider.

collaboration is pursuing the superconducting approach. Within the warm category the S-band linear collider (SBLC) represents the lowest frequency option based on SLC technology developed about 30 years ago while the NLC ($\sim$ JLC) parameters shown in Table 7 push the accelerating frequency up by a factor of 4, which allows higher acceleration gradient.

Table 6: Accelerator Physics Workshops.

General Workshops:

Date	Workshop	Location
1988	LC88	SLAC
1990	LC90	KEK
1991	LC91	Protvino
1992	LC92	Garmisch
1993	LC93	SLAC
1995	LC95	KEK

Topical Meetings:

Date	Meeting	Location
1992	Final Focus and Interaction Region	SLAC
1992	Damping Rings	KEK
1993	Emittance Preservation	KEK
1994	e^+e^- Sources	DESY
1994	RF Power Sources	BNL

Although all designs shown here have 0.5 TeV lowest energy, it is, in general, not a problem to drop the energy to the Top threshold with some modest loss of luminosity. Note that a single 1.5 TeV parameter set is given for NLC. This parameter set was prepared at the request of other working groups and evolved from studies several years ago by the SLAC parameters group to evaluate several designs as to energy flexibility. This analysis showed that the NLC/JLC designs can cover the 0.5–1.5 TeV range. Another high-energy design was produced by CLIC for 2 TeV; however, recently they have focused on the 0.5 TeV design. The other designs shown in Table 3, however, have not been extended to 1.5 TeV.

Before beginning it is useful to note that as we have learned more from the technology development and SLC experience, the designs in this report have evolved. Although this evolution is still taking place, the designs and the enabling techniques are reaching maturity now.

To begin the detailed discussion, we present a summary of the SLC to set a baseline for future designs. Although at first look the SLC is a rather specialized hybrid collider, in fact, the SLC is addressing all of the key issues that are being studied for future colliders. The third section presents an overview of linear collider R&D and introduces the parameters sets and main features of all the designs. The next five sections each discuss a particular design and/or the associated test facilities that are under development. Finally, the report summary discusses world collaboration.

3.2 The SLAC Linear Collider Program

3.2.1 The SLAC Linear Collider [5]

The SLAC Linear Collider (SLC) holds a unique place in the field of high energy accelerators as the world's first electron-positron linear collider. By the early 1980's, it was evident that the previous generation of electron-positron storage ring accelerators could not practically be extended beyond the energy range of LEP and LEP 2. The circumference, and hence cost, of an electron storage ring scales like the square of the energy making a 0.5–1.0 Tev machine prohibitively expensive. Linear colliders scale linearly with energy and can potentially provide higher energy at lower cost. The SLC was proposed as both a prototype for a new generation of electron accelerators and as an inexpensive entry into the physics available at the mass of the Z. Because it was the first example of a new accelerator technology, it was a particularly difficult machine to understand and commission. It was almost 2 years from the completion of construction in 1987 until the first Z event was seen by the Mark II detector on April 11, 1989. Several more years of effort were required to develop the knowledge and experience to operate the first linear collider reliably. Today the success of the SLC has generated widespread interest in linear colliders throughout the international accelerator community. Close collaboration between the SLC and other linear collider groups has been of great benefit to both efforts.

A cycle of the SLC begins with two electron bunches stored in the north damping ring and two positron bunches stored in the south ring. Each 1/120 second, one positron bunch and both electron bunches are kicked out of the rings and accelerated down the linac. The positrons, followed by the first electron bunch, are accelerated to about 47 GeV before they are separated by a dipole magnet at the end of the linac, transported around the collider arcs and brought into collision. These bunches then travel back through part of the beam line until they are ejected into extraction lines to beam dumps. The second electron bunch is deflected onto a target two-thirds of the way down the linac to produce the next bunch of positrons. The positrons from the target are collected, accelerated to 200 MeV and transported back to the beginning of the linac. There they are joined by two bunches of electrons from the source, accelerated to 1.2 GeV and stored in the two damping rings for the next cycle.

Table 7: Comparison of Linear Collider Parameters

Parameter	TESLA		SBLC		NLC (~JLC)		
Center-of-mass energy (TeV)	0.5	1	0.5	1	0.5	1	1.5
Luminosity (10^{33})	7	10	4	6	8	20	19
Repetition rate (Hz)	0	5	50	50	180	120	120
Bunches/rf pulse	800	4180	125	50	90	75	75
N (10^{10})	5.15	0.91	2.9	2.9	0.65	1.3	1.3
x/y emittance (10^{-8} m)	2000/100	520/6.3	1000/5	1000/10	500/5	500/5	500/5
x/y beta at IP (mm)	25/2	20/1	22/0.8	32/0.8	10/0.1	40/0.1	40/0.1
x/y sigma at IP (nm)	1000/64	325/8	670/28	572/9	300/3	425/2	425/2
Bunch length (microns)	1000	500	00	500	100	100	100
Upsilon	0.029	0.058	0.055	0.091	0.096	0.28	0.42
Pinch enhancement	2.3	2	1.6	1.7	1.4	1.4	1.3
Beamstrahlung delta	0.03	0.03	0.03	0.07	0.03	0.08	0.10
Number photons/e^-	2.7	1.3	2	2.3	0.9	1.1	1.1
Loaded gradient (MV/m)	25	25	17	34	37	74	74
Linac length (km)	20	40	29.4	29.4	14 1	4	21
Number klystrons	1202	2404	2450	4900	1890	3850	5820
Klystron peak power (MW)	3.25	3.25	150	150	96	105	105
Pulse compression gain	—	—	—	2	3.6	6.6	6.6
Power/beam PB (MW)	16.5	15.2	7.3	5.8	4.2	9.4	14
AC power PAC (MW)	137	159	114	200	92	144	218
$2P_B$/PAC	0.24	0.19	0.13	0.06	0.09	0.13	0.13

3.2.2 Polarized Beams at the SLC

The first SLC polarized source, using a bulk GaAs cathode illuminated by a dye laser operating at 715 nm, produced a polarization of 27–29%. In 1993 both the cathode and the laser were upgraded for higher performance. A new Ti:Sapphire laser pumped by a doubled YAG laser has a hundred times more power and a tuneable wavelength.

The present cathode uses a "strained lattice" technology to achieve higher polarization. In a bulk GaAs photocathode the laser light excites electrons from two degenerate energy levels. The electrons from these two levels have different polarizations which partially cancel, giving a maximum achievable polarization of 50%. For the strained lattice, a 100–300 nm layer of GaAs is grown on a GaAsP substrate. Because the GaAs layer is thin, its lattice distorts to match the GaAsP lattice. With long wavelength laser light (typically 865 nm), only one of the two energy levels is excited and polarization up to 90% can be attained.

The major surprise was a large anomalous spin precession through the north arc. At the energy of the Z, the arc vertical betatron tune is very nearly equal to the spin tune causing a resonance-like behavior. Small changes in the vertical trajectory can strongly affect the spin orientation. This effect also causes a few percent depolarization due to the different paths followed by particles of slightly different energy. New feedback systems to better control the beam trajectory through the arc have allowed stable delivery of polarized electrons. The extreme sensitivity of polarization to vertical orbit has also allowed the development of techniques for using controlled oscillations to optimize the spin orientation and to minimize the depolarization effects. The polarized source has proven very reliable with over 95% availability during more than two years of operation. For the 1994 run, a thin (100 nm) strained lattice cathode was installed to provide about 80% polarization. A larger area cathode configuration has also been developed to support higher beam intensity.

Further research on cathode technology is aimed at even higher polarization for future runs.

3.2.3 Luminosity

The luminosity of a linear collider is given by the formula

$$L = \frac{N^+ N^- f H_d}{4\pi\sigma_x\sigma_y}\eta \qquad (1)$$

where N^+ and N^- are the number of positrons and electrons per bunch, f is the collision frequency, σ_x and σ_y are the horizontal and vertical beam sizes at the collision point. The H_d is an enhancement factor due to disruption which describes the additional focusing of one bunch by the field of the other bunch during the collision. The efficiency factor, η, relates the peak luminosity to the average luminosity delivered over a period of weeks. This formula clearly indicates the strategy for increasing the luminosity: for fixed collision frequency, one must increase the number of particles per bunch and/or decrease the beam area. To translate gains in peak luminosity into an increase in total events, the efficiency is a critical factor. Progress at the SLC has been made in all parameters but the biggest payoff has come from smaller bunch size and more efficient operation.

The SLC polarized electron source and the positron production system are capable of producing 4–5 $\times 10^{10}$ particles per bunch but the maximum operational intensity has been limited by constraints on stability and on non-linear effects which increase the beam size. In 1993, the intensity was limited by the onset of a bunch length instability in the damping rings at a threshold of 3×10^{10} particles per bunch. The instability is caused by the interaction of the intense bunches with the impedance of the vacuum chamber, primarily due to sharp transitions in its shape. For the 1994 run, smoother vacuum chambers have been constructed for both rings. This has raised the threshold for turbulent bunch lengthening by a factor of 1.5 to 2, well beyond other practical limits on the SLC beam intensity. Stable operation at 4×10^{10} particles per bunch has already been established.

Producing high-current, low-emittance beams and transporting them without increasing the emittance is one of the most difficult challenges of a linear collider. The history of emittance reduction in the SLC is an excellent example of the development of the technology required for linear colliders through successive refinements in measurement and control techniques. Throughout the SLC, the basic strategy is to match the bunch as well as possible theoretically, then measure the residual error and compensate for it by introducing a deliberate error to cancel it exactly. Precision, non-invasive diagnostics are required to measure and match the beams through the transport lines and to accelerate them down the Linac with minimal emittance growth. Careful alignment and orbit control are required to prevent wakefield growth in the copper accelerating structure, where the fields induced by the head of the bunch can deflect the tail and distort the bunch along its length. Residual wakefield tail can be minimized by using a controlled oscillation to generate errors which exactly cancel the measured tails. Emittances better than the original design value have been achieved for both electrons and positrons at the end of the linac

At the end of the 1992 run, a new operating mode with flat beams (vertical size much smaller than horizontal) was tested in preparation for the Final Focus Test Beam, a linear collider development project. Flat beams have the advantage of a smaller beam area, which produces higher luminosity. It is easy to create a small vertical beam size in the damping rings; the challenge is in preserving it. The original SLC design had relied on operation with round beams because of the difficulty of controlling the vertical-to-horizontal coupling in transport lines such as the arcs. In the brief test run, beam size ratios of 10:1 were easily achieved out of the damping rings and the long years of effort in developing techniques to control the beams throughout the SLC allowed the very small vertical size to be delivered for collisions. The luminosity doubled almost immediately. Flat beam operation was successfully commissioned early in the 1993 run. In this configuration, the beams can be focused to a size of 2.1 μm horizontally and 0.8 μm vertically was achieved. The corresponding bunch area of 1.7 square μm is significantly smaller than the SLC design goal of 2.9 square μm. (See Fig. 4.)

For the 1994 run, the final focus was upgraded further to take full advantage of flat beam operation. With roughly equal horizontal and vertical emittances as in the original SLC design, the beam size at collisions is dominated by the linear size due to the emittance. With a very small vertical emittance, this is no longer true and the contribution to the beam size from residual chromatic aberrations is much larger than the linear size. Additional quadrupole and sextupole magnets have been installed to reduce these aberrations and improve tuning flexibility. New wire scanners in each final focus provide more precision diagnostics. With these improvements, the SLC has been able to focus a low current beam to a vertical beam size of less than 500 nm, one third of the original design. This provides a factor of two increase in the peak specific luminosity. The challenge is to reach and maintain these small beam sizes at operating intensity. (See Fig. 5.)

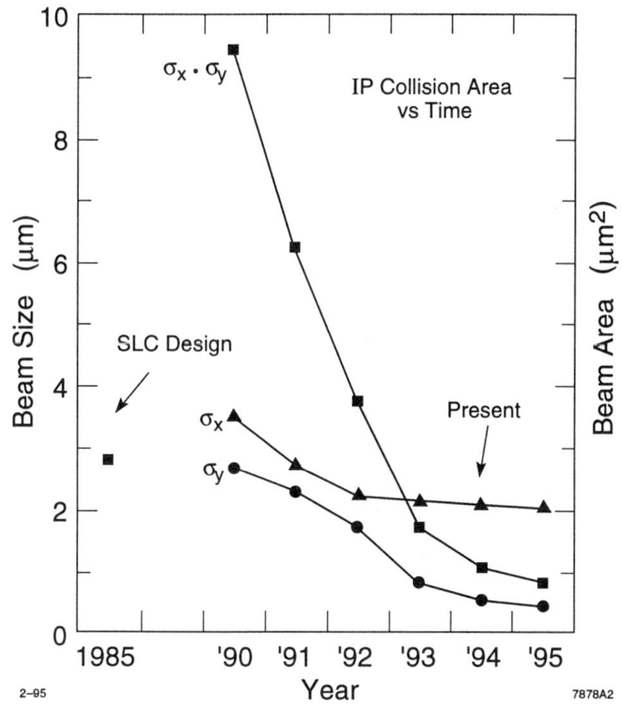

Figure 4: The evolution of SLC IP beam size versus time.

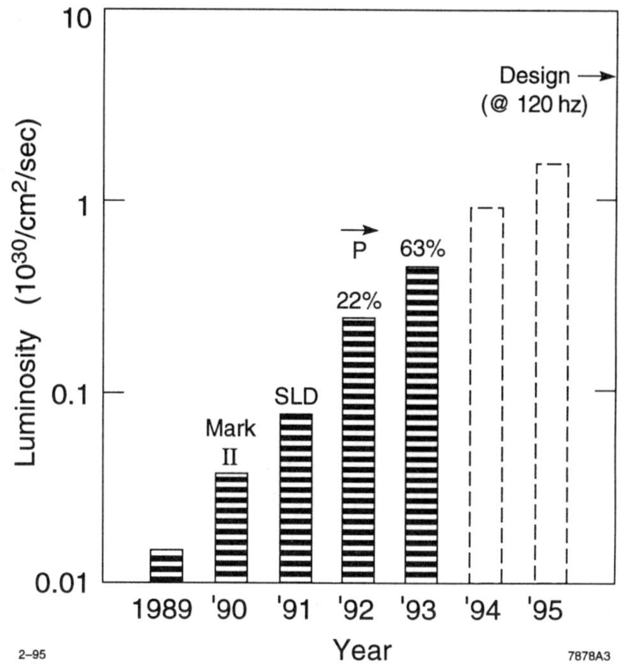

Figure 5: SLC peak luminosity.

3.2.4 Efficiency

The final important factor affecting the event rate is the machine operating efficiency, which has steadily improved to an average of 60–70%. While hardware reliability is essential, the major advance in SLC operations has come from the development of an extensive network of feedback systems to stabilize the bunches throughout the machine. In contrast to a storage ring, a linear collider is not inherently stable and is extremely dependent on feedback and other controls for reliable operation. The basic SLC feedback systems provide pulse-to-pulse control of beam energy and trajectory with special applications for the polarized source and other critical components. There are more than 50 feedback systems active during SLC operation, controlling over 200 parameters.

3.2.5 Future Plans

Motivated by the success of the polarized electron source and by the increase in luminosity, SLAC has completed two major improvement projects aimed at maintaining a vital SLC physics program for the next few years. For the 1994 run, new damping ring vacuum chambers and a major upgrade of the final focus magnet system have been installed to support higher beam intensity and smaller beam size. For these parameters, the focusing of one bunch by the field of the other can become significant, producing up to 30% higher luminosity. The first observation of the long-predicted disruption enhancement would be an exciting achievement.

These upgrades are expected to produce an overall factor of 2 to 4 increase in luminosity compared to 1993 after they are fully commissioned. In the 1994 run, SLD should log more than 100K Zs with 80% polarization to improve the measurement of the weak mixing angle by a factor of two. In future runs with the new SLD vertex detector installed, the goal is more than 500 thousand Zs with over 80% polarization in the next four years. This will provide the highest precision measurement of the Weinberg angle and important B physics results.

In the ten years since construction began on the SLC, the interest in linear colliders has grown enormously. Today most of the major high-energy physics laboratories in the world have active linear collider development programs. Tremendous progress has been made in understanding the techniques required to build and operate this new kind of particle collider. The SLC remains in the forefront of these efforts as the only existing linear collider where new ideas and techniques can be tested. Its recent success with high polarization and increased luminosity has given new impetus to plans for the next linear collider. The SLC has conclusively demonstrated that linear colliders are the way to reach the TeV scale in electron-positron accelerators.

3.3 *An Overview of Future e^+e^- Linear Collider Designs [6]*

The parameters of linear colliders with $E_{\rm c.m.} = 0.5$ TeV based on a compilation made by G. Loew at the LC–93 Conference is given in Table 8. These colliders are:

- TESLA which is based on superconducting rf. All the others would use room temperature rf.
- {SBLC which uses S-band (3 GHz) rf where there is extensive operating experience.
- NLC which uses higher frequency X-band (11.4 GHz) rf in a modulator-klystron-accelerator configuration similar to S-band linacs.
- JLC–I which has three frequency options, S-band, C-band (5.7 GHz), and X-band. Multiple bunches are accelerated in each rf pulse as they are in TESLA, SBLC, and NLC.
- VLEPP which employs a single high-intensity bunch rather than multiple bunches.
- CLIC which is a two-beam accelerator with klystrons replaced by an rf power source based on a high-current, low-energy beam traveling parallel to the high-energy beam.

3.3.1 Beam Power and Spot Size

The luminosity is given by

$$L = \frac{1}{4\pi}\frac{N^2 f_c}{\sigma_{x0}\sigma_{y0}} H_D = \frac{1}{4\pi}\frac{N^2 f_c}{\sigma_x \sigma_y}, \qquad (2)$$

where N is the number of particles/bunch and fc is the collision frequency. Focusing during the collision, disruption, can be accounted for by an enhancement factor H_D in the left-hand expression where the beams sizes without disruption σ_{x0}, σ_{y0} are used or by using the disrupted beam sizes σ_x, σ_y as in the right-hand expression.

The electromagnetic fields at the collision point are paramaterized by

$$\Upsilon = \frac{5r_e^2}{6\alpha}\frac{\gamma N}{\sigma_L(\sigma_x+\sigma_y)}, \qquad (3)$$

where field enhancement is approximately accounted for by using the disrupted sizes. The energy in units of mc^2 is denoted by γ, and σ_L is the bunch length. The mean beamstrahlung energy loss $\delta_B \propto \Upsilon^2$ and backgrounds from beamstrahlung, e^+e^- pairs, and hadronic events depend on Υ. When $\Upsilon \ll 1$ and $\sigma_x \gg \sigma_y$, the mean number of beamstrahlung photons per incident particle is

$$n_\gamma \approx \Upsilon\frac{5\alpha\sigma_L}{2r_e\gamma} \approx \frac{2\alpha r_e N}{\sigma_x}. \qquad (4)$$

n_γ serves as an approximate measure of backgrounds.

The luminosity can be rewritten in terms of four parameters: γ, n_γ, σ_y, and the beam power, $P_B = N f_c \gamma m c^2$

$$L \approx \frac{1}{8\pi\alpha r_e mc^2}\frac{P_B n_\gamma}{\gamma\sigma_y}. \qquad (5)$$

Detector backgrounds fix n_γ, and the center-of-mass energy determines γ. The trade-off is between beam power and spot size. Roughly speaking, TESLA and SBLC would have high beam powers and large spots while the others would have small beam powers and small spots. "Large" and "small" are relative to each other; all of these colliders have large beam powers and small spots compared to present day practice at the SLC.

3.3.2 Multiple Bunches

All designs except CLIC and VLEPP use multiple bunches to achieve good energy efficiency. The cost of filling the accelerator structure with rf energy is amortized over a large number of bunches rather than a single bunch. Using multiple bunches has implications for both the fundamental and higher modes. Each bunch needs roughly the same energy profile down the linac to avoid emittance blow-up from dispersive effects, and they need to have the same energy in the final focus to minimize chromatic aberrations. Bunch train lengths and accelerator structure filling times are comparable, and the accelerator must be prefilled and the rf amplitude ramped so that each bunch gains the same energy

The long-range transverse wakefields from higher modes can cause emittance blow-up that is in addition to that from short range wakefields. The higher modes must be damped or detuned (spread out in frequency) to reduce these wakefields to acceptable levels.

The dominant multibunch effect in TESLA is associated with chromatic effects from the bunch-to-bunch energy spread. A systematic spread could be caused by effects such as Lorentz detuning where the cavity dimensions change during the pulse due to the pressure from the stored electromagnetic energy. With a bunch-to-bunch energy spread of 10^{-3}, the emittance increase is a factor of ten. This results in a tight tolerance of a few times 10^{-4} on the energy spread.

Two of the designs, VLEPP and CLIC, are not designed for multiple bunches. VLEPP has a large, single bunch, and requires stringent tolerances on the linac for emittance preservation and a novel final focus that employs a traveling focus. The possibility of using up to

Table 8: Selected Linear Collider Parameters for $E_{\rm c.m.} = 0.5$ TeV (G. Loew, Linear Collider 93), where VLEPP employs a "traveling focus." The loaded gradient is calculated before applying further reductions for off-crest running, BNS damping, etc., VLEPP excluded. The AC power is calculated only for the linac damping ring, detector, utility power, etc., and power for klystron focusing are not included [SBLC bases its number on a combined klystron-modulator efficiency of 45%, JLC and NLC have assumed this number to be closer to 35%, and SLED–I (for JLC–I(S)) and SLED–II (for JLC–I(C), JLC–I(X), NLC, and VLEPP) are assumed to be 65% efficient].

Parameter	TESLA	SBLC	JLC-I (S)	JLC-I (C)	JLC-I (X)	NLC	VLEPP	CLIC
L (10^{33} cm^{-2} s^{-1})	7	4	4	7	6	8	15	2–9
rf frequency (GHz)	1.3	3.0	2.8	5.7	11.4	11.4	14	30
Repetition rate (Hz)	10	50	50	100	150	180	300	1700
Bunches per rf pulse	800	125	55	72	90	90	1	14
N (10^{10})	5.15	2.9	1.30	1.0	0.63	0.65	20.	0.6
BPM precision (μm)	10.	10.	NA	NA	1.	1.	0.1	0.1
$\gamma\epsilon_x$ / $\gamma\epsilon_y$ (10^8 m)	2000/100	1000/50	330/4.5	330/4.5	330/4.5	500/5	2000/7.5	180/20
β_x^* / β_y^* (mm)	25/2	22/0.8	10/0.1	10/0.1	10/0.1	10/0.1	100/0.1	2.2/.16
σ_{x0} / σ_{y0} (nm)	1000/64	670/28	300/3	260/3	260/3	300/3	2000/4	90/8
σ_L (mm)	1000	500	80	80	67	100	750	170
IP crossing angle (mrad)	0	3	7.3	8	7.2	3	—	1
Υ	0.029	0.055	0.24	0.21	0.16	0.096	0.074	0.35
Disruptions, D_x / D_y	0.54/8.5	0.36/8.5	0.13/13	0.13/11.7	0.07/6	0.08/8.2	0.4/215	1.3/15
H_D	2.3	1.6	1.6	1.5	1.7	1.4	1.3	3.3
δ_B	0.03	0.03	0.10	0.08	0.05	0.03	0.13	0.36
n_γ	2.7	2.0	1.6	1.4	1.0	0.9	5.0	4.7
Unloaded gradient (Mv/m)	25	21	22	40	40	50	108	80
Loaded gradient (MV/m)	25	17	19	33	31	38	96	78–73
Active linac length (km)	20	29.4	28	16.7	17.7	14	6.4	6.6
Section length (m)	1.04	6	3.6	2	1.3	1.8	1.01	0.273
Number of sections	19232	4900	7776	8360	13600	7778	5200	24000
Number of klystrons	1202	2450	1944	4180	3400	1945	1300	2
Klystron peak power (MW)	3.25	150	85	45	70	94	150	700
Klystron pulse length (μs)	1300	2.8	4.5	3.6	0.84	1.5	0.7	0.011
Pulse length to section (μs)	1300	2.8	1.2	0.6	0.21	0.25	0.11	0.011
Pulse compression/gain	—	—	2.4	4.2	3.2	4	4.22	—
a/λ (input/output cavity)	0.15	.15/.11	0.13	.16/.12	.24/.14	.22/.15	.14	.2
P_B (MW)	16.5	7.3	1.4	2.9	3.4	4.2	2.4	.4–1.6
AC power (MW)	137	114	106	193	86	141	91	175
$2P_B/P_{AC}$	0.24	0.13	0.03	0.04	0.09	0.06	0.05	0.02

four bunches in CLIC is still under study, with the issues being energy compensation and control of intrabunch transverse wakefields.

3.3.3 Radio Frequency Power

Most of the colliders in Table 8 require a large number of high peak-power klystrons. S–band klystrons delivering 80 MW peak power for 4.5 μs are in operation at the KEK Accelerator Test Facility. This is essentially the performance assumed for JLC–I (S). A 150 MW, 3.0 μs klystron developed by SLAC and DESY for the DESY SBLC and S–band preaccelerators of the NLC has recently operated at full power and will be used in a test facility at DESY. X–band klystrons have operated at KEK with the latest tube reaching an output power of 70 MW in 50 ns pulses with 35% efficiency. At SLAC, two X–band klystrons have reached a power over 50 MW with an output pulse duration of 1.5 μs and an efficiency of about 40%.

The efficiency for transforming AC wall plug power to rf power is one of the major factors determining the economics of linear colliders. This efficiency has two major contributors, the modulator and the klystron. The need to improve modulator efficiency has led to a number of ideas, including a Blumlein configuration for the pulse-forming network that has been successfully tested at KEK and the use of a high voltage switch tube rather than a pulse transformer being pursued for SBLC. This has proven to be more difficult than anticipated, and switch tubes are now a backup to conventional modulators in the SBLC test linac.

It is impractical to directly generate the short rf pulses needed for the high-frequency colliders. A substantial fraction of modulator inefficiency comes from the rise- and fall-times of the pulse transformer that steps-up the output voltage. Pulse compression matches a long klystron pulse at a reduced peak power to the short bunch trains and accelerator structures that are appropriate for high-frequency rf. SLED–II pulse compression has been tested with 50 MW, ~ 1 μs input pulses, and power gains of 3.6 to 4.1, with efficiencies of 60–68% have been demonstrated. The intrinsic efficiency of this design is 70–75%. With improved components, this is expected to reach a power gain of 4.2 to 4.6 with 66–70% efficiency.

TESLA has unique power source requirements. The high-Q cavities and long pulse length reduce the peak power to 3.25 MW, but the modulator must be capable of delivering that power for over a millisecond.

The two-beam accelerator, personified by CLIC in Table 8, avoids the thousands of individual rf power sources by replacing them with a high-current, low-energy drive beam. This low-energy beam has a time structure appropriate for generating 30 GHz rf. It is accelerated by a 350 MHz superconducting rf system, and energy is extracted with transfer structures spaced roughly 1.5 m apart. Drive beam generation is under study at the CLIC Test Facility, where single, 8 bunch, and 24 bunch pulse trains have been produced and—using a CLIC section as the transfer structure—56 MW of rf power has been generated. This corresponds to a peak decelerating field in the last cell of 107 MV/m. When this power was transferred to the accelerating section an average accelerating field of 71 MV/m was seen with no signs of breakdown.

3.3.4 Accelerator Structures

Room temperature accelerators are performing with gradients close to those listed in Table 8. Precision machining is being used for tight fabrication tolerances and for the surface qualities needed for high gradients. Structures are being made at KEK and CERN using direct copper-to-copper diffusion bonding of precision machined cells. A CERN–made 11 GHz structure has been tested at KEK and exceeded 100 MV/m accelerating field after a reasonable amount of conditioning. Its performance was limited by the available rf power. A full-length NLC section has reached over 55 MV/m in a high-power test, and, as mentioned above, a 30 GHz CLIC structure has shown excellent performance. Costs for mass fabrication are not excessive.

High-power pulsed processing is having continuing success in raising the gradient of superconducting cavities, with a gradient of 25 MV/m being reached in a five-cell, 1.3 GHz cavity. Demonstrating this type of behavior in a larger scale linac, together with reducing costs, are major goals of the TESLA Test Facility under construction at DESY.

3.3.5 Emittance Preservation

The vertical invariant emittances $\gamma\epsilon_y$ are small, and emittance preservation during acceleration is a central consideration. Emittance growth caused by the combination of injection jitter and wakefields must be controlled by tight tolerances on injection elements and BNS damping.

Misalignments of the accelerator sections, quadrupoles, and beam position monitors in the main linac cause emittance growth through wakefields and dispersion different central trajectories for different energies. Beam based orbit correction procedures, where optical elements are varied and orbit changes measured, have become almost universally adopted as the way to loosen alignment tolerances. Experience is being gained at the SLC where these technologies are used to control emittance dilution.

3.3.6 Center-of-Mass Energy ($E_{\rm c.m.}$) greater than 1 TeV

The discussion above concentrated on $E_{\rm c.m.} = 0.5$ TeV, but there is substantial interest in higher center-of-mass energies. Ideally the luminosity would increase as ($E^2_{\rm c.m.}$) to reflect the decrease in the cross section for production of point-like particles. Table 7 in the Introduction gives recent parameters for upgrades of TESLA, SBLC, and NLC to 1 TeV (and 1.5 TeV for NLC).

The two ways to reach 1 TeV are to double the length, or double the gradient by quadrupling the peak power using a combination of increasing the number of klystrons, the klystron peak power, and the pulse compression gain. The former is used in TESLA since 50 MV/m would be near the fundamental gradient limit of superconducting rf in Nb. The latter is used in SBLC and NLC. The threshold for rf capture of dark current is about 20 MV/m at 3 GHz and about 80 MV/m at 11.4 GHz. The SBLC gradient is well above that threshold, but may be reached with appropriate attention to surface preparation and with RF processing. The NLC gradient is below the dark current limit and has been easily exceeded in the test mentioned above.

A straightforward application of either method of doubling the energy leads to unacceptable AC power consumption, so the parameters have changed to reflect this. First, the luminosity has not been scaled as $E^2_{\rm c.m.}$. Second, the trade-off between high beam power and small spots (Eq. (5)) is no longer the central theme it was in the 0.5 TeV discussion. All of the parameter lists have evolved in the direction of small spots, with nearly identical invariant emittances. The underlying assumption for TESLA and SBLC are that after gaining experience with correction and optimization procedures, the vertical emittance can be reduced by an order of magnitude. This is possible if the accelerator complex has been designed and constructed with the goal of producing small emittances.

The 1 TeV NLC parameters show a third approach to the problem of AC power consumption. They are based on improved efficiency. Some of that improvement has come from evaluating the modulator, klystron, and pulse compression efficiencies for the $E_{\rm c.m.} = 0.5$ TeV collider. The current estimate is that the AC power for the linac would be 92 MW versus the 141 MW at LC–93. Additional efficiency improvements are counted on for 1 TeV. These may include elimination of modulators through using a gridded klystron or a cluster klystron.

3.4 The Japanese Linear Collider (JLC) and Accelerator Test Facility (ATF) [7]

3.4.1 Basic Design Approach

The Japanese Linear Collider (JLC) design is based on normal-conducting linac technology in order to save R&D time for construction of a TeV linear collider. The wall plug power was assumed to be a few hundred MW at largest. The linac gradient was set to be at almost 100 MV/m. In order to achieve a luminosity of the order of 10^{34} cm^{-2} s^{-1} with those parameter constraints, X–band ($\sim$ 10 GHz) was chosen for the main linac frequency. Since the X–band linac technology was quite immature, the top priority was to develop it by extending the well established S-band technology.

An Accelerator Test Facility (ATF) was founded at KEK in 1988 in order to promote linac R&D work. At first an S–band rf system was established with several 5045 klystrons sent from SLAC. High gradient tests were carried out for S–band structures. Based upon this infrastructure, R&D work for X–band linac technology was at once started. Thereby efforts have been devoted mostly to develop X–band klystrons of 100 MW class. High-gradient tests of X–band structures also were carried out, as klystrons of a few tens of MW became available.

In the course of the X–band work and miscellaneous studies, the main purpose of the ATF gradually became construction of a test damping ring. Its energy was chosen to be 1.54 GeV. The injector is an S–band linac of the same energy. Construction of the accelerator complex was started in 1991. The ring is expected to provide electron bunches with their normalized emittance as small as a required value of 5×10^{-8} rad m.

Theoretical design work has been developed in order to optimize general parameter sets for the JLC, along with the above experimental work. The basic guideline has been to adopt a multibunch scheme per linac pulse and a small-crossing-angle collision. Major results in this field are: 1) study of the beam behavior at the collision which enables accurate evaluations of the luminosity, 2) study of achromatic final-focus optics which enables a large momentum acceptance, 3) detailed estimation of background noises, and 4) study of beam break-up and beam loading in the main linac.

3.4.2 Critical Issues for the Main Linac

Total Wall-Plug Power. The total power would certainly be within the above constraint for $E_{\rm c.m.} = 0.5$ TeV with parameter sets presently under consideration, but it would be necessary to reduce the repetition rate and the expected luminosity for $E_{\rm c.m.} = 1$ TeV or higher, since the total power is approximately proportional to $E_{\rm c.m.}$

Radio Frequency Power Source. Desirable ratings for the klystron output are 130 MW, 500-ns-long pulses per tube. Presently available tubes provide $\sim$ 70 MW, 100 ns pulses. Development of reliable output windows is the key issue in order to achieve these ratings. Due to large power consumption, the conventional focusing solenoid should be replaced—*e.g.*, by a superconducting one.

In order to attain required gradients in the main linac, at least one-fold compression of the klystron output power is necessary, with the peak power doubled thereby. It should be tested to see whether amplitude or phase modulations brought about in the process are tolerable for an identical acceleration of each bunch.

The high-voltage pulse should have rise/fall times short enough compared with the pulse length of 500 ns in order to improve the overall power efficiency. A modulator with a Blumlein PFN seems promising, since the step-up ratio of a pulse transformer can then be halved, leading to a considerable shortening of its time constants. Capacitors with the smallest possible intrinsic inductance are also a necessary ingredient for the PFN in order to ensure fast ramping times.

Structure. High-gradient tests were carried out to verify the feasibility of required gradients 50 to ~ 100 MV/m with an X–band structure. The gradients seem attainable, but the conditioning time was an order of magnitude longer than SLAC and CERN structures. Study for treatment of the structure surface is an immediate problem.

The X–band structure needs to be fabricated very accurately. Fine machining of structure pieces must be well established in order to attain dimensional accuracies on the order of 1 μm or better. Progress is yet only midway, however, toward bonding the pieces into a structure with a tolerable loss of the accuracies and with good electric and vacuum contact. A diffusion bonding technique is being developed for this purpose.

In order to damp the HOM effect, simulation studies have been carried out for both damped and detuned structures. A reasonable set of dimensions was found for a damped structure with four circumferential slots, although its fabrication may be cumbersome. Concerning a detuned structure, a new method was developed to calculate the dispersion relations of HOM modes up to the order of 20 or higher. Wakefields would then be calculated with a good accuracy for the multibunch case. It is urgently necessary to establish some facility (such as the Next Linear Collider Test Accelerator (NLCTA) at SLAC) to carry out beam measurements of the wake and check the validity of simulations.

Technology for accurate beam monitoring and alignment system for miscellaneous linac elements should be developed in this experimental work.

3.4.3 Injector for the Main Linac

Electron sources. Conventional cathodes can provide sufficient electrons per bunch for the JLC design, but it is not yet clear how uniform the population should be among the bunches in a train. This should be estimated carefully in order to control the BBU in the main linac. In this respect, a conventional grid-controlled thermionic gun might be inadequate, and an rf gun might be needed.

As a polarized electron source, GaAs photocathodes have come into practical use. The issue at present would be to incorporate the cathodes in an rf gun system.

Damping Ring. The damping ring must provide beams with the small vertical emittance given above with a repetition rate as high as 150 Hz. A lattice of FOBO type seems promising to attain the low emittance according to simulations. A long wiggler magnet section is necessary to ensure a fast damping time. The specifications are much more stringent than those for conventional rings; therefore, it is quite desirable to verify the feasibility by constructing a test ring. A fine alignment system for magnets is a key ingredient, since the coupling between vertical and horizontal betatron oscillations should be minimized to achieve the low emittance. The emittances, both transverse and longitudinal, should be as uniform as possible among the bunches for the beam dynamics in the main linac. For this sake, the ring rf system must comprise damped cavities and energy-compensating cavities. Those technologies will be developed in the ATF damping ring .

Beam Transport and Bunch Compression Every bunch should be equally extracted from the ring. Hence, two kicker magnets, spaced by a π phase and driven by a common power unit, will be used to cancel the effect of ripples in a driving pulse.

The bunch length σ_z should be about 60 μm, while it is about 3 mm; in the ring, therefore, a bunch compressing line should be very carefully designed. Emittances should not grow in the line. Furthermore, a uniformity among the bunches is necessary also for this compressing process.

3.4.4 Interaction Region

Final Focus. The optics should be such that: 1) it includes a collimator to reject too many off-momentum particles, which might otherwise cause background noises at the collision point; 2) the achromatic correction should be sufficient to keep the beam size small at the collision point, and 3) the bore aperture of the final quadrupole magnet should be large enough to avoid an emittance growth due to image currents on the pole surfaces.

The final quadrupole magnet is the most technically difficult element. The poles should be machined with an accuracy of about 1 μm. A 1 m model was fabricated, and has been successfully used at the FFTB line. However, for the JLC, it would have to be at least a few meters long, making the machining much more

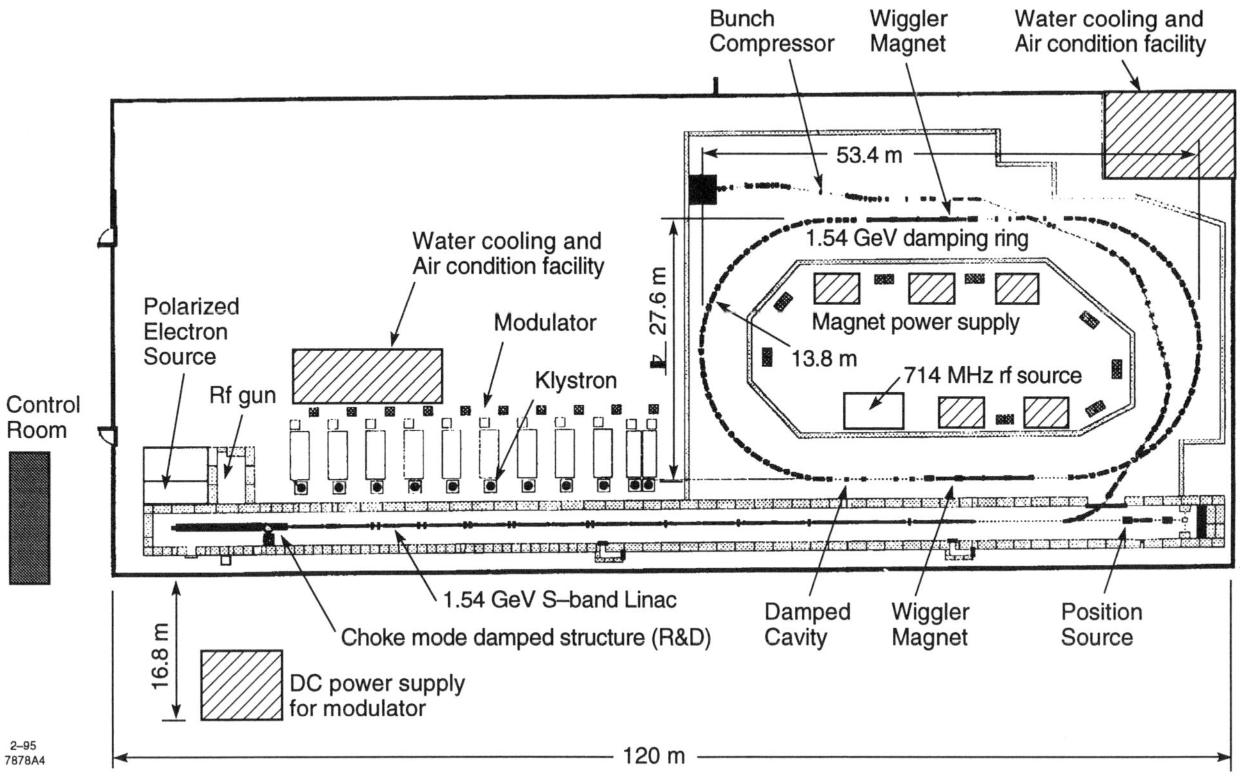

Figure 6: Layout of the ATF at KEK.

difficult. Moreover, it is beyond the present technology to measure the quadrupole and higher order fields of such a long magnet with a sufficient accuracy.

Collision Point. It is essential to measure the position and profile of incoming bunches. The vertical dimension σ_y is as small as several nanometers. A laser-compton profile monitor was developed and successfully tested at the FFTB line for beam sizes of several tens of nanometers by using 1.06 mm laser beams. For the JLC, this scheme depends critically on the feasibility of intense laser beams of one tenth of the wave length.

The background problem is another serious issue at the collision point. Pair creation due to strong electromagnetic fields of the colliding bunches is a major cause of the background. The creation process is being analyzed in detail by use of a newly developed code, ABEL. In order to cope with noise particles, the solenoid field of the detector may be higher than 2 T. A shield must be designed to confine this field within the detector.

3.4.5 Status of the ATF Construction

Shield walls for both the injection linac and the test ring were completed in 1993. Radiation safety systems were also completed. Construction of high-power AC lines and cooling water pipings for the ring was completed by the end of 1994. (See Fig. 6.)

The 1.54 GeV S-band Linac. The first 3-m section of the linac has been in commission since August 1993. Various beam monitors have been tested with the beams. A choke cavity structure also will be tested in a few months. In parallel with the commission, construction of the linac has been continued. About half of the nineteen 3-m accelerating sections have already been installed. Nine out of ten modulators have been fabricated. A sufficient number of 80 MW klystrons are in stock.

The 1.54 GeV Test Damping Ring. One unit section of the arc has been fabricated, comprised of a defocusing dipole magnet, a quadrupole, a sextupole, and steering magnets set on a common table with precision movers. A few wiggler magnets have also been fabricated. Prototype vacuum chambers are under fabrication. Cold test of a damped cavity was completed. A 50 kW, 714 MHz klystron has been successfully tested.

3.5 CLIC: The Basic Scheme, Critical Issues, and Status of Test Facilities [8]

3.5.1 The Basic Scheme

CERN is studying the feasibility of building a 0.5–2.0 TeV e^+e^- linear collider, using classical normal-conducting traveling-wave rf accelerating sections, powered by a superconducting drive linac.

The 3–5 GeV high-intensity electron drive linac runs in parallel with the main linac. The bunched drive beam is decelerated in so-called transfer structures where 30 GHz rf power is generated and fed via standard waveguide to the accelerating structures.

Periodic replacement of the energy lost by the beam to the transfer structures is made by short sections of 6 MV/m superconducting cavities driven by 350 MHz 1 MW klystrons. The superconducting cavities and klystrons already developed by CERN for LEP are ideal for this application. Generation of the rf power by a drive linac, rather than by thousands of individual klystrons, is a distinctive feature of the CLIC design and results in a particularly simple tunnel cross section.

An obvious design aim is to make the linacs as short as possible to keep the cost down. This implies high accelerating gradients, which would normally result in a higher power requirement, but is compensated for by operating at a high frequency to maximize the rf-to-beam power conversion efficiency. CLIC has chosen to work at 80 MV/m and 30 GHz. At this frequency, wakefield effects, alignment tolerances, and fabrication problems seem just manageable, but demand state-of-the-art technology.

For a 2 TeV center-of-mass collider, each of the high-energy linacs is made up of 45,750 30-cm-long iris-loaded traveling wave sections. A relatively large aperture-to-wavelength ratio of 0.2 has been chosen for these sections to maintain the destructive effect of the single-bunch transverse wakefields within reasonable limits. The 35 mm outer diameter, which is machined to a precision and concentricity with the beam aperture of ± 1 μm, serves as the reference for alignment. Individual cells are pumped by four vacuum manifolds through radial holes. The unit cost of mass-producing the discs for these sections has been estimated by a recent in-depth cost-estimate study carried out by four independent firms to be about $12 per cell.

A stringent test of the ability of CLIC–type sections to withstand high accelerating gradients was made at KEK in September 1993, when an 11.4 GHz traveling wave section built by CERN using the same special techniques developed for CLIC sections reached an average gradient of 85 MV/m after 50 hours of conditioning (the peak gradient in the first cell was 138 MV/m). The dark current level at 50 MV/m was < 1 μA.

Limiting the emittance blowup along the main linacs in the face of strong transverse wakefields is a concern, and sets tight tolerances on the transverse alignment of the components (typically 5 μm on BPM's and accelerating sections for 50 μm initial offsets of the quadrupoles). Such tolerances can only be achieved with a beam-based active alignment system and requires in particular micron resolution BPM's. Prototype BPM's based on simple E_{110} cylindrical cavities have been built and their ability to resolve such small displacements has been evaluated on the bench by exciting them with an antenna to simulate the passage of an off-axis electron bunch. The output E_{110} signal as a function of antenna position over a range of ± 1 μm clearly shows a resolution capability of $<$ 10 nm.

3.5.2 Critical Issues

Drive-beam generation. Generation of trains of short, high-charge bunches at 1 cm spacing is an essential and difficult element of the CLIC two-beam scheme. Three basic types of solution are being studied.

The first scheme creates the trains by combining the outputs of an array of pre-injector linacs of slightly different energies using a magnetic "switch-yard" system.

The second scheme uses the bunching capability of a Free Electron Maser (FEM) to produce the 1 cm bunch structure directly. The primary beam comes from an induction linac, after bunching in the FEM bunch trains are created by a 350 MHz chopper.

The third scheme generates the total required charge of 7 mC at low energy in long trains of long, widely spaced (6.8 ms), relatively low intensity bunches. These are subsequently compressed to 1 mm by two stages of conventional magnetic compression, accelerated to 3 GeV in a recirculating superconducting linac, and finally stacked in an isochronous ring having a circumference equal to the initial bunch spacing plus 1 cm to give them the correct bunch spacing.

Alignment and Emittance Blowup. Limiting emittance blowup in the main linac is a serious concern and requires the use of beam-based active alignment schemes and sophisticated alignment correction algorithms. Although results from computer simulations indicate that this is in fact feasible, it does require a measurement of the emittance at many locations along the linac. At this moment no suitable emittance measurement device is available.

Luminosity. Further studies of schemes using damped or detuned structures are required to evaluate the feasibility of running the CLIC machine in a multibunch mode to either increase the luminosity-to-power ratio or to reduce (δ) the energy-loss parameter.

3.5.3 Status of Test Facilities

CERN has built, and is currently operating, a test facility for linear collider studies (the CLIC Test Facility or CTF) to:

- study the production of short, high-charge electron bunches from laser illuminated photocathodes in rf guns,
- generate high-power 30 GHz rf pulses by passing bunch trains through transfer cavities for testing CLIC prototype components, and
- test beam position monitors.

Significant amounts of 30 GHz power can only be extracted from the CLIC section by using bunch trains. 60 MW has been generated by a 24-bunch train containing 67 nC of total charge or 2.8 nC per bunch. This power level corresponds to a decelerating field in the section of 110 MV/m (more than the CLIC nominal accelerating gradient). The charge per bunch which can be transmitted through the structure is at this moment limited by wakefields in the traveling wave accelerating section. The accelerating field in the second CLIC structure is determined from the difference between maximum and minimum energy gain of the beam as its phase with respect to the beam-induced rf accelerating field is varied. An average accelerating gradient of 73 MV/m was measured in this way for the best performance so far. Plans to increase the 30 GHz power generation include:

- raising the gradient in the accelerating section to reduce the effect of wakefields,
- making shorter bunches using a magnetic bunch compressor, and
- building a new 100 nC/bunch rf gun.

CLIC Alignment Test Facility. An active alignment test facility has been built in an unused underground tunnel at CERN to study the feasibility of making controlled submicron displacements and to try out alignment systems.

The structures to be aligned, dummy accelerating sections for the moment, are supported by V–blocks on 1.4-m-long silicon carbide girders. The ends of two adjacent girders sit on a common platform which ensures continuity of position between units.

The platforms are activated by three stepping-motor-driven precision jacks (two in the vertical plane for vertical displacement and axial rotation, and one in the horizontal plane). The setup is equipped with linear and angular displacement transducers (0.1 μm and 10 μ-rad resolution respectively) and is piloted remotely from a small computer. After deliberate misalignments of 1 mm, the system which is programmed for automatic alignment with respect to the transducers, settles back to nominal positions within < 1 μm.

The set-up is also being used to test a new optical pre-alignment system, developed by NIKHEF in Amsterdam, for use before injection of the beam. The image of a square-shaped red-light source is focused on a light-detecting four-quadrant cell by a thin lens. Displacements of the source, lens or four-quadrant cell out of the optical axis of the instrument produce an imbalance at the detector. This system has been incorporated into the six hollow support girders of the test module and enables the relative positions of the far ends of two adjacent girders to be maintained in position with respect to the ideal straight line to < 2 μm.

The CESTA Test Facility. A collaboration to study the use of an FEM to create the CLIC drive beam exists between CERN and the Centre dEtudes Scientifiques et Techniques dAquitaine (CESTA) in Bordeaux. The aim is to use a helical wiggler to bunch a beam from an induction linac, and in a later phase to use this beam to generate power in a CLIC transfer structure. A preliminary experiment to measure the bunching produced by a helical wiggler using the beam from a gun-diode is already underway.

The MIT Test Facility. CERN is collaborating with MIT to test CLIC prototype components. MIT has a gun diode driven FEL that produces 20-ns-long, 60 MW power pulses at 33 GHz. This facility is at present being used to test a 25-cell prototype CLIC accelerating section.

3.6 The S-Band Linear Collider [9]

Among the different design studies for a next generation e^+e^- linear collider, the SBLC approach follows the concept of a relatively low rf-frequency $\omega_{\rm rf}$ and a moderate accelerating gradient g. This allows for a high overall efficiency and relaxed tolerances in combination with reduced wake field effects in the linac (the wake fields scale approximately as $W_{\|} \propto \omega_{\rm rf}^2, W_{\perp} \propto \omega_{\rm rf}^3$). In the SBLC, conventional traveling wave accelerating structures at 3 GHz are used running at a gradient of 17 MV/m. To a certain extent the SLC can be considered a 20% model for the SBLC and thus the SBLC can greatly make use of the wealth of experience from running this first linear collider.

The SBLC linear collider study is pursued at DESY in the frame of an international collaboration with institutes in China, France, Germany, Japan, Netherlands, Russia, and USA contributing to the technical R&D and/or the design of the 500 GeV collider.

3.6.1 Parameters

The achievable luminosity of a linear collider is determined by the following basic parameters:

- average power per beam P_b, which is limited by a reasonable AC-power limit and the overall AC-to-beam transfer efficiency η,
- normalized vertical emittance ε_y, limited by tolerances, and
- maximum tolerable beamstrahlung energy loss $\langle \Delta E/E \rangle_{\rm rad}$ limited by background considerations and the energy resolution required by the high-energy physics experiment.

Using basic relations for the luminosity and the beamstrahlung, and assuming an optimum beta-function β_y at the IP equal to the bunch length σ_z the luminosity is in good approximation given by

$$L = \text{const.} \times \frac{P_B}{\gamma} \frac{\left\langle \Delta E \; E_{\rm rad}^{1/2} \right\rangle}{\varepsilon_y^{1/2}}. \tag{6}$$

With P_B the beam power in MW, $\left\langle \Delta E \; E_{\rm rad}^{1/2} \right\rangle$ the energy loss due to beamstrahlung in percent, and ε_y the vertical emittance in 10^{-6} m, we find the const/γ $\approx 2.8\times10^{32}$ cm^{-2} s^{-1} at $E_{\rm c.m.}$ = 500 GeV. A large efficiency is achieved by accelerating long bunch trains per linac pulse allowing for high P_B, at the same time keeping beamstrahlung at a low level. The required emittances and beam sizes at the IP are close to what has been achieved recently at the FFTB experiment. The main S–Band Linear Collider parameters are given in Table 9.

3.6.2 Interaction region, Final Focus, and Collimation

Keeping beamstrahlung at a low level is an essential prerequisite for acceptable background conditions and good energy resolution for the high energy physics experiment. The most important parameters concerning beam-beam effects are summarized in Table 10.

In case of the SBLC, beams have to cross at an angle (θ_c = 3 mrad) in order to avoid the multibunch kink-instability. A reduction of luminosity caused by an effective increase of the horizontal beam size is avoided by employing a simple crab-crossing scheme with finite dispersion at the interaction point), making use of a coherent energy spread within the bunch of about σ_E = 0.5%.

The lattice between the interaction region (IR) and the main linac consists of the final focus system (FFS) for beam size demagnification and chromatic corrections, a collimation section to protect the IR quadrupoles from large amplitude particles, and bending sections for creating a sufficient separation between two beam lines if the collider has to serve two experiments. The bend between collimation and the FFS also helps to reduce background due to muons originating at the collimators.

The momentum acceptance of ±2.0% ($\sigma_{E,r,\rm beam}$ = 0.5%) with an optimized sextupole distribution of the

Table 9: Main Parameters of the S–band 500 GeV (c.m.) Linear Collider.

Parameter	Value
Active length	29.4 km
$t_{\rm pulse}$	2 μs
n_b/pulse	125
Δt_b	16 ns
$f_{\rm rep}$	50 Hz
$\varepsilon_x/\varepsilon_y$	10/0.5 $\times 10^6$ m
β_x^*/β_y^*	22/0.88.8 mm
σ_x^*/σ_y^*	670/28 nm
σ_z	0.5 mm
$\left\langle \Delta E E_{\rm rad}^{1/2} \right\rangle$	3.2%
P_b	2 × 7.2 MW
$P_{\rm AC}$ (2 L's)	113 MW
$\eta_{\rm AC-to-beam}$	13%
L (including H_D)	3.6 × 10^{33} cm^{-2} s^{-1}

Table 10: Results of Beam–Beam Simulations for the SBLC.

Parameter	Value
$\left\langle \Delta E E_{c.m.;rms}^{1/2} \right\rangle$	2.7%
Disr. D_x/D_y	0.4/8.5
angle $\theta_{\gamma x/y}$	1.28/0.55 mrad
$N_{\rm pair}$/bunch	7
$N_{\rm hadr}$/bunch	0.2

FFS for the SBLC design is far in excess of the beam energy spread. The requirements for beam collimation are determined by the condition that synchrotron radiation generated in the doublet before the IP has to pass freely through the aperture of the final quadrupole on the opposite side. This means that particle amplitudes have to be restricted to $6\sigma_x \times 8\sigma_y$. Following concepts developed at SLAC, a beam optics design for simultaneous collimation in x, y and dE/E has been worked out. The entire lattice from the linac to the IP will require approximately 1.1 km on either side of the IP.

3.6.3 Main Linac

As any other Linear Collider, the SBLC consists of a large number of basic cells. In the SBCL such a cell is made of two 6-m-long traveling wave, constant-gradient copper structures. They are powered by one 150 MW klystron. A modulator flat top pulse width of 2.8 μs is required to handle the current pulse of 2 μs.

A total of 2450 klystrons and modulators is necessary for the two 250 GeV linacs. Different focusing schemes are still being considered, while in general the focusing strength should vary with $\beta \propto \gamma^{1/2}$ in order to keep the BNS energy spread constant. After the damping ring the basic FODO cell length with $\beta_{\text{initial}} = 18$ m is determined by the length of one accelerating structure. Further upstream the β–γ scaling could be realized in good approximation by distributing fewer quadrupoles of the same focal length along the linac. This scheme has the advantage of keeping the number of different units down.

In order to minimize the civil engineering costs it is planned to have the klystrons and the linac housed in one single tunnel. Preliminary radiation simulations have shown that legal restrictions can be met by employing suitable concrete shielding between the accelerator and the klystron gallery.

One of the most important accelerator physics issues in a linear collider is the preservation of a (very) small emittance along the linac. Emittance dilution caused by chromatic effects (dispersion, filamentation) due to energy spread in the bunch, short range wake fields and long range deflecting modes are investigated extensively. Recent computer simulation results for SBLC show that single bunch beam break-up due to injection oscillations can be well suppressed by applying BNS damping only over the first 30% of the linac employing a coherent energy spread of $\sigma_E = 0.9\%$. In the remaining part of the linac σ_E can be reduced again to 0.3% by shifting the bunches 12 degrees off the rf wave crest. Taking into account transverse positioning errors of quadrupoles, accelerating structures and beam position monitors (BPMs) of 0.1 mm (rms), it is shown that after applying various correction algorithms (wakefield-free (WF) orbit correction, bumps for dispersion and wakefield compensation), the dilution can be kept as small as $\Delta\varepsilon_x/\varepsilon_y = 3\%$. For a BPM resolution of 5 μm this correction scheme requires four emittance measurement stations with an accuracy of 3%.

Higher order modes, induced by bunches passing off-axis through the accelerating structures, cause oscillations growing over the entire bunch train and along the linac (cumulative multibunch beam break-up). In order to keep this effect within limits, a frequency spread for the deflecting modes is introduced (the maximum variation is assumed to be 40 MHz) by changing the resonant behavior of the rf structure for the HOM modes and at the same time keeping it identical for the accelerating mode. Since detuning alone is not sufficient, the quality factor of the modes with the strongest coupling to the beam have to be lowered. This is achieved by adding three HOM-couplers per structure resulting in a Q-profile of typically $\approx 3 \times 10^3$ for the harmful dipole modes. With these assumptions and tolerances, computer simulations yield an effective multibunch emittance dilution of $\Delta\varepsilon_x/\varepsilon_y = 20\%$. This dilution can be further reduced by "beam-based alignment" of the cavities (zeroing of the HOM power by moving the structures towards the beam axis) and/or by using fast kickers to place all bunches back on the same orbit from pulse to pulse. The latter method requires pulse to pulse stability of the beam break up behavior.

3.6.4 Injection

The emittances required for the SBLC beams are provided by two damping rings of 650 m circumference operating at 3.15 GeV. A beam optics delivering $\varepsilon_x = 5 \times 10^{-6}$ m (50% of the design value at the IP) has been designed. The normalized dynamic aperture of 2.4×10^{-2} m is sufficient to accept the beam delivered from the e^+ source. Positrons are produced by converting γs in a thin $(0.4x_0)$ target. The required intense photon source is realized by passing the e^- beam after collision through a 30-m-long wiggler. This method drastically reduces the heat load on the target and opens up the possibility to produce polarized positrons by using a helical undulator.

3.6.5 Upgrade Potential

The SBLC design is very well suited for pushing the vertical emittance to lower values in a later upgrade program. After gaining experience with various correction and optimization procedures, reducing the vertical emittance ε_y by an order of magnitude seems to be conceivable. This reduction would allow for higher luminosity and at the same time lower AC–power consumption. Achieving such small emittances becomes very important (if not inevitable) for an energy upgrade to 1 TeV. In the SBLC design such an upgrade could be realized without extending the overall tunnel length. It would require doubling the number of klystrons and compressing the rf-pulse with a SLED system. Both means together provide the accelerating gradient of 34 MV/m needed to reach 1 TeV within the existing linac length. The price to pay though is a doubled overall power consumption.

3.6.6 Test Facilities

The goal of the SBLC test facility under construction at DESY is to construct and test the basic components required for the 2×250 GeV linear accelerators. The test linac consists of an injector providing bunch trains similar to those to be used in the collider. Two 150 MW klystrons (built by SLAC) power four 6-m-long accelerating structures. A beam diagnostics station is foreseen to measure bunch to bunch offsets as well as single and multibunch energy-spread. The injector provides a 6 A pulse out of a 90 kV gun with a duration of 2 ns. This pulse is compressed longitudinally by more than a factor of 200 resulting in a bunch length of ≈ 3 mm. Although longer than the final design bunch length, this value is sufficient to study most multibunch and beam dynamics effects. The subharmonic cavities and the vacuum system will be assembled and installed during 1994 and be commissioned early 1995.

Recently, the first klystron has reached its full design parameters in rf tests at SLAC. Conventional line-type modulators are foreseen for pulsing the klystrons. As an alternative solution a hard tube switching device is also under study at DESY. A klystron test stand for the 150 MW klystron with water loads and further required infrastructure is under construction.

The design of the accelerating sections concentrates on the first series production of 5.2-m-long structures for the DESY injector linac, which are very similar to the test facility structures. Nine-hundred cells for 6 sections (from overall 14) have been ordered from industry and are being brazed at DESY. Horizontal or vertical brazing, vacuum- or hydrogen-atmosphere ovens or inductive heating are still being investigated. A cup-tuning machine to match the structure for the accelerating wave after brazing and before final installation is now operating.

Low power test models of the symmetric high-power couplers have been manufactured and matched to an accelerating structure while the high power versions are scheduled autumn 1994. Different types of additional couplers required for the HOM damping and/or measurement of the beam induced higher order mode power are being investigated. After brazing and tuning, the sections for the test facility will be mounted on different types of temperature insensitive girders (glass ceramics, carbon fiber composite and heat shielded stainless steel) to keep the six meters of copper waveguide straight within 15 μm rms. Every girder is equipped with micromovers at both ends to allow for ± 1.5 mm offset in both directions.

Magnets, structure supports, and precision movers, as well as methods to compensate ground vibrations, are investigated. The design of the linac quadrupoles completely decouples the coil windings from the iron yoke, which automatically minimizes the vibration effects due to cooling water flow. Although the effects which have been described before will not affect the final emittance in the test accelerator, feedback systems and control loops will be tested on the supports and girders, which have been designed to fit the tunnel requirements and the minimum height for final installation.

3.6.7 Conclusion

The low-frequency approach of SBLC is well suited to achieve the performance goals of a next generation linear collider based on extensions of available technology. At the same time, it is also well suited for later upgrade programs to either higher luminosity or higher energy without enlarging the overall size of the machine.

3.7 TESLA: The Superconducting Radio Frequency Approach[10,11]:

3.7.1 The Basic Approach

All electron-positron linear collider designs are constrained by the need to limit the number of photons emitted during each beam-beam interaction. The practical result of this is that the luminosity is simply proportional to the ratio of the beam power to the beam area at the crossing point.

$$L = \text{const} \cdot [\mathrm{P_{beam}}/\mathrm{Area}] \tag{7}$$

The constant is determined by the number of photons deemed acceptable in a particular design. Copper-based designs need to minimize the required beam power by working with the smallest feasible beam cross section area. In this case, the AC power is naturally large because of the significant I^2R losses required to establish the accelerating gradient and the relatively low efficiency in converting mains power to beam power. There is also a premium in using as high a frequency as possible since that reduces the wall loss for a given accelerating gradient. At higher frequency, the conductors are closer to the beam making tolerances tighter and the emittance diluting wake effects stronger.

This situation is radically altered if the accelerating cavity walls are superconducting. In this case the wall losses are reduced by several times 10^5 and the efficiency of conversion of mains power to beam power can be relatively high, with the added dividend that conventional rf power supplies can be used. The low loss permits use of a lower frequency for the superconducting version with small power penalty, thereby ameliorating the wake effects. Further, because of the higher relative efficiency, it is possible to use more of a given, total, facility power in the beam, thereby allowing use of a relatively larger beam area at the crossing. A concomitant of the low wall losses and relatively large beam area is the ability to deliver the beam in long pulses with bunches separated on the order of a microsecond in headon collisions, with relatively long free space on either side of the IP, $L*$ being of the order of 3 m.

The needed emittances are achievable with damping rings in existence today and the demagnification required of the final focus has already been demonstrated at the FFTB experiment at SLAC. It should be noted that even in the TESLA case, the beam cross section at the crossing point will be smaller than in present practice at the SLC and of greatly reduced aspect ratio. A proposed layout for TESLA is shown in Fig. 7.

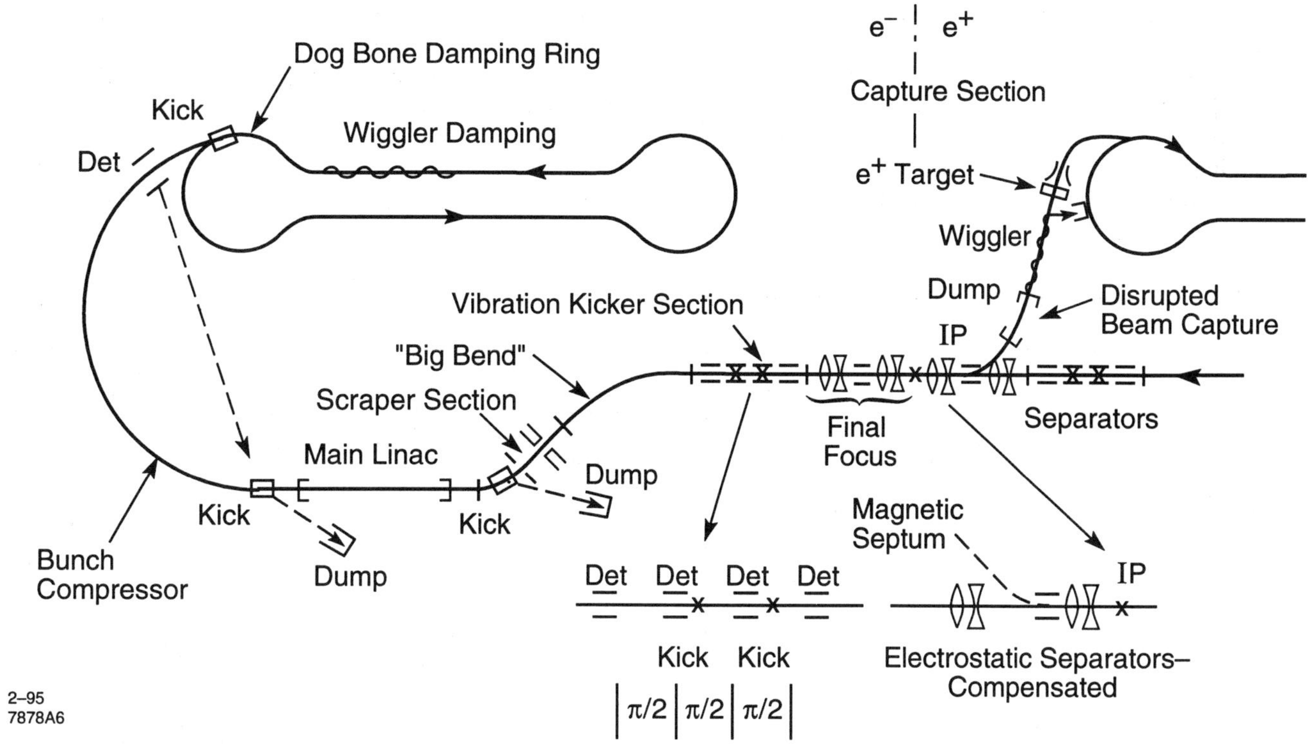

Figure 7: Shown is an example of a positron damping ring and a collision region that takes advantage of the long pulses and microsecond separation to give head-on collisions with an in-pulse feedback on bunch position and compensated electrostatic separation of the beams after the collision.

3.7.2 Critical Issues

While the potential advantages are manifest, technical realization will be challenging. There are both cost and technical factors to be considered. On the cost side the achievable gradients need to be raised from the 5 MeV/m typical of existing storage ring cavities to 20 or 25 MeV/m. The cost per unit length of the accelerating cavity plus cryostat needs to be lowered a factor of five to bring the price to about 50k\$/m to give a competitive cost per MeV figure. Another source of concern is the relative delicacy of the superconducting state. How reliable would such a system be in the face of the realities of vacuum and cryo system failures? Will it be possible to recover from such a failure by in situ processing? At the long rf wavelength, which is a strength of the SC approach, electrons present in the accelerator vacuum can be more easily captured from rest to make parasitic beams. These parasite beams could perhaps sap energy from the main beam and perhaps cause unwanted cryo losses and beam background in addition to instabilities of the wanted beam. Because of the very high Q of the cavities, small changes in dimension can make large changes in amplitude and phase of the accelerating wave. The pressure caused by the growing stored energy during the pulse can make such changes, the so-called Lorentz detuning. Can an economical cavity stiffening with active control scheme be found? Finally, the large beam power which is on the one hand a virtue implies the need for the most powerful positron source of all the approaches and gives the challenge of safe disposal of the high beam power in addition. Progress in all of these areas has been made and a plan to demonstrate solutions to many of them put in place.

3.7.3 Progress of the TESLA Test Facility

A 500 MeV TESLA Test Facility Linac (TTFL) is now under construction at DESY. It will consist of four cryomodules of a type that could be used in a linear collider. Each module contains eight 1 m, nine-cell cavity units, plus a focusing doublet assembly with beam monitors. Each cavity subunit has its own couplers. Two cryomodules are driven by one klystron/modulator set. An injector section brings the initial beam up to 15 MeV nominal before introduction into the TTFL proper. A beam analysis station will be installed both after the injector and at the high-energy end of the TTFL. The linac will be installed in Halle 3 at DESY adjacent to the recently completed cavity chemical processing area, with cleanrooms, vertical test cryostats, and rf power supply for cavity testing and high-power pulse processing (HPP). The cavities and the cryostat have been ordered for the first cryomodule, with delivery expected in early 1995. Initial cavity tests have begun. In a prototype processing and test setup at Cornell, several multicell cavities have been chemically processed and subjected to HPP.

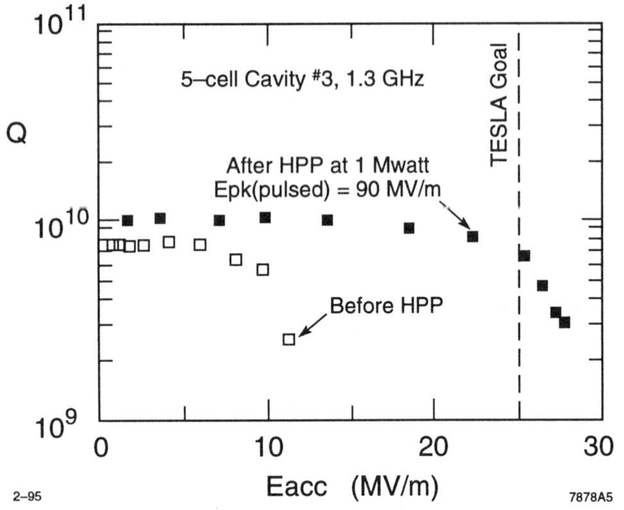

Figure 8: Q versus accelerating field before and after HPP.

Figure 8 shows the Q versus Field, CW, before and after HPP. It can seen that HPP is quite effective in raising the achievable gradient. After exposure to air, HPP was successful in recovering the gradient. How this will apply in a linac environment must await the completion and operation of the TTFL now expected in the 1997–1998 time frame. Experience with the facility will yield information about many of the critical issues cited above; for example, dark current, Lorentz detuning, cost, robustness against vacuum or cryogenic failures, and so on.

3.8 Results from the Final Focus Test Beam (FFTB) [12]

One of the challenges to the development of TeV-scale e^+e^- linear colliders is to make particle beams with extremely small sizes. Whereas the bunches in the SLC are millimeter-long needles a micron across, those in future machines will need to be ten times shorter and up to a hundred times narrower. Producing and colliding tightly focused beams requires careful control and stabilization of magnetic elements, and places considerable emphasis on accurate measurement of the properties of the beam itself. The FFTB Collaboration [13] has completed construction of a prototype focussing system for a future linear collider (see Fig. 9). This FFTB, which occupies 200 m in the straight-ahead channel at the end of the SLAC linac, is designed to accept the SLC electron beam as input and to produce a focal point at which the beam height is demagnified by a factor of 380, to a size smaller than 100 nm. Similar compression factors will be required for the final focus of TeV-scale linear colliders.

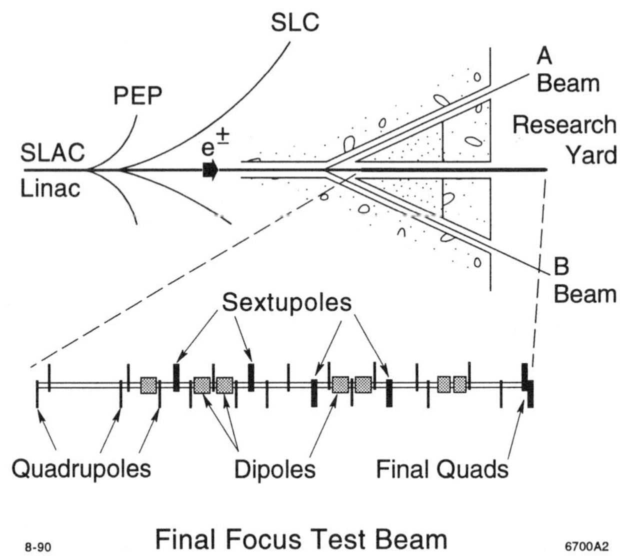

Figure 9: Location of the FFTB at the end of the SLAC 50 GeV inac.

Results presented here were obtained with 0.65×10^{10} electrons per pulse transported to the end of the linac with a vertical emittance of typically 2×10^{-6} rad m.

The nominal beam energy was 46.6 GeV, and the spread in particle energies was maintained at $\pm 0.05 - -0.1\%$. The optics of the FFTB are corrected to third order for geometric and chromatic aberrations, and are designed to reduce this beam to a spot with vertical height of 52 nm.

The FFTB contains five optical sections. The beam at the end of the linac is first matched to the lattice of the FFTB in a section that controls the launch of the beam into the FFTB, and contains quadrupole lenses, both normal and rolled, that are able to fully adjust the betatron space of the beam. Two sections that contain sextupole magnets at points of high dispersion allow the chromaticity of the lattice to be tuned separately in the horizontal and vertical planes. The chromaticity introduced by the focussing quadrupoles must be cancelled with approximately 1% accuracy by that generated in the sextupoles. Geometric aberrations are controlled with pairs of sextupoles placed at points of equal dispersion but spaced exactly π radians apart in betatron space. These aberrations must also be cancelled to ≈ 1%, so it is important to maintain the proper phase advance between the sextupole magnets. The lattice includes a "β–exchanger" to match the optics from one chromatic correction section to the other. This section contains an intermediate focal point at which the vertical beam height is reduced to 1 μm. The overall demagnification of the beam is determined by the focal lengths of the initial matching section and the final telescopic section. The basic principles of this scheme have been successfully demonstrated at the SLC, but the demagnification of the beam in the FFTB is an order of magnitude greater.

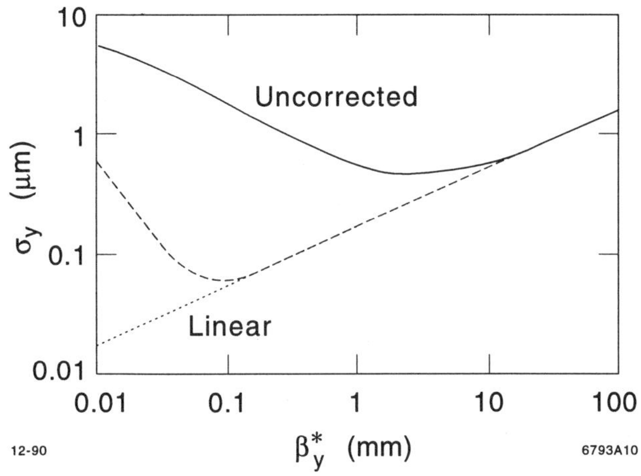

Figure 10: Vertical beam height at the FFTB focal point. The curves are uncorrected for chromaticity (solid), corrected for chromaticity (dashed), and the ideal monochromatic behaviour (dotted).

Two major third-order aberrations remain in the spot produced by the FFTB. Geometric aberrations are introduced as the beam envelope changes within the finite length of the sextupoles. Also, synchrotron radiation in the bend magnets destroys the exact cancellation between the chromaticity introduced by the sextupoles and that by the quadrupole lenses. An optimal design that minimizes the spot dilution from these two effects is used. Its performance is shown in Fig. 10. A similar optimization that includes the emittance growth due to synchrotron radiation in the bend magnets is done for the horizontal plane.

The optical design is tailored so that 28 of the 31 quadrupoles required to focus the beam are constructed as identical solid iron-core magnets each with an effective length of 46 cm and bore diameter of 2.3 cm operated with pole-tip fields below 10 kG. The design of the pole-tip contour limits the non-quadrupole field to less than 0.1% of the primary field at 70% of the full aperture. Tolerances on the lenses of the final doublet are more stringent with restriction on their harmonic content of 0.03%, and in some cases these magnets must operate with pole-tip fields as large as 14 kG. Permendur is used in the fabrication of the pole-tips of these magnets. There are four sextupole magnets in the chromatic correction sections of the FFTB. The field in each need only be pure to 1%. All FFTB magnets meet or exceed the requirements for strength and harmonic content.

Errors in the position or orientation of magnetic elements in the FFTB (with respect to the ideal beam line coordinates) can introduce anomalous dispersion or coupling into the beam phase space and can change the focusing of the optics. Alignment errors also introduce linkage between the correcting elements of the beamline.

Optical effects created by errors in the alignment of the main FFTB magnets can be corrected with vernier tuning elements as long as the alignment is within certain tolerances. Simulations showed that spot sizes could be created at the focal point that differ from the design value by only several percent, as long as the magnetic centers of the quadrupole and sextupole magnets are initially placed within 100 μm of their ideal horizontal and 60 μm of their ideal vertical positions.

Each quadrupole and sextupole magnet in the beamline is placed on a remotely-controllable support capable of translating laterally over a range of 1 mm in steps of 0.3 μm. The design uses a set of cam shafts driven by precision stepping motors. Shown in Fig. 11 is the response of a typical magnet mover. These devices are linear to a few parts per mil over their full range of motion, and are able to move quarter-ton magnets with submicron precision and little or no backlash. The precision of these movers makes it possible to use the beam to accurately align the FFTB magnets, and to simultaneously calibrate the response of the beam position monitors. Magnet-to-magnet alignments of 10–50 μm are estimated to be achieved with application of beam-based procedures.

Carbon filaments used in wirescanners are destroyed by thermomechanical stresses induced by submicron beams, so new instrumentation is required to measure beam profiles at the final focal point of the FFTB. Two devices were built for this purpose. One monitor uses the coherent interaction of the electron bunch with a gas-jet. Atoms of helium or argon, injected into the path of the beam, are ionized and trapped in the potential-well created by the passing electron bunch. Plasma oscillations, excited in the plane transverse to the beam direction, effectively transfer energy to the ions. The amplitude of the oscillations in the horizontal and transverse planes are proportional to the corresponding dimension of the electron bunch, so the azimuthal distribution of ejected ions reflects the aspect ratio of the bunch. The distribution and time-of-flight of the ejected ions are detected in a ring of multichannel plates that surrounds the focal point.

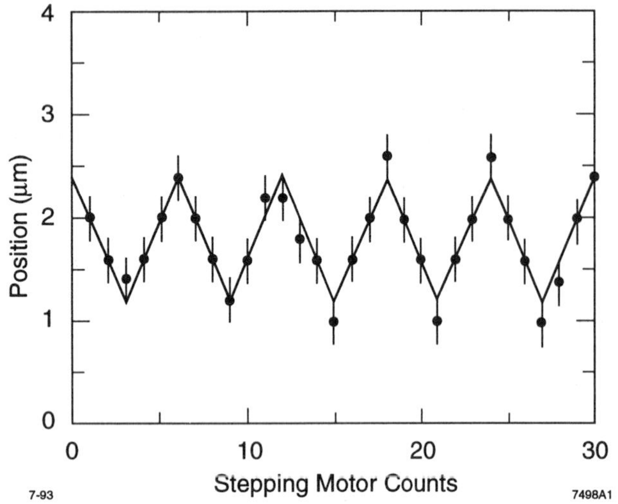

Figure 11: Measured response of a magnet mover. The solid line traces the input command value, and the points are the measured response of the mover. The errors on the points are the estimated resolution of the position readout.

A second monitor built for the FFTB focal point uses the concept of an optical cavity. A laser beam is split, and folded onto itself to produce an interference fringe pattern in space. The electron beam is scanned across this pattern to yield a modulated rate of Compton-scattered photons in the forward direction. The ratio of peak to minimum Compton rates depends on the size of the beam and the fringe spacing.

Time-of-flight signals and azimuthal distributions from the gas-jet monitor agreed well with theoretical expectations, and were used to optimize the position of the beam waist, and to adjust skew quadrupoles to remove astigmatic coupling introduced by the final quadrupole lenses. This monitor was able to measure beam heights from a few microns down to 100–200 nm, and was an important tool to make intial adjustments of the beamline optics.

Precise tuning of the smallest spots was achieved with the laser-Compton monitor. Iterative tuning of vernier knobs to minimize residual dispersion and coupling of the beam phase space resulted in beam heights $\sigma_y \approx 100$ nm. Further reduction was made by adjustment of trim sextupoles just upstream and at the same betatron phase as the final quadrupoles. An example of the measurement is shown in Fig. 12. The beam size measured by the laser-Compton monitor must be corrected for the mismatch between the length of the electron bunch, its divergence at the focal point, and the extent of the laser field along the beamline. This is estimated to be a 10% correction. Repeated measurements taken at the focal point over a period of several hours were distributed approximately Gaussian with corrected mean 70 nm and standard deviation 6 nm.

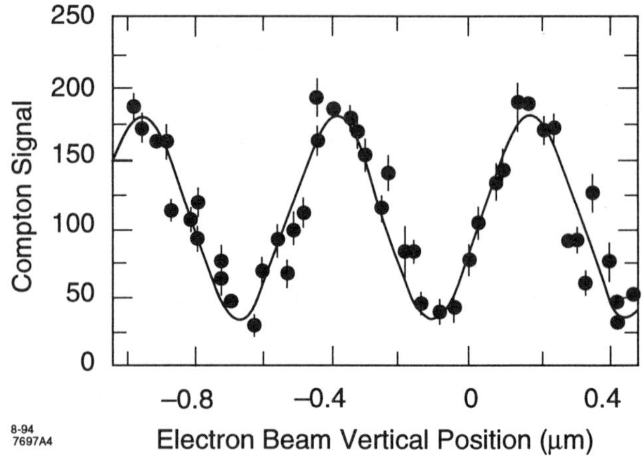

Figure 12: Measurement of the vertical height of the beam at the FFTB focal point with the laser-Compton spot monitor. Observed fringe spacing agrees well with the 0.5 μm expected from the wavelength of the laser. In this case, the beam height is 73 nm.

We conclude that we have focused the SLC 47 GeV electron beam through a demagnification $\approx$ 320 to a vertical height of $\sigma_y \approx 70$ nm. This represents a significant advance of technologies and accelerator physics required for the design and implementation of a future TeV-scale electron-positron collider.

3.9 The Next Linear Collider Test Accelerator [14]

3.9.1 Introduction

During the past several years, there has been tremendous progress on the development of the radio frequency (rf) system and accelerating structures for a Next Linear Collider (NLC). Developments include high-power klystrons, rf pulse compression systems and damped/detuned accelerator structures to reduce wakefields. In order to integrate these separate development efforts into an X-band accelerator capable of accelerating the electron beams necessary for an NLC, SLAC is building an NLC Test Accelerator (NLCTA). The goal of the NLCTA is to bring together all elements of the entire accelerating system by constructing and reliably operating an engineered model of a high-gradient linac suitable for the NLC. The NLCTA will serve as a test-bed as the design of the NLC evolves. In addition to testing the rf acceleration system, the NLCTA is designed to address many questions related to the dynamics of the beam during acceleration. In this section, we will report on the status of the design, component development, and construction of the NLC Test Accelerator.

Table 11: NLCTA rf System Parameters.

Parameter	Design	Upgrade
Linac unloaded energy gain	540 MeV	1080 MeV
Linac active length	10.8 m	10.8 m
Unloaded accelerating gradient	50 MV/m	100 MV/m
Injection energy	90 MeV	90 MeV
rf frequency	11.424 GHz	11.424 GHz
Number of klystrons	3	6
Klystron peakpower	50 MW	100 MW
Klystron pulselength	1.5 μs	1.5 μs
rf pulse compression power gain	4.0	4.0
Phase advance/cell	2 π/3	2 π
HOM suppression technique	Detuning	Detuning

In order to control the linac length of the NLC, current designs at SLAC and KEK use acceleration gradients which begin at 50 MV/m for the 0.5 TeV linear collider and finish with 100 MV/m in the upgraded 1 TeV or 1.5 TeV collider. These gradients are provided by an 11.4 GHz rf system (X-band). Although there has been experience with short X-band accelerators in industrial and medical applications, there are presently no high-gradient X-band accelerators in operation.

Much experience has been gained during the past several years with this radio frequency at SLAC and KEK. We have powered 11.4 GHz structures to reach peak surface fields in excess of 500 MV/m. Short traveling-wave accelerating structures have been powered to accelerating fields in excess of 100 MV/m. High-power klystrons have been constructed which reach 50 MW in pulses 1.5 μs long and 85 MW in pulses 200 ns long. We have constructed high-power rf pulse-compression systems that achieve a factor of four to five in peak-power multiplication. More efficient modulators are being developed. Finally, we are developing low-loss components for manipulation of high-power pulses of 11.4 GHz rf.

The NLCTA is primarily a high-gradient X-band linac consisting of six 1.8 m-long accelerator sections. These sections are fed by three 50 MW klystrons, which make use of SLED–II pulse compression to increase the peak power by a factor of four. This yields an acceleration gradient of 50 MV/m, so that the total unloaded energy gain of the beam in the X-band linac is 540 MeV. The NLCTA parameters are listed in Table 11. The right-hand column of Table 11 lists the parameters for an upgrade of the X-band linac to 100 MV/m by the use of six 100 MW klystrons.

3.9.2 Injector

The NLCTA injector will consist of a 150 kV gridded thermionic cathode gun, an X-band prebuncher, a capture section with solenoid focusing, and a rectangular chicane magnetic bunch compressor.

3.9.3 Radio Frequency System

The high-gradient accelerator will be fed with rf power through overmoded circular waveguides which penetrate the shielding blocks above the accelerator. Four 50-MW klystrons will be positioned along the accelerator, outside the shielded enclosure. Each klystron is powered by an independent modulator, allowing the flexibility needed for multibunch energy control and adequate power for an upgrade to a 100-MV/m accelerating gradient with six 100-MW klystrons, as indicated in Table 11. Each klystron feeds a SLED–II pulse compressor. The pairs of delay lines of the SLED–II pulse compressors are overlapped, parallel to the accelerator, outside the shielding. The output of each SLED–II is split to feed two accelerator sections. In the case of the injector, the SLED–II output is split to feed the two short injector sections to provide overhead for beam loading. The first two klystrons have exceeded NLCTA specifications (50 MW, 1.5 μs), and one is being used to test a complete NLCTA pulse compression system.

3.9.4 Accelerator Structure

To increase the luminosity of an NLC well beyond the minimum levels necessary for high-energy physics experiments, a train of bunches must be accelerated on each rf pulse. The primary impact of this choice is in the design of the rf structure. As each bunch traverses the structure, it excites wakefields which can remain until the next bunch passes. If this happens, each bunch resonantly drives all the bunches behind it. This leads to transverse multibunch beam breakup. However, beam breakup can be eliminated by choosing an rf structure in which the wakefields damp significantly between bunches.

There are two methods that can be used to achieve this damping. In the first, the higher order modes in the structure can be damped by coupling them to radial waveguides that are terminated with matched loads. This causes the energy to radiate out of the structure between bunches. The second technique involves changing the frequency of the higher order modes (HOMs) of each cell. Qualitatively, the total wakefield is then composed of a sum of wakefields, one from each cell. Behind the driving bunch, the wakefield decoheres because of the differing frequencies, and the net effect is a reduction of the wakefield. With this technique, the frequency distribution is important in determining the subsequent decay of the wakefield behind the driving bunch. Both of these techniques have been tested experimentally.

For the NLCTA, we plan to use a detuned structure which is a 2/3 "constant-gradient-like" structure modified every half meter to include four symmetric pumping holes. These holes lead to parallel vacuum manifolds, which provide sufficient pumping speed despite the small beam aperture. The cavities are machined to provide a precise mechanical reference from the inside dimensions to the exterior of the structure.

In order to achieve the reduced wakefield, the structure is configured to be very nearly constant gradient. The decoherence of the wakefield between bunches will be achieved by a Gaussian distribution of HOM frequencies with a standard deviation of 2.5%, which results in a Gaussian decay in time for the initial wakefield. This distribution can be obtained by tailoring a constant-gradient section so that more cells are near the central frequency, while fewer are near the ends of the frequency band. This choice results in a structure in which the iris size along the structure first decreases rather quickly, then decreases slowly in the middle, and finally decreases quickly along the structure towards the output end.

With this distribution of HOMs, the wakefield decoheres to less than 1% of its peak value. his decoherence is sufficient to eliminate beam breakup in the NLC or NLCTA. Because of the low injection energy, the NLCTA has a sensitivity to transverse wakefields comparable to the much longer NLC linac. The NLCTA will permit the verification that detuned structures can indeed suppress wakefields to the levels necessary for stable acceleration.

The first detuned accelerator structure has been completed and tested successfully at high gradient. In addition, the resulting wakefield was measured in ASSET in the SLC and was found to agree quite well with theoretical predictions. The second structure plus the two injector structures are presently in fabrication.

3.9.5 Beam Analysis

A magnetic spectrometer has been designed that will analyze the bunch train after acceleration in the linac in order to determine beam energy, beam-energy spread, and bunch-to-bunch offsets. The optics in the beam analysis region allow for the measurement of emittance in both transverse planes. A vertical kicker magnet upstream of the spectrometer provides a method for separating the bunches vertically so that the energy, energy-spread, and horizontal offsets can be independently measured along the bunch train. After initial commissioning, an extensive set of experiments is planned to verify that the NLCTA can indeed stably accelerate trains of low-emittance bunches suitable for a full-scale NLC.

3.9.6 Summary and Plans

The NLCTA is proceeding on schedule. The accelerator shielded enclosure is complete and all infrastructure is in place (girders, cable trays, water, lights, racks, and control room). All the magnets are complete and installed in the enclosure, power supplies are installed. The prototypes for the klystron and accelerator structure have performed at NLCTA specifications. Additional klystrons and structures are in fabrication. The prototype of the pulse compression system has been tested up to full power. The injector will be tested in the summer of 1995, and the full NLCTA should be complete by end of 1996.

3.10 Summary

The Electron Linear collider Facilities section gives a snap shot of the designs for a future e^+e^- linear collider. The detailed designs are developing very rapidly as the various test facilities are being constructed. Most designs and test facilities will reach completion around the end of 1996. In order to facilitate communication and cooperation in the above process, a Linear Collider World Collaboration has recently been established. The following is an excerpt from the Memorandum of Understanding:

3.10.1 Purpose and Guidelines

This Memorandum of Understanding (MoU) describes the Collaboration formed by the Signatory Institutions for the following purpose:

- to coordinate international Research and Development (R&D) efforts towards linear colliders,
- to review and monitor the status and progress of technical approaches towards linear colliders, and
- to facilitate the exchange of personnel among the participating Institutions, and to promote the sharing of research facilities among the Participating Institutions.

This Collaboration will carry out its business according to the following guiding principles:

- The Collaboration is open to any Institution which agrees to make significant Contributions toward the development of linear colliders.
- This Collaboration is not intended to displace or supersede any existing organization.
- The Collaboration recognizes the requirement that R&D towards future linear colliders be conducted without prejudice to potential sites of construction.

- The Collaboration anticipates that future linear colliders will be constructed by international coalitions.

The Linear Collider World Collaboration is now in the process of conducting its first technical review of linear colliders. The following is the charge to the Technical Review Committee:

Charge to the 1994/1995 Technical Review Committee

The Technical Review Committee is to consider the goal to design, build, and operate a TeV-scale linear electron-positron collider capable of satisfying the need to explore the particle physics of this energy range. Specifically the Committee is to examine accelerator designs and technologies suitable for a collider that will initially have center of mass energy of 500 GeV and luminosity in excess of 10^{33} cm^{-2} s^{-1} and be built so that it can be expanded in energy and luminosity to reach 1 TeV center of mass energy with luminosity 10^{34} cm^{-2} s^{-1} The Committee should consider construction and operation of both the initial facility and the upgrade path to 1 TeV. The Committee is also asked to comment on the potential of technologies to reach higher energies and luminosities, and to provide alternative physics capabilities, for example gamma-gamma collision.

The Technical Review Committee is to identify the accelerator physics issues and technological requirements for each approach to provide particle physics opportunities at the energy and luminosity goals stated above. The report of the Committee should contain a brief summary of the status of and expected progress toward understanding and achieving the most important of these requirements. The Committee should attempt to identify areas of possible further collaboration in the world-wide linear collider R&D program.

A draft of the Committee report should be submitted to the Collaboration Council shortly after the LC95 meeting scheduled for March 1995 in Japan.

We note that the Technical Review Committee is beginning its work now and will have a draft report around the time that the Proceedings of this Workshop is published. We anticipate their report will give a more detailed and indepth assessment of the status of the world linear collider effort.

4 Electron–Positron Circular Collider Facilities

4.1 Operating Colliders and Future Plans

Currently, there are five e^+e^- circular colliders (VEPP–2M at the Budker Institute in Novosibirsk, Russia, BEPC at the Institute of High Energy Physics in Beijing, CESR at Cornell University in Ithaca, New York, TRISTAN at KEK in Tsukuba, Japan and LEP at CERN in Geneva, Switzerland) operating in the world. The total center-of-mass energy for the beams in these colliders ranges from the ϕ mass (1.020 GeV) to the Z mass (91.2 GeV). For circular colliders, the emphasis is on luminosity and on providing a very clean background environment for the detectors. Upgrade plans for VEPP–2M, BEPC, and CESR are to increase their luminosities while the operating energy of LEP will be increased to above W^+W^- threshold by the installation of approximately 200 multicell superconducting rf accelerating cavities. The luminosity of LEP will also be increased by a more aggressive pursuit of multibunch operation with pretzels with initial operation planned for 1996.

Future plans are for the completion of the construction of PEP II, a B Factory at SLAC in Stanford, California and KEK B, a B Factory at KEK. Both machines are being constructed in existing tunnels at these laboratories and will operate with asymmetric beam energies to provide a significant Lorentz boost for the center of mass in the laboratory frame. This makes it possible to observe the time evolution of the B^0 decays from the $\Upsilon(4S)$ and thereby the violation of CP for the first time outside the K system. At Frascati in Italy, a ϕ Factory, DaΦne is under construction with initial operation expected in early 1996. At Cornell, CESR's luminosity is being increased to the region of $10^{33}\,\mathrm{cm}^{-2}\,\mathrm{s}^{-1}$ to provide a large data sample at the $\Upsilon(4S)$ for the study of rare B decays in an upgraded CLEO detector. The remainder of this short chapter will be a small expansion of these remarks on each of the colliders.

4.2 Facilities

4.2.1 Phi Factories

VEPP–2M at the Budker Institute in Novosibirsk has been operating for the past several years with a peak luminosity of 5×10^{30} cm^{-2} s^{-1} and currently has the highest luminosity of any collider in this energy region. An upgrade in luminosity is planned for the end of 1995 by installing a solenoidal final focusing system. This will permit operation with round beams and has the potential of increasing the luminosity by a factor of 20. It will be the first extended operation of round beams in a circular e^+e^- collider.

Table 12: DaΦne Main Features.

Initial luminosity	1.3×10^{32} cm $^{-2}$ s^{-1}
Luminosity in 3 years	$\times3$ increase
Interaction regions	KLO, CP violation
	FINUDA, hypernuclei
Circumference	97.69 m
Bunches	initial: 30, final:120
Particles/bunch	8.9×10^{10}
Radio frequency	368.25 MHz
Cavity design, single cell	Low HOM impedance
Feedback	DSP-based longitudinal similar to PEP II
Initial operation	Early 1996

DaΦne is being constructed at Frascati in buildings previously occupied by ADONE, one of the first successful modern e^+e^- colliders, which was retired in April 1992. The main features of DaΦne are listed in Table 12.

4.2.2 Tau Charm Factories

The collider at the High-Energy Institute in Beijing, BEPC continues to be the premier facility in this energy region. An upgrade of the luminosity of the collider is planned and an improvement program for the detector is currently underway. The projected peak luminosity performance is 3×10^{31} cm^{-2} s^{-1}, which will be an increase of approximately a factor of 5 over present peak performance.

A project to construct a high-luminosity collider in Spain by a consortium of European and US physicists is currently on indefinite hold and is likely to remain so. As a result, BEPC will continue to be the facility for this energy region for the foreseeable future.

4.2.3 CESR Luminosity Upgrade

The phased upgrade of CESR to peak luminosities of 10^{33} cm^{-2} s^{-1} is making good progress. Operation with ±2 mrad horizontal crossing angle for the two beams is now routine with beam-beam tune shifts of 0.036 compared with 0.042 for the best head-on operation. Peak luminosities of 2.5×10^{32} cm^{-2} s^{-1} are common at the start of collisions and improving steadily as unwanted multipoles from field errors in special magnets are corrected. With head-on collisions, the best peak luminosity was 3×10^{32} cm^{-2} s^{-1}. Installation of large-aperture final-focus quadrupoles and beam tubes along with the new silicon strip vertex detector in CLEO in early 1995 will make operation much easier and allow for bunch train operation with low backgrounds. Initial test runs with two bunches per train (position of original single bunch) spaced by 28 ns yielded peak luminosities of 2.4×10^{32} cm^{-2} s^{-1} with acceptable detector backgrounds. The total beam current in these tests was comparable to the normal operating currents. For the last month of $\Upsilon(4S)$ running in 1994, two bunches per train was the normal operating mode and the monthly integrated luminosity was the second highest recorded at CESR.

The key to the success of the upgrade plan is the lowering of the higher-order mode impedance of the ring. An initial beam test in CESR in August 1994 of a 500 MHz single cell superconducting rf cavity with all higher-order modes well damped went well. The cavity delivered 155 kW of power to the beam and achieved an accelerating gradient of 6 MV/m. A current of 220 mA was stored with the cavity powering the beam in conjunction with the conventional rf system. By operating CESR at 4.5 GeV the single cell cavity was able to store beam with the conventional rf system detuned. Operation of superconducting cavities in the harsh environment of a high-current storage ring should be similar to that of conventional cavities on the basis of this very successful test.

The elements of the upgrade appear to be in hand, and what remains to be achieved is successful integration of these elements into a smoothly working system. A peak luminosity of 10^{33} cm^{-2} s^{-1} by 1998 seems possible.

4.2.4 *B* Factories

The luminosity frontier will be the main thrust of these new colliders operating with a center-of-mass energy at the $\Upsilon(4S)$. The goal is to produce as many B mesons as possible in the clean environment of an e^+e^- collider. These colliders have the added advantage of operating with asymmetric beam energies to boost the B's from the $\Upsilon(4S)$ decay so that the time dependence of the B decays can be determined to study CP violation. The clean environment will enable very sensitive tests of rare B decays. The two projects represent the future of circular colliders and promise a rich physics return.

PEP II at SLAC in the US is well underway with completion of the machine scheduled for fall 1998. The Japanese project at KEK, KEK B, is receiving significant funding this fiscal year with an agreed upon plan that calls for commissioning of the storage rings in late 1998. Both machines use as much of existing infrastructure as possible to minimize the cost of the facilities. Because of the large currents and large numbers of bunches, both designs work hard to minimize the higher order mode impedance of the beam environment and feature state of the art feedback systems. A short summary of the main machine parameters is given in Table 13. For details, the reader is referred to the latest versions of the design reports. Achieving these parameters will be a challenge.

Table 13: Selected PEP II and KEK B Parameters.

	PEP II		KEK B	
Parameter	e^+	e^-	e^+	e^-
Beam energy (GeV)	3.1	9.0	3.5	8
Circumference (m)	2199		3016	
Current (A)	2.14	1.48	2.6	1.1
No. bunches	1658	1658	5120	5120
Bunch length (cm)	1.0	1.0	0.32	0.4
β_V/β_H (IP, cm)	$\frac{1.5}{37.5}$	$\frac{3.0}{75}$	$\frac{1.0}{33.0}$	$\frac{1.0}{33.0}$
Crossing angle (mrad)	0		±10	
Beam-beam tune shift	0.03	0.03	0.05	0.05
Luminosity (10^{33} cm^{-2} s^{-1})	3.0	10.0		

Table 14: Summary of machine parameters for LEP 200.

Luminosity	7×10^{31} cm^{-2} s^{-1}
Circumference	26.7 km
Bend radius	3.28 km
Maximum beam energy	90 GeV
Superconducting rf cavities	192
Radio frequency voltage	1900 MV
Radio frequency power	15.6 MW
Bunches	8
Bunch current	1 mA

4.2.5 LEP 200

Adding 192 superconducting cavities to LEP to achieve a center-of-mass energy above the W^+W^- threshold will approach the upper limit of what is sensible for the maximum energy of a circular collider using electrons and positrons. Above this energy, the trade off between circumference and total length and voltage gain of the rf system forces a very large cost for the storage ring. Linear colliders are the sensible approach above LEP 2 energies.

The present plan for installation of the superconducting cavities in LEP calls for operation at energies above the W^+W^- threshold in 1996. Because the cross section for W pair production is significantly smaller than the cross section at the Z, a luminosity upgrade is also planned. This will make use of pretzel orbits to increase the number of bunches from 4 to 8. The integrated luminosity goal is to accumulate 500 pb^{-1} (1 pb^{-1} = operation of the collider at a luminosity of 10^{32} cm^{-2} sec^{-1} for 10^4 seconds) by the end of 1999. A short summary of the main machine parameters is given in Table 14. The latest accelerator conference reports should be consulted for details.

5 Advanced Accelerators

Cost played a pivotal role in the demise of the SSC and is a major issue in the LHC. Its importance will increase in the future. As the energy scale of the phenomena being studied rises, it is more than likely that the luminosities needed for precision measurements will rise as well. No technical schemes for producing energies and luminosities much beyond LHC and NHC are in hand. Thus we are being challenged both economically and technically to find acceptable vehicles for carrying us to the multi-TeV energy and precision measurement frontiers.

Confronted with this situation in which clear directions have not emerged, it behooves us to proceed on as broad a front as possible in seeking solutions for both accelerator and detector systems at the next frontiers. In carrying out such a program there are three principal directions in which to go. First there is the approach that has served us well for almost half a century: development of improvements to proton synchrotrons and microwave electron linear accelerators. Next on the list is the introduction of completely new systems for acceleration such as those employing lasers, high-current relativistic beams, etc., in which one harnesses technical components that are available from collateral fields or must be especially developed for the purpose. The third approach which has also received attention of late is to consider the collisions of different particles, for which the physics goals and design constraints can be different. Muon and gamma-gamma collisions have been suggested as possible niche accelerators for different purposes and are under continuing investigation.

As we take stock of the current situation and try to lay out strategies that will take us to the frontiers beyond LHC and NHC, we need to keep in mind that our challenge involves a system which must meet cost and many technical requirements simultaneously. What this tells us is that proceeding along any new avenues needs to be disciplined by system studies, including the detectors, from early on. In this way the strengths and weaknesses of potential approaches, including operating and capital cost dimensions, can be confronted and used to guide the evolving R&D.

5.1 Improvements In Conventional Acceleration

5.1.1 Introduction

A natural approach to an advanced accelerator is to scale down the dimensions and rf wavelength (λ) of a conventional structure. This could be a good idea for several reasons. First, at fixed peak power per unit length (p) or fixed rf pulse energy (u) there is a gain [15] in accelerating gradient (E_a): $E_a \propto p^{1/2}\lambda^{-1/4}$ $E_a \propto u^{1/2}\lambda^{-1}$ Second, the limit imposed by electrical breakdown of the structure walls (E_b) increases [16] : $E_b \propto \lambda^{-1/2}\tau^{1/4} \propto \lambda^{-7/8}$ where τ is the filling time ($\leq 1\ \mu$s).

Many designs based on scaling linacs to higher frequency are being pursued for the next linear collider; they include collaborations on DLC (DESY Darmstadt), JLC (KEK, Japan) and SLAC–NLC (USA). (The TESLA collaboration is exploring a superconducting structure at lower frequency and gradient [17]).

Two big issues for extending such schemes to the ≥ 100 MeV/m level are: 1) developing efficient power sources matched to the structure's requirements on peak power, frequency, and pulse length; and 2) avoiding breakdown/dark current.

5.1.2 Power Sources

There is a world-wide effort on alternative power sources including klystrons, gyroklystrons, EIKs (extended interaction klystrons), TWTs, twystrons, magnicons, gyrotrons, and FELs (free electron lasers) [18]. 11.4 GHz klystrons producing > 50 MW for 0.5 μs with 44% efficiency have already been used to generate 100 MeV/m gradients in an unloaded structure at SLAC [19,20].

Novel concepts such as the cluster klystron [21] and the sheet-beam klystron [22] may significantly improve the overall rf system efficiency by operating at lower beam current densities, thereby permitting the use of highly optimized extraction circuits which should yield high (approaching 60–70%) rf efficiencies. Moreover, in some systems, modulating grids or nonintercepting modulating-anodes (magnicon injection guns) would permit rapid modulating of the klystron beam pulses, thereby permitting the use of DC power supplies and a concomitant increase in modulator efficiency.

5.1.3 Two Beam Accelerators (TBA's)

An alternative power source that might represent a significant simplification is the two-beam accelerator (TBA) concept [23]. In this scheme a high-current (drive) beam runs parallel to the accelerated beam in a separate structure. The drive beam produces radiation via a free electron laser mechanism [Lawrence-Berkeley Laboratory, USA approach] or a relativistic klystron interaction (CLIC, CERN and LBL approach) [24]. This radiation is diverted via output couplers to the high-gradient structure.

The drive beam is periodically re-accelerated (for example, via induction accelerator units or superconducting cavities). This scheme takes advantage of the high peak power and efficiency possible with a relativistic driver (*e.g.*, 1000 MW at 34% efficiency were demonstrated in an FEL at LBL) [25]. The re-acceleration of the drive beam overcomes the need for thousands of separate power sources.

Recent work on TBA's has included extensive modeling of the tolerance on drive beam quality in order to maintain control of rf phase [26], fabrication of scaled high-gradient structures for testing with conventional power sources [27], and development of a new theoretical model that allows comparison of the relativistic klystron and FEL approaches to a TBA [28]. Simulations have shown that an FEL in a standing wave structure enables adequate control of rf phase, if the structure is also designed to suppress BBU instabilities. It has been shown that since the frequency generated by the relativistic klystron interaction is dependent on beam speed rather than energy, the tolerance on energy spread ($\delta\gamma/\gamma$) is relaxed at high γ compared to the FEL interaction [24]. The final choice will depend on how such advantages trade off against the added difficulty of accelerating a bunched drive beam to high γ. Experiments at the CLIC test facility have already reached a 50 MeV/m gradient with a relativistic klystron TBA.

5.2 Advanced Acceleration Methods

5.2.1 Wakefield Accelerators

One step simpler than the TBA's, conceptually at least, are the wakefield accelerator schemes. Here the drive beam excites a wakefield in a structure or a medium (*e.g.*, dielectric or plasma), and the high-energy beam is directly accelerated by this wakefield (*i.e.*, without shunting the wakefield off to a separate structure). In order to overcome a fundamental wakefield theorem [29] and obtain a large transformer ratio (ratio of energy gained per trailing particle to energy per driving particle), various geometries have been proposed. These include noncollinear or hollow drive beams and shaped beams [30] with slowly rising and sharply falling current profiles [31]

Various proof-of-principle experiments were performed in the late 1980's [26,32,33,34]. At Argonne National Laboratory (ANL, USA) Jim Simpson's group used test particles to successfully confirm theoretical predictions of wakefields in hollow dielectric tubes [28] and in plasmas [29]. A multibunch driver experiment at KEK (Japan) led by A. Ogata accelerated trailing particles by 10 MeV over .75 m in a plasma [30]. Currently experiments are planned at ANL and UCLA (USA) to extend the experiments to high-gradients (up to 100 MeV/m) and non-negligible transformer ratio. The early ANL work showed that BBU instability and charging of the walls are critical issues in dielectric tubes.

To overcome these, a design has been developed by which the driving and accelerating beams travel through different tubes. This scheme then resembles the TBAs of the previous section.

5.2.2 Laser Near Field Accelerators

A natural approach to achieving acceleration at laser wavelengths is with miniaturized conventional or unconventional slow-wave structures. Although new micromachining technology using lithographic techniques or open structures may make these feasible to build, some novel approach is still required to make the transverse wakefields (scaling as $1/\gamma^3$) small enough to avoid instabilities and degradation of beam quality. Furthermore, electrical breakdown is obviously an issue. Assuming for the moment that these problems could be overcome, R. Byer has described a scenario for a 1-km-long TeV accelerator with 10^{10} electrons and a rep rate of 120 Hz [35]. To accommodate the beam load and avoid surface damage requires a rectangular slow wave structure 3 cm wide by 1 μ high. The laser needed to fill a 10 cm length of this structure would have a kilowatt of average power. Byer points out that such a laser could be commercially available in four years; and diode pumped solid state lasers already operate with wall plug efficiency of 10%.

5.2.3 Inverse Cerenkov Accelerators

An alternate approach to coupling particles to parallel component of laser electric field is the Inverse Cerenkov Accelerator (ICA). In the ICA the parallel component of E is created by tilting the laser at a slight angle Θ to the particle beam direction in a gaseous medium. By choosing Θ to be the Cerenkov angle ($\Theta = \cos[1/n]$, where n is the refractive index of the gas), the particles remain in phase with the (slowed) laser and can gain energy at a rate $q\vec{v}\vec{E} \approx qcE\sin\Theta_c$. The perpendicular component of E can be nearly canceled (leaving a modest net focusing force) by using an axicon geometry [36]. In a recent experiment led by W. Kimura at Brookhaven National Laboratory (USA) [37], this technique was used to accelerate test particles with a .7 GW CO_2 laser ($\lambda = 10.6\ \mu$) in an H_2 gas at 2.2 atm (Θ = 20 mrad). Energy gain from 40 MeV to 43.7 MeV over 12 cm was measured for a gradient of 31 MeV/m in agreement with their theoretical model.

Critical issues for this scheme are how to avoid gas ionization by the laser (limiting E), how to maximize n, and hence Θ and $E_{\|}$, without increasing gas pressure and consequent particle beam scattering (*e.g.*, by taking advantage of atomic resonances), and how to stage (by re-focusing the laser beam).

5.2.4 Inverse FEL

There is no limit from breakdown of fields at a focus in a vacuum, but such fields do not accelerate an unperturbed particle. If the particle is bent or wiggled by an external magnetic field, then acceleration is possible.

For fixed field it would be Inverse Synchrotron Acceleration; for a wiggling field it is an Inverse Free Electron Laser (IFEL) [38]. Since, in either case, there need be no structure near the beam, there are no structure induced wakefields; and, since the magnetic fields can be essentially invariant with transverse position, there are no other transverse wake fields. So, unlike most other advanced acceleration schemes, an IFEL is peculiarly suitable for accelerating very small emittances. Unfortunately loss of energy from direct synchrotron radiation sets an effective limit on the maximum energy achievable. For electrons, this limit is at only about 50 GeV [39]; *i.e.*, at an energy already obtained, but for muons the limit would be much higher. An experiment to demonstrate significant IFEL Acceleration is in preparation [40] at the BNL Accelerator Test Facility (ATF) using a CO_2 laser and a pulsed magnetic wiggler.

5.2.5 Plasma Acceleration

Another approach to utilizing the high peak power of lasers is to use them to drive longitudinal space charge waves in a plasma. Plasma waves are accelerating structures that support large parallel electric fields and are immune from electrical breakdown. The plasma waves can be driven resonantly by the radiation pressure of a train of laser pulses [41,42], separated approximately by a plasma period $2\pi/\omega_p$, where $\omega_p = [4\pi n_0 e^2 / m]^{1/2}$ as in the beat wave accelerator or by a single short ($\approx \pi/\omega_p$) pulse as in the laser wakefield accelerator [37,43]. In the beatwave scheme the accelerating gradient scales,

$$E_{\|} = \frac{\sqrt{n_0}}{4} \int_0^{\tau} \alpha_1 \alpha_2 \omega_p \, dt, \tag{8}$$

where $\omega_{1,2} = eE_{1,2}/m\omega_{1,2}c$ is the normalized oscillatory velocity in the laser field, $\omega_1 - \omega_2 = \omega_p, n_0$ is in cm^{-3} and τ is the smaller of the laser pulse length, the relativistic detuning time ($\approx 7[\alpha_1\alpha_2]^{-2/3}/\omega_p$) and the time scale for ion instabilities (a few times the ion plasma period $\approx 43 \times 2\pi/\omega_p$).

Several groups including C. Joshi *et al.* at UCLA (USA), A. E. Dangor *et al.* at Rutherford (UK), F. Amiranoff *et al.* at Ecole Polytechnique (France) and Y. Kitagawa *et al.* in Osaka (Japan) have performed beat wave experiments in the last decade, successfully demonstrating the generation of plasma waves with longitudinal fields of 1–3 GeV/m and phase velocity [44] $\approx c$. Recently, UCLA [45], Osaka [46], and a group at the National Research Council (Canada) [47] have also reported acceleration of injected particles.

The UCLA group accelerated $\sim 10^5$ test particles (approximately 1% of the randomly phased injected particles from 2 MeV to up to 30 MeV over 1 cm in a plasma of density 8×10^{15} cm^{-3}. This corresponds to a gradient of 3 GeV/m. The energy gain was limited by the focal depth of their CO_2 lasers. The experiment confirmed earlier simulations and theoretical work predicting that competing instabilities could be avoided by employing short laser pulses (compared to an ion plasma period $\approx 43\times 2\pi/\omega_p$) of sufficient amplitude ($eE/m\omega c \geq 0.1$). Based on these results, it is possible to extrapolate the UCLA design to a 1 GeV experiment over ~ 10 cm using T-class lasers beating in a 10^{17} cm^{-3} density plasma.

The critical issues for laser-plasma accelerators beyond 1 GeV include diffraction and staging; long time scale instabilities and pump evolution [48], dephasing [49], beam loading; beam quality, and efficiency [50]. Preformed plasma channels have been explored recently both theoretically [51,52], and experimentally [53] as a means of confining the laser for many diffraction lengths as well as for creating accelerating and focusing fields that are optimal for high beam quality.

5.3 Alternative Particles

5.3.1 Linear Colliders with γ–γ Collisions

There exist strong arguments that the particle physics one can study with a linear electron-positron collider is special and complementary to that of a hadron supercollider. Similarly the study of electron-photon and photon-photon collisions would provide unique access to some areas of fundamental physics plus some not unwelcome redundancy with measurements obtained from electron-positron collisions. Physics goals for a gamma-gamma collider are significant indeed, and are presented below.

Initial studies reveal no show-stoppers to γ–γ and e–γ colliders. Therefore, it would be prudent that the conceptual design of a linear collider include multiple interaction regions, one of which is dedicated to e–γ and γ–γ collisions. Given that e^+e^-, e^-–γ, and γ–γ colliders are the same except for the interaction region, the incremental cost of adding gamma collisions to the capabilities at one of the interaction regions is relatively small.

The R&D required for e–γ, and γ–γ colliders, in addition to that for the e^+e^- machine would be:

- detectors and masking;
- high-power lasers, including FEL's;
- special final focus components;
- bright sources of polarized electrons; and
- high-power, low-loss optical components

One initial step, for example, could be a Photon Linear Collider with a center-of-mass energy of 80–180 GeV. This collider will be less expensive than other variants (less energy, positrons are not necessary), but it would have high discovery potential in Higgs physics, etc.

5.3.2 Physics Goals for a γ–γ Collider

The structure functions of the photon, probed by deep inelastic scattering from a photon target, is a fundamental and largely unresolved area of investigation in quantum chromodynamics. Clearly the electron-photon collision option would provide the paramount facility for these studies. In another important area of QCD, photon-photon collisions would allow studies of the op quark threshold region that would complement studies performed in electron-positron collisions. Here some unique measurements are possible using polarized photon beams: large circular polarization will allow direct observation of p-wave toponium, not possible in electron-positron collisions, while with linear polarization it may also be possible to make very sensitive measurements of the strong coupling constant.

Study of W boson pair production in photon-photon collisions provides the most sensitive tests for quartic anomalous interactions of the electroweak gauge bosons. The photon-photon option also provides some unique advantages for Higgs boson studies. The two photon width of a Higgs boson is most directly measured here; it is a fundamental probe both of the electroweak theory and for electrically charged ultraheavy quanta which would affect the two photon rate if their mass is generated by the Higgs boson. The search for supersymmetric Higgs bosons is also enhanced by the photon-photon option. The heavy scalar and pseudoscalar of the minimal supersymmetric model must be produced together in the same event in electron-positron collisions, requiring energy greater than the sum of their masses, but in photon-photon collisions they can be produced and observed individually. The circular polarization of the photon beams is an important asset in these studies, both enhancing the signal and suppressing the background. Linear photon polarization may also be useful, since it would allow direct measurement of the properties of Higgs bosons, which might not be directly measured in any other way.

5.3.3 μ^+–μ^- Colliders

Lepton colliders produce simple interactions essential in the exploration of new particle states. Extension of e^+e^- colliders to multi-TeV energies is performance-constrained by radiation and "beamstrahlung" effects, which increase as $(E_c/m_c)^4$ and cost-constrained by the need for two full-energy linacs [54]. However muons

Table 15: Parameter list for a 4 TeV $\mu^+\mu^-$ Collider.

Parameter	Symbol	Value
Energy/beam	E_μ	2 TeV
Luminosity	$L = \dfrac{f_0 n_s n_b N_\mu^2}{4\pi\sigma^2}$	$3\times10^{34}\,\mathrm{cm^{-2}\,s^{-1}}$
Source Parameters		
Proton energy	E_p	30 GeV
Protons/pulse	N_p	$2\times3\times10^{13}$
Pulse rate	f_0	10 Hz
μ-production acceptance	μ/p	0.15
μ-survival allowance	$N_\mu/N_{\mathrm{source}}$	0.33
Collider Parameters		
Number of μ/bunch	$N_{\Delta\pm}$	1.5×10^{12}
Number of bunches	$4n_B$	1
Luminosity turns	n_s	900
Normalized emittance	ϵ_N	3×10^{-5} m-rad
μ-beam emittance	$\epsilon_t = \epsilon_N/\gamma$	1.5×10^{-9} m-rad
Interaction focus	β^*	0.3 cm
Beam size at IP	$\sigma = \sqrt{\epsilon_t}\,\beta_0$	2.1 μm

(heavy electrons with $m_\mu = 200\ \mathrm{m}_e$) have negligible beamstrahlung and can be accelerated and stored in rings. The liabilities of muons are that they decay, with a lifetime of $2.2\times10^{-6}\ E_\mu/m_\mu$ s, and that they are created through decay into a diffuse phase space. But a study [55] suggests that a $\mu^+\mu^-$ collider, with an energy of $E_{\mathrm{cm}} = 2E_\mu = 4$ TeV, and luminosity of $L = 3\times10^{34}\,\mathrm{cm^{-2}\,s^{-1}}$, might be possible using only existing technical capabilities.

The possibility of muon ($\mu^+\mu^-$) colliders has been introduced by Skrinsky *et al.*, [56] and Neuffer [57]. More recently, several mini-workshops have greatly increased the level of discussion [51,58,59,60]. Palmer and Neuffer [51] have introduced improvements, and developed a complete scenario. Table 15 shows their parameters. The design consists of a muon source, a muon collection, cooling and compression system, a recirculating linac system for acceleration, and a full-energy collider with detectors for multiturn high-luminosity collisions.

The μ-source driver is a high-intensity rapid-cycling (10 Hz) synchrotron, similar in scope to the proposed 30 GeV facility, KAON [61]. Two bunches of 3×10^{18} protons each are accelerated and extracted into separate lines for μ^+ and μ^- production. (Separate lines permit use of higher-acceptance $\pi\to\mu$ capture lines.) Each bunch collides into a target, producing π's ($\sim 1\pi$/interacting p) over a broad energy and angular range. The target is followed by Li lenses which collect the πs into a large-aperture, strong-focusing transport line, which has a large energy width and angular acceptance. This line is sufficiently long to insure $\pi\to\mu$ decay, plus bunch lengthening. Bunch lengthening is followed by a nonlinear rf acceleration system which implements an rf rotation, in which the bunch is lengthened while the energy spread decreases. The μ beam is then matched into a beam cooling system. From Monte Carlo simulations, they obtain ~ 0.2 captured μs per initial proton.

For collider intensities, the phase-space volume must be reduced. This is obtained by "ionization cooling" of muons which is described in detail in references [52,53,62]. In ionization cooling, beam transverse and longitudinal energy losses in passing through a material medium are followed by coherent reacceleration, resulting in beam phase-space reduction [63]. The coherent cooling is opposed by the randomizing (heating) effects of multiple scattering and energy straggling. To minimize energy straggling, cooling is done at low relativistic energies ($E_\mu \sim 300$ MeV), in a strong-focusing lens which maintains small beam size over extended lengths, and a low-Z material. A series of about 20 Li current-carrying rods, where the high current provides strong radial focusing, alternating with linear accelerators provide adequate transverse cooling.

Phase-space exchange in wedge absorbers is used to balance the longitudinal and transverse cooling to design parameters. In this scenario, they cool only with ionization cooling in conducting rods, and that is sufficient. However, other techniques, such as ionization cooling in focusing transports or rings, or optical stochastic cooling [64], may permit improvements.

Following cooling and preacceleration, the beams are accelerated to full energy (2 TeV). A full-energy linac would work, but it would be expensive and does not use our ability to recirculate μ's. A recirculating linac similar to CEBAF [65] could obtain acceleration to full energy in 10–20 recirculations, using only 200–100 GeV of linac. After acceleration, the μ^+ and μ^- bunches would be injected into the 2-TeV superconducting storage ring ($\sim$ 1 km radius), with collisions in two interaction areas. The beam size at collision is focused to $\sigma \approx 2\ \mu$m, similar to hadron collider values. For a mean bending field of 6 T the bunches circulate for ~ 900 turns before decay.

The greatest uncertainty in such a concept is whether the detector background can be made acceptable. Electrons, from decay of muons in the storage ring, collide with the walls and collimators yielding massive electromagnetic showers, hadronic particles and significant numbers of lower energy muons. If this problem can be solved, the $\mu^+\mu^-$ collider may well provide the only economically and technically feasible access to energies beyond the LHC.

6 Conclusions

The Accelerator Physics, Technologies, and Facilities Working Group has, over the course of this study, come to several conclusions with regard to the need for continuing support of accelerator-directed research and development in the United States. It is evident from past experience that research into accelerator technologies is required for the long-term health of the field of high-energy physics. Our conclusions in this regard are most easily grouped into three categories: hadron colliders, electron linear colliders, and novel technologies.

For high-energy hadron colliders, superconducting magnets are the enabling technology. As such, reassembly of a United States superconducting magnet research and development program is critical to the future. The key to increasing the luminosity of the Tevatron proton-antiproton collider has been, and will continue to be, increasing the antiproton production rate. Development of this capability deserves continuing support. Significant participation in the Large Hadron Collider project is expected to be a valuable and important component of a United States program for hadron collider research and development. Such participation would be most valuable if targeted toward those areas that represent the greatest challenges in moving toward the future; for example, superconducting magnet and beam-tube vacuum technologies. Finally, there are new ideas for overall systems designs for a 60 TeV collider facility that could be constructed sometime following completion of the LHC. These ideas deserve to be pursued.

For electron linear colliders, the enabling technologies are the high-gradient acceleration system: radio frequency power sources, accelerator structures, etc. The technology required to support a next-generation linear collider has made significant advances over the past several years, and is expected to reach maturity during the next few years. There are linear collider test facilities now under construction that should be completed as expeditiously as possible. Finally, conceptual designs of next-generation linear colliders based on proven technologies should be developed over the next two to three years.

To conclude, we expect that new novel technologies and/or different types of collider may be required to allow high-energy physicists to explore the multi–TeV scale (a muon collider is a possible example). Continued investment in generic accelerator research is required if we are to continue to open new experimental opportunities at an ever expanding energy frontier.

Acknowledgments

The conveners wish to express their gratitude to the Working Group members for their contributions, and to the conveners of the other working groups for their useful suggestions and comments. We also appreciate the work of participatants of the Indiana Workshop on Future Hadron Facilities for input into Section 2.

References

[1] Proc. Workshop on Future Hadron Facilities in the US, Bloomington, Indiana, 1994; Fermilab–TM–1907.

[2] SSC Central Design Group, "Conceptual Design of the Superconducting Super Collider," SSC–SR–2020 (1986).

[3] A. Piwinski, Proc. 9th Int. Conf. High Energy Part. Accel., **405** (1974).

[4] Convener's Note: Following the workshop, a promising concept '(multibatch coalescing) has been proposed for the Main Injector. If viable, this would reduce the need for new accelerators to one.

[5] Contributed by N. Phinney.

[6] Contributed by R. Siemann; see also 8/94 Linac Conf. paper, SLAC–PUB–6646 (1994).

[7] Contributed by K. Takata

[8] Contributed by I. Wilson, CERN/SL 94–89 (RF) CLIC Note 253.

[9] Contributed by R. Brinkmann, N. Holtkamp, and T. Weiland.

[10] The TESLA Collaboration: CERN, Cornell, CE–S DAPHNIA, DESY, Fermilab, Gesamthochschule Wuppertal, IHEP Beijing, INFN Frascati, LASA Milan, IPN Orsay, Johan-W. Goethe U. Frankfurt, KFK Karlsruhe, LAL Orsay, SEFT Helsinki, TH Darmstadt, TU Berlin, TU Dresden, UCLA Los Angeles.

[11] Contributed by M. Tigner for the TESLA Collaboration (Ref. [10]); see also H. T. Edwards, " TESLA Parameters Update, a Progress Report to be published in Part. Accel. by Gordon and Breach.

[12] Contributed by P. K. Tannebaum for the FFTB Collaboration (Ref. [13]); see also V. Balakin *et al.*, *Phys. Rev. Lett.* **74**, No. **13**, 2479 (1974).

[13] The FFTB Collaboration: Deutsches Elektronen–Synchrotron, DESY, Hamburg, Germany; Fermi National Laboratory, Batavia, IL, USA; The Budker Institute for Nuclear Physics, Russia; Laboratoire de l'Accelerateur Lineaire, Orsay, France; Max–Planck–Institute, Munich, Germany; National Laboratory for High Energy Physics, KEK, Tsukuba, Japan; and Stanford Linear Accelerator Center, Stanford University, tanford,CA, USA.

[14] Contributed by R. Ruth; see also R. D. Ruth *et al.*, "The Next Linear Collider Test Accelerator," Proc. PAC93, Washington, DC., and in SLAC–PUB–6252.

[15] P. Wilson, "Pulsed RF Technology for Future Linear Colliders," invited talk at the April 21, 1992 APS meeting in Washington DC.

[16] P. Wilson, private communication.

[17] R.H. Siemann, in Proc. 1993 Part. Accel. Conf., Washington DC, (IEEE, Piscataway, NJ, 1993) p. 532.

[18] V. Granatstein and C. Striffler, Proc. Advanced Accelerator Concepts, Port Jefferson, NY, 1992, AIP Conf. Proc. **279**, ed. J. Wurtele (AIP, NY, 1993) p. 1.

[19] T. Lavine, private communication; and R. Ruth *et al.*, Ref. [8], p. 543.

[20] "NLC Test Accelerator, Conceptual Design Report," SLAC Report–411, August 1993 (available from National Technical Information Service, US Dept. of Commerce, 5285 Port Royal Road., Springfield, VA 22161).

[21] R.B. Palmer, W.B. Herrmannsfeldt, K.R. Eppley, "An Immersed Field Cluster Klystron," Proc. Int. Conf. High Energy Accel., Tsukuba, Japan, 1989; and SLAC–PUB–5026 (1989).

[22] D.U.L. Yu, J.S. Kim, and P.B. Wilson, "Design of a High-Power Sheet-Beam Klystron," Proc. Advanced Accelerator Concepts, op. cit., p. 85–102.

[23] A.M. Sessler in Laser Acceleration of Particles, ed. P.J. Channell (AIP, NY, 1982) p. 154; A.M. Sessler, D.H. Whittum, W. M. Sharp, M.A. Makowski, and J.S. Wurtele, *Nucl. Inst. and Meth.* A **306**, 592 (1991).

[24] S. Van der Meer, *Part. Accel.* **30**, 127 (1990); W. Schnell, Proc. ICFA Workshop on e^+e^- Linear Colliders, ICFA 93–154, ed. R. Settles, 1993, p. 267; CLIC Study Group Report in Ref. [3], p. 540.

[25] T. Orzechowski *et al., Phys. Rev. Lett.* **57**, 2172 (1986).

[26] W.M. Sharp *et al.*, in Proc. Intense Microwave and Part. Beams II, O/E Lase 91, Los Angeles, in Proc. Int. Soc. OptĖng. (SPIE) (1991).

[27] T. Lavine, private communication.

[28] J.S. Wurtele *et al.*, in Ref. [4], p. 143.

[29] S. Heifets and S. Kheifets, *Rev. Mod. Phys.* **63**, 631 (1991).

[30] W. Bialowons *et al.*, in New Developments in Part. Accel. Techniques, Orsay, France, ed. S. Turner, (CERN 87–11, Geneva, 1987) p. 298.

[31] K.L.F. Bane, P. Chen, and P.B. Wilson, *IEEE Trans. Nucl. Sci.* **32**, 3524 (1985).

[32] W. Gai *et al., Phys. Rev. Lett.* **61**, 24 (1988).

[33] J. B. Rosenzweig *et al., Phys. Rev. Lett.* **61**, 98 (1988).

[34] A. Ogata *et al.*, in Proc. HEACC 89, Tsukuba, Japan.

[35] R. Byer, in Proc. Adv. Accel. Concepts. Workshop, Lake Geneva, WI, 1994.

[36] G. Fontana and R. Pantell, *J. Appl. Phys.* **54**, 4285 (1983).

[37] W. Kimura, in Proc. Adv. Accel. Concepts. Workshop, Lake Geneva, WI, 1994.

[38] R.B. Palmer, *J. Appl. Phys.* **43**, 3014, 1972.

[39] C. Pellegrini, P. Sprangle, and W. Zakowicz, Int. Conf. on High Energy Accelerators, 1983.

[40] IFEL Exp. at BNL

[41] T. Tajima and J.M. Dawson, *Phys. Rev. Lett.* **43**, 267 (1979).

[42] Vi Berezhiani and Ig. Murusidze, *Physica Scripta* **45**, 87 (1992); D. Umstadter *et al., Phys. Rev. Lett.* **72**, 1224 (1994).

[43] P. Sprangle *et al., Appl. Phys. Lett.* **53**, 2146 (1988).

[44] C.E. Clayton *et al., Phys. Rev. Lett.* **55**, 1962 (1985); A.E. Dangor *et al., Physica Scripta* T **30**, 107 (1990); Y. Kitagawa *et al., Phys. Rev. Lett.* **68**, 45 (1992); F. Amiranoff *et al., Phys. Rev. Lett.* **68**, 4780 (1992).

[45] M. Everett *et al., Nature* **368**, 527 (1994).

[46] A. Ogata, private communication.

[47] N. Ebrahim, private communication.

[48] T. Antonsen and P. Mora, *Phys. Rev. Lett.* **69**, 2204 (1992); P. Sprangle *et al.*, **69**, 2200 (1992); W. B. Mori *et al., Phys. Rev. Lett.* **72**, 1482 (1994); J.S. Wurtele and G. Shvets, submitted to *Phys. Rev. Lett.* (1993); P. Gibbon and A.R. Bell, *Phys. Rev. Lett.* **61**, 1599 (1988). These works examine long time scale instabilities that may be both deleterious and beneficial. Recent simulations have shown that forward Raman scatter, or self-modulations of a single frequency laser, can be used to accelerate large numbers of self-trapped plasma electrons to order 100 MeV in a millimeter. Such an accelerator could be an economical tool for applications that do not require high beam quality.

[49] T. Katsouleas and J.M. Dawson, *Phys. Rev. Lett.* **51**, 392 (1983).

[50] T. Katsouleas *et al., Part. Accel. Conf. Proc.* , 81 (1987).

[51] T. Krall *et al., Phys. Rev.* E **48**, 2157 (1993).

[52] T. Katsouleas *et al.*, Ref. (14), p. 480.

[53] G. Durfee and H. Milchberg, *Phys. Rev. Lett.* **71**, 2409 (1993).

[54] M. Tigner, Ref. [14], p. 1.

[55] R.B. Palmer, David Neuffer, and J. Gallardo, Proc. Advanced Accelerator Concepts, Lake Geneva, WI, 1994.

[56] E.A. Perevedentsev and A.N. Skrinsky, Proc. 12th Int. Conf. on High Energy Accel., 485 (1983); A.N. Skrinsky and V.V. Parkhomchuk, *Sov. J. Nucl. Physics* **12**, 3 (1981).

[57] D. Neuffer, *Part. Accel.*, **14**, 75 (1983); Proc. 12th Int. Conf on High Energy Accel., **481**, 1983); Advanced Accelerator Concepts, AIP Conf. Proc. **156**, 201 (1987).

[58] R.J. Noble, in Ref. (14), p. 949.

[59] Proc. Muon Collider Workshop, Los Alamos National Laboratory Report LA-UR-93-866, ed. H.A. Thiessen, 1993.

[60] Proc. Mini-Workshop on $\mu^+\mu^-$ Colliders; Particle Physics and Design, Napa CA, 1994, ed. D. Cline, to be published in *Nucl. Instánd Meth.*

[61] *KAON Factory proposal, TRIUMF*, KAON internal report (Vancouver, Canada, 1990).

[62] D. Neuffer, to be published in *Nucl. Inst. and Methods* (1994).

[63] Ionization cooling is not practical for protons and electrons because of nuclear scattering (ps) and bremstrahlung (*e*s) effects, but is for μs, and the necessary energy losses are easily obtained within the μ lifetime.

[64] A.A. Mikhailichenko and M.S. Zolotorev, *Phys. Rev. Lett.* **71**, 4146 (1993).

[65] CEBAF Design Report, CEBAF internal report (CEBAF, Newport News, VA, 1986).

DETECTORS

JAMES R. BENSINGER

Brandeis University, Waltham, MA 02254

ANDREW J. LANKFORD

University of California at Irvine, Irvine, CA 92717

DANIEL R. MARLOW

Princeton University, Princeton, NJ 08544

Working Group Members

C.M. Bromberg	*Michigan State University*	J. Paradiso	*MIT*
P.R. Burchat	*Stanford University*	S.I. Parker	*University of Hawaii*
P. Cushman	*University of Minnesota*	M. Piccolo	*INFN Frascati*
S.C. Eno	*University of Maryland*	V. Polychronakos	*Brookhaven National Laboratory*
H.A. Gordon	*Brookhaven National Laboratory*	J. Proudfoot	*Argonne National Laboratory*
W.T. Ford	*University of Colorado*	S. Reucroft	*Northeastern University*
M.C. Goodman	*Argonne National Laboratory*	H.F. Sadrozinski	*U.C. Santa Cruz*
A. Korytov	*MIT*	S. Schnetzer	*Rutgers University*
J.F. Krizmanic	*Johns Hopkins University*	M.C. Sheaff	*University of Wisconsin*
J.G. Learned	*University of Hawaii*	A.F. Sill	*Texas Technical University*
N.S. Lockyer	*University of Pennsylvania*	J.G. Smith	*University of Colorado*
S. Mani	*U.C. Davis*	F.E. Taylor	*MIT*
K.T. Mcdonald	*Princeton University*	S.E. Willis	*Northern Illinois University*
H. Montgomery	*Fermilab*	C.R. Wuest	*Lawrence Livermore*
H.B. Newman	*Caltech*		

1 Introduction

Detectors play a crucial role in the study of elementary particles. The rapid advance of particle detector technology has allowed experimentalists to take advantage of corresponding improvements in particle accelerators. Notable examples from the past few decades include the multi-wire proportional chamber and its numerous variants, large-scale calorimeters, silicon strip detectors, large water-Čerenkov detectors, and the automation of data acquisition through online computers. Without these and other important developments, our current knowledge of elementary particles and their interactions would no doubt be far less advanced than it is today. Much of this technology finds immediate application in other areas of physics and in fields such as medicine, where there is a direct impact on everyday life.

To be able to take advantage of new accelerator facilities in the U.S. and abroad, it is essential that particle physicists continue to make progress in the development of detectors. Such advances will not only be of value at new facilities, but will allow us to exploit further existing machines. One can expect that as in the past, advances will take the form of refinements and improvements in existing techniques as well as completely new inventions.

A number of areas where progress can and should be made have been identified. These include the development of:

- radiation hard materials and electronics;
- advanced vertex detectors such as strip detectors, pixel detectors, and detectors based on new materials such as gallium arsenide and diamond;
- improved photodetectors;
- fast vertex triggers;
- high-precision and/or large-area arrays for muon detection;
- fast, bright, radiation-hard crystals;
- high-speed noble-liquid calorimeters;
- high-density, low-cost electronics;
- large bandwidth data acquisition systems and switching networks; and

- hadron identification devices for the study of B physics.

These technologies are of importance not only to accelerator-based experiments but to the growing number of particle astrophysics experiments as well.

In this chapter, we present an overview of the state of the art in particle detectors and consider in more detail the areas in which further progress should be made. In our conclusions we propose a detector R&D program that will help stimulate work in this important area. Although the current funding situation is such that it is not realistic to think in terms of replacing the generic detector R&D programs sponsored by the SSC Laboratory and the Texas National Laboratory Research Commission, we believe that even a considerably more modest program would have an important impact.

2 Tracking

2.1 Requirements

Tracking of charged particles, usually in a magnetic field, is a standard function of most HEP detectors. The past years have seen an extension of trackers designed mainly for momentum determination and pattern recognition to devices capable of specific ionization measurements for particle ID and, more recently, the incorporation of precision vertex measurements to identify long-lived particles. In the future, high rate operation will be a major challenge. This applies to both gaseous and solid state tracking detectors. High occupancy and radiation damage have to be dealt with in nearly all future applications:

- B Factories at Cornell, SLAC and KEK;
- FNAL collider upgrades of CDF and DØ;
- SLD upgrade;
- RHIC experiments;
- LHC experiments (Atlas, CMS, ALICE);
- Fixed Target experiments at FNAL and CERN;
- Experiments at HERA; and
- Space applications.

Engineering questions will play an increasingly important role in all of these applications as outlined below.

2.2 Semiconductor Detectors

Semiconductor tracking detectors play a crucial role in precision vertex reconstruction. They range from silicon microstrip detectors, which are essentially one-dimensional devices, to inherently two-dimensional devices such as pixel detectors and charged-coupled devices (CCD's). There is also an emerging diamond detector technology. For storage rings, pixel detectors could offer significantly better intrinsic resolution than silicon strips because of their high signal-to-noise. Multiple scattering could, in principle, be decreased if the electronics are kept thin and if the system can be supported and cooled with minimum material. At the B Factories, where typical track momenta are quite low (a few hundred MeV), these improvements will result in improved impact parameter resolution only if the beam pipe is also significantly thinner than those being designed and constructed today.

The development of semiconductor detectors goes hand in hand with the development of ASIC (application specific integrated circuits) front-end electronics and readout systems. Modern designs are fully integrated mechanically and electrically, so as to achieve a match between the channel pitch of the detector and its electronics. The resulting channel counts are very large and require special care in the design of data storage and readout systems.

2.2.1 Silicon Strip Detectors

In the arena of silicon microstrip detectors, two areas of development stand out: forward silicon tracking in colliding beam experiments and improved detector spatial resolution. The benefits of improved forward tracking coverage are obvious. Research efforts have demonstrated that "wedge" silicon microstrip detectors with constant or variable strip pitch work. Further investigation of double sided designs and radiation hardness need to be performed. The high spatial resolutions of microstrip detectors sharpen impact parameter determination, event pattern recognition, and momentum resolution. Several groups have obtained position resolutions under 5 μm with microstrip detectors by incorporating passive intermediate strips. Further improvements are possible.

Another challenge for vertex detectors at many collider experiments, especially asymmetric B Factories, is the measurement of both coordinates in the plane of the detector. Currently this is being addressed by double-sided silicon strips.

In future applications at LHC and in space, large-scale use (tens of square meters) of silicon strips is foreseen. The engineering issues associated with such large arrays are much the same as for smaller systems and include: cooling, mechanical supports, thinning of electronics and detectors, radiation hardness, and the production of reliable and uniform detectors. Clearly unit cost will be a prime consideration. Reliable electronics will also be essential for progress in developing the planned large and precise systems.

2.2.2 Diamonds

Detectors based on diamond material constitute a rapidly emerging new technology. In this approach diamond sheets are grown by the chemical vapor deposition (CVD) process [1], which allows large areas of high quality diamond to be grown at reasonable cost. Diamond has many properties that make it an excellent material for use in particle detectors: high resistance to radiation damage, large resistivity, saturated carrier velocities, low dielectric constant, high thermal conductivity, large mechanical strength, and potentially low cost. The most important of these is diamond's radiation hardness relative to conventional detector materials. This makes diamond detectors highly attractive for coping with the intense radiation environments that will exist in planned future experiments at the LHC, the Fermilab Collider in the era of the Main Injector, and high-intensity fixed-target experiments.

Since diamond is an excellent electrical insulator, there is negligible leakage current in diamond detectors. In addition to effectively eliminating shot noise, this property also simplifies the fabrication of detector devices since there is no need for a *p-n* junction. Instead, an electric field, typically 10^4 V/cm, is directly applied across the thickness of the diamond. A charged particle traversing the diamond creates electron-hole pairs, which then separate in the applied field, inducing charge on the surface electrodes.

The quality of detector-grade diamond is characterized in terms of collection distance: the average distance an electron-hole pair separates under an applied electric field. Over the past five years, the value of the collection distance of the best available diamond has increased at the rate of about an order of magnitude every 18 months. The highest quality diamond currently tested has a collection distance of 85 μm, corresponding to an average collected charge of 3000 electrons for a minimum ionizing particle. A twenty-layer twenty-radiation length diamond/tungsten sampling calorimeter was tested and found to have the same performance as a silicon/tungsten calorimeter and performed as predicted by an EGS calculation [2]. This demonstrates that current diamond quality is sufficient for employing diamond calorimeters in experiments. A series of diamond microstrip detectors have also been tested. One of these, with a collection distance of 50 μm, has a signal to noise of 6:1 and a position resolution of 25 μm for a 100 μm strip pitch [3]. A diamond tracker with suitable signal-to-noise for use in experiments will require a collection distance of about 200 μm, corresponding to 7000 collected electrons; a factor of two-to-three improvement over current diamond.

The radiation hardness of natural diamond has been known for many years [4]. Recently a set of radiation hardness measurements of CVD diamond has been made. The tests revealed no degradation of pulse height nor any measurable leakage current after exposures of up to 10 MRad of ^{60}Co gamma rays, fluences of $10^{14}/\text{cm}^2$ of 300 MeV pions, and fluences of a few $10^{13}/\text{cm}^2$ of 500 MeV protons. Further studies are planned to extend the above exposures by an order of magnitude and to include exposures to a fluences of few $10^{14}/\text{cm}^2$ neutrons [5].

2.2.3 Pixel Detectors

Pixel detectors represent a natural next step in the evolution of position-sensitive semiconductor detectors. Although strip detectors have been highly successful in applications where high precision and good double track separation are required, pixel detectors offer a number of important advantages:

- Each measurement plane provides a complete space-point measurement, serving the same function as two planes of strips.
- The tracking combinatorics are greatly reduced. With two sets of strips, n simultaneous hits results in $n!$ possible solutions, whereas pixel detectors naturally correlate the x and y views.
- The per-element occupancy in pixel detectors is much lower, allowing operation at much higher beam intensity.
- The low capacitance of pixels results in lower noise and a correspondingly better signal-to-noise ratio.

The first use of pixel detectors was by Damerell who showed that CCD's could detect minimum ionizing particles [6]. The CCD's proved to have great advantages in pattern recognition, but were not widely used in other experiments due to their slow readout and thin sensitive region (a few tens of μm). These problems do not occur in the pixel detectors now under development, which can have sparse field readout and in which nearly the full thickness of a chip is sensitive. The simplest uses a PIN diode with an array of collection electrodes bump-bonded to a readout chip [7]. Monolithic detectors, first fabricated in 1990-1991, with tests described in a series of publications (see reference [8]), incorporate the readout electronics on the same substrate as the charge-collection silicon and consequently are thinner and have five times lower input capacity than bump-bonded devices. Fully depleted CCD's have also been fabricated, but still have the slow readout of the original versions [9].

Monolithic Detectors In a monolithic device, ionization charge goes to a two-dimensional array of collection electrodes that are in contact with the substrate. The electrodes, which are typically several microns across, are connected to the gates of readout transistors. These, and the rest of the pixel circuits are in a single implanted n well that surrounds each collection electrode and occupies the top few microns of the wafer. It provides an optimum environment for the transistors, shields the substrate from transistor switching transients, and, together with a diode covering the bottom of the chip, sets up the collection field that spans the wafer and focuses the charge on the small collection electrodes. The charge remains on the nearby MOS gates where it controls currents flowing to the edge of the chip. It can send current in more than one direction, can be read out at once, and can be stored and read out later as well. Signals from minimum ionizing particles in silicon strip detectors, before amplification, are about 0.01 to 0.1 mV; here they are about 0.1 V. Additional readout electronics are in n and p wells along the chip edges.

Infrared tests show that more than 99.995% of the positive ionization charge generated under the n well, but away from the array edges, goes to the collection electrodes rather than to the n well, which occupies over 90% of the area. In beam tests of an array of 34×125 μm^2 pixels with 500 GeV/c muons, signal to single-channel noise ratios of 65 to 1 (with much of the noise due to off-chip electronics) were observed and spatial resolutions of $\sigma \simeq 2$ μm were obtained for 90°-incident tracks. Hits with large pulse heights had a larger error due to Landau fluctuations, as did small ones due to their reduced signal-to-noise ratio. For the central 63% the spatial resolution was in the 1.3 to 1.5 μm range. Angled tracks also had poorer resolution, typically in the range of 3 to 6 μm. However, the separation of their entrance and exit positions could be used to give a direction vector having an accuracy of 1° to 3° though usually with a two-fold pointing ambiguity.

Charge sharing near the border of a large pixel (in this case the long direction of the 34×125 μm^2 pixel), can improve the position accuracy, although the errors here have a significant non-gaussian tail. If long pixels are used in a staggered "brick" like arrangement, charge spreading to adjacent rows provides two sharing regions in each length: two at the ends of each pixel and one in the middle next to the dividing line in the adjacent rows. The central non-sharing region is, in addition, split into two shorter ones, so the overall resolution in the long direction is greatly improved. Alternately, square pixels with sides of 60–70 μm or less would provide sets of space points with small errors in both directions and, for angled tracks, with a pointing vector attached uniquely to each point. This could provide significant help in tracking complex events.

These first-generation detectors have also been used to detect and measure a 0.1 -mm-wide calcification in a tissue sample from a breast tumor, about half the diameter of the smallest ones normally seen in clinical practice [10].

The second-generation detectors now in fabrication will have sparse-field readout. A proposed quantum mammography system, using monolithic detectors, would not only be fully digital, but would have higher resolution and efficiency than present systems and would count each x-ray quantum to produce the maximum possible information from the available radiation. Such chips would use a high-speed data-push readout [10].

Hybrid Pixel Detectors The main advantage of these detectors is that two known commercial components are combined with a method that is commercially available, if not commonly so. Bumps have been made with solder, gold, tin, indium, and other materials. They have been made with ball bonders that do not leave the usual wire on the ball, by plating, and by thermal evaporation followed by lift-off above the masked regions around the bumps. The Hughes Aircraft Company used indium bumps that were made this way to bond one of their readout chips to a detector made by Micron Semiconductor [7]. Indium bumps are also available from Rockwell International, among other sources [11]. Small indium bumps, 25 μm or less in diameter can be made and reliably bonded using commercial aligner bonders [12] without needing high temperatures.

A SLAC/UC Berkeley Space Sciences Lab/Hughes collaboration tested indium bumped hybrid detectors in a beam at Fermilab. An analysis of that data showed a resolution of 2.6 μm [7]. Hybrid detectors also are used in Omega-WA97 and will be used in the LEP-Delphi experiments at CERN [13].

Several groups are now designing readout chips for use in LHC that will uniquely identify tracks from one of perhaps 20 events in one beam crossing while also avoiding confusion from the tracks of the 80 or more following beam crossings that occur before the arrival of a Level 1 trigger. Due to the expected high rates, all designs keep the information and readout control within individual columns, which operate independently of each other prior to the Level 1 trigger.

The CERN-Omega group plans to store hit information within the pixel while a digital delay within each pixel identifies those with the proper hit times by a coincidence with the Level 1 trigger [13].

The Marseille group plans to shift the address of each hit pixel down its column, one shift per beam crossing, and to calculate the hit time from the address and the arrival time at the column end [14]. They have also placed integrated busses for the readout chips on the detector chip, which spans a group of readout units.

The LBL group plans to set a hit latch and store a buffer address and analog charge information in each hit pixel. The buffers, located at the ends of each column, store only the hit times.

All of the readout architectures will have one form or another of dead time, which is not expected to be too serious. Indeed, radiation damage is expected to be the main limitation on how close the detectors can be placed to the beams.

2.3 Gaseous Detectors

Gas-filled wire chambers continue to be a mainstay of charged particle tracking systems. They provide moderate-to-high rate capability and excellent spatial resolution over large volumes at a manageable cost. Although at the LHC, where the rates will be extremely severe, traditional wire-cathode drift chambers are likely to be displaced by semiconductor detectors, straws, fibers, and/or gas microstrip detectors, gaseous detectors will continue to play an important role at the e^+e^- B Factories and elsewhere.

2.3.1 Small-Cell Wire-Cathode Devices

The main challenge for tracking at the B Factories concerns improving the momentum resolution for tracks with momenta less than 3 GeV/c. This involves operation in high magnetic fields (1.5T) and reducing the amount of material traversed by tracks. The latter requirement has implications for the inner wall of the tracking device, the composition of the wires and the choice of gas. These considerations preclude pressurized devices and point to a low-mass drift chamber as the detector of choice. Thin end-plate supports are also useful to minimize material in front of the particle ID devices and the EM calorimeter.

Many studies during the past five years have shown helium-based gases to be an attractive choice. The radiation length of the gas can be reduced by a factor of five or more. The resulting gas has a much smaller Lorentz angle (and hence much improved high-B-field operation), while still providing spatial and dE/dx resolution comparable to argon-based gases.

2.3.2 Jet-Cell Chambers

The CDF central tracking chamber (CTC) was designed to provide isolated charged-particle tracking, but has since been used to reconstruct on a routine basis nearly all of the charged particles in the very complex, high-rate events at the Tevatron collider down to a tracking threshold of only a few hundred MeV. This advance was made possible by the enormous increase in online and offline computing available to the experiment since it was designed in the 1980's.

2.3.3 Straw Chambers

Arrays of detectors comprising large numbers of individual thin-walled mylar or kapton cylindrical straws offer important advantages for the high rates anticipated at the LHC. In particular, comparatively small ($r \simeq 2$ mm) drift distances are possible (this minimizes the pulse-pair resolution time and the current draw per anode wire) and the wire tension load is greatly reduced since there are no cathode wires. Moreover, the straws can be arranged into mechanically robust modules and individual cells electrically isolated from one another, minimizing the impact of a single broken wire. These devices received much attention as part of the SSC detector R&D program, where the state of the art was considerably advanced. Position resolutions of order 100 μm were measured in 4-mm-diameter straws and straw modules as long as 4 m were constructed.

Serious challenges remain, however, and the particular problems of pileup, occupancy, and detector aging will be even more severe at the LHC. A large volume must be filled with small detector elements, resulting in a high density of signal lines emanating from the detector's readout surfaces. Signal time shaping and recovery are critical. Signal dynamic range and cross-talk mitigation are also a challenge. High drift velocity gases with adequate gain, primary ionization rate and efficient collection and benign chemistry are required. Low mass structures capable of 20 μm alignment tolerances need to be developed.

2.3.4 Time Projection Chambers

Time projection chambers (TPC) are powerful tracking devices that find application in collider experiments where minimization of material is not critical and where time-integrated particle fluxes are fairly low. The direct association of axial with azimuthal information at each radius renders pattern recognition very straightforward. The feasibility of operation at elevated pressure enhances position resolution and dE/dx precision.

2.3.5 Gas Microstrips

This technology is getting very serious attention in Europe, and was the baseline choice for SDC for the forward disks. The small charge collection distance permits fast response. That property and the implementation on a solid substrate allows fine pitch (good position resolution) and flexible geometry. The devices are a potentially lower cost alternative to silicon strips with application to larger area detectors. Work is still needed on substrate choices, where suitable low-mass materials with the correct resistivity must be identified, and on techniques for etching.

2.4 Scintillating Fibers

Much work has been done within SDC and DØ on this technology, which offers a good solution to the need for fine granularity and fast time response and provides natural basis for fast triggering. Sophisticated light detection techniques are under development. In particular, the visible light photon counters (VLPC's) developed by Rockwell International Science Center, offer quantum efficiencies of $\sim 75\%$, good single photoelectron resolution, gains in excess of 10^4, and risetimes under 10 ns [15,16]. Moreover, they are capable of operation in high magnetic fields. With high light-detection efficiency the granularity, position resolution, and mass can be pushed to make fiber trackers a very attractive technology. Better understanding of radiation damage may be needed for very high-radiation applications.

2.5 Areas for Future Development

Research on semiconductor devices has yielded fruitful results in the areas of radiation damage, novel strip geometries, and inherent sensor design and operation. At the same time, bipolar and CMOS ASIC's have been developed for optimizing the readout. In light of the importance of silicon strip technology across all of experimental particle physics, we need to develop U.S. suppliers of these devices, and the opportunity for R&D efforts in close collaboration with industry.

Strong R&D efforts by effective groups in pixel detector development are needed to realize a working system. To take complete advantage of gains in vertex detector technology, technologies for building thin beam pipes and low-mass support structures also need to be pursued.

For collider geometries good solutions combining high speed and granularity with low mass are still elusive. Detector mass exacerbates the problem of high background rates. Monte Carlo calculations of background rates are inadequate—*e.g.*, CDF and DØ experience substantially larger backgrounds than the programs predict.

Drift chamber technology in general would benefit from further systematic study and classification of gas properties: ionization potentials, electron attachment rates, drift in electric and magnetic fields, diffusion, and the chemistry of aging. Work with helium-based gases is quite new and substantial improvements may be possible. More studies are needed regarding the suitability of these gases in high-radiation environments. The capability of Boltzmann-simulation code to predict detailed properties of gases has improved enormously during the past five years and further improvements in the knowledge of the cross sections of popular components are needed.

Similar improvements in materials are being considered for the wires and end plates. Carbon-fiber and magnesium are possible alternatives to aluminum for the end plates. Small-diameter gold-plated aluminum wires require development of special feedthroughs, and the possibility of using gold-plated, carbon-fiber wires should be explored.

3 Calorimetry

3.0.1 Requirements

The primary requirement for electromagnetic calorimetry is energy resolution, which can be parametrized as

$$\frac{\sigma_E}{E} = \frac{A}{E} \oplus \frac{B}{\sqrt{E}} \oplus C \tag{1}$$

In detectors for e^+e^- storage rings and certain fixed-target experiments, where detection of photons with energies below 100 MeV is important, the first term, which is mainly a consequence of electronic noise, is often crucial, whereas at LHC, where $e^\pm$ and γ energies ranging well above 100 GeV will be encountered (*e.g.*, in the decays $H \to \gamma\gamma$ or $Z \to e^+e^-$) the third term, usually called the "constant term," can dominate. The stochastic term $(B/\sqrt{E})$ plays an important role in both regimes. For LHC detectors, typical target values are $B = 5\%$–10% (E in GeV) and $C = 0.5\%$–1.0%. At the B Factories, noise contributions of 0.5 MeV or less are desirable, as are stochastic terms corresponding to $B = 1\%$–2%. For hadronic calorimetry, the energy resolution requirements are typically relaxed to $\sigma_E/E = 65\%/\sqrt{E} + 5\%$.

3.1 Liquid Ionization Calorimeters

The precision measurement of the ionization in noble liquid calorimeters started with the pioneering work of Willis and Radeka [17]. Many large scale liquid argon calorimeters have led to productive physics programs: Mark II at SLAC, R806 at the CERN ISR, E706 at Fermilab, HELIOS at CERN, and more recently, DØ at Fermilab [18], SLD at SLAC [19], and H1 at HERA/DESY. All of these devices displayed attractive features intrinsic to the technique: stability, unit gain, flexible longitudinal and transverse segmentation, uniformity, and (in some cases) hermeticity. The DØ detector employs a 50,000 channel liquid argon calorimeter that includes both electromagnetic (EM) and hadronic sections. It is hermetic for missing energy studies and has proven very effective for triggering on and measuring electrons, jets, and photons. The SLD lead liquid-argon calorimeter (LAC) includes an EM section and a short hadron section, followed by an Iarocci-tube calorimeter.

At the LHC liquid calorimetry will be pushed to new frontiers in the following areas: 1) energy resolution, 2) γ pointing, 3) π^0 rejection, 4) dynamic range, and 5) radiation hardness. The energy resolution can be improved by increasing the sampling fraction by using thinner absorbers or using denser active liquids such as krypton. However, good energy resolution requires an excellent calibration system, as has recently been developed at BNL [20].

Achieving the requisite 16–18 bits of dynamic range while maintaining excellent energy resolution presents a serious challenge to the designers of readout electronics. There has been R&D in switched capacitor arrays (SCA) and ADC's to handle more than 12 bits. This work must continue.

The large number of channels being considered for future experiments requires the development of a compact feedthrough that will be cost effective. Another challenge is to integrate the preshower function (pointing and π^0 rejection) into the main calorimeter. Still further R&D is required for the extreme radiation environment at the LHC. There has been a long program of certifying which materials will survive in this environment—but it must continue with the added direction of considering the possible poisoning of the noble liquid. There is also important R&D going on for the radiation stability of the electronics, which is often in close proximity to the calorimeter.

The ATLAS [21] experiment is planning a barrel EM calorimeter and endcap EM and hadron calorimeter with an integrated forward liquid calorimeter. The electrode design is based on the accordion design first introduced by the RD3 Collaboration at CERN [22]. That R&D program has progressed over the last four years to include the demonstration of a 2-m prototype readout with fast (20 ns) shaping that achieved an energy resolution for electromagnetic showers of (9-10)%$/\sqrt{E}$ with a constant term of less than 0.4%. Building on this work, R&D using liquid krypton has produced energy resolution of 6.7%$/\sqrt{E}$, timing resolution of 280 ps for 20 GeV [23], and the first time-sampled waveforms (every 18 ns).

Another use for liquid ionization calorimetry is being pursued by the NA48 experiment, which is designed to measure CP violation in K-decays. They have developed a large almost totally liquid krypton EM calorimeter. This device has achieved an energy resolution of 3%$/\sqrt{E}$, and exhibits excellent timing performance [24].

3.2 Crystal Calorimeters

The superb energy resolution and detection efficiency of total absorption shower counters composed of inorganic scintillating crystals have been known for decades. In high-energy physics, the use of large arrays of scintillating crystals for the precision measurement of the energy and angle of photons and electrons has been a key factor in many fundamental discoveries [25]. This was first demonstrated in the Crystal Ball detector's study [26] of radiative transitions and decays of the charmonium family. Over the last decade, larger crystal calorimeters have been constructed, and their use has been a key factor in the successful physics programs of the L3 experiment at LEP [27], CLEO II at CESR [28] and the Crystal Barrel at LEAR [29]. Crystal detector arrays of this type also have been designed and are under development for the next generation of high-energy physics experiments aimed at the study of CP violation, including KTeV at Fermilab [30]), and the SLAC [31] and KEK [32] B Factory detectors.

In addition, crystal calorimeters containing 10^4 to more than 10^5 elements have been designed and studied intensively by a large sector of the high-energy physics community planning experiments for multi-TeV hadron colliders, including the Superconducting SuperCollider (SSC) in the U.S. [33,34] and the Large Hadronic Collider (LHC) at CERN in Europe [35,36].

The unique physics capability of crystal calorimeters is a result of their good electron and photon energy resolution (typically $2\%/\sqrt{E} \oplus 0.5\%$) over a wide energy range, uniform hermetic coverage, fine granularity over a large solid angle, and clean electron and photon identification. A well-designed crystal calorimeter has greater sensitivity than sampling calorimeters for detecting multibody final states containing low-mass resonances (such as π^0, η, and η'), or yet-to-be-discovered particles such as the Higgs in the "Intermediate Mass" range between 80 and 170 GeV [37].

In order to maintain the high resolution and the resultant sensitivity to new physics, the key detector requirements are radiation hardness and precision calibration, in addition to high speed, mechanical robustness, and a stable readout with high linearity and large dynamic range. For experiments that are required to operate in the high-rate high-radiation environment of multi-TeV hadron colliders, very high speed (triggering within the typical beam-crossing time of 25 ns) and radiation resistance up to the 10 MRad range are among the principal design requirements. Following an intensive R&D program for GEM at the SSC, it was demonstrated that mass-produced BaF_2 crystals from China could be rendered radiation hard by optical bleaching *in situ* with visible light [38,39].

For the larger radius crystal calorimeters being considered for CMS at the LHC, CeF_3 and $PbWO_4$ —which are denser than BaF_2—are the leading candidate crystals. Encouraging recent progress has been made in the production of both large CeF_3 and $PbWO_4$ crystals with improved radiation resistance, where crystals of appropriate size and quality have been produced in a number of different laboratories around the world.

At lower energies, high-luminosity experiments searching for rare Design B, K, or τ decays, or performing precise CP-violation measurements through the study of B and K final states containing e^+e^- pairs, single or multiphotons (including π^0's and η's), have chosen CsI(Tl) (CLEO II, Crystal Barrel, SLAC and KEK B Factories) or undoped CsI (KTeV). This offers a mechanically robust, relatively low cost solution to the requirements of high light output, a minimum of material in front of and between the crystals, radiation hardness up to the 10 krad range, and moderate to high speed.

3.3 Scintillator Calorimeters

The use of plastic scintillator as the sensitive medium in sampling calorimeters has been a common approach in experimental high-energy physics for much of the past three decades. Almost all possible sampling geometries have been employed—scintillator tiles with waveshifter bar or waveshifter fiber readout in varying geometries; pure scintillating fiber detectors with fibers running both parallel and perpendicular to the incident particle; and hybrids of the above. The outstanding questions relevant to future high-energy experiments are well defined, although their relative importance depends on the particular application.

Maximizing the fiducial volume is of obvious importance. The use of waveshifting fiber readout can reduce the uninstrumented region by as much as a factor of five relative to the conventional approach using waveshifter bars. However, fabrication labor is increased in the fiber technique, and further refinements are needed. One possibility is the use of injection molding to form the scintillator tile simultaneously with a ready-made slot to contain the readout fiber. At present, however, injection molding does not produce tiles with adequate optical properties. Alternatively, one could use a "spaghetti" calorimeter approach for which no waveshifter is needed and the scintillating fibers run parallel to the direction of incoming particles. This approach is also labor intensive, needs large volumes of scintillating fiber (for which cost reduction is an issue), and is difficult to implement if depth segmentation is necessary. These are all areas appropriate for additional R&D.

For many future applications, radiation damage is still an issue for scintillator calorimetry. Doses of one MRad can be tolerated under normal conditions, and doses up to about 5 MRads can be accommodated if reinstrumentation/recycling of the scintillator is possible. Beyond this level there are no realistic solutions in hand. This is one area where R&D is still needed both on the development of new scintillator and waveshifter fluor systems (for example at longer wavelengths) as well as more radiation tolerant plastic bases suitable for commercial production.

The signal response of scintillator is also an issue for high speed sampling. At the present time it appears that a minimum integration gate width of about 40 ns is needed to fully contain the signal, although improvements brought about by R&D on fluors might make shorter gates possible. Signal-width effects might also be mitigated by through waveform recording.

The calibration of scintillator calorimeters is also of importance. Typical test-beam calibrations (*e.g.*, CDF and ZEUS) achieve of order 2% precision. With *in situ* charged particle tracking, this can be improved to better than 1%. Achieving this precision in the absence of a magnetic spectrometer (*e.g.*, by a light flasher system) is an area where future R&D might make some inroads.

There are still open questions with respect to hadron/jet calorimetry, in the details of hadron shower development. Continuing R&D in this area both with test beams and simulations is desirable to form a basis for improved optimization of calorimeter designs. Moreover, in the long term the use of fine-grained calorimetry may allow (via for example neural networks) good discrimination from background QCD processes of W and Z decay to jets, hadronic τ decays and perhaps even B jets.

Finally, the granularity of a scintillator calorimeter is mainly determined by cost at the present time. As noted, labor costs can be high for many fabrication approaches. The most popular readout device is the photomultiplier tube, which is a costly element as well as one that can be sensitive to environmental factors. R&D to develop a cheaper and/or more robust alternative should have high priority for applications demanding high granularity.

3.4 Preshower and Shower Maximum Detectors

Hadron collider physics requires good lepton and photon identification; *e.g.*, good single-γ/π^0 discrimination in intermediate mass Higgs searches ($H \to \gamma\gamma$) and the ability to identify electrons in complicated events. Although electron identification can be achieved with standard calorimeters by using global properties of the shower, this is not sufficient to tag partially isolated electrons such as those from $b \to ce\nu$, or background processes such as c-decay or e^+e^- pairs from conversions in beam pipe and tracker. Additional electron identification can be provided by the introduction of a special-purpose detector to sample the distribution of shower energy in finer detail. The type of device selected depends on the calorimeter and physics priorities, *e.g.*, sampling preshower detectors are not appropriate in front of high resolution, homogeneous calorimeters. Moreover, the degree of segmentation is a compromise between cost and performance.

Beam tests show that while shower maximum devices obtain better e/π separation than preshower detectors as standalone devices, their information is highly correlated with global calorimeter cuts. Preshower detectors give a much larger additional uncorrelated pion rejection (factor of 5-15) when used in combination with calorimeter data [40,41]. If the calorimeter follows a substantial amount of dead material, such as a 1 X_0 solenoidal coil, the electromagnetic resolution will be degraded. Since the energy deposited in an active layer of a preshower detector immediately following the coil is related to the amount of energy lost therein, a preshower detector can be used to correct for this degradation. This function goes under the name "massless gap" in LAr calorimetry [42]. It was also modeled and explored in a test beam for the SDC preshower detector [41].

Highly articulated preshower detectors provide additional benefits. They can take advantage of the submillimeter size of the shower in the first few radiation lengths, providing a long lever arm to trackers and excellent background rejection of events with overlapping charged hadrons and π^0's or γ's. If the preshower detector is finely segmented longitudinally as well, and is deep enough to convert photons with a reasonable probability (>3 X_o), then it can distinguish between single γ's and π^0's on an event-by-event basis.

The most widely studied technologies are silicon strips and scintillating fibers or tile/fiber arrangements, although work is also currently proceeding in the LHC RD-3 group on forming a preshower detector from specially instrumented layers of a LAr calorimeter. Work on silicon preshower detectors has been supported by the LHC RD-2 program and by major new detectors such as Zeus [43]. The advantages of silicon preshower detectors are their excellent timing properties, thin profile, and natural ease in adapting to any pixel geometry. Scintillating fiber preshower detectors were pioneered by UA2 [44]; the outer layers of their fiber tracker were essentially an imaging preshower detector, read out by image intensifiers and CCD's.

The high channel count of preshower/shower maximum detectors means that the development of cost effective readout is the driving R&D issue. For both silicon and LAr, combining radiation resistance with low noise and low power dissipation while maintaining a large enough dynamic range provides a challenging readout design problem. Photodetector development for fast fiber readout (CCD's used in UA2 are too slow for LHC applications) resulted in new low cross-talk designs of conventional multi-channel tubes manufactured by Hamamatsu and Philips and in the development of new hybrid tubes with PIN diodes [45] or avalanche photodiodes (APD's) [46]. More development is needed to make segmented hybrid tubes reliably and cheaply. Bare APD arrays operated in linear mode can now be made with gains of 500-1000, but work must continue to improve the yield and manufacturing process to reach the goal of $3. per channel. Visible light photon counters, which offer quantum efficiencies of about 75% are also a promising option for preshower detector readout.

4 Muon Detectors

4.1 Overview of Requirements

A number of large-scale muon systems are envisioned both as upgrades and new projects within the US HEP program during the next decade. Muon systems will likely play a key role in the detection and study of massive objects, such as the top and bottom quarks, where the sign of the electric charge of a high-P_t muon serves as an efficient quark flavor tag, and the Higgs, W', or Z' bosons, where the multi-muon signature will be distinctive above the large background expected. Muon systems will play a key role in neutrino and cosmic ray physics, where the flavor of the incident neutrino is labeled by the presence or absence and sign of the muon.

An overview of these projects where there is, or is likely to be, a large involvement within the US HEP community is listed below:

- LHC muon systems: ATLAS and CMS;
- Upgrades of CDF and DØ muon systems;
- B Factories at Cornell, SLAC, and KEK;
- Neutrino Oscillation experiments at CERN, BNL, and FNAL;
- Dedicated B-physics initiatives at LHC and HERA;
- Detectors for RHIC; and
- Cosmic ray experiments.

Muon detectors frequently require excellent spatial and time resolutions in order to achieve the necessary momentum resolution in high-rate environments. Moreover, the distinct signatures produced by high-energy muons makes them very useful at the trigger level and it is generally desirable to have the muon system produce signals for this purpose. This is typically done either by extracting some sort of timing signal from the primary muon detector or by augmenting the muon detectors with dedicated trigger devices.

Since muons are identified by their penetrating power, muon systems generally lie on the outside or end of detectors. The resulting systems are therefore generally quite large, making cost an important consideration and pointing the way to strong industrial involvement.

4.2 Current Status of Muon Detector-Trigger Technologies

For muon track reconstruction and/or triggering needed in the projects envisioned above, several attractive technologies have recently been developed or improved. All the technologies under review here are based on gas-ionization technologies where either the drift time and/or the induced charge are detected. Spatial resolutions of order 50 to 200 μm and time resolutions of a few nanoseconds have been achieved. Further improvements in economy and performance can be expected with additional R&D.

4.2.1 Cathode Strip Chambers

Cathode Strip Chambers (CSC's) are based on the multiwire proportional chamber (MWPC) concept. These devices provide high resolution measurements of both position and time, making it possible to combine tracking and triggering into a single detector [47]. Spatial resolutions of order 50 to 100 μm and time resolutions of order 24 ns have been achieved. The precision spatial resolution is obtained by measuring the charge induced on cathode strips— typically 0.5 to 1.0 cm pitch—to of order 1% with a precision integrating amplifier (t > 300 ns). Timing signals and a second coordinate are furnished by the system by reading out the anode wire. By "ORing" layers of chambers, a time resolution was demonstrated adequate to tag the beam crossing time of 16 ns efficiently at the SSC [34]. By measuring the charge of both the anode and cathode signals the chamber can furnish a true space point—a distinct advantage in a high rate environment.

These devices were chosen as the tracking and trigger technology for the GEM muon system. They are being considered for the very high $|\eta|$ region of the ATLAS muon system and for the entire endcap system of the CMS detector. The present R&D on these devices is focused on developing mass production methods to mitigate the high costs, small chamber sizes, and exacting tolerances of the technology.

Honeycomb Strip Chambers (HSC) share many of the properties of the CSC concept [48]. The cell structure is fabricated out of mylar folded into a hexagonal tube of 11-cm width. Cathode pickup strips are placed on a mylar substrate which runs perpendicular to the anode wires. Drift times measure the distance to the wire, and the center of gravity of the charge distribution induced on the cathode strips measures the coordinate along the wire. The chambers can be mass-produced cheaply by folding sheets of mylar with etched cathode strips to form the upper and lower halves of the chamber hexagon. A spatial resolution using the cathode strips of 80 μm has been measured in RD5 at CERN [49].

4.2.2 Drift Tubes

Drift tube tracking chambers have been in use for a number of years in a variety of experimental settings. Owing to their simplicity and time-proven properties this technology continues to be attractive in the LHC era. They have been operated in both limited-streamer and proportional mode, and can be pressurized to several atmospheres for better resolution resulting from slower drift times, smaller Lorentz angles, and more ionization clusters per unit length in the gas.

There are many variations of the basic construction of the cathode surface surrounding the anode wire. The round drift tube geometry is especially simple in that the spatial information is independent of the perpendicular angle of crossing. This technology has variously been called PDT for pressurized drift tubes, RDT for round drift tubes (RDTs were considered for GEM [50]), or MDT for Monitored Drift Tubes. The latter is the baseline technology for ATLAS barrel muon spectrometer. The "monitored" concept is based on the detection of the last electron to drift to the anode wire as well as the first to give a self-calibrating drift time and a discriminator against neutron and soft-gamma backgrounds [51].

In the presence of an axial magnetic field, the isochrones of the cylindrical drift tubes remain circles centered on the anode wire with a modified distance-to-time relation. A more complicated Lorentz angle compensation is needed for transverse magnetic fields. The round drift tube devices are especially suited for muon spectrometers based on solenoidal magnetic fields where the anode wires can be oriented along the axial field component, although they are being considered for the ATLAS toroid system [52]. To provide a bunch crossing capability using mean timers, the DTBX (Drift Tube with Beam Crossing Identification Capability) has been considered by CMS [35].

Some care in the choice of gases is needed for the RDT technology applied to LHC. Because of the expected high backgrounds, small drift times are needed of order < 500 ns. This requires the use of a "fast" gas. Since the E-field drops as $1/r$, non-flammable gas mixtures, such as Ar-CO_2 (80-20), can not be used since they would lead to very non-linear distance-time relation, which would be difficult to control. A small Lorentz angle is also desirable imposing further requirements on the gas mixture [53].

Limited streamer drift chambers (LSDT) considered for the GEM muon system [54] are constructed out of 2.54-cm open drift cells made from thin Al sheets on three sides and are operated in the limited streamer mode. By measuring the drift time, spatial resolutions of order 100 μm have been obtained. An orthogonal coordinate is measured by cathode strips placed above the open side of the cathode drift cell. By using the correlation of times of the wire pulses with those of the cathode, a two-dimensional space point is measured.

4.2.3 Plastic Streamer Tubes

PST's are modular chambers made from extruded PVC plastic cells coated on the inside with graphite paint to form the cathode. They can be operated in either the limited streamer mode or the proportional mode. The anode-cathode wire spacing is typically 0.5 cm in 1×1 cm^2 cells. A 100-μm-diameter wire is usually used. Cathode strips parallel to the wires and pads are frequently employed to given orthogonal coordinate readouts.

The technology is now quite well understood and has been employed extensively in LEP-CERN experiments (ALEPH, OPAL, and DELPHI), in the SLD-SLAC, as well as in FERMILAB, and various cosmic ray experiments. Several manufacturing concerns in the US and Europe now mass-produce the chambers and associated strip electronics.

4.2.4 Jet Cell Chambers

Jet cell chambers are essentially multiwire chambers in which the anode wire plane is oriented radially. In this configuration the drift cells are staggered to resolve the L-R ambiguity. In another configuration the resolution of the L-R ambiguity is accomplished by arranging the jet cells at a finite angle with respect to the infinite momentum trajectories. Each track is on both the left and right sides of the anode plane and the chamber is self-calibrating. The open wire drift chamber concept has been used with success in the L3 muon system at CERN and was proposed for the L* detector at the SSC [55].

The merit of this technology is that a large number of measurement points along the trajectory are possible, giving good determination of the local track vector. Spatial resolutions of the order 200 μm have been achieved using a non-flammable gas mixture of Ar-CF_4-CO_2. Double-track resolution of 2-3 mm is possible.

4.2.5 Parallel Plate Chambers

Parallel plate chambers (PPC) have been considered for triggering in the very forward region of the CMS detector [35]. The PPC is a single-gap gas detector with planar electrodes flat to 5 μm separated by 1 to 2 mm. The chamber is operated in the avalanche mode and has been tested up to rates of 108 cm^{-2}s^{-1}. Sizes up to 50×50 mm^2 can be fabricated in any shape to meet the spatial requirements of the large-η region of a muon system at a hadron collider.

4.2.6 Thin Gap Chambers

Thin gap chambers have been used for the OPAL Pole Tip Calorimeter [56] and have properties which make them suitable for the trigger of the ATLAS forward muon system. Constructed as multi-wire chambers with a 1.6-mm anode-cathode distance and 2-mm anode wire pitch, the chambers have good rate capabilities and resistance to radiation damage. The inner cathode surfaces are coated with carbon thereby allowing both strip and pad readout. The chambers are operated in the saturated gas-gain mode and are not very sensitive to mechanical deformations. Typical drift times of < 8 ns are obtained and the chambers have been operated in environments of up to 500 MIP-mm^{-2}s^{-1}. Early work on this chamber concept is described in reference [57].

4.2.7 Resistive Plate Chambers

The RPC is essentially a narrow gap (2 mm) avalanche chamber with a strongly quenched streamer [58]. The chambers can be configured to give a two-coordinate readout. The plates of the chamber are coated with thin sheets of semi-conducting plastic or glass (*e.g.*, bakelite) of high resistance (typically 10^{10} to 10^{12} Ω-cm) and are operated with a gas mixture containing a strong electronegative component such as freon. Both gas and electrode structure lead to the strong quenching of the avalanche. The merits of this technology are its low cost and fast timing (< 5 ns).

The RPC is a good candidate for a muon trigger system [both the ATLAS and CMS barrel regions] and for providing the second coordinate for a drift tracking system. RPC's are also under consideration for a muon tracking system in the SLAC B Factory project. They have been used in a number of applications, including the L3 endcap system.

The R&D effort on this technology is concentrated on improving the RPC rate capability. At present, the efficiency of the chamber sags and time jitter degrades with increasing rate burden. Promising results have recently been obtained by lowering the gas-gain with more freon and by using plastics of lower resistivity [59,60] the chambers can be operated at count rates approaching 1,000 MIP-cm^{-2}s^{-1}.

4.3 Precision Alignment of Detectors

In order to achieve their design resolution, most muon systems require precision alignment over large areas and distances. Typical requirements depend on the design concept of the system but the requirements can be as exacting as 25 μm over distances of order several meters. This is quite a challenge and several techniques are under development.

A few of the techniques for both global and local alignment presently under consideration for both the ATLAS and CMS muon systems are listed below:

- stretched wire with capacitive readout,
- LASER straightness monitors,
- LED–Photodiode optical straightness systems,
- bar-code optical systems, and
- proximity monitors

Many of the properties of these technologies were explored in the development of the SSC experiments (see references [34] and [16]).

5 Particle Identification

The e^+e^- B Factories at CESR, SLAC and KEK and the Fermilab Tevatron collider program will require good hadron identification for CP-violation physics. In particular, the ability to separate π's from K's over a wide range of momenta will be of paramount importance. For the most part this section will deal with techniques devised for this purpose. However, transition radiation detectors, which are uniquely suited for electron identification and have a wide range of applications in both B physics and elsewhere, will also be considered.

5.1 Requirements

5.1.1 e^+e^- Colliders

The generation of e^+e^- B Factories now under construction employs K-π separation in two distinct ways: flavor tagging of "opposite-side" particles and particle identification in exclusive decays. The former application is primarily of interest to the experiments at asymmetric colliders.

In the flavor-tagging application, the sign of the b quark is inferred from the sign of the charged kaon that emerges from the end result of the Cabibbo-favored $b \to c \to s$ cascade. This technique offers relatively high tagging efficiency ($\sim 35\%$) and a false-tagging probability of only ($\sim 10\%$), which arises mainly from Cabibbo-suppressed decays in the cascade. The momentum spectrum of the tagging kaons is reasonably soft, extending upward to only ~ 1.5 GeV/c.

Particle ID is also expected to play a crucial role in the reconstruction of exclusive decay modes such as $B^0 \to \pi^+\pi^-$ and $B^\pm \to D^0 K^\pm$. In these decays the charged hadrons are characterized by a much harder momentum spectrum and coverage up to about 4 GeV/c is needed (3 GeV/c in the case of symmetric machines). Since the lower end of the momentum spectrum for decay daughters extends down to 1.5 GeV/c, the ideal system would provide coverage over the full range up to 4 GeV/c.

The level of rejection required for unwanted particles varies depending on the application at hand, but in some cases is as stringent as 100:1. If this is to be accomplished without significant loss of acceptance the π-K separation should be of order 3–4σ.

The problem of particle identification is made more difficult in the e^+e^- environment by the need to fit the system in a relatively small radial gap (≤ 25 cm) between the central detector and the electromagnetic calorimeter. Moreover, since it is important to be able to detect low-energy photons with high efficiency, the amount of material in front of the calorimeter must be limited to $\sim 20\%$ of a radiation length.

5.1.2 Hadron-Beam Experiments

As is the case at e^+e^- colliders, the principal need for hadron identification in contemporary experiments with hadron beams is for studies of B-particle decays with emphasis on CP violation and neutral B-meson mixing.

The π's and K's from B decays in hadron experiments have only slightly larger transverse momenta than at e^+e^- colliders; good efficiencies would be obtained in detectors with coverage up to $P_t = 1.5$ GeV/c for tagging of opposite-side B's, and up to 2.5-3 GeV/c for two-body B decays. Thus a central detector for B physics at a hadron collider has almost identical kinematic requirements for hadron identification as does an e^+e^- collider. An important difference is that a substantial fraction of relevant particles occurs at small angles to the beam in hadron experiments, so the total-momentum spectrum of particles produced at polar angle θ is scaled by $1/\sin\theta$ times the transverse-momentum spectra discussed above. For small polar angles, particle identification via Čerenkov radiation in gases is a useful technique that is less relevant in central detectors.

The B decays of interest for CP violation are not associated with high-transverse momentum jets, so the π's and K's to be identified are typically not accompanied closely by other particles. The local density of B-decay products per interaction is only slightly higher in hadron experiments than in e^+e^-, when angular distributions are discussed in terms of the pseudorapidity, $\eta = -\log\tan\theta/2$, a logarithmic angular variable. Namely $dN/d\eta \approx 4-6$ in hadron experiments, where N is the number of charged particles per interaction. This kinematic feature leads to the requirement that the detector segmentation be roughly uniform in pseudorapidity rather than polar angle. Further, a detector with nearly full angular acceptance should cover $\Delta\eta \approx$ 8–10 at the Tevatron collider, and 10–12 and LHC.

Another distinguishing feature of hadron-beam experiments is the interaction rate. An ambitious (but typical) goal is to operate at 10^7 interactions per second, compared to only about 20 (hadron producing) interactions/second at an e^+e^- B Factory. The electronics of the particle ID systems must have a higher rate capability for hadron experiments, although the minimum time between interactions (16-200 ns) is actually longer in hadron experiments than at e^+e^- B Factories (2-4 ns).

5.1.3 Dedicated Electron Identification

It is difficult to quantify the requirements for dedicated electron identification systems based on transition radiation, since these system often work in conjunction with other systems (calorimeters and preshower detectors) that also provide e/π separation. Generally speaking, however, one seeks to maintain high ($\geq 90\%$) electron acceptance while achieving hadron rejections on order of 100:1.

5.2 Technology Options

Pions and kaons are most readily distinguished from one another through a measurement of their velocity, which when combined with a momentum measurement allows one to deduce their mass. Thus virtually all dedicated hadron identification devices are based on either a time-of-flight (TOF) measurement or on the detection of Čerenkov light. In one novel scheme [61] these two approaches are combined as described below.

5.2.1 Time of Flight

A particle's velocity can be directly determined by measuring its flight time along a known path length. For relativistic particles the maximum momentum at which a specified time separation, Δt can be obtained is given by

$$p_{\max} = \sqrt{\frac{L\,(m_K^2 - m_\pi^2)}{2c\Delta t}} \qquad (2)$$

where L is the path length, typically of order 1.0–1.5 m.

The next generation of scintillator-based timing systems are expected to provide timing resolutions of $\sigma_t \simeq$ 80–100 ps (see *e.g.*, reference [32] pp.. 73–80), which will yield a 3σ separation to momenta of about 1.3 GeV/c. This is adequate to cover nearly all of the tagging region in e^+e^- experiments and the low part of the momentum range needed for hadron colliders.

The 100 ps figure represents an advance over previous systems, which realized resolutions of order 150 ps [28]. It is at least in part due to the use of magnetic field tolerant fine mesh photomultiplier tubes [62] coupled directly to the ends of the scintillator. Although the basic technique was one that was well established, it has benefitted from technological advances.

Scintillator-based TOF systems have long been used in particle and nuclear physics experiments and represent minimal technological risk. Moreover, apart from the need to maintain rather stringent timing stability, they are robust and straightforward to operate. Since limited momentum coverage can be extended with the addition of a Čerenkov based system as discussed below, TOF systems are serious contenders for the particle identification systems of B-physics experiments.

5.2.2 Threshold Čerenkov Detectors

In a threshold Čerenkov system one chooses a radiator medium with a refraction index n such that $\beta_K < 1/n < \beta_\pi$ in the range of momentum of interest. When that is the case, pions will emit Čerenkov light whereas kaons will not.

In the momentum range relevant to the e^+e^- B experiments, one requires a refractive index in the range $1.01 < n < 1.02$ (larger indices may be required if one wishes to extend the coverage to lower momentum), which can be achieved using silica aerogels or pressurized gases. Systems based on both aerogel [31,32,63] and high pressure gas are under study [63].

Aerogel The aerogel technique is not new but has enjoyed a resurgence in recent years in anticipation of the turn-on of the B Factories. Although the primary light yield is reasonable, a significant fraction of the photons are lost due to Rayleigh scattering and absorption in the aerogel. Even in the most favorable cases only a few photons will be detected. Moreover it is not practical to transport them any distance from the radiator. One therefore requires a high-gain photodetector that provides good single photoelectron resolution while operating in a strong magnetic field. Thus in addition to the work on the aerogel itself, attention has been paid to the application of new photodetectors.

A number of lab and industry groups have been engaged in the manufacture of aerogel material. Although some of these efforts have been aimed specifically at high-energy and astrophysics applications, others originally were driven by aerospace applications.

The photodetectors under consideration as detectors of aerogel light are fine mesh photomultipliers with conventional bialkali photocathodes, micro-channel plate photomultipliers using GaAsP photocathodes and a hybrid device, which consists of an avalanche photodiode coupled to a conventional photocathode via a vacuum acceleration gap [64,46]. As with the TOF systems, high performance photodetectors will be crucial to the success of this technology.

High Pressure Gas Pressurized gas has been used as a Čerenkov radiator for many years (for reviews see references [65] and [66]). There are a number of gases that can be used to achieve refractive indices in the desired range at workable pressures (one to several atmospheres). Many of these exhibit good UV transparency and are readily available. A typical gas radiator Čerenkov counter for a fixed target experiment consists of a large volume of gas viewed by photomultipliers, often with the assistance of light-collecting mirrors. Most such approaches, however, are not suitable for the barrel regions of colliding-beam experiments, where space is at a premium. Workers at CLEO [67] have investigated a variety of novel schemes to address this difficulty. Generally, they employ an array of thin-walled 10-cm-diameter tubes, which the particles traverse at right angles. The light is redirected to end-mounted photodetectors (*e.g.*, high-field fine-mesh PMT's) using specially designed reflector arrays or waveshifters.

5.2.3 Ring Imaging Čerenkovs

Ring Imaging Čerenkov (RICH) detectors extend the momentum typically attained with threshold counters by operating in a regime where both π's and K's are above threshold. Particle species are distinguished by measuring the opening angle of their Čerenkov cone, θ_C which is given by

$$\theta_C = \cos^{-1}\left[\frac{1}{n}\sqrt{1+\left(\frac{m}{p}\right)^2}\right] \qquad (3)$$

Solid, liquid, or gaseous radiators can be employed, depending on the momentum range being covered. (For a review see reference [68].) Although the extended momentum coverage is an obvious plus, RICH systems are typically more expensive and complex than threshold devices.

RICH's Based on Gas-Filled Photodetectors The rings must be detected using highly-segmented two-dimensional photodetectors. Although conventional photomultiplier tubes could in principle be used for this, the number of detector elements needed is generally prohibitive—channel (pixel) counts of several tens of thousands, and even hundreds of thousands, are not uncommon. Thus workers have been led to develop photodetectors based on wirechambers filled with photosensitive gases such as TMAE or TEA [69].

An important technical advance of the past decade was the successful deployment of two large scale ring-imaging detectors (the DELPHI RICH [70] and the SLD CRID [71]). Although these devices are well matched to their current application, they are limited in their ability to handle high rates. This has led to efforts to develop improved photocathodes. One promising approach is to form a photocathode using a thin layer of CsI. The CsI layer can be deposited directly on a two-dimensional array of metal readout pads. Since the drift time for all photoelectrons is roughly the same, and since the pads can be individually instrumented with readout electronics, the intrinsic rate handling capability of the devices is greatly increased.

Solid CsI photocathodes have been the subject of an extensive R&D program over the past few years [72] and much progress has been made in understanding their capabilities and limitations. The work is now entering the critical phase of determining whether the successes achieved with table-top devices can be realized in full-scale detectors.

The DIRC A novel implementation of the ring-imaging approach, called the DIRC (**D**etection of **I**nternally **R**eflected **Č**erenkov light), was recently proposed by workers at SLAC [73]. This device uses the total internal reflection of light in long thin quartz bars to form a velocity dependent image. That image is transported to the edges of the detector where it is proximity focussed onto an array of photomultiplier tubes. An attractive feature of this approach is that while the idea is a new one, its implementation is based on largely conventional technology. The DIRC requires little space in the active region of the detector. However, the requirement that the quartz bars pass through to the outside of the detector imposes some constraints on the overall detector design. Moreover, a large phototube array ($\sim$ 10–20 K PMT's) is needed.

Since the HEP community has relatively limited experience with this sort of detector, its proponents have set out upon a program to experimentally measure important quantities, such as light yield and the quality of the quartz bars that can be obtained from industry. These studies have begun to produce results (see reference [31] pp.. 138–151.) and thus far support the design assumptions. A number of engineering details having to do with the magnetic flux return (for a solenoidal detector) and the support of the bars and other detector components remain to be worked out.

5.2.4 Čerenkov Correlated Timing

The RICH technique can be extended to implement an improved time-of-flight system. As noted by workers at CLEO [61] the light produced in a quartz radiator will propagate in way that increases the intrinsic difference in arrival times for signals from particles of different velocities. Since the time scale for emission of Čerenkov light is very short ($\sim 10^{-15}$ s) excellent timing resolution and therefore greatly enhanced particle separation is possible.

The effective propagation velocity for light trapped by total internal reflection in long quartz bars is such that the leading photons from faster particles (larger Čerenkov angles) arrive sooner than those from slower particles (smaller larger Čerenkov angles). This effect enhances the intrinsic time separation that arises from the difference in particle velocities. In this way, the $\sim$ 1.3 GeV/c upper limit on momentum coverage from conventional scintillator-based time-of-flight systems can be extended by a substantial factor.

5.2.5 Transition Radiation Detectors

Transition Radiation Detectors (TRD's) have been used successfully for particle identification for approximately the last fifteen years. They have been employed in a variety of experimental environments, including the Intersecting Storage Rings at CERN [74], hadron collider experiments at the CERN $Sp\bar{p}S$ and at the Fermilab Tevatron, an internal gas jet target experiment at the $Sp\bar{p}S$, as well as fixed target experiments at both laboratories [74,75,76,77]. The primary application has been electron identification [74,75], but more recently they have been used to identify hadrons as well [76,77]. These versatile detectors show great promise for use in the identification of B decay products in future heavy quark experiments at hadron accelerators. While their anticipated role in the central or moderately forward region of a collider would be to identify electrons, they would be expected to discriminate among hadron species in fixed target experiments or in the very forward regions of collider experiments. Moreover, they can be used simultaneously for particle identification and for precise tracking [78,79,80] by subjecting the output signals to multiple thresholds.

The total transition-radiation (TR) energy radiated is proportional to the Lorentz factor, γ, of the traversing charged particle, so that a TRD that turns on and reaches saturation for electrons between 1 and 2 GeV demonstrates the same response to pions only when they reach an energy of 250-500 GeV. Thus, the range over which TRD's are useful for electron identification spans more than two orders of magnitude in momentum.

Because space is so limited, the design of a TRD for electron identification in the barrel region of a collider experiment presents the biggest challenge. Such detectors must be compact and must present as little material as possible to the detectors placed behind them. TRD's make optimum use of the material they contain when radiator and detector are interleaved more frequently into a "fine-sampling" detector [81,82]. Thin detector volumes are desirable not only to minimize the amount of material, but also because they reduce the time over which the signal is collected. Drift times in mixtures containing xenon are long, although there is some evidence [83,84] that this can be improved by approximately a factor of two with the admixture of "fast" gases like CF_4. Further study of aging properties of TR detectors using this gas is needed.

The method of "cluster counting" has been demonstrated to be more effective in separating species than the method of total charge collection [82,85]. This technique entails shaping the pulses so that capture of an x-ray, which deposits a large amount of ionization in a small volume, yields a significantly larger pulse than the signal due to ionization loss along the path length sampled within the same integration time. For small cell depth, the difference in the two techniques is less significant than for a cell of larger depth, where the total ionization loss signal is larger and competes favorably with signals from x-ray capture.

Most research over the last few years has been in the study of radiator materials. The goal has been to find bulk materials that compete with regularly spaced foil sets in the amount of radiation produced. Such materials make the detector much simpler to construct and should be capable of supporting themselves without frames or spacers. They might even be sufficiently rigid to form the support structure for the detectors. Foam radiators have been compared to foils containing approximately the same amount of material in radiation lengths for both very thin [86], $\sim$ 0.04 g/cm^2, radiators and some [79,84,87] of a thickness $\sim 0.2 - 0.7$ g/cm^2, more typical in terms of detectors which have been built in the past. The foams are seen to yield of order 80% of the TR that is produced by the foils, independent of the thickness at which they are compared. Preliminary data indicate that a somewhat better yield can be obtained from mats made from CH_2 fibers and possibly also from Rohacell foam [87].

6 Large-Area Detectors for Non-Accelerator Physics

Large area muon detectors are needed for non-accelerator experiments are measured in sizes of tens of thousands of square meters. The two main usages are in underground detectors of low- and high-energy neutrinos and in Extensive Air Shower (EAS) experiments.

6.1 Underground Water Čerenkov Detectors

The first of the large underground water Čerenkov experiments was the IMB detector, located in a salt mine near Cleveland, Ohio [88]. It was soon followed by the Kamioka detector in Japan [89] and by the HPW experiment in the Silver King Mine in Utah [90]. All three were built in the early 1980's with the hope of detecting nucleon decay. Although these experiments did not detect nucleon decay, they did eliminate the simple SU(5) model of grand unification. The designs, which were similar, employed photomultipliers immersed in ultra-pure water. They registered contained events (few MeV to $\sim$ 1 GeV) and throughgoing muons (typically 20 GeV). Effective volumes ranged from 600 to 10,000 tons, and effective light-collection areas ranged from 100 to 400 m^2. The detectors used of order 10^3 photomultiplier tubes and achieved energy sensitivities in the range of one to six photoelectrons per MeV.

While IMB and HPW have shut down, Kamioka continues to operate, and two new large detectors are under construction. The SuperKamioka [91] tank, which employs 11 K twenty-inch Hamamatsu PMT's around its walls, is roughly 40 m in diameter and 40 m tall. It is scheduled to begin operation in 1996, and will be able to make order-of-magnitude improvements in measuring solar neutrinos, searching for nucleon decay, studying atmospheric neutrinos, and keeping a watch for supernovae from within or near our galaxy. Another experiment, called SNO [92], is being built in Sudbury, Canada, for operation in 1996. It will be filled with 1000 tons of D_2O for studying solar neutrinos. Both experiments are pushing down to thresholds in the several MeV range, and both will need to reduce radioactive isotope content to extremely low levels.

Light collecting optics are of importance as well. Kamioka and SNO employ light cones of a modified Winston type to achieve significant (tens of percent) increases in light collection. IMB employed 0.75 m square wavelength shifter (WLS) plates and achieved about a 40% increase in collection area over bare 5-inch PMTs. While WLS plates are cost effective, they add a few nanoseconds to the photoelectron arrival-time spread [88], significantly degrading the timing capabilities of the instruments.

6.2 Underwater Neutrino Detectors

DUMAND [93], Baikal [94], AMANDA [95] and NESTOR [96] are all projects aimed at underwater or ice detection of high-energy (typically TeV) neutrinos. They have sensitive muon detection areas of order a few $\times 10^4$ m^2 and angular resolutions of about 1°. Although these instruments primarily detect muons that traverse their volumes, they also are intended to detect high-energy cascades that occur throughout the array volume and even from distances up to 100 m away at PeV energies.

The water Čerenkov technique currently provides the only economical means of achieving muon detection areas at or beyond those now planned. The universally acknowledged goal for high-energy neutrino astronomy is a 1-km-scale instrument [97]. There are discussions of acoustical and radio detection techniques for sensing neutrino induced cascades of energies beyond 1 PeV. It is likely that these techniques will initially run parasitically on the Čerenkov instruments, which employ well-known technology and low thresholds.

6.2.1 DUMAND

DUMAND will be 4.7 km deep in the ocean near Hawaii and will employ 15-inch PMT's arrayed on a 10×40 m^2 horizontal lattice. The optical detectors are "smart" units, with local microprocessor control. Signals are sent to a "string-level" controller that contains digitizing and communication circuitry. The DUMAND group has designed and built a 200 K-gate GaAs LSI chip that digitizes 27 channels of pipelined data with 1 ns least count. Data are transmitted to shore on single mode fiber at 625 Mbaud, where trigger finding and processing takes place in digital hardware.

6.2.2 Baikal NT-200

The instrument in Baikal employs similarly large PMT's. However, because of the relatively shallow (1 km) depth and a background level that is 1000 times higher than DUMAND, the tubes are arranged in tighter clusters. The Baikal detector readout uses more traditional local coincidences, which are formed underwater and transmitted to shore via coaxial cable at low speeds. Eighteen dual-PMT units are currently installed and counting muons. This group has developed techniques for deploying detectors through the winter ice and laying cables in slots running from the site to the shore station.

6.2.3 AMANDA

The AMANDA array is being installed at the South Pole, with PMT's placed in holes drilled through the ice using hot water. In an initial deployment, 80 eight-inch PMT's were placed at depths between 800 m and 1000 m. It appears that air bubbles in the ice produce a 10- to 20-cm attenuation length, so that light progresses by diffusion, rendering fast timing over a useful distance range impossible. There are plans are for further PMT installation at a depth of 1500 m. Thus far this experiment employed coaxial cable communication to the surface and traditional signal processing using NIM and CAMAC electronics.

6.2.4 NESTOR

The NESTOR collaboration is planning an array similar to DUMAND but placed in the Mediterranean, off Pylos in the Southwest of Greece at a depth of 3.7 km. This collaboration will also employ large PMT's, fiber optics, and shore-based data filtering. The array is intended to be somewhat more dense than DUMAND and consequently will have a lower energy threshold (a few GeV). The NESTOR engineering design is still evolving.

6.2.5 KM3, NAO

Several meetings have now taken place to form the beginnings of a world consortium for building a km-scale instrument, which would have of order 10^4 sensors. The cost of such a device, which might be built around the turn of the century, would likely be of order \$ 150 M, comparable to that of a colliding beam experiment. The choice of venue has yet to be made for this collaborative effort. It will depend upon engineering experience with the present instruments. Nonetheless, many of the engineering requirements are already known, and work has already begun at JPL and Saclay.

6.3 Water Čerenkov Detectors at the Earth's Surface

Extended air shower (EAS) detectors have traditionally employed shielded liquid or plastic scintillators for muon detection. Generally the cost has been driven by civil engineering, since the cost of piling up absorber soon dominates. There is one exception, however. In recent years it has been realized that the water-Čerenkov technique can be applied to produce inexpensive large-area arrays for EAS detection.

The MILAGRO detector, which is aimed at studying air showers induced by gamma rays of cosmic origin, is now under construction near Los Alamos. It covers an active area of about 5000 m^2 with a single layer of photomultiplier tubes and will be sensitive to incident photons with energies extending to below 100 GeV. The detector consists of an 8-m-deep pond with a lattice of eight-inch photomultipliers looking upward to an opaque cover 2 m above. The top layer detects electromagnetic energy, and another smaller and deeper layer detects penetrating muons.

We note that seven proposals in the last few years (notably the GRANDE group in the U.S.) have aimed at building a larger and thicker water Čerenkov arrays than MILAGRO, with dimensions up to 10^5 m^2 by 50 m deep, with three layers, some looking down as well as up.

6.4 Technology Issues

We have outlined above the larger efforts in this field, with emphasis upon the technological issues. Here we will discuss what will be needed to achieve a fifty-fold increase in detector size.

6.4.1 Optical Detection

It appears that large-area photomultipliers will continue to be the transducer of choice for Čerenkov radiation for some years. Large-area solid state detectors are currently too expensive and suffer from high dark currents in comparison to PMT's. Work on pressure tolerant detectors, such as liquid-filled photosensors, could result in a breakthrough, but there is no program currently underway. One new approach is being explored at LBL [98], which envisages a cylindrical PMT with WLS collection down the axis. Modest improvements might also made through the use of mirrors and WLS plates to enhance light collection in selected directions. Both approaches currently suffer from degraded timing resolution and substantial mechanical problems when deployed in either ocean or ice environments and are therefore not being actively considered.

6.4.2 New Techniques of Detection

As noted, there are groups interested in acoustical [99] and radio detection [100] of ultra-high-energy neutrino-induced particle cascades, both in ice and in the ocean. If first measurements of *in situ* backgrounds and actual attenuation lengths are favorable, then these techniques may be pursued more vigorously in the next few years.

6.4.3 Fast Pipelined Electronics and Data Transmission

All cosmic ray experiments must collect and analyze their data in real time. This has caused some difficulties in employing off-the-shelf electronics, and has led to some new solutions which have applications elsewhere. A prime example is the DUMAND digitizer chip mentioned above, now being applied in the BNL $(g-2)$ experiment, and considered for other accelerator applications. Some experiments, such as MILAGRO, will face trigger and data handling problems which begin to approach those of future collider experiments.

Data transmission will surely move to fiber optics. While the HEP experiments have taken advantage of technology advances driven by the communications industry, some novel problems persist. For example, PMT pulses to be transmitted as time-over-threshold pulses are not well matched to off-the-shelf links, which generally require a 50% duty factor.

6.4.4 Fast Pipelined Data Processing

As with large accelerator experiments, massive underearth instruments must employ multi-layer trigger schemes to achieve the requisite reduction in rate. Although the flexibility of software-based trigger systems is an obvious attraction, CPU speeds are not yet high enough to handle data directly from the detectors (the gap is currently less than a factor of ten, however). Much experience with the architecture and algorithms for such systems will be gained in the next few years

6.4.5 Calibrations and Survey

All the Čerenkov detectors discussed require calibration for amplitude and timing. Large neutrino detectors, which span a very large dynamic range in energy, require only logarithmic calibration. Although presently the dynamic range is typically only 10^3, improvements will be needed in the future.

The pulsed light sources that are presently most frequently used (typically N_2 lasers), work either by illuminating the PMT's via a scattering ball or by illuminating a scintillator that reradiates the pulse. While this works well enough, lasers are awkward and not completely reliable. Improvements in fast blue LED's would be a great help, though they are presently limited in stability, time and output amplitude.

For the solar neutrino detectors, low-level calibration is quite delicate, and several techniques will be employed. These will use radioactive sources and possibly even small electron accelerators to achieve calibration accuracies of few percent at 5 MeV.

Surveying is a special problem in both the ice and ocean. In the ice it must be done well once. In the ocean, due to a tidal excursions of a few meters for bottom-tethered instrument strings of several-hundred-meter height, one must monitor the position on the time scale of hours. All three deep water experiments plan to use acoustical surveying. Indeed, the DUMAND Seattle group has advanced the field somewhat already with an acoustical chirp technique using DSPs for running correlations. Their system achieves a resolution of a few centimeters, which is orders of magnitude better than commercial deep-ocean systems.

6.4.6 Reliability

As for all massive detectors, quality control and reliability assurance are of paramount importance since the instruments can be serviced only with difficulty or not at all. A major challenge therefore is to achieve the sort of mean time between failure statistics typical of the aerospace industry without adding significantly to the cost of the instruments.

6.4.7 Service and Robotics

Service is a serious issue for these experiments. For example, the SuperKamioka detector will take nearly a year to drain, service and refill with adequately pure water. The PMT's in the ice at the South Pole are not removable once installed. SNO plans to use a robot for in-tank service. The deep ocean experiments also plan to use robots and submarines. For a future deep water detector, adequate robotic capability not only will allow repair and replacement, but will also allow reconfiguration of the array to respond to changes in the physics program.

6.4.8 Deep Ocean Technology

DUMAND's unique requirements for deep underwater operation have led to substantial technological innovation in a number of areas as mundane, but vital, as connectors and cable terminations. The development of pressure-tolerant electronics suitable for direct immersion in fluid would be of great benefit to future efforts since it would obviate the need for heavy, expensive, and unreliable housings and feedthroughs. Corrosion is always a concern, but simple procedures and the use of non-metals and titanium eliminate that risk.

7 Electronics

Historically, application of rapidly developing electronics technology to electronics for HEP experiments has enabled more sophisticated and more powerful physics experiments. In general, the HEP community needs an ongoing R&D program to further develop application of existing technologies, to study application of new technologies, and to become aware of developing technologies. For example, circuits using existing technologies to deliver the large dynamic range required for future calorimeters must be developed. In addition, the emerging technology of multiquantum well devices is promising for high-speed analog and digital data transmission in compact, low-power units and should be studied for HEP application. The rapidly evolving nature of electronics technology requires a strong R&D program in order to exploit technological advances in extending physics capabilities. In the following sections, some of the current challenges needing the attention of an active R&D program are discussed under the general headings of Front-End Electronics and Trigger Systems. Challenges to data acquisition and computing systems, which also exploit rapidly evolving electronics technology, are discussed in the Computing section.

7.1 Front-End Electronics

Front-end electronics extract physical quantities of interest, usually time and/or charge, from detector signals. They perform analog signal processing, digitization, and buffering of detector signals. The capabilities of the HEP community to develop its own custom analog integrated circuits for front-end electronics have made it possible to instrument detectors with hundreds of thousands of channels. This has in turn made it possible to build highly granular detectors that offer better resolution and the ability to reconstruct and study ever more complicated events. For example, the development of integrated circuits with complete readout channels on a pitch of 50 μm has enabled us to read out silicon microstrip detectors directly, in turn enabling the detection and measurement of secondary vertices from the decay of short-lived particles. The use of custom integrated front-end electronics is still in its infancy, now being used in only a small fraction of the detector systems in most experiments. However, all large new experiments planned or under construction depend critically on custom IC's for all detector systems, and the use of these techniques will largely define physics capabilities in the future.

Future front-end electronics require circuits that are high speed, low noise, low power, and high density, requirements which are sometimes at odds with one another. Circuits for high rate environments also require pipelined architectures allowing signals from many events to be buffered while trigger processing occurs. They must be simultaneously read/write such that signals from new events can be processed while data from a previous event is read out. Analog integrated circuit development requires considerable experience and expertise. Requirements that are specific to the front-end electronics for certain detector types are outlined below.

7.1.1 Electronics for Calorimeters

Calorimetry requires pulse height measurement with high dynamic range, seldom less than 15 bits. This requirement places great demands on readout electronics. Preamplifiers, amplifiers, ADC's, and analog memory cells of sufficiently high dynamic range are needed. The performance needed for the next generation of high-energy accelerators, where calorimetry requires even greater dynamic range (about 18 bits), is not yet in hand. R&D of both the building blocks, such as analog pipelines, and of complete integrated systems should be continued. Such R&D can provide not only the enabling technology for read-out of calorimetry, but also production circuits that can be used on more than one experiment.

7.1.2 Electronics for Drift Chambers

Drift chambers require time measurement circuits with resolution of one nanosecond or less. They also often require high density, and they sometimes require radiation hardness. Circuits which meet these requirements now exist; however, fully integrated systems have not yet been produced. Drift chambers also often require pulse height measurement. Circuits for this purpose, and circuits for combined measurement of time and pulse height are also needed.

7.1.3 Electronics for Particle Identification Systems

In general, electronics for particle identification systems, such as ring-imaging Čerenkov detectors and for time-of-flight counters, is generally not as far along in its development as electronics for calorimeters and drift chambers. The requirements for a particular particle identification technique are often unique. For example, electronics for high resolution time of flight systems must provide time resolution of 50 ps, along with a pulse height measurement of several bits resolution. Such performance has not yet been achieved in integrated circuits.

7.1.4 Electronics for Semiconductor Trackers

The development of electronics for silicon microstrip detectors is a good example of how custom integrated circuits can play a key role in detector technology. Although progress in this area has been impressive, circuits with the complete functionality needed for future experiments, in particular circuits with pipelined memory for simultaneous read/write, are not yet complete. In addition, development of readout electronics for pixel detectors, affording high resolution, two-dimensional position measurement, is in its infancy. The design of signal-processing and data-storage circuitry with the ultra-high density required for pixels is a great challenge. The size of pixel readout circuitry will limit pixel size, and hence resolution, for the foreseeable future. The readout electronics for microstrip gas chambers (MSGC's), can be similar to silicon microstrips; however, MSGC's will impose some unique signal processing requirements if they are to achieve their ultimate performance in a high-rate environment.

7.1.5 Radiation hard electronics

One of the greatest R&D challenges in the electronics area is the development of circuitry with sufficient radiation hardness to withstand the very high fluxes of particles associated with high-luminosity colliders. For instance, doses of 10 MRads and fluences of 10^{14} n cm^{-2} will not be unusual in future experiments. Development of radiation-hard custom circuits is not yet sufficiently advanced. Although radiation hard semiconductor processes are commercially available, vendors often know little about the radiation hardness of analog circuits manufactured in their processes. Much further study of the radiation hardness properties of available processes for a variety of circuit types is needed. In addition, R&D on this subject is needed to study new processes, monitor changes in processes, and to study suitability of processes to new circuit designs.

7.2 Trigger Systems

Although trigger systems are generally experiment specific, the architectures and tools used in their implementation are generic. The techniques available at the trigger level often define the physics accessible to a given experiment. Consequently the lack of sufficiently powerful tools can limit physics potential, particularly for high-rate hadron physics experiments. As the technological capabilities of electronics have grown, so has the ability to design trigger systems with greater and greater physics capability. Remarkable progress has been made in a few short years since the days when trigger systems were constructed from modular logic units that provided two to four simple logic functions per module, which were cabled together to select event candidates. Trigger systems are now "cabled up" using commercial programmable logic, which provides thousands of logic functions per integrated circuit. Moreover, custom integrated circuits capable of providing tens of thousands of functions, including computational functions, are often implemented. Modern trigger systems also harness the power of hundreds of commercial processors in trigger "farms," greatly enhancing the ability to implement sophisticated trigger algorithms. Further developments along these lines will be needed for full exploitation of hadron colliders.

Trigger architectures for high-luminosity experiments now generally consist of multiple levels employing pipelined and parallel processing techniques and high bandwidth data transfer. Accordingly, the discussion of existing technical challenges and needed R&D is by trigger level.

7.2.1 First Level Triggers

First level triggers are characterized by high input rates (for instance 40 MHz at the LHC) and short latencies (typically 2 to 4 μs). At the first level, algorithms must be provided by dedicated logic, sometimes referred to as hard-wired or special-purpose processors. This logic must be pipelined to accept input rates with the time between event candidates less than the fastest imaginable processing times, and it must be massively parallel in order to simultaneously process all trigger inputs. For instance, processing of hundreds of thousands of input channels in parallel is not unusual. To address these requirements in a cost-effective and reliable manner, and in a way that offers adequate flexibility and programmability, custom and semi-custom (*e.g.*, field-programmable gate arrays) integrated circuits are required. Such IC's are the only way to provide sufficient logic to address the requirements. R&D into the development of high-density digital integrated circuits for high-energy physics is needed to make commercially available techniques accessible to the HEP community, to gain experience with available techniques from outside HEP, to track rapidly evolving technologies, and to learn how to apply them to HEP. This R&D should also include study of synthesis tools for facilitating IC design.

Another important challenge in the design of first-level trigger systems is the need to bring a very large number of input high-rate signals to the processor hardware. With the use of application-specific IC's (ASIC's) to perform the processing, the performance and cost of first level triggers can be severely limited by the required connectivity, *e.g.*, by the bandwidth of data links or by the number of input pins on IC's. R&D on high-speed data paths is needed to enable effective first-level triggers for future experiments. Topics for R&D include the use of commercial very high-speed fiber optics (greater than a GigaByte per second) in synchronous operation and serial data receiver circuit blocks for custom IC's.

7.2.2 High level triggers

High level triggers, often referred to as Level 3 triggers, are characterized by the use of commercial, general-purpose processors, such as RISC processors found in high-end workstations. These processors offer the advantage of cost-effective, flexible, programmable, and upgradeable implementations of trigger algorithms. Nevertheless, the use of these processors in new environments with higher rates and more demanding physics requirements presents several challenges. These challenges relate to the use of larger and larger numbers of processors in parallel. For instance, LHC experiments postulate the use of "farms" of as many as one thousand processors. Mechanisms for delivering data at very high rates (1 to 10 kHz), such as high-speed switching networks require development, particularly as applied to the HEP trigger environment, where the nature of the data flow differs from existing commercial applications. Software tools for managing and debugging code running in such a large distributed system, and software techniques for production of reliable code for programs executing a trillion operations per second, are also required.

7.2.3 Intermediate Level Triggers

Triggers at levels intermediate between the first level and high levels are characterized by input rates (*e.g.*, 100 kHz) that approach those of the first-level trigger yet require selection algorithms that are comparable in complexity to those of higher levels. Investigations into numerous processing techniques are warranted in this domain, for instance, studies of digital signal processors (DSP's), neural networks, commercial image processors, and advanced custom processors. Studies of architectures, including embedding of commercial CPU chips into custom architectures as well as use of commercial CPU boards or workstations, should also be pursued. Finally, the ability to guide and deliver very high bandwidth data to multiple processors is also crucial in intermediate levels; however, the challenges differ from those at the first level or at higher levels and deserve separate study.

7.2.4 Trigger Algorithms

Although most trigger algorithms are experiment-specific, several techniques and challenges are rather generic. For instance, calorimeter triggers for identifying electrons and jets are similar in many experiments. A general challenge is fast tracking triggers, driven in large measure by the desire to study heavy-quark physics. Trigger algorithms capable of identifying secondary vertices in high-rate large-multiplicity environments are of particular interest in the study of B physics. Development of such algorithms could lead to a breakthrough in physics potential in both the fixed target and hadron collider fields.

8 Conclusions and Recommendations

8.1 Importance of Detector R&D

From the preceding sections it is clear that the upcoming generation of accelerators will place severe demands on detector technology. Although in recent years impressive progress has been made in most areas of detector R&D, much remains to be done.

To maintain the health of our field in the future, it is important that new ideas for detection and measurement of the observables of particle physics nurture and bloom. These ideas are the "gene pool" from which future detectors will evolve. Without such activity it will become increasingly difficult to take full advantage of the physics opportunities presented by the new generation of facilities. In short, detector R&D is an important part of the HEP program and must be protected.

8.2 Generic vs. Detector-Specific R&D

When considering how best to proceed in ensuring that progress in detector development continues to be made, one encounters the question of whether detector R&D should be funded through a separate program dedicated to this end—*i.e.*, a "generic" R&D program—or within the context of experiments—*i.e.*, "detector-specific".

Until recently, most detector development work was detector-specific at least in a formal sense. The situation changed around 1990, when the SSC laboratory established a program of generic detector R&D as a means of stimulating early work on the SSC's experimental program. At about the same time, the Texas National Research Laboratory Commission (TNRLC), began to fund to research proposals of a generic nature. (The TNRLC also funded on-going physics research and infrastructure.) Unfortunately, with the demise of the SSC both the generic detector R&D and the TNRLC programs were cancelled, leaving much excellent detector development work without support.

At present there are several other pressures that work against a healthy program for developing new detection techniques. The scale and cost of detectors has grown enormously. With that growth comes an increased conservatism in making technological decisions. The scientists charged with managing large detector projects understandably look for a high degree of assurance that a technology will perform before committing substantial funding. Moreover, the time and resources required for a technology to progress from the conceptual phase to a mature state has increased in accordance with increases in complexity. Finally, in an atmosphere of limited funding, "discretionary" funds (*e.g.*, director's funds at laboratories and endowed funds at universities) are increasingly likely to be used to address short-term crises rather than to provide seed money for the sort of innovative but speculative ideas that lead to new techniques.

There are, however, valid considerations that tend to favor a detector-specific funding mechanism. Perhaps the most important of these is that with a fixed overall budget, any funds devoted to generic R&D must ultimately come from other parts of the HEP program, none of which currently have enough support. Perhaps the best arbiters of which detection techniques ought to be developed are the detector collaborations themselves. A large purely generic program runs the risk of diverting precious HEP resources to support the development of technologies that are "interesting" but of no particular benefit to particle physicists.

From the preceding paragraphs it follows that while some funding of generic detector R&D is warranted, any program to do so should be of modest scope and should be aimed primarily at nurturing new techniques just long enough to provide proof-of-principle demonstrations. Beyond that "market forces"—*i.e.*, decisions made by detector collaborations—should determine which technologies receive the level of support necessary for the deployment of practical devices.

8.3 Features of the Proposed Program

The details of any such program are best determined by members of the HEP community working in conjunction with the funding agencies. In this section, we present only general considerations.

8.3.1 Level of Support

Most new techniques will benefit enormously from relatively modest levels of support. Generally speaking, amounts between $50 K and $250 K per project per year will suffice. Indeed, funding on a scale larger than that runs the risk of creating large programs that are not particularly conducive to innovation. If one assumes that six projects each averaging $150 K/year are started each year and continued for three years, the steady state outlay for the total program would be about $3 M per year or about 0.5% of the current HEP budget. Even in a scenario with flat or declining budgets, the budgetary impact of such a program should be manageable since funds "taxed" away from other HEP programs will be redistributed, easing the pain, at least on average.

8.3.2 Duration of Support

As noted, the intent of the proposed program is to provide an incubation period for new or improved detector technologies. Generally speaking, three years should suffice to develop new techniques to the point where detector collaborations can seriously consider their use. Thus the typical development project might be funded in three one-year installments, with the second- and third-year renewals being conditioned primarily on demonstrated progress. In exceptional circumstances, longer periods might be allowed, although this would obviously have an adverse impact the ability to launch new starts. In any event, projects extending beyond three years would not be "renewed" in the usual sense but instead would be subjected to the same sort of review applied to new applicants.

8.3.3 Method of Peer Review

Since the number of funded projects will be small and many of the proposals are likely to be highly technical in nature, the selection of proposals for funding will most efficiently be done by a panel of experts. The panel should comprise physicists (and where appropriate engineers) with expertise in accelerator- and non-accelerator-based detector technology as well as high-energy physicists representative of the "end-user" community. Outside experts might also be included to provide additional perspective or technical expertise. By asking panel members to serve in three-year staggered terms, the panel would retain some "memory" while at the same time being continually refreshed.

8.3.4 Eligible Applicants

Individual or group proposals from laboratories, universities, and industry should be eligible. Collaborative combinations should also be permitted, although the relatively modest levels of funding will naturally tend to limit the size of collaborative efforts. In exceptional cases, larger collaborations may be formed if this is deemed to be the most efficient approach to a particular development effort.

8.3.5 Criteria for Evaluation

To be successful this program should have as its primary goal the development of new and improved detector technologies. Thus this should be the main criterion in evaluating proposals. Of course, the usual technical criteria (capability and/or track record of the investigator(s), feasibility of the proposed plan, appropriateness of the budget, etc.) should be taken into account, since such considerations are important to the success of any research venture.

8.3.6 Use of Funds

Since the amount of funding will be limited, it is essential that there be a minimum number of restrictions on their use. Investigators should be free to use the funds in a way that is most effective. Since second- and third-year renewals will not be automatic, there will be a strong incentive for investigators to use their funding in the most effective way.

References

[1] J. Angus and C. Hyman, *Science* **241**, 688 (1990); W. Yarbrough and R. Messier, *Science* **247**, 688 (1990).

[2] R.J. Tesarek *et al.*, *Nucl. Instrum. Methods A* **349**, 96 (1994)

[3] F. Borchelt *et al.*, CERN preprint PPE/94-113 submitted to *Nucl. Instrum. Methods* (1994).

[4] S.F. Kozlov *et al.*, *IEEE Trans. Nucl. Sci.* **22**, 160 (1975).

[5] The RD-42 Collaboration, CERN/DRDC 94-21, DRDC/P56 (1994).

[6] M. Algranati *et al.*, *Nucl. Instrum. Methods* **164**, 615 (1979); C.J.S. Damerell *et al.*, *Nucl. Instrum. Methods* **185**, 33 (1981); C.J.S. Damerell, *Nucl. Instrum. Methods A* **342**, 78 (1994); C.J.S. Damerell *et al.*, Proc. XXVI Conf. on High Energy Physics, Dallas 1862 (1992).

[7] S. Gaalema, *IEEE Trans. Nucl. Sci.* **32**, 417 (1985); S. Shapiro *et al.*, *Nucl. Instrum. Methods A* **275**, 580 (1988); S. Shapiro *et al.*, SLAC Pub 6249, June 1993.

[8] S. Parker, *Nucl. Instrum. Methods A* **275**, 494 (1989); C. Kenney *et al.*, *Nucl. Instrum. Methods A* **342**, 59 (1994); W. Snoeys *et al.*, *IEEE Trans. Electr. Dev.* **41**, 903 (1994).

[9] B. Heinemann *et al.*, *Nucl. Instrum. Methods A* **305**, 517 (1991).

[10] S. Parker *et al.*, *IEEE Trans. Nucl. Sci.* **41**, 2862 (1994).

[11] Rockwell International, Microelectronics Process Center (PO Box 3105, Anaheim, CA 92803). A survey of companies in this field has been done by G. Gilchriese, J. Millaud, and D. Nygren and presented to an ATLAS collaboration meeting at the Lawrence Berkeley Lab (7 March 1995).

[12] Research Devices, 121 Ethel Road West, Piscataway, N.J. 08854.

[13] M. Campbell *et al.*, *Nucl. Instrum. Methods A* **342**, 52 (1994).

[14] P. Delpierre *et al.*, *Nucl. Instrum. Methods A* **342**, 233 (1994).

[15] M.D. Petroff and M.G. Stapelbrock, *IEEE Trans. Nucl. Sci.* **36**, 158 (1989); M.D. Petroff and M. Atac *IEEE Trans. Nucl. Sci.* **36**, 163 (1989). M. Atac preprint UCLA-PPH0040-2/92 (1992).

[16] Solenoidal Detector Collaboration Technical Design Report, SDC-92-201, SSCL-SR-1215, April 1992.

[17] W.J. Willis and V. Radeka, *Nucl. Instrum. Methods* **120**, 221 (1974).

[18] S. Abachi *et al.*, *Nucl. Instrum. Methods A* **324**, 52 (1993).

[19] D. Axen *et al.*, SLAC-PUB-5354(1992).

[20] O. Benary *et al.*, *Nucl. Instrum. Methods A* **344**, 363 (1994).

[21] ATLAS Letter of Intent, CERN/LHCC/92-4, October 1992.

[22] B. Aubert *et al.*, *Nucl. Instrum. Methods A* **309**, 438 (1992); *Nucl. Instrum. Methods A* **321**, 467 (1992); *Nucl. Instrum. Methods A* **325**, 116 (1993); *Nucl. Instrum. Methods A* **330**, 405 (1993).

[23] O. Benary *et al.*, to be published in *Nucl. Instrum. Methods A* (1994).

[24] V. Fanti *et al.*, *Nucl. Instrum. Methods A* **344**, 507 (1994); G.D. Barr *et al.*, *Nucl. Instrum. Methods A* **323**, 393 (1992).

[25] G. Gratta, H. Newman and R.Y. Zhu, Crystal Calorimeters in Particle Physics, CALT-68-1922, March 1994, to be published in *Ann. Rev. Nucl. and Part. Sci.* **44** (1994).

[26] M. Oreglia *et al.*, *Phys. Rev. D* **25**, 2295 (1982).

[27] L3 Collaboration, *Nucl. Instrum. Methods A* **289**, 35 (1990).

[28] Y. Kubota *et al.*, *Nucl. Instrum. Methods A* **320**, 66 (1992).

[29] E. Aker *et al.*, *Nucl. Instrum. Methods A* **321**, 69 (1992).

[30] K. Arisaka *et al.*, KTeV Design Report, FN-580, January 1992.

[31] PEP-II Collaboration Letter of Intent, June 1994.

[32] BELLE Collaboration Letter of Intent, KEK Report 94-2 April 1994.

[33] L* Letter of Intent to the SSC Laboratory, November 1990.

[34] GEM Letter of Intent, SSCL SR-1184, November 1991.

[35] Compact Muon Solenoid Letter of Intent, CERN/LHCC 92-3, October 1992.

[36] L3P Letter of Intent, CERN/LHCC 92-5, 1992.

[37] R.Y. Zhu and H. Yamammoto, GEM TN-92-126 and CALT 68-1802, 1992.

[38] D.A. Ma and R.Y. Zhu, *Nucl. Instrum. Methods A* **332**, 113 (1993).

[39] See papers by Y. Dafinei. D. Hitlin, H. Newman, B. Winstein, C.L. Woody, R.Y. Zhu, in Heavy Scintillators for Scientific and Industrial Applications, Proceedings of the "Crystal 2000" International Workshop, Editions Frontieres Volume C58, edited by F. Nataristefani *et al.*, September 1992.

[40] D. Acosta *et al.*, *Nucl. Instrum. Methods A* **302**, 36 (1991).

[41] M. Hulbert *et al.*, *Nucl. Instrum. Methods A* **335**, 427 (1993).

[42] G.S. Abrams *et al.*, *IEEE Trans. Nucl. Sci.* **27**, 59 (1980).

[43] A. Dwurazny *et al.*, DESY Report 88-070.

[44] D.R. Wood, *Nucl. Instrum. Methods A* **289**, 331 (1990).

[45] R. DeSalvo *et al.*, *Nucl. Instrum. Methods A* **315**, 375 (1992).

[46] P. Cushman and R. Rusack, *Nucl. Instrum. Methods A* **333**, 381 (1993).

[47] E. Mathieson and G.C. Smith, *IEEE Trans. Nucl. Sci.* **36**, (1989).

[48] H. Van der Graaf *et al.*, *Nucl. Instrum. Methods A* **307**, 220 (1991).

[49] A. Bohrer *et al.*, Status Report of the RD5 Experiment, CERN-DRDC-91-53, January, 1992.

[50] C. Bromberg *et al.*, GEM TN-93-288 (1993).

[51] ATLAS MDT Concept ATLAS Internal Note Muon-NO-029, October, 1993.

[52] ATLAS Letter of Intent, CERN/LHCC/92-4, LHCC/12, October, 1992.

[53] M. Dold *et al.*, ATLAS Internal Note Muon-NO-043, June 1994.

[54] L. Osborne *et al.*, GEM Technical Note TN-92-203, October 1992.

[55] U. Becker *et al.*, *Nucl. Instrum. Methods A* **253**, 15 (1986).

[56] G. Mikenburg, *Nucl. Instrum. Methods A* **265**, 223 (1988).

[57] S. Majewski *et al.*, *Nucl. Instrum. Methods A* **217**, 265 (1983).

[58] R. Santonico and R. Cardarelli, *Nucl. Instrum. Methods A* **187**, 377 (1981).

[59] R. Santonico, private communication.

[60] C. Wuest, R. Bionta, and I. Pless, Appendix of GEM Technical Note TN-93-288 (1993).

[61] M. Selen and K. Honscheid, CLEO Preprint CBX 92-116 (1992); K. Honscheid, M. Selen, and M. Sivertz, *Nucl. Instrum. Methods A* **343**, 306 (1994).

[62] H. Kichimi *et al.*, *Nucl. Instrum. Methods A* **235**, 451 (1993).

[63] CLEO Staff, CLNS Note 93-1265 (1993).

[64] S. Fagan, Laser Focus World, November 1993, 125; M. Szawiowski *et al.*, Advanced Photonics preprint, December 1993.

[65] A.S. Vovenko *et al.*, Soviet Physics Uspekhi **6**, 794 (1964).

[66] J. Litt and R. Meunier, Ann. Rev. Nucl. Sci. **23**, 1 (1973).

[67] Various options have been discussed by members of the CLEO collaboration; see *e.g.*, talk by K. Kinoshita at the *B* Factory detector meeting at KEK, November 1993.

[68] J. Séquinot and T. Ypsilantis, *Nucl. Instrum. Methods A* **343**, 1 (1994); T. Ypsilantis and J. Séquinot, *Nucl. Instrum. Methods A* **343**, 30 (1994);

[69] J. Séquinot and T. Ypsilantis, *Nucl. Instrum. Methods* **142**, 377 (1977).

[70] See *e.g.*, R. Arnold *et al.*, *Nucl. Instrum. Methods A* **252**, 188 (1986).

[71] D.W.G.S. Leith, *Nucl. Instrum. Methods A* **265**, 120 (1988).

[72] P. Mine, *Nucl. Instrum. Methods A* **343**, 99 (1994); D.F. Anderson, S. Kwan and V. Peskov, *ibid* p. 109; C. Lu and K.T. McDonald, *ibid* p. 135; A. Breskin *et al.*, *ibid* p. 159.

[73] B. Ratcliff, BaBar Collaboration Note #92 (Dec. 1992); P. Coyle *et al.*, SLAC Report SLAC-PUB-6371, (1993). See also P. Coyle *et al.*, *Nucl. Instrum. Methods A* **343**, 292 (1994).

[74] J. Cobb *et al.*, *Nucl. Instrum. Methods* **140**, 413 (1977).

[75] G. Barr *et al.*, *IEEE Trans. Nucl. Sci.* **36**, 66 (1989).

[76] D. Errede *et al.*, *Nucl. Instrum. Methods A* **309**, 386 (1991).

[77] M. Baake *et al.*, *Nucl. Instrum. Methods A* **281**, 325 (1989).

[78] V. Bondarenko *et al.*, *Nucl. Instrum. Methods A* **327**, 386 (1993).

[79] E. O'Brien *et al.*, *IEEE Trans. Nucl. Sci.* **40**, 153 (1993).

[80] J.T. Shank *et al.*, *Nucl. Instrum. Methods A* **309**, 377 (1991).

[81] B. Dolgoshein, *Nucl. Instrum. Methods A* **252**, 137 (1986).

[82] T. Ludlam *et al.*, *Nucl. Instrum. Methods A* **180**, 413 (1981).

[83] L. G. Christophorou *et al.*, *Nucl. Instrum. Methods* **171**, 491 (1980).

[84] KTeV TRD group, KTeV technical notes 13, 146, 150-6, 158, 164, 166, 172, 178, 180, 185, 197, 225-6, 232-3, and 243. These can be obtained from Ms. Laura Sedlacek, MS 231, Fermilab, P.O. Box 500, Batavia, IL 60510.

[85] C. Fabjan *et al.*, *Nucl. Instrum. Methods A* **185**, 119 (1981).

[86] V. Chernyatin *et al.*, *Nucl. Instrum. Methods A* **325**, 441 (1993).

[87] M. Sheaff *et al.*, Design Studies for Compact, Low-Mass, High-Rate Transition Radiation Detectors, SSC Generic Research proposal #432.

[88] R. Becker-Szendy *et al.*, *Nucl. Instrum. Methods A* **324**, 363 (1993).

[89] K. Hirata *et al.*, *Phys. Rev. D* **45**, 2170 (1992).

[90] J. Negret, Ph.D. thesis, Purdue Univ., UMI 88-14570-MC, 1987.

[91] K. Nakamura, Proceedings of the 3rd International NESTOR Workshop, edited by L. Resvanis, Pylos, Greece, Oct. 1993.

[92] H. Chen *et al.*, *Nucl. Instrum. Methods A* **264**, 48 (1988).

[93] C. Wiebusch, DUMAND II Project Report, HDC-5-94 (1994).

[94] G. Domogatsky, *et al.*, 6th International Workshop on Neutrino Telescopes, edited by M. Baldo-Ceolin, Venice, Feb. 1994.

[95] A. Goobar *et al.*, Proceedings of Neutrino 1994, XVI International Conference on Neutrino Physics and Astrophysics, Eilat, Isareal, June 1994, edited by A. Dar and M. Gronau (1994).

[96] various papers in the Proceed 3rd International NESTOR Workshop, Pylos, Greece, Oct. 1993, edited by L. Resvanis (1994).

[97] J. Learned, Phil. Trans. R. Soc. Lond. A **346**, 99 (1994).

[98] D. Nygren, LBL, private communication (1994).

[99] J. G. Learned and R. J. Wilkes, Proceed. 23 ICRC, Calgary, Canada, 1993, for acoustical detection in water; B. Price, *Nucl. Instrum. Methods A* **325**, 346 (1993).

[100] See reports on radio detection in ice by J. Ralston and G. Smoot from Snowmass 1994.

Computing

KENNETH MCFARLANE

CEBAF, Newport News, VA 23606, USA

Working Group Members

S.C. Loken,	*Lawrence Berkeley Laboratory*	C. Watson,	*CEBAF*
R.P. Mount	*California Institute of Technology*	J. Womersley	*Fermilab*

1 Introduction

High-energy physics has made heavy use of computational technology throughout the development of both fields. Common roots include the first "AND" gate using vacuum tubes, and the development of Monte Carlo simulation techniques for neutron transport during the Manhattan Project. Now, "computing in HEP" is taken to mean application of (primarily) digital computer hardware and software in a variety of overlapping areas, including DAQ/event building, control and monitor systems, on-line and off-line filter and reconstruction systems, networks and interconnections, data handling and storage systems, user interfaces and visualization, information systems and communications, simulations, and analysis systems.

We discuss these various aspects of computing under three headings:

- software systems and tools,
- computing hardware, and
- information systems and communications (including both hardware and software).

The remarkable progress in computing, along with early HEP adoption and adaptation of hardware developments in computing and electronics, has enabled the HEP program to be successful. There are, however, issues in each of the above areas which will require significant R&D and capital investment to meet the future needs of HEP. Within the U.S., HEP laboratories have not produced a coherent body of software with scope anything close to that produced by CERN. With the cancelation of the SSC project, the U.S. has lost both a source of funding for, and a vehicle for coordination of, computing R&D. We believe that the proportion of the HEP effort devoted to computing, and the level of coordination, must be increased if HEP computing technology is to keep pace with other detector developments.

In the following, we discuss past progress and the current status, review likely technology needs, and make recommendations for the future.

2 Past Progress and Current Status

Advances in computing made within HEP have been critical to the success of the HEP program, and some developments have had, and are having, wide influence on computing in general, despite the fact that HEP is a now a relatively small part of the user community. There have been notable HEP-industry associations, from which both have benefited. HEP-industry projects have influenced IBM (HEPVM, REXX, storage management, tape robots, real-time AIX, WDSF), DEC (VAX/VMS to STK silo connection), STK (tests of high-capacity cartridges). An L3-IBM project has implemented a high-performance system using HiPPI. HEP representatives sit on computing industry Customer Councils and Advisory boards, as well as standards bodies. Computer "farms" (*e.g.*, ACP at FNAL, SHIFT at CERN, PDSF at SSCL) have influenced commercial products.

The advent of the SSC and LHC created new challenges to computing systems. After some study, it appeared that (Level 1) trigger systems could reduce the trigger rate for general-purpose detectors by about three orders of magnitude, from the collision rate of 60 MHz. Given expected event sizes, the task of DAQ and computing was to take a data rate of about 30 GB/s of event fragments and make that consistent with storage technology (a rate to storage of order 100 MB/s seemed possible) and analysis requirements, in a cost-effective way. Most of the resulting issues re-appear in near-future experimental environments (*e.g.*, Tevatron collider upgrades and high-rate fixed-target experiments such as KTeV) and future colliders (*e.g.*, LHC, SLAC PEP-II, and RHIC). To address the challenges, a variety of R&D projects in DAQ and computing were begun: in the SSCL Generic R&D Program, in the SDC and GEM R&D programs, within the SSCL, and in concert with the DOE part of HPCCI. (Some funding also came from the TNRLC.)

Projects carried out in the SSC context included:

- studies of massively-parallel closely-coupled computing systems for DAQ (for B physics),
- studies of CPU farms for on-line filters and production,

- trials of commercial network components for DAQ,
- design of DAQ/event builders in combination with on-line filtering,
- R&D on data-base technology for event storage (the PASS project),
- development of detailed software concepts for off-line computing, and
- donstruction and operation of heterogeneous computing systems based on UNIX (*e.g.*, PDSF at SSCL).

A notable characteristic of these efforts was the use of modeling and discrete simulation of the hardware systems, to an extent new in our field. Several vendors participated in projects, and the PASS (Petabyte Access Storage Solutions) project (in which SSCL was one participant) is continuing. Also, there was a significant effort to involve computer professionals in the SSC program. Below, we give some examples of past progress and achievements (this is not an exhaustive list and we hope that any omissions are drawn to our attention), and describe the current status of computing in HEP. Next, we comment on the mode of operation and funding approaches.

2.1 Software Systems and Tools

The character of HEP has resulted in many software developments designed to enhance the productivity of physicists who are not computer specialists; these developments have been freely available and shared internationally. Notable ones include:

- standard packages, beginning with such programs as SUMX, KIOWA, and TVGP, which were crucial in the era of resonance studies,
- early use of interactive/time-share computing (Wylbur),
- HEPVM, with priority batch and tape-handling systems,
- tools built to extend the capabilities of FORTRAN (Mortran and ZEBRA),
- EGS, for electromagnetic shower simulation, and
- the CERNLIB package, with many tools, including detector simulation (GEANT), analysis (PAW), fitting (MINUIT), numerical analysis, and the ZEBRA memory management/data structure system, that were ported to many systems.

EGS has been widely adopted for simulation of medical radiation treatment systems, and CERNLIB has been used outside the HEP community. Industry software standards committees, such as those for POSIX and FORTRAN) have had HEP representatives. The current status of software technology in HEP can be summarized as follows:

- Tools such as GEANT and PAW are the HEP standard (*e.g.*, CDF's primary tool for their top analysis was PAW).
- database technology has been clearly shown to be necessary for Large experiments.
- UNIX has gained acceptance in HEP (for many experiments, the bulk of the CPU power is in UNIX systems).
- Beginning at the detector end (DAQ, slow control), C and C++ are supplanting FORTRAN as the language of choice; many projects using an object-oriented approach are underway.
- Software methodologies and tools (particularly SA/SD) have been used with some success; tools for developing object-oriented systems are coming into use.

The software situation is less than satisfactory: there have been several cases where insufficient attention to software has led to loss of results. HEP is behind the computing industry in the use of new operating systems and languages, and software engineering approaches. Whereas 10 years ago HEP graduate students received a valuable training in computing, many potential students are now repelled by the need to learn and use outdated and inefficient software techniques.

2.2 Computing Hardware

The HEP program has always had a high demand for cost-effective CPU power, and has needed access to very large data sets. Data acquisition from electronic detectors is also a critical application of computing and storage hardware technologies. These needs have driven a number of innovations and adaptations:

- emulators as an inexpensive source of CPU cycles, for analysis and on-line filters;
- pioneering use of computers for real-time DAQ (PDP-1, 8, 11), in conjunction with standards such as CAMAC (CAMAC was adopted by IEEE as a U.S. standard for DAQ and control);
- early trial of videotape for storage;
- farms, adopted by industry for some applications; and
- massively-parallel computing systems (MPP) developed for lattice QCD calculations (following the demonstration that problems can be decomposed onto MPP) have been crucial to the development of industry MPP systems.

HEP representatives have been involved in developing hardware standards: SCI and CAMAC are examples.

The current technical status of computing hardware in HEP can be summarized as follows:

- Data-compression at the front end is increasingly used to reduce bandwidth requirements, at the expense of more CPU power being needed later.

- The improving cost, performance, and power consumption of VLSI CPU's is leading to their use closer and closer to the front end of the DAQ chain.
- CPU farms, using volume-produced commercial boxes, are established as a cost-effective and manageable approach to computing power.
- Modern multiprocessor "mainframes" with I/O capacity of hundreds MB/s (megabytes/second) are used for data-intensive physics analysis.
- Tape, and tape robots, can handle the present generation of experiments.
- High-performance tapes (*e.g.*, Ampex DD-2) are being used for data storage.

The availability of hardware resources to U.S. HEP is nearly adequate: fortunately the extremely rapid cost and performance improvement of industry products has more or less kept pace with (even unanticipated) HEP needs and budgets. There is presently a CPU resource shortage (of a factor of two to four) which affects the ability to run simulations of existing and proposed detectors and to carry out needed reconstruction and DST production of existing data. Up to its demise in mid-1994, the PDSF at SSCL was still heavily exploited by U.S. HEP.

2.3 Information and Communication Systems

Information and communication systems have been critical to the rapid progress in HEP; HEP has been a model for "getting online." The experiments run by large collaborations could not be managed without e-mail. File and information sharing over networks is also critical to rapid analysis of data and dissemination of results.

Notable developments include:

- early extensive use of e-mail for collaborative efforts;
- the SPIRES data bases and the QSPIRES no-login access system;
- World-Wide Web; this hypertext information system, which does not require logging into the computing system carrying the information, was originated at CERN and has now spread far beyond HEP;
- networking installed in Europe for HEP has been extremely influential in developing the infrastructure there, and links to China, Russia, and Brazil have had important impacts; and
- HEP is an active participant in development of desktop video-conferencing over the network.

The current technical status of information and communication systems in HEP can be summarized as follows:

- Video-conferencing is in increasing use and was essential in the development of SSC collaborations, including links to Japan.
- World-Wide Web is being used extensively to support collaborative efforts.
- Commercial packages for documents (text and figures) are being used; there are incompatibilities which inhibit collaborative efforts.

Information and communication systems are currently adequate within the U.S. and parts of Europe. International links have been restricted by financial and political considerations, and by lack of communications infrastructure within regions that are active in HEP, such as China and the former Soviet Union.

3 Mode of Operation

The current administrative status for computing in HEP is unsatisfactory from a number of perspectives. The demise of the SSC has removed an important source of funding. While the CERN LHC RD program continues with its support of computing projects, the U.S. has lost a central focus and vehicle for coordination. Budgets for computing organizations in HEP are flat, and many costs are hidden and dispersed, making cost/benefit optimization and coordination of programs difficult. R&D funds for computing in HEP are small compared with other R&D funds: in the SSC Generic R&D program, funding for computing was 3% of the total.

Besides funding of R&D work by DOE and other agencies, another avenue for making progress is through joint work with industry. This has been quite successful at CERN, which is seen by vendors as a display point for Europe; U.S. HEP laboratories have not been viewed as demonstration sites to the same extent.

For the U.S. high energy physicist, computing is not seen as an advantageous career path, though computing systems need to be built just as detectors need to be built. This has had disadvantages: problems are solved only in the immediate context, new languages and operating systems are adopted slowly, and software is poorly documented and maintained. Within the U.S., HEP laboratories have not cooperated to produce a coherent body of software with scope anything close to that produced by CERN.

4 Technology Needs

Technology needs of near-term and future experiments can be measured by the Level 1 trigger rate, event size (which is a measure of complexity), and acceptable rate to storage. Current experiments operate at a DAQ rate (product of L1 rate and event size) of about 30 MB/s with data rates to storage of order 3 MB/s. In the future, fixed-target experiments will have DAQ rates of 2 GB/s (*e.g.*, KTeV) and high-luminosity hadron collider experiments, 40 GB/s (see Table I, taken from various sources). Rates to storage will be about 3 MB/s and 100 MB/s respectively. For typical runs, this implies data samples up to 1 PB. The rates anticipated for LHC detectors very likely will be the maximum for the foreseeable future: the next generation of colliders will likely trade higher energy for lower luminosity.

The high data rates and large data samples place demands on hardware, and the large reductions from DAQ rates to storage rates, along with the need to simulate and analyze, place demands on software. In addition, large collaborations need improved information and communication systems. Both simulation and reconstruction will require the computing power available to an experiment to be increased by three orders of magnitude. We must fully exploit available technologies so that a shrinking or stable community can maximize its productivity.

Opportunities for HEP to contribute to the computing industry may include the systems that can handle DAQ and analysis of large samples of "events." The analogy of a shopping cart full of items to an HEP event has been suggested. A large chain of stores might generate up to 100 Hz of customer checkouts, with perhaps 2 kB of data per checkout, a rate quite comparable with LEP experiments, but occurring day in and day out. Each event will have correlations between purchases; these are already being exploited by retailers with point-of-sale terminals to increase the market impact of coupon campaigns, but the ability to record and analyze large data sets efficiently may give industry more sophisticated capabilities.

4.1 Software Systems and Tools

Software systems must be improved so as to enhance physicist productivity. An "engineering" approach should be taken: the practices of the most successful experiments and the best tools of industry should become standard. A way must be found within this framework to overcome the inevitable conservatism of HEP software: the very success of the HEP tools in use inhibits a shift to new methods. Complex systems with many tools that depend on detailed information about the data structures become very hard to change.

Issues to be addressed are:

- software management for distributed efforts;
- certification and control of code, especially for online filter systems;
- flexibility, to take advantage of evolving technology and requirements without disruption of on-going work;
- the use of commercial tools (*e.g.*, software development environments, analysis packages, distributed system management);
- analysis frameworks that facilitate the posing of physics questions;
- data bases and data access tools;
- computing professionals as part of teams, or hired to produce specific software systems; and
- interoperability, to allow the use of heterogeneous systems.

4.2 Computing Hardware

Computing hardware provides CPU power, interconnections in distributed systems (including DAQ systems), data storage, and data access systems. The technical needs of near-term experiments appear to be satisfiable with existing hardware, or hardware available in the near term. Future experiments are more demanding and extrapolations of computing cost and performance less certain.

The main areas of need for high-luminosity experiments are:

- specialized systems for trigger and DAQ,
- data storage and data access, and
- CPU power for physics simulations and event reconstruction.

Trigger and DAQ systems will use some combination of off-the-shelf items and special-purpose items, and will be tailored to fit available technology and the specific needs of experiments. Estimates of CPU power needed for experiments vary from 1-10 kMIPS for KTeV in the next two or three years, to 1,000 to 10,000 kMIPS for LHC in ten years. This includes CPU power needed for simulation, which is likely to equal or exceed that needed for reconstruction. It should be noted that the Physics and Simulation Facility (PDSF) at SSCL was a 4 kMIPS facility and was heavily used, after the cancelation of SSC, for about a year for HEP work (unrelated to SSC). Rates in data-storage and data-access systems can be very high; systems currently being envisaged will require many links operating at 100 MB/s. For access to data and distributed analysis, network bandwidths of order 1 GB/s will be needed. New data base technologies will impose new demands on hardware, as will desktop videoconferencing.

There will be two major needs in HEP that will likely require substantial adaptation of what we can expect from industry:

- data compression and pipelining in the front end and DAQ, which will require CPU power and high-speed communications in specially-configured systems to build events and unpack the data;
- storage of data and access for analysis. The data-access requirements will dominate the bandwidth and structure of the storage systems.

Computing systems have been evolving especially rapidly since 1980, in performance and cost, as a result of a number of factors: VLSI technology, hardware-software optimization (RISC), portable operating systems (UNIX), volume production, competition, and widespread use (from less than 200 systems on networks fifteen years ago, there are now millions). Current trends are that:

Table 1: Estimated rates and event sizes for various experiments

Experiment	L1 Rate	Event size	Data rate (peak)	Rate to storage (mean)
CDF				1 TB/yr
KTeV	500 kHz	4 kB	2 GB/s	3.3 MB/s
KEK B	200 Hz	30 kB	6 MB/s	
OPAL	3 Hz	100 kB	0.3 MB/s	0.3 MB/s
DØ	150 Hz	400 kB	60 MB/s	1 MB/s
GEM	100 kHz	300 kB	30 GB/s	1 MB/s
LSND	100 Hz	2 kB	0.2 MB/s	0.2 MB/s
CMS	50 kHz	500 kB	25 GB/s	20 MB/s
FNAL Run 3				
EOI	5 kHz	300 kB	2 GB/s	60 MB/s

- The power of a single CPU increases by a factor of 1.4–1.6 per year.
- CPU power drops in cost by a factor of 1.6–1.8 each year.
- Fast memory drops a factor of 1.3 per year.
- Slow memory drops by a factor of 1.4 per year.
- Magnetic disks drop a factor of 1.2–1.7 per year.
- The cost of bandwidth into mass storage drops by a factor of 1.1 per year.
- The speed of communication links is expected to change in steps: perhaps a factor of three every four years.

These trends mean that a constant expenditure of funds on equipment will buy double the power every two years. Measuring computing power in VAX-11/780 equivalents (VUPs), a million VUPs in workstations cost about $130M in 1993, and will cost $10M in 1999. The cost of bandwidth to mass storage is declining more slowly; nevertheless, from 1993 to 1999 a factor of three decrease can be expected. I/O bandwidth will likely evolve in a more step-wise fashion, *e.g.*, by ATM or Fibre Channel replacing Ethernet and SCSI.

These trends appear to put the CPU power required, and storage for pre-LHC era experiments, within an acceptable cost range. Even for LHC, computing hardware costs may be a small fraction of the total detector cost, provided that volume-produced systems continue to use chip area for CPU's and caches, and not for signal processors for special tasks such as speech recognition or image processing. Network technology appears to be advancing at an appropriate rate. The area of data base technology is also advancing, but LHC experiments may have difficulty in meeting their cost and performance requirements in this area.

4.3 Information Systems and Communications

With the large collaborations and international character of HEP, providing consistent data and rapid communication will be of prime importance for our field's productivity in future years. During the SSC era, physicists were added to the U.S. HEP community, and there were new university groups, resulting in more groups, smaller groups, and scattered groups, increasing the need for effective communication. Several collaborations have found that publication of results is delayed by the pace of communication within a widely dispersed group.

Communications within collaborations will need to be supported in the following areas:

- videoconferencing at every location;
- systems to support "virtual meetings" across widely separated time-zones'
- document standards, including graphics, to permit electronic sharing, and group editing, of documents; compatible document processing software licensed at all HEP sites; and
- inter-regional and international networks to pull together world-wide collaborations.

Bringing communications to the desktop of the active participant is critical to get the most productivity out of a widely-dispersed community. Currently, the Internet is a bottleneck in fully exploiting computing capabilities. Trends in information systems (World-Wide Web, satellite systems, desktop videoconferencing, BISDN and SONET) are favorable, as the commercial world develops similar needs. There are still national and regional barriers to fully open inter-regional and international communications, since communications are now so linked to national and regional competitiveness that progress may be slower than purely technical considerations would require. As recently as last year, the primary link for HEP between the U.S. and Europe was connected with a specific experiment, and links to Russia could not be direct, and went via a 9600 baud connection. These arrangements are being changed but are indicative of the problems that must be addressed.

There are also very significant network needs for data serving and data analysis, and for management of distributed computing facilities. With petabytes of data being contemplated, the issues of centralizing data and production computing, versus having regional centers are inextricably tied up with those of network capacity, reach, and cost. Within the U.S, the HEP community has benefited from having connectivity provided without direct cost on ESNET and the Internet; it is very likely that increasing commercial exploitation will change this, especially as the bandwidth needs increase.

Much of the international connectivity needed for participation in the CERN and DESY physics programs is directly funded by U.S. HEP. Extrapolation of the historical downward evolution of international tariffs (10–15% per year) yields a highly constrained model for the collaborative analysis of petabyte data samples. Deregulation and a massive growth of commercial traffic could result in one or two orders of magnitude more bandwidth per unit cost.

5 Recommendations

The computing environment has changed greatly over the past decade, in ways that dictate new approaches for our discipline: HEP is not a driver of general computing technology, we must position ourselves to move with industry and to take advantage of industry developments for other customers. Overall, the field must find ways to do more with less; this to us implies greater coordination of computing efforts, but also an investment in new software and hardware to maximize physicist productivity. Particularly in the U.S., computing efforts have been fragmented and have not had the usefulness or scope of the coordinated effort in Europe. In the future, it is clear that the development of computing systems for HEP will be driven by the needs of LHC, and hence will be focused at CERN. Nevertheless, the U.S. HEP community must participate in these developments, to provide training for students and support for U.S. efforts in HEP.

There are several paths that could be followed. One would be to make the bulk of computing support a professional activity, by employing software and hardware professionals to do most of the work (this has budget implications, but may increase physicist productivity). Another approach, in which physicists would continue to develop computing systems, would be to raise the status of contributions to HEP computing, and define satisfying career paths for those who choose this area, partly by tying the work to a coherent body of HEP developments (thus in effect professionalizing the work inside HEP). That is, either regard it as engineering to be paid for and managed, or as part of the discipline to be valued as it builds a productive technology within a coordinated framework.

- Our main recommendation is that the U.S. HEP community should formulate a coherent approach to the use of computing technology. A mechanism should be created to foster these discussions, involving the HEP and Nuclear Physics laboratories and user groups.

Specific recommendations that we make here, and that we expect would be part of a coordinated approach, are:

- Make policy choices lying between the extremes of
  - contracting out all computing and software responsibilities and
  - assigning these responsibilities to university and laboratory personnel.

  The policy must be formulated with the agreement of funding agencies. Where responsibilities are assigned to universities and laboratories, appropriate career structures may have to be created, particularly if the universities are to retain a major rôle in this key area of HEP.
- Re-capitalize U.S. HEP institutions (universities and laboratories) in a consistent way, compatible with current and anticipated industry development toward open systems. This means consistent hardware, consistent operating systems, consistent application software, and good networks. If necessary, provide training in workshops and summer schools to speed transition. The goal is to provide access to adequate CPU power, good networks, and videoconferencing.
- Provide centers of CPU power and storage for simulation, reconstruction, and analysis, accessible to any HEP institution.
- Fund R&D in computing:
  - software systems and tools: data bases and data access methods, software methodologies;
  - computing hardware: data handling and analysis systems (parallel I/O, disk farms, robots), trigger and DAQ systems; and
  - information systems and communications: esktop video conference, document preparation systems.

  In view of the rapid changes in computing and software technologies, R&D projects should be regularly reviewed by experts from within and from outside HEP.

There are questions which must be debated, areas where there is advantage to be gained in taking a specific path. As one example, one operating system could be chosen and full use made of the services provided, or we could continue the CERN approach of providing applications that run on top of any OS.

We conclude by emphasizing increasingly important to the HEP community, and a coherent and well-designed approach is needed to maximize the productivity of a distributed community with high demands.

Structural Issues In High-Energy Physics

Raymond L. Brock

Michigan State University, East Lansing, MI 48824

Robert N. Cahn

Lawrence Berkeley Laboratory, Berkeley, CA 94720

A.J. Stewart Smith

Princeton University, Princeton, NJ 08544

Michael Barnett

Lawrence Berkeley Laboratory, Berkeley, CA 94720

1 Introduction

In light of such events as the SSC cancellation, the controversy surrounding the decision on siting the B Factory, the declining budgets, and other problems facing U.S. high-energy physics, the DPF Executive Committee decided in late 1993 to add a Working Group on Structural Issues to the Long Term Planning Study already underway. At a time of such great change, it seemed reasonable to ask, for example, if the present HEP advisory structure needed improving. The time frame of this DPF Study overlapped the deliberations of the HEPAP Subpanel [1] chaired by S.D. Drell. To keep the Subpanel informed of our activities and to avoid interference, the DPF conveners participated in town meetings of the Drell Panel, and several Subpanel members attended meetings of the DPF Working Group. Interactions between the DPF and the Drell Panel were highly constructive, and the section on Governance in the Drell Panel's Report, issued in May 1994, reflects much of the input from DPF. The Drell Report has served as an important resource for subsequent DPF deliberations.

The DPF Working Group on Structural Issues began its work by posing to the community a number of questions and issues in February 1994 (See Appendix A). We received many messages by e-mail in response to this, and conducted three open meetings to get further input: May 6/7 at Johns Hopkins University; July 8/9 at the Lawrence Berkeley Laboratory; and August 7 in Albuquerque, following the 1994 DPF meeting.

It was decided early to separate the issues to be covered into three distinct tasks, each with a separate subgroup: Governance and Advising; Career Issues; and Education and Outreach. The reports of the subgroups follow.

2 Governance and Advising

2.1 The Existing Systems

For many years HEPAP has served as the primary advisory body on high-energy physics, and has long been cited as a successful model for communicating technical advice and, through HEPAP subpanels, making important decisions affecting the directions of the U.S. HEP program. The primary charge to HEPAP is to advise on the health of the High Energy Program. HEPAP members are chosen from among practicing experimental and theoretical high-energy physicists and accelerator physicists, in such a way to ensure diverse expertise and knowledge of the problems of each substantial constituency.

Recently, however, concerns have been raised regarding the effectiveness, limitations, and flaws of HEPAP in an era of constant or shrinking budgets and increasing demands on available funds. Since HEPAP is a DoE Institution it is constituted to advise on questions raised by DOE, with the result that the dominant concern is necessarily the operation of the DOE accelerator laboratories. Participation by the NSF is asymmetric and secondary. Scientific issues are discussed only infrequently, and there is no role for an advocate for University programs equivalent to that of the Director of a National Laboratory. It has become difficult for HEPAP to cope with the interlaboratory strife resulting from inadequate budgets.

HEPAP's effectiveness is compromised because by law it can meet only in public session. Because of this it is denied access to confidential information including preliminary budgets, and is unable to conduct the frank, confidential discussions that are essential to a thorough analysis of policy and program alternatives. As a consequence HEPAP is often reduced to reacting after the fact to decisions to which it has had little input.

Whenever important issues arise demanding advice based upon confidential deliberations and privileged information, DOE requests HEPAP to convene a Subpanel that is not bound by the public-meeting constraint. This process has usually resulted in well-thought-out reports and recommendations. However, the recommendations have often not been fully implemented by DOE and occasionally they have even been contradicted. As a result a certain cynicism has set in throughout the high-energy physics community. Even under the best of circumstances, the Subpanel process can be slow and divisive.

2.2 Deliberations toward a New Role for DPF

The Structural Issues Working Group focused on the advisory structures in HEP, and how to improve them. After several meetings at which the issue was discussed, draft proposals were prepared for a fuller discussion and adoption at the DPF94 meeting in Albuquerque last August. The proposals centered around the issue, raised often in e-mail and discussion, of whether DPF should institute a panel with a somewhat broader, complementary role to that of HEPAP. Such a group could concentrate on physics questions with less attention to institutional issues, address issues of concern to either or both DOE and NSF programs, and be a resource for governmental agencies and institutions, and for the public generally, providing information on particle physics.

One possible proposal for debate called for a STANDING COMMITTEE, commissioned by the DPF Executive, to examine and report on scientific progress and opportunities in Elementary Particle Physics. The committee would be composed of nine active researchers in the field, chosen for their knowledge of the program and insight into the scientific problems that the program addresses. More specifically, such a committee could perform the following:

- study physics questions and explore future directions by commissioning committees and workshops to address specific questions of science;
- monitor the modes of research in HEP, with a view toward ensuring that resources are optimized to produce the strongest possible scientific program, giving particular attention to the needs of universities;
- explore and enhance the relationship between HEP and related fields;
- serve as a resource to HEPAP, governmental bodies, and the general public; and
- serve as an advocate on behalf of high-energy physics and related science.

What sort of science questions with major impact on the HEP program might this committee consider? Examples of current interest might include: new experimental initiatives on neutrino oscillations, neutrino astrophysics, or rare decays; the need for overlap in various B-physics programs; upgrades of major accelerators and detectors, *e.g.*, the Tevatron energy or intensity; and the Next Linear Collider.

It was strongly emphasized that this committee should focus on the science, so as to complement and not directly compete with HEPAP. However, this proposal if adopted would institute an additional standing advisory body, whose role would have to be carefully crafted to avoid confusion or confrontations.

Promising developments during the spring and summer of 1994 led the Working Group to consider modifying the above proposal. First, the Drell Subpanel, which had expanded its charge to address governance and management, clearly took into account the ideas discussed by the DPF. Next, the report was well received in Congress, and by the funding agencies. The DoE and NSF have both declared their commitment to increase communication and cooperation, and to devise a more formal, expanded role on HEPAP for the NSF. A third development was the agreement of DoE to implement the Panel's recommendation for more direct input by scientists in budget planning.

As a result, an alternative proposal was presented by the Working Group at the Albuquerque meeting. Working within the present DPF structure, it builds on the experience and success of the SNOWMASS concept, invented by DPF in 1982, which greatly influenced HEP planning for many years. The main difference with the proposal described above is that, instead of commissioning a new standing committee, the DPF Executive Committee would itself take on an expanded role to take care of the responsibilities of the proposed standing committee.

2.3 Our Major Recommendation

During the debate in Albuquerque a consensus formed against forming a new standing committee, and in favour of the following recommendation:

- The DPF Executive Committee should commission workshops and studies on issues of great importance to the field. The Executive Committee would not carry them out, but instead appoint ad-hoc panels to do so.
- Each study should be of short duration, so as to make its conclusions and recommendations available in a timely manner.
- The reports resulting from workshops and studies should be disseminated promptly to the HEP community, DoE, NSF, the Laboratory Directors, and the public.

We are confident that this proposal will greatly reduce any potential conflict with HEPAP. The expanded responsibilities and roles will change the nature of the DPF Executive Committee, and in fact of DPF itself. However, these new responsibilities fall within the DPF charter, which reads:

> "The objective of the Division shall be the advancement and diffusion of knowledge of the fundamental particles and fields, their structure, their interactions and interrelationships, the design and development of high-energy accelerators, and the design and development of instrumentation techniques for high-energy physics."

2.4 Discussion

The recommendation only addresses items (1) and (2) of the new responsibilities proposed in the previous section for DPF. During the course of discussion, many members of the working group felt that items (3) (exploring relationships among HEP and other related fields) and (4) (acting as a resource to governmental bodies and the general public) have not diminished in importance and should also be addressed by the next DPF Executive Committee as part of its new expanded role. For example, item (3) could be dealt with by a study commissioned by the Executive Committee. However, for items (4) and (5), a satisfactory solution has not yet been found. Some feel that it would be a healthy component in the advisory apparatus to have a voice for the field which is not bound by a direct relationship with a single federal agency when so much of the government (Congress, OMB, OSTP, NSF) is directly involved in matters which very directly impact the planning and execution of a balanced program. (Here, "balanced" means proportioned according to science and not necessarily to institutional priorities.) While the Executive Committee could take it upon itself to represent the Field to these other important players in the management of HEP, this could also be delegated to an ad hoc group.

Finally, we suggest that the Executive Committee begin a tradition of preparing an annual, written "white paper" for the DPF membership on the State of the Field. This document should integrate over the panorama of the recent scientific accomplishments as well as current scientific needs and preparation for future scientific projects. It should include an appraisal of the progress made, or not, by the relevant governmental partners in meeting those needs. Given its intended audience it is best that this document should be rather brief. It could be prepared for the Executive Committee through an ad hoc review panel, which would broaden its base. This document would be distributed to the DPF membership, discussed with HEPAP, and generally shared with the other relevant agencies as an assessment of how well the scientific and administrative necessities are being met to keep this fragile science on course to continued success. The purpose would be to be constructive, to help keep policy in line with the scientific priorities.

3 Career Issues

3.1 The Problem

The demise of the SSC immediately put more than 100 physicists into the job market. The resulting shortage of positions and loss of physics opportunities convinced some of the SSC displaced-persons as well as many other promising people in HEP to leave the field, and also demoralized scientists with short-term positions as they contemplated searching for their next job in such a hostile climate. Calls for action came from many sources; the Users' Groups at the National Laboratories were particularly concerned. The DPF responded by including a study of careers as one of the Structural Issues in HEP, and the working group has begun to study the situation.

3.2 Establishing a Data Base

As soon as the discussions about career issues was begun it clear that reliable demographic data would be a crucial prerequisite for meaningful deliberations. Unfortunately very little useful data exists. The latest relevant survey was done by the Sciulli Subpanel of HEPAP in 1991. Hence the first action of the working group was to ascertain what plans there were, if any, to compile the necessary information and to analyze it. We learned in June 1994 that the DoE had proposed two years ago that the Particle Data Group (PDG) should undertake a comprehensive manpower survey, and that The PDG had submitted a draft questionnaire to the DoE for approval. No action had been taken on it, however, because several well-known crises with higher priority for DoE attention had intervened. The Structural Issues group decided to try to catalyze the process, with the result that the DoE and the PDG have scheduled the survey for Fall 1994. We are glad to report that the NSF has agreed to join the effort, and to contribute to its financial support. (Such a survey could not be complete without the close involvement and cooperation of the NSF.) The survey is now being prepared for broad circulation, its prime goals being to obtain least a 90 per cent response rate, and to reach all persons who have been active in HEP for any significant period within the past 15 years.

The survey will answer demographic questions such as the following, and then will search for trends:

- How many students, postdocs,faculty members, permanent laboratory staff members, etc., are supported by the U.S. HEP program?
- How long do typical scientists spend at each level as they progress from new PhD to senior researcher?

- What classes of positions do new PhD's choose in HEP, industry, medicine, law, finance, etc.?

Obtaining accurate data is only the first step toward improving the advising and utilization of high-energy physicists. Even more important, the DPF must inform its scientists of all the various career paths open to them both within HEP and without; it must educate them as to the needs of employers in industrial laboratories, government, finance, etc so that they can optimize their qualifications and résumés; and it should show employers how high-energy physicists can contribute to their enterprises. As a first attempt in this direction, the DPF co-sponsored a Career Workshop at the Albuquerque meeting along with the AIP and APS.

3.3 A Recommendation

To produce lasting, significant improvements in the career situation, a long-term, planned effort will be required. Therefore, the Structural Issues group recommends:

- The DPF Executive Committee should organize a Subcommittee on Careers to arrange meetings, prepare literature, consider the appropriateness of, and possible improvements in the training of graduate students and postdocs, etc.

Several people willing to serve have already been identified, and have begun by assisting the DoE and the PDG with the Career Survey. The subcommittee would optimally be expanded to include representatives from each major constituency: universities, national laboratories, and industry.

4 Education and Outreach

4.1 Introduction

In the post-Cold War, post-SSC era, it became clear to many high-energy Physicists that we need to redouble our efforts to communicate what we do and why we do it, even though we feel that time is a scarcer resource than money. Nonetheless very substantial efforts are already underway, and we just need to build upon these existing activities. Ongoing efforts range from those done by individuals, to those done by the labs and other organized groups. Inevitably there are questions about whether such efforts are or should be done for selfish or altruistic motives. Such discussions do not seem to be useful. Many physicists have spent many years in education and outreach; the public and our field have both benefitted.

The Division of Particles and Fields of the APS, as part of its current study of future directions for our field, has created a working group on "Structural Issues." This working group formed a subgroup to examine ongoing outreach and education efforts and seek proposals for new directions. Named the "Public Outreach and Education Team" (or POET), this group has initiated several activities so far, including:

- An evening meeting at Fermilab on February 14, 1994 at which general issues were discussed.
- A survey of ongoing ideas and of proposals was conducted via e-mail. The results of the survey and the Fermilab meeting were distributed via e-mail.
- An electronic bulletin board was set for discussion of these issues (but it has not been used much).
- At the DPF 1994 meeting in Albuquerque on August 2, 1994, a plenary session was held on Public Outreach and Education, at which four reports on these issues were given. Later in the meeting, the POET group met with conference attendees and had an intensive discussion of proposals for future directions.

The DPF 1994 meeting was the first general meeting at which a plenary session was devoted to outreach and education in high-energy physics. Despite the late hour of the session (4:30-6:00 PM) half of the attendees remained for the entire session. The moderator of the session was Geoffrey West (LANL). The speakers included: Malcolm Browne (Pulitzer-Prize-winning writer for the New York Times) on "High Energy vs. Low Education: A National Challenge," Julia Thompson (Pittsburgh) on "Outreach to Women and Minorities," Ernest Malamud (SciTech and Fermilab) on "Using Science Centers to Expose the public to the Microworld."

A general them of the many comments we have received is the need to convey the excitement of physics — not just the big discoveries but the controversies too. There seems to be a consensus that we need to find means to recognize and reward outreach and education activities by physicists. The holding of a plenary session at a major conference was a good first step. A list of possible proposals for future action is given in Appendix B.

In the following sections, we shall briefly review selected activities and proposals concerning general lay audiences, students, the government, the news media, etc. Of particular interest are the use of the World-Wide Web, the creation of a catalog, and activities by collaborations. We are also working on activities targeted on underrepresented groups in HEP, and on enhancing the offerings at science museums.

4.2 Reaching Lay Audiences

Many people have told us about their talks and classes for lay audiences. In a later section the resources that are available to people interested in presentations to any non-technical audience will be discussed. There are a variety of books available to the public about Particle Physics topics. There have been suggestions that someone should produce a large-size "coffee table" photo book showing some of the detectors, accelerators, events, and even people of high-energy Physics.

Two recent books brought to our attention were: Cindy Schwarz's *A Tour of the Subatomic Zoo*, which was written for the interested layperson/ undergraduate/ high school teacher or student. It assumes no prior physics background and can serve as an introduction to the basics. It was published by AIP. Lawrence Krauss' *Fear of Physics*, tries to reach out to a broad popular audience, in order to explain what physicists are interested in and why.

The Florida State University Physics Department has been producing mall exhibits for a number of years and report that they are quite popular. CERN has set up its own science museum, MICROCOSM, and has now built a separate building for it. They say its purpose is "to let the public see the research work carried out by the physicists in their quest to understand the laws of Nature." The number of visitors has remained consistently very high.

4.2.1 News Media

In connection with the announcement of the initial top quark candidates, Fermilab carried out an excellent program to inform the news media about the physics and the experiments in a manner that allowed the media to report the news accurately. They put together a substantial package of information well-suited for the target audience. They had excellent results in getting good coverage in many media outlets throughout the country.

We do not always have big news to report. However, many people have reported success at getting media coverage for aspects of their experiments. Some sample comments:

> "We had a press conference when we did the last touches on the experimental hardware...
>
> It works, but takes a lot of work and courage...
>
> We were on the evening news. One key point was good contact to the public relations office. One has to be extremely careful about the scope of their press release...
>
> I think a press conference after the publication of key result or a press conference when an experiment takes first data is the best approach."

One comment was "It is often easy to interest journalists in (well-defined) stories, but it needs a significant effort to establish the 'networking' links to them." Some thought should be given on how to do this.

A common problem felt by many physicists "is the NEGATIVE peer pressure to go public." The culture of our field has equated talking to the press about one's research to "publishing in the New York Times." Clearly one should continue to publish in the standard journals, but in the world we now live in, we are obligated to communicate our results, our conclusions, and the benefits of our work to a broad audience. Physicists should be *encouraged* to describe what we are doing and why. We are excited by the theories and experiments of our field, and we should not be ashamed to share that excitement. These lessons have not been lost on the astrophysicists; their stories appear weekly in the press (even the less-glamorous stories).

A number of people at Fermilab have proposed a national meeting of science writers and physicists to discuss the reporting of science. Clearly many of us feel that both the quantity and quality of reporting about particle physics are not adequate. It is a difficult subject about which many writers may feel insecure. Such a meeting might not only help break down some these barriers, but would help foster contacts between writers and physicists.

4.2.2 Radio, TV, and Cable TV

In general it is difficult to present science on television because of the cost. However, Bernice Durand (Wisconsin) teaches modern physics for nonscientists very successfully on Madison area cable TV where watchers know her as the "physics lady."

PBS has recently begun a new television series called the *Magic School Bus*. Several people has asked whether the producers could be interested in an episode on Particle Physics. It is a fully animated children's educational series, featuring a teacher named Ms. Frizzle (played by Lily Tomlin), who takes her students on a magically powered bus for scientific field trips into the human body, around the solar system, or back to the time of dinosaurs. "Children's interest in science starts to erode in the elementary grades," project organizers say. *The Magic School Bus* project is designed to keep children's curiosity alive.

We note that other sciences seem to be featured in 60-second science profiles on the radio — why not HEP?

4.2.3 Government

It is generally agreed that our field could do a better job of informing Washington officials about what we do and why we do it, about the benefits of our research, and about the excitement of particle physics. Other areas of physics participate in APS' congressional visits programs much more than we do. Many Members of Congress and their aides have never seen an HEP physicist and are happy for the opportunity. The recent Drell Panel report had a significant impact, in part because of significant followup in Washington by members of the panel and others.

It has been suggested that the DPF should sponsor

occasional Congressional Fellows similar to those from the APS and AAAS. The cost is about $50,000 each in salary, moving expenses, etc. Unfortunately it is doubtful that the DPF can afford this. However, one should not underestimate the impact of Congressional Fellows, who are often regarded by lay congressional staffers as "gurus" on science issues. Unfortunately these fellows have rarely been from our field, and in fact, they have even campaigned *against* our interests. It has been suggested that the DPF should simply push to end APS's program, which we pay for and which some believe may have done more harm than good with respect to HEP. In no sense do they represent our field, nor is it clear that we get the best qualified people to accept such positions. A more positive view would hold that DPF should identify and recruit better qualified fellows.

A former Congressional aide has suggested that we would benefit more directly by sponsoring quarterly receptions for Members and aides (from the House and Senate) at which leading figures in HEP would discuss HEP physics issues and developments. He estimated that these evening receptions would typically attract 15-20 people (assuming food was provided), and felt that such numbers were well worthwhile. This is already done by other fields including chemistry and biological sciences.

4.2.4 Science Community

One of the lessons of the SSC debacle is that we could benefit from better relations with the rest of the science community. A British correspondent reported that particle physicists have made great strides in improving their relations with other communities in Britain, and that it has greatly benefitted them.

A proposal has been made to hold meetings in Washington on the benefits of basic research for America, cosponsoring it with biologists, chemists, medical researchers, geologists, astronomers, etc. Leading researchers from each field would speak about the importance of basic research. Reporters would be invited to attend. Later, participants in the meeting could visit the Capitol to relate this message to whatever committees or individuals are interested. The purpose of such meetings would be general and not to promote any particular projects. They would serve the dual purpose of reaching out to these other fields and explaining to the public the value of basic research.

4.2.5 Documenting the Value of Basic Research in HEP

A number of people have urged a new effort to document the impact of basic research in areas ranging from education to technology transfer to medical benefits to economic impact. One suggestion is to trace the history of particular technologies. We have not received any specific proposals on how to coordinate this.

4.2.6 College Students

A recurrent theme from many respondents is that there are enormous numbers of young people taking introductory physics courses in our own universities and that we are wasting a tremendous opportunity by not turning them on to physics and basic research as much as possible. These people will be the congressional aides, opinion leaders, etc. in a few years.

Others have proposed that we should spend more time giving talks at neighboring colleges.

4.2.7 Teachers and School Children

Many physicists are currently active in bringing particle physics to high school students. This can be done through presentations, workshops, open houses, the creation of materials, etc. The national laboratories all have such programs which I will discuss later. One very active national group is the Contemporary Physics Education Project (CPEP) which consists of teachers,educators, and physicists (among the physicists are M. Barnett, R. Cahn, G. Goldhaber, H. Quinn, M. Riordan, C. Schwarz). This group has created the "wall chart" on *Fundamental Particles and Interactions* (in three sizes) and distributed more than 100,000 copies of it. It also has very popular color software for high school/college students in both Mac and PC versions. Packets of classroom activities about particle physics have been mailed to every high school physics teacher in the US, and a a book on the subject of particle physics, detectors, accelerators, and astrophysics is nearing completion. CPEP conducts many workshops for teachers on how to use CPEP materials to teach particle physics. CPEP has been featured in *Science*, *Physics Teacher*, and even on the BBC World Service.

The American Chemical Society together with AIP periodically publishes booklets for students with cartoons, etc. The April 1993 issue was on particle accelerators. The book published by Cindy Schwarz with AIP (described earlier) is intended for high school students.

A popular suggestion has been the idea of creating a catalog of resources, materials, workshops, etc. on particle physics. This would be made available (for free) not only to teachers but to physicists to aid and stimulate them in joining education/outreach efforts. The catalog would be available both in printed form and on the World-Wide Web. Some people propose mailing it to all high school teachers, but others feel that would not be useful.

A number of people are currently making presentations and giving workshops at teachers meetings such as the American Association of Physics Teachers (AAPT) and the National Science Teachers Association (NSTA). These organizations have national, regional, and state meetings. Those involved in these presentations find them well received and advocate that more people do it.

Another proposal is that we set up a national referral service (via telephone and e-mail) that would direct high school and college teachers with HEP questions to physicists who are willing to answer questions. The idea would be to refer the teachers to physicists in or near their own state. They might call a number such as 1-800-PARTICLE (extra digits are ignored). This service may also provide a list of speakers.

Finally, physicists can and do work together with local school districts and state agencies. In addition, there are university, college, high school alliancc programs (organized via the APS).

4.3 Resources Available to Physicists

Many of the national laboratories such as Fermilab, SLAC, Brookhaven, and CEBAF have substantial education departments that sponsor workshops and programs for both students and teachers, and material development. They are anxious for the involvement of additional physicists.

Fermilab opened the Leon M. Lederman Science Education Center in September 1992 in a dedicated building with many exhibits. They have 45 precollege programs serving over 40,000 teachers and 8,000 teachers per year, as well as many college programs. They sponsor workshops for Latin American countries and create Spanish versions of instructional materials. Physicists are involved in Fermilab programs as research mentors, seminar speakers, role models, question & answer sessions with school children, consultants on science content, hands-on-science in the classroom, museum volunteers, and SBIR proposals. Programs at CEBAF emphasize "Teach science by doing science."

Existing materials include transparencies, slides, comics, software, etc. These will be included in the catalog discussed above in the section on teachers and school children. The public relations staff at laboratories and universities often have resources available for physicists.

We should continue to report on outreach/education at DPF meetings to inform physicists about resources and ongoing activities. Many have suggested that we should work through the DPF and other organizations. We can also communicate about these activities via Internet bulletin boards and newsgroups.

4.4 Using the Information Superhighway

More and more public schools are gaining access to the Internet. One suggestion is that the labs should set up files from which events pictures, detector designs, accelerator pictures, etc. can be obtained by anonymous ftp. These should be appropriately annotated.

The World-Wide Web (WWW) presents tremendous opportunities as use is growing by 300 percent a year. Major news media are searching the Web for stories, among them the New York Times. Even the sheriff of Tulsa, Oklahoma has listed Tulsa's most wanted criminals on the Web

An example of the impact of WWW can be seen in the interest generated by LBL's "Whole Frog" link-up, with which users can examine many three-dimensional images of the frog with or without skin, from any angle. Different organs can also be seen separately. In half a year 160,000 users from 56 countries have connected to it: (http: //george.lbl.gov/ITG.hm.pg.docs/dissect/info.html).

Other organizations are already well underway in exploiting the Web for educational purposes. As a prominent example, CERN organized a major WWW Workshop on Teaching and Learning with the Web in May 1994. They had speakers and participants from throughout Europe but few from the US.

NASA has placed on WWW tremendous numbers of images from the Hubble Space telescope and elsewhere including pictures of supernova, comets, galaxies, planets, etc. These are annotated and sometimes very useful for education. There are also a variety of animations. A prime focus of NASA pages is always on hot and current topics. They have coordinated the efforts of their many different labs and facilities.

Fermilab has made great strides in making a major presence on the Web with some excellent educational pages and a coherent, organized approach. They cover the physics, the detectors, the accelerators, the benefits, and more. I suggest you look at it.

Clearly HEP (like NASA) should have a coordinated approach to the Web with a single homepage for the public that points to the labs and other relevant sites. This effort may require a meeting of the interested groups. This page should contain short items summarizing the current excitement and controversies in particle physics and point to lab and university homepages for more information.

Physicists may also need to make some effort to aid schools and libraries getting onto WWW. Many are already on the Web (even some elementary schools classes have their own pages), but most are not.

Other suggested approaches are to create multimedia CD-ROM programs about particle physics or even Nintendo-type games.

The AIP has an e-mail news service on physics education. It summarizes information on resources, national initiatives, outreach programs, grants, publications, etc. To subscribe to AIP's PEN, send an e-mail message to listserv@aip.org . Leave the "Subject" line blank. In the body of the message, enter the following command: <add pen>.

4.5 Outreach by Experimental Collaborations

An interesting suggestion is that experimental collaborations should be creating WWW and ordinary printed materials about their experiment. These should describe the physics motivations of their experiment and explain how the experiment might accomplish these goals. There are people who believe that any experimental collaboration that cannot explain these basic concepts to the public should not be funded.

Several people have suggested that experimental collaborations can do much more. A very interesting proposal is one under which traditionally non-research colleges (and possibly high schools) could become "affiliates" of experimental collaborations. Arrangements would be made whereby they would "participate" in research activities. Their work might involve a small scale hardware study (table top) or a simulation study. They would need computer time or the loan of some small hardware system for a few months. An incentive for these schools would be very important: some degree of recognition of being part of the experiment. The institution names might be listed on scientific papers under the banner "educational institutions." One possibility suggested for a college senior lab experiment would be to do some data analysis and event reconstruction for particles such as Z, W, and top, using CDF or DØ data. Clearly it would take a significant amount of development work for this to be feasible, but once such educational material is developed, it could be distributed to other colleges. Later it might even be distributed to high-level high schools as a test.

Astrophysicists have already developed such a program, and it has been very successful. It is called "Hands on Universe." The organizers feel it gives high school teachers and their students the opportunity to become collaborators on real scientific research. The program provides them with access to professional grade telescopes, analytical tools and the training to use them. It is currently delivered to high schools across the United States. Students can request telescope time to obtain images of the moon, planets, galaxies, or supernovae.

The program recently made national news (ABC Nightly News, Associated Press, etc.) when two 17-year-old juniors at a Pennsylvania high school while searching for a galaxy photographed a supernova (1994I). While they did not, of course, recognize this, their photograph was the earliest one taken and therefore quite valuable. Both the publicity for science and the impact on young people were also valuable.

4.6 Summary

There is no doubt that there are some exciting things happening in high-energy physics outreach and education, carried out by educators and by physicists. However, the reality is that extremely few physicists spend any time at all on these efforts. They heartily endorse these programs, but find that they lack either the time or the inclination to join in.

Perhaps, as we hope, attitudes are changing, since it is important to show by our actions that we value public awareness. We should make communication a priority and reward it. We need a mechanism to make this happen, and motivation for people to do it.

Appendix A: Questions on Governance

During the spring of 1994, the HEP community was encouraged to express its views on the following issues, responding to the Working Group by letter or e-mail.

- The only formal mechanism for advising the federal government is the HEPAP organization. Does HEPAP work effectively? If "yes", why. If "no", why not?
- What additional mechanisms do you think should be considered by DPF to more adequately plan for the future of HEP?
- Should structured consultative mechanisms be created between HEP and Congress as a whole or in part, NSF, DoE, OSTP?
- How should the support levels be optimized between university and national laboratory programs, domestic and offshore experiments, etc., and unnecessary duplication be avoided.
- Should interface subjects such as nuclear physics, astrophysics, etc., have separate, identifiable mechanisms for consultation with the government?
- Should HEP seek a structured public education role? If so, how broad should such an activity be and how should it be organized?

Appendix B: Proposals for Outreach Activities

DPF94 Conference participants who attended the POET meeting seemed especially interested in the following proposals:

- Create a catalog of HEP resources (materials, workshops, etc.) for teachers and for physicists. It would be printed and on the World-Wide Web.

- Together with basic researchers from other fields, organize a meeting in Washington on the impact and importance of basic research.
- Organize a unified approach to presenting Particle Physics on the World-Wide Web, presenting the highlights and controversies of our field.
- Begin a program of educational affiliates of experimental collaborations who would perform specially designed analysis or experiments.
- Find means to better inform Washington staff and officials about HEP (quarterly receptions at the Capitol, congressional fellows, etc.).
- Organize a national science writers meeting with physicists.
- Encourage more HEP participation in science museum programs and find means to present our subject in museum-type settings.

For these and other efforts to succeed, the DPF needs to give them some priority and provide vital organizational support. Moral support is welcome, but if we wish for outreach and education activities to progress, meaningful action by the DPF would be more beneficial.

References

[1] S.D. Drell *et al.*, "High Energy Physics Advisory Panel's Subpanel on Vision for the Future of High-Energy Physics", May 1994, DOE/ER-0614P (U.S. Department of Energy, Office of Energy Research, Washington, D.C. 20585).